International Conference on Defects in Insulating Materials 2008

Journal of Physics: Conference Series
Volume 249

Aracaju, Sergipe, Brazil
24-29 August 2008

ISBN: 978-1-61782-078-6

Printed from e-media with permission by:

Curran Associates, Inc.
57 Morehouse Lane
Red Hook, NY 12571

Some format issues inherent in the e-media version may also appear in this print version.

Printed by Curran Associates, Inc. (2010)

Institute of Physics
Dirac House, Temple Back
Bristol BS1 6BE UK

Phone: 44 1 17 929 7481
Fax: 44 1 17 920 0979

techtracking@iop.org

Additional copies of this publication are available from:

Curran Associates, Inc.
57 Morehouse Lane
Red Hook, NY 12571 USA
Phone: 845-758-0400
Fax: 845-758-2634
Email: curran@proceedings.com
Web: www.proceedings.com

TABLE OF CONTENTS

FOREWORD

These proceedings represent a sample of the scientific works presented during ICDIM2008, the 16th International Conference on Defects in Insulating Materials, held at the Federal University of Sergipe, Aracaju, Brazil from 24–29 August 2008. The conference was the latest in a series which began at Argonne in 1956, and which has been held most recently in Riga, Latvia (2004) and Johannesburg, South Africa (2000). The conference was also related scientifically to the EURODIM series, which have been held most recently in Milan, Italy (2006), Wroclaw, Poland (2002) and Pecs, Hungary (2010).

The aim of the conference was to bring together physicists, chemists and materials scientist to discuss defects in insulating materials and their effect on materials, including their optical, mass/charge transport, energy storage and sensor properties. The conference featured 6 plenary lectures, 60 contributed lectures and about 130 posters. The posters were displayed for the whole conference, but discussed in two three-hour sessions.

We are grateful to the International Advisory Committee for suggesting invited speakers and to the Programme Committee for their help in refereeing all the abstracts and choosing the contributed oral contributions. We would also like to thank the Local Organising Committee and the Brazilian Physical Society for their help with local organisation and the online registration/payment process respectively. The chairpersons would like to specially thanks all the sponsors listed below for financial support.

The Federal University of Sergipe, one of the public and 'free tuition' Universities of the Country, run by the Brazilian Ministry of Education, were pleased to host this 16th meeting, the first one in Latin America.

Mario E G Valerio *Conference Chair*
Robert A Jackson *Programme Chair*

Conference Scope

Scope of the Conference was the presentation of the latest investigations on point and extended defects in bulk materials and thin films. Technological applications will be presented alongside fundamental measurements and theories.

The main scientific areas included:

1 Fundamental physical phenomena
Point and extended defects in wide band-gap systems: oxides, fluorides, nitrides, alkali- and silver-halides, perovskites, minerals, ceramics, nano-structures, organic molecular crystals, glasses, high-k and low-k materials, photonic crystals.

2 Defects at surfaces and interfaces
Thin films and low-dimensional systems. Colloids, nano-crystals, and aggregates. Defects and material preparation technology. Defects modelling and computational methods. Radiation effects, radiation induced defects, colour centres. Luminescence of excitons, impurities, and defects. Electronic excitations, excited state dynamics, radiative and non-radiative relaxations. Scintillation, energy transfer and storage,

carrier trapping phenomena. Non-linear optical phenomena. Laser active centres. Phonons and defects, electron–phonon interactions. Defect diffusion, ionic relaxations, ionic transport.

3 Technological applications

Radiologic imaging and detection, scintillators, and dosimeters. Optical devices and photonics, photorefractive electro-optics, optical fibres, lasers. Materials for micro-electronics. Solid electrolytes, fuel cells, electrochemical sensors, fast ionic conductors.

Conference chairpersons:
Mário E G Valerio (Conference Chairman), Physics Department, Federal University of Sergipe, SE, Brasil
Robert A Jackson (Programme Chairman), School of Physical and Geographical Sciences, Keele University, Keele, UK

Conference committees:
International Advisory Committee
R Capelletti, Italy
A V Chadwick, UK
J Corish, Ireland
J D Comins, South Africa
H W den Hartog, The Netherlands
K Funke, Germany
Robert A Jackson, UK
O Kanert, Germany
A A Kaplyanskii, Russia
A Lushchik, Estonia
F Lüty, USA
M Moreno, Spain
P E Ngoepe, South Africa
M Nikl, Czech Republic
S V Nistor, Romania
Ch Pedrini, France
O F Schirmer, Germany
J-M Spaeth, Germany
A M Stoneham, UK
M Suszynska, Poland
I Tale, Latvia
M E G Valerio, Brasil
R T Williams, USA

Programme Committee
Robert A Jackson (Chair), University of Keele, UK
R M Montereali, ENEA C.R. Frascati, Rome, Italy
M Moreno, University of Cantabria, Spain
Ch Pedrini, University Lyon, France
Klaus W H Krambrock, UFMG, MG, Brasil
Volkmar Dierolf, Lehigh University, USA
Laszlo Kovács, Hungarian Academy of Sciences, Hungary
M E G Valerio, UFS, SE, Brasil

Local Organizing Committee
M E G Valerio, UFS, SE, Brasil
Sonia L Baldochi, IPEN, SP, Brasil
Klaus W H Krambrock, UFMG, MG, Brasil
Livio Amaral, UFRGS, RS, Brasil
Ana R Blak, USP, SP, Brasil
Marco Cremona, PUC-RJ, RJ, Brasil
Anderson S L Gomes, UFPE, PE, Brasil
Spero Penha Morato, LaserTools, SP, Brasil
Alejandro Ayala, UFC, CE, Brasil

ICDIM2008 Sponsors:

PEER REVIEW STATEMENT

All papers published in this volume of *Journal of Physics: Conference Series* have been peer reviewed through processes administered by the proceedings Editors. Reviews were conducted by expert referees to the professional and scientific standards expected of a proceedings journal published by IOP Publishing.

Journal of Physics: Conference Series **249** (2010) 012001 doi:10.1088/1742-6596/249/1/012001

Ultraviolet luminescence and creation of $(WO_4)^{3-}$-type centers under UV irradiation of $PbWO_4$ crystals doped with trivalent rare-earth ions

A Krasnikov[1], T Kärner[1], V Kiisk[1], V V Laguta[2,3], M Nikl[3], I Sildos[1], Y Usuki[4] and S Zazubovich[1,5]

[1,5] Institute of Physics, University of Tartu, Riia 142, 51014 Tartu, Estonia

[2] Institute for Problems of Material Science, Ukrainian AS, Krjijanovskogo 3, 03142 Kiev, Ukraine

[3]Institute of Physics, AS CR, Cukrovarnicka 10, 162 53 Prague, Czech Republic

[4]Material Research Laboratory, Furukawa Co. Ltd., Tsukuba 305-0856, Japan

E-mail: svet@fi.tartu.ee

Abstract. Luminescence characteristics and creation of $(WO_4)^{3-}$ and $(WO_4)^{3-}$-A^{3+} centers were studied in $PbWO_4$ crystals doped with trivalent rare-earth A^{3+} ions (A^{3+}: La^{3+}, Lu^{3+}, Y^{3+}, Ce^{3+}, Gd^{3+}, Eu^{3+}) under selective irradiation in the 3.4-5.0 eV energy range at 5-80 K. Optically created centers were detected by the TSL and ESR methods. Creation spectra of the mentioned centers were measured and the activation energies needed for their creation were calculated. It was found that both the $(WO_4)^{3-}$ and $(WO_4)^{3-}$-A^{3+} centers can be created not only in the band-to-band transitions region but also in the 3.8 ± 0.4 eV range. The activation energy for their creation at T<40 K is close to zero. Under irradiation in the energy range around 3.8 eV at T>60 K, the activation energy for $(WO_4)^{3-}$-A^{3+} centers creation varies from 10 to 25 meV, depending on the irradiation energy. A new UV emission, peaking in the 3.05-3.20 eV range, was found whose thermal quenching takes place with the same activation energy. Radiative decay of defect-related states, as well as the processes of their non-radiative decay, resulting in creation of $(WO_4)^{3-}$-A^{3+} centers and thermal quenching of the UV emission, are considered to explain the effects observed.

1. Introduction

Single crystals of lead tungstate ($PbWO_4$) are well known as a fast and heavy scintillation material for high-energy physics experiments. Radiative and non-radiative decay of various exciton- and defect-related states under selective UV excitation at T>80 K was systematically studied for many $PbWO_4$ crystals (see, e.g., [1-4] and references therein). It was found that the radiative decay of the self-trapped and various localized excitons of the $(WO_4)^{2-}$ type, arising from the regular crystal regions, is responsible for the complex blue B emission band. The complex low-temperature green G(I) emission band of undoped crystals was ascribed to excitons of the same type, but arising from the lead-deficient

[5] To whom any correspondence should be addressed.

crystal regions. Photo-thermally stimulated non-radiative decay of the exciton- and defect-related states results in the appearance in the crystal lattice of various electron and hole centers. Tunneling recombination of the electron centers connected with oxygen vacancies and the hole centers connected with lead vacancies is accompanied with the green G(II) emission. It was concluded that lead and oxygen vacancies play the most important role in the optically and thermally stimulated recombination processes. Co-doping with stable trivalent rare-earth A^{3+} ions which substitute for Pb^{2+} ions in the $PbWO_4$ crystal lattice [5] suppresses the lead-deficient structure and reduces the number of the isolated vacancies [2, 6, 7]. This leads to a strong suppression of both the G(I) emission and slow (μs-ms) tunnelling recombination G(II) emission and to an enhancement of the fast (2-4 ns at RT) B emission. As a result, considerable improvement of scintillation characteristics of $PbWO_4$ crystals is obtained. However, the peculiarities of the luminescence and defects creation processes in A^{3+}-doped $PbWO_4$ crystals were not studied in detail. Therefore, the aim of the present work was to study the processes of luminescence and creation of electron $(WO_4)^{3-}$ and $(WO_4)^{3-}$-A^{3+} centers in $PbWO_4$:A^{3+} crystals (A^{3+}: La^{3+}, Lu^{3+}, Y^{3+}, Ce^{3+}, Gd^{3+}, Eu^{3+}) under their selective irradiation in the 3.4-5.0 eV energy range at 5-80 K.

2. Experimental procedure

Single crystals of $PbWO_4$:A^{3+} were grown in Japan by Furukawa Co. Ltd from 4N purity raw material, using the Czochralski method in air with a Pt crucible and following the third crystallization method [8]. The concentration of A^{3+} ions was about 135 ppm in the melt. The crystals contain less than 1 ppm of other impurities [8] and also very few isolated (non-compensated) lead and oxygen vacancies [2-6].

Luminescence characteristics were studied with the use of the set-ups and procedures described in [1, 3]. Decay kinetics were measured with a modified Spectrofluorometer 199S (Edinburgh Instruments) under excitation with a nanosecond coaxial hydrogen-filled flashlamp (IBH) or microsecond Xe lamp (IBH) and using two single grating monochromators. The detection was performed with a IBH-04 photomultiplier module using the method of time-correlated single photon counting. A deconvolution procedure (SpectraSolve software package) was applied to extract true decay times from multiexponential approximation. Time-resolved emission spectra were measured under excitation with a Nd:YAG laser-pumped optical parametric oscillator NT 342/1/UVE (EKSPLA, Lithuania). It allows to obtain pulsed monochromatic radiation in the 0.6-6.0 eV energy range with a pulse duration of 5 ns.

Optically created centers were detected by the thermally stimulated luminescence (TSL) and electron spin resonance (ESR) methods. The TSL glow curves were measured with a heating rate of 0.2 K/s after selective UV-irradiation of a crystal for 15-30 min through a monochromator with spectral width of the slits ≈5 nm. The spectral integrated emission of a crystal was detected with a photomultiplier or a Hamamatsu (6240) photon counting system. The ESR spectra were measured at 18 K using a ERS 231spectrometer (Germany) after irradiation of a non-oriented $PbWO_4$:Y^{3+} crystal at different temperatures (10-80 K) with the Nd:YAG laser. After each TSL or ESR measurement the crystal was heated up to 150 K to destroy the optically created centers.

The experiments at low temperatures were carried out with the use of immersion helium cryostats and vacuum nitrogen cryostats.

3. Experimental results

3.1. Photoluminescence characteristics

In the $PbWO_4$:A^{3+} crystals studied, besides the intense B emission, the weak ultraviolet - UV, green - G(I) and red - 1.9 eV emissions are observed (figure 1a). The position and FWHM of the UV band are slightly different in different crystals, varying from 3.05 eV to 3.20 eV and from 0.56 eV to 0.64 eV, respectively. This difference can be caused by a complex structure of this band and by its overlap with the B band. The B/UV emission intensity ratio depends on the crystal (table 1). The maximum position of the G(I) emission band varies in different crystals from 2.25 eV to 2.40 eV. The 1.9 eV emission is different from the red emissions (1.57 eV and 1.48 eV) ascribed in [9] to complex lead-vacancies-containing centers. These centers are absent in most of the crystals studied in the present work. Under 3.85 eV excitation, the intensities of the UV, G(I) and 1.9 eV emissions are comparable (figure 1a). The intensity ratio of the UV and G(I) bands depends on the crystal (table 1).

In the excitation spectrum of the UV emission, the exciton band located at about 4.1 eV and a complex band located in the 3.50-3.95 eV range are observed (figure 1b). The 4.1 eV band appears partly due to the overlap of the UV and B emission bands. Its maximum is shifted to lower energy with respect to the exciton band located at 4.13-4.15 eV in the excitation spectra of the other emissions.

Table 1. Intensity ratios obtained from corrected emission spectra measured at 80 K under E_{exc}=3.85 eV (for the UV and G(I) emissions) and E_{exc}=4.5 eV (for the B emission). The activation energies E_q for thermal quenching of the UV emission. The activation energies E_a for the creation of the TSL peak at 97-107 K under irradiation at T_{irr}>55 K in the exciton (4.1 eV) and in the defect-related (3.7 eV) energy range.

	Emission intensity ratios		E_q, meV	E_a,meV	
Crystal	UV/G(I)	B/UV	UV	4.1 eV	3.7 eV
$PbWO_4$:Lu	1.38	84	15	11	
$PbWO_4$:Y	1.26	108	16	11	25
$PbWO_4$:Ce	1.19	540	12	10	
$PbWO_4$:Gd	0.96	108	10		
$PbWO_4$:Eu	0.84	288	15		

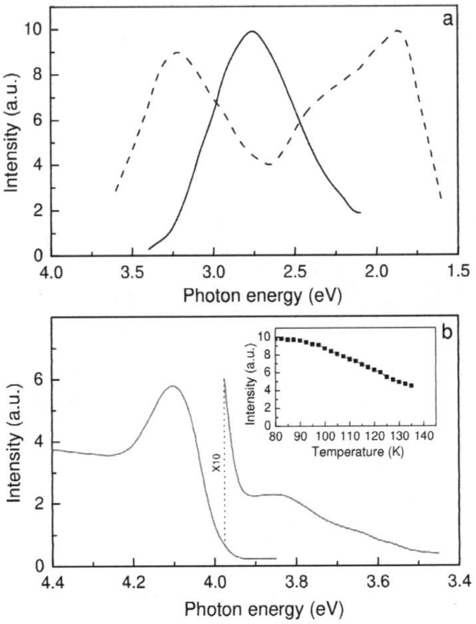

Figure 1. (a) Corrected and normalized emission spectra of the $PbWO_4$:135 ppm Y^{3+} crystal measured at 80 K under excitation at 4.5 eV (solid line: the B band) and 3.85 eV (dashed line: the UV, G(I), and R bands). The B/UV emission intensity ratio is equal to 108 (see table 1).
(b) Corrected excitation spectrum of the UV emission at 80 K. Temperature dependence of the UV emission intensity (in the inset).

Thermal quenching of the UV emission takes place at much lower temperatures (T_q≈130 K, see the inset in figure 1b) as compared with the B emission (T_q≈160-180 K [1]) (T_q is the temperature where the emission intensity has decreased by a factor of 2). The activation energies E_q for thermal quenching of this emission, calculated from the lnI(1/T) dependences, are about 10-16 meV and slightly depend on the crystal (see table 1).

At 80 K the decay kinetics of the UV emission of the $PbWO_4$:Y^{3+} crystal consists of three components with decay times of 35 ns, 350 ns (figure 2a) and 11-14 μs (figure 2b, curve 1). The time-resolved emission spectra, shown in the inset of figure 2a, indicate that both the ns-components arise from the 3.15 eV band. In the emission spectrum of the μs-component, the UV band located at about 3.05 eV and the G(I) band are observed (see the inset in figure 2b). The decay time of the G(I) emission (figure 2b, curve 2) is close to the value (≈20 μs) observed in other undoped crystals [3]. The time-resolved emission spectra measured at 10 K at different time gates (figure 3) also indicate that the spectra of the fast (curves 1, 2) and the slow (curves 4, 5) decay components of the UV emission are located at about 3.15 eV and 3.05 eV, respectively. At 10 K, the decay time of the slow component is about 100 μs.

A very fast weak ≈3.25 eV emission band was detected at RT in [10] and ascribed to the radiative decay of the non-relaxed self-trapped exciton state. If so, the B/3.25 eV emission intensity ratio should be sample-independent. However, this is not the case for the UV emission studied (table 1). Besides, according to [10], the intensity of the 3.25 eV emission decreases with decreasing temperature. Thus, the UV emission reported in the present paper is different from that observed before.

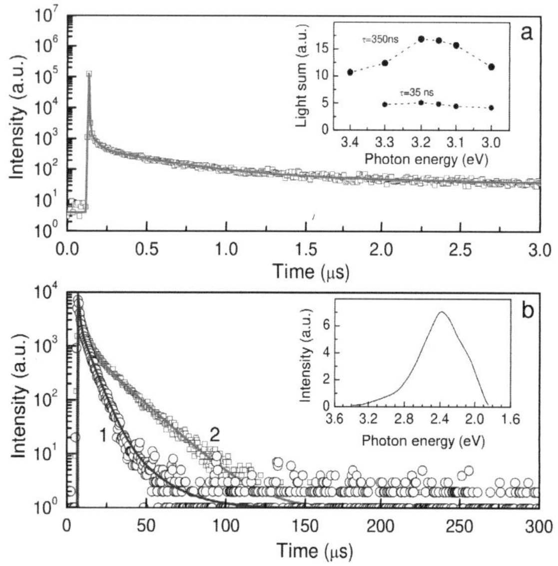

Figure 2. Luminescence decay kinetics in the $PbWO_4:Y^{3+}$ crystal measured at 80 K for the (a) 3.15 eV; (b) 3.0 eV (curve 1) and 2.35 eV (curve 2) emission bands. In the insets: (a) the emission spectra of the ns-components; (b) the emission spectrum measured at t=40 μs after the excitation pulse. E_{exc}=3.8 eV.

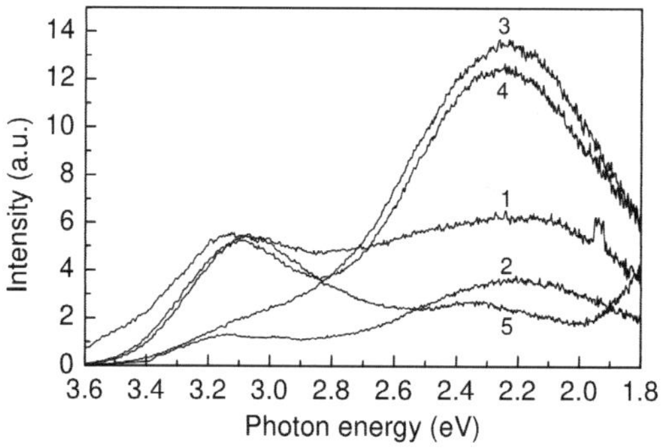

Figure 3. Time-resolved emission spectra of the $PbWO_4$:135 ppm Y^{3+} crystal at 10 K. Time gates: 0-0.1 μs (curve 1), 0.1-1.1 μs (curve 2), 1-11 μs (curve 3), 10-110 μs (curve 4), and 100-1100 μs (curve 5). Excitation with the Nd:YAG laser at 3.87 K; for emission detection spectral width of slits 10 nm.

3.2. Thermally stimulated luminescence studies

Thermally stimulated luminescence of X-ray irradiated $PbWO_4:A^{3+}$ crystals was studied in [5, 11-15]. The TSL peaks at about 50 K and 95-110 K were ascribed to the thermal destruction of $(WO_4)^{3-}$ and $(WO_4)^{3-}$-A^{3+} centers, respectively. The TSL intensity in the UV-irradiated crystals is weak. The characteristics of all the crystals studied are similar. In the present paper they are illustrated on the example of the $PbWO_4:Y^{3+}$ crystal. After irradiation at T_{irr}<40 K, peaks appear at 53 K and 107 K (figure 4a). They are most effectively created in the band-to-band transitions region and by about two orders of magnitude less effectively, in the 3.8 ± 0.4 eV range. The creation spectra of both TSL peaks coincide (figure 4b), and around 3.8 eV they are close to the excitation spectrum of the UV emission (compare figures 4c and 1b).

From the dependences of the TSL peak intensity on the irradiation temperature T_{irr}, the activation energies E_a were determined for the TSL peaks creation under irradiation in the exciton (4.05 eV) and defect-related (3.7 eV) regions. For $(WO_4)^{3-}$ centers creation, E_a=0 (figure 5a). At T_{irr}<40 K, the E_a value for the creation of $(WO_4)^{3-}$-A^{3+} centers is also close to zero (figure 5b) which means that these centers are created only at thermal destruction of $(WO_4)^{3-}$ centers. At T_{irr}>60 K, $(WO_4)^{3-}$-A^{3+} centers are created with E_a=10-25 meV (table 1) depending on the irradiation energy E_{irr} (see the inset in figure 5b).

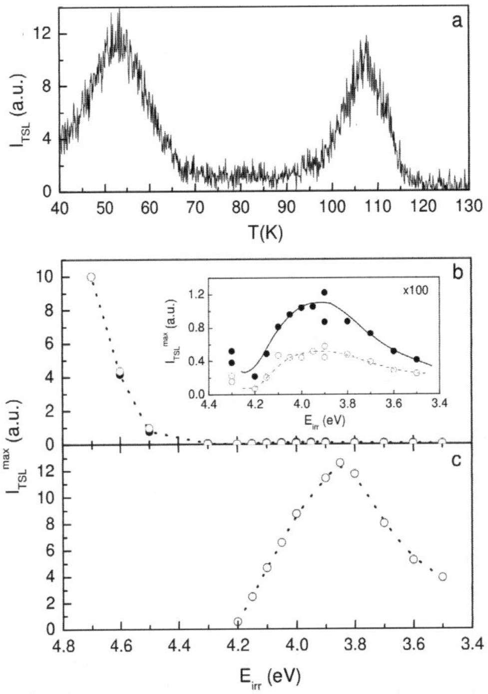

Figure 4. (a) The TSL glow curve measured after irradiation of the $PbWO_4:Y^{3+}$ crystal at 26 K with E_{irr}=3.8 eV. (b, c) Creation spectra of the TSL peaks at 53 K (solid circles) and 107 K (open circles) (normalized at 4.7 eV) measured under irradiation (b) at 26 K and (c) at 75 K.

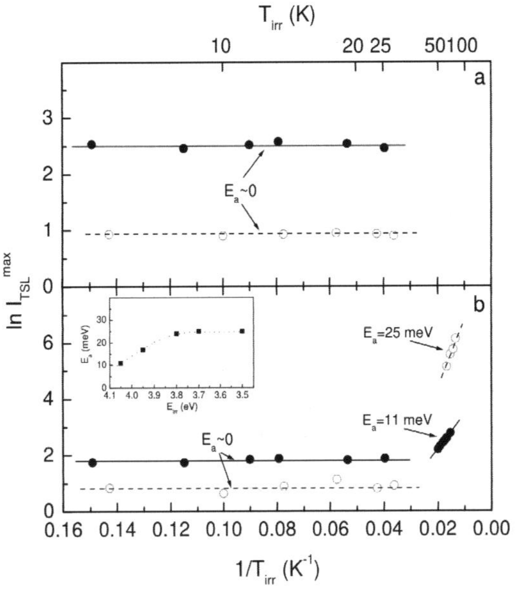

Figure 5. Dependence of the TSL intensity on the irradiation temperature T_{irr} measured for the TSL peaks located at (a) 53 K and (b) 107 K after irradiation of the $PbWO_4:Y^{3+}$ crystal with E_{irr}=4.05 eV (solid circles) or E_{irr}=3.7-3.8 eV (open circles). In the inset, dependence of the activation energy E_a for the 107 K peak creation on E_{irr} obtained under irradiation at T_{irr}>55 K.

3.3. Electron spin resonance studies

The ESR spectra of $(WO_4)^{3-}$ and $(WO_4)^{3-}$-La^{3+} centers in $PbWO_4$ crystals were studied in [16, 17] and [5, 13], respectively. The ESR characteristics of $(WO_4)^{3-}$-Lu^{3+} centers were described in [17]. In the present paper, the ESR signals from $(WO_4)^{3-}$ and $(WO_4)^{3-}$-Y^{3+} centers were detected. After irradiation of $PbWO_4:Y^{3+}$ with E_{irr}=4.05 eV at T_{irr}<40 K, $(WO_4)^{3-}$ centers are mainly created, while $(WO_4)^{3-}$-Y^{3+} centers appear only after heating of the irradiated sample up to 60 K or after irradiation at 60-80 K (figure 6). Similar results were obtained in [5, 17] under irradiation of $PbWO_4:A^{3+}$ crystals with the He-Cd laser (3.82 eV). From the dependence of the ESR signal intensity on the irradiation temperature, the activation energies E_a were calculated for the creation of $(WO_4)^{3-}$-Y^{3+} centers. Under irradiation at T_{irr}<40 K, the E_a values are close to zero. Under irradiation at 4.05 eV in the 60-80 K temperature range, a value of E_a=11 meV was obtained. This value is close to the E_a value obtained for the creation of the TSL peak at 107 K under the same irradiation conditions, as well as to the value of E_q for thermal quenching of the UV emission in the $PbWO_4:Y^{3+}$ crystal (table 1).

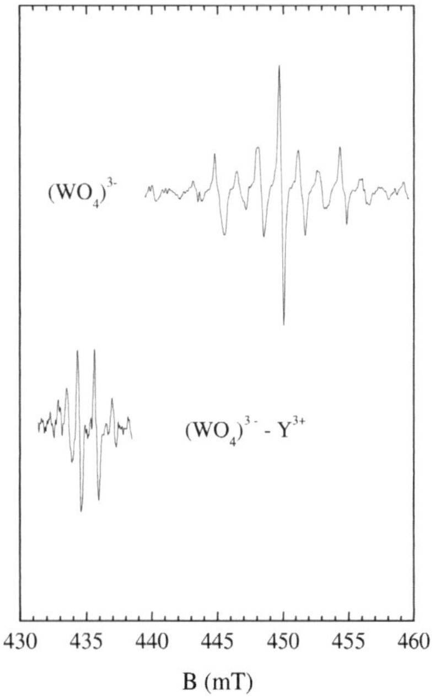

Figure 6. The ESR spectra of $(WO_4)^{3-}$ and $(WO_4)^{3-}$-Y^{3+} centers measured at 18 K after irradiation of the $PbWO_4:Y^{3+}$ crystal at T_{irr}=20 K and T_{irr}=60 K, respectively. E_{irr}=4.05 eV. The orientation of the magnetic field was close to the [100] axis of the crystal.

4. Discussion

The data obtained in [5, 17, 18] and in the present paper indicate that electron $(WO_4)^{3-}$ and $(WO_4)^{3-}$-A^{3+} centers in $PbWO_4$ crystals can be created by the photons of energies 3.8 ± 0.4 eV which are about 1 eV smaller than the band gap energy. The creation of $(WO_4)^{3-}$ centers was explained before by a tunnelling process of the electron from an excited Pb^{2+} ion [17] or by the ionisation of $(MoO_4)^{2-}$ groups and the subsequent trapping of the released electrons at $(WO_4)^{2-}$ groups [18]. However, the $PbWO_4:A^{3+}$ crystals studied do not contain the Mo impurity. In the undoped $PbWO_4$ crystals, the absorption in the energy range around 3.8 eV is usually ascribed to the excitons from Pb-deficient (e.g., $Pb_{7.5}W_8O_{32}$ [19]) crystal regions or to some defect-related centers. In [1-4] it was shown that at T_{irr}>120 K, the photo-thermally stimulated disintegration of the mentioned exciton- and defect-related states takes place with the activation energies of about 0.2-0.4 eV which are much larger than the values of E_q and E_a obtained in the present paper.

The UV emission band, whose excitation spectrum coincides with the creation spectrum of electron $(WO_4)^{3-}$ and $(WO_4)^{3-}$-A^{3+} centers in the energy range around 3.8 eV, is relatively wide. Its FWHM=0.60±0.04 eV, like for the B and G(I) emission bands which both are of exciton origin. However, the Stokes shift (≈0.7 eV) is smaller than that observed for the other exciton emission bands in $PbWO_4$ crystals. The presence in the decay kinetics of the UV emission of two (fast - 35 ns and slow - 350 ns at 80 K) components with practically coinciding emission spectra is characteristic for an emission, arising from the triplet relaxed excited state (see, e.g., [20] and references therein).

We assume that the effects observed are due to the radiative and non-radiative decay of an excited state of some defect-related center. Under excitation around 3.8 eV, not only the radiative decay of this state, accompanied with the weak UV emission, but also ionisation of the defect center is possible. At T_{irr}<40 K, the electrons released into the conduction band are trapped at $(WO_4)^{2-}$ groups and electron $(WO_4)^{3-}$ centers are created with the activation energy E_a=0. The energy level of $(WO_4)^{3-}$ is located at 50 meV below the bottom of the conduction band [16]. We assume that tunneling transitions of electrons from the $(WO_4)^{3-}$ centers into the excited state of the mentioned defect center can take place. This process results also in the appearance of the UV emission. At T_{irr}>60 K, when the $(WO)_4^{3-}$

centers are thermally unstable, the optically released and trapped at $(WO_4)^{2-}$ electrons are immediately thermally released from the created $(WO_4)^{3-}$ centers into the conduction band and retrapped at more deep $(WO_4)^{2-}$-A^{3+} traps. As a result of this dynamical process, electron $(WO_4)^{3-}$-A^{3+} centers are created with the activation energy which is smaller than 50 meV. This process leads to the decrease of the population of the excited state of the defect center and, consequently, to the thermal quenching of the UV emission with the same activation energy of 10-25 meV. Further studies are needed to make a justified conclusion on the origin of the defect center responsible for the effects observed and on the structure and dynamics of its excited states.

5. Conclusions

The ESR and TSL studies of $PbWO_4$:A^{3+} crystals (A^{3+}: La^{3+}, Lu^{3+}, Y^{3+}, Ce^{3+}, Gd^{3+}, Eu^{3+}) irradiated in the 3.4-5.0 eV energy range at 5-80 K indicate that electron $(WO_4)^{3-}$ and $(WO_4)^{3-}$-A^{3+} centers can be created under irradiation of the crystals not only in the band-to-band transitions region but also in the energy range around 3.8 eV. The activation energies for their creation are E_a=0 and E_a=10-25 meV, respectively. Under excitation in this range, a new UV emission band peaking at 3.05-3.20 eV is found. At T>60 K, both thermal quenching of the UV emission and the photo-thermally stimulated creation of $(WO_4)^{3-}$-A^{3+} centers take place with the same activation energy. The radiative and non-radiative decay of a defect-related state are proposed as possible reasons of the effects observed.

Acknowledgements

This work was partly supported by the Estonian Science Foundation Grants No. 7507, 6999 and Czech Institutional Research Plan AV0Z10100521.

References

[1] Krasnikov A, Nikl M and Zazubovich S 2006 *Phys. Status Solidi* B **243** 1727
[2] Laguta V V, Nikl M and Zazubovich S 2007 *Rad. Measur.* **42** 515
[3] Babin V, Bohacek P, Krasnikov A, Nikl M, Stolovits A and Zazubovich S 2007 *J. Lumin.* **124** 113
[4] Laguta V V, Nikl M and Zazubovich S 2008 *IEEE Transactions on Nuclear Science* **55** issue 3 part 1 1275
[5] Laguta V V, Martini M, Meinardi F, Vedda A, Hofstaetter A, Meyer B K, Nikl M, Mihokova E, Rosa J and Usuki Y 2000 *Phys. Rev.* B **62** 10109
[6] Laguta V V, Martini M, Vedda A, Nikl M, Mihokova E, Bohacek P, Rosa J, Hofstaetter A, Meyer B K and Usuki Y 2001 *Phys. Rev.* **B 64** 165102
[7] Han Baoguo, Feng Xiqi, Hu Guanqin, Wang Pingchu and Yin Zhiwen 1998 *J. Appl. Phys.* **84** 2831
[8] Kobayashi M, Ishii M, Harada K, Usuki Y, Okuno H, Shimizu H and Yazawa T 1996 *Nucl. Instr. Meth. Phys. Research* A **373** 333
[9] Bohacek P, Senguttuvan N, Kiisk V, Krasnikov A, Nikl M, Sildos I, Usuki Y and Zazubovich S 2004 *Rad. Measur.* **38** 623
[10] Tamulaitis G, Burachas S, Martinov V P, Ryzhikov V D, Gutbrod H H, Manko V I and Terekhin M 1997 *Phys. Status Solidi* A **161** 533
[11] Martini M, Meinardi F, Spinolo G, Vedda A, Nikl M and Usuki Y 1999 *Phys. Rev.* B **60** 4653
[12] Böhm M et al. 2000 *Proc. of the 5th Int. Conf. on Inorganic Scintillators and Their Applications (Moscow, Russia, August 16-20, 1999)* ed V V Mikhailin (Moscow: Moscow State University Press) p 619
[13] Hofstaetter A et al. 2000 *Proc. of the 5th Int. Conf. on Inorganic Scintillators and Their Applications (Moscow, Russia, August 16-20, 1999)* ed V V Mikhailin (Moscow: Moscow State University Press) p 128
[14] Nikl M, Bohacek P, Mihokova E, Martini M, Meinardi F, Vedda A, Fabeni P, Pazzi G P, Kobayashi M, Ishii M and Usuki Y 2000 *J. Appl. Phys.* **87** 4243

[15] Nikl M, Bohacek P, Mihokova E, Solovieva N, Martini M, Vedda A, Fabeni P, Pazzi G P, Kobayashi M, Ishii M, Usuki Y and Zimmerman D 2001 *J. Cryst. Growth* **229** 312
[16] Laguta V V, Rosa J, Zaritskii M I, Nikl M and Usuki Y 1998 *J. Phys.: Condens. Matter* **10** 7293
[17] Böhm M, Henecker F, Hofstaetter A, Luh M, Meyer B K, Scharmann A, Kondratiev O V and Korzhik M V 1999 *Rad. Eff. & Defects Solids* **150** 21
[18] Hofstaetter A, Korzhik M V, Laguta V V, Meyer B K, Nagirnyi V and Novotny R 2001 *Rad. Measur.* **33** 533
[19] Moreau J M, Galez Ph, Gladyshevski R E, Peigneux J P and Korzhik M V 1999 *J. Allows and Comp.* **284** 104
[20] Fukuda A 1970 *Phys. Rev.* **B1** 4161

Journal of Physics: Conference Series **249** (2010) 012002 doi:10.1088/1742-6596/249/1/012002

Visible photoluminescence in coloured Lithium Fluoride under ultra-violet continuous wave excitation

H J Kalinowski[1], **R M Monterali**[2], **M A Vincenti**[2] **and R N Nogueira**[3]

[1] Federal University of Technology – Paraná (UTFPR), Av. Sete de Setembro 3165, 80230-901 Curitiba, Brazil

[2] ENEA, Physical Technologies and New Materials Dept., C.R. Frascati, via Enrico Fermi 45, 00044 Frascati (RM), Italy

[3] Instituto de Telecommunicações and University of Aveiro, Campus Universitário de Santiago, 3410-193 Aveiro, Portugal

E-mail: `hjkalin@cpgei.ct.utfpr.edu.br`,`montereali@frascati.enea.it`,`rnogueira@av.it.pt`

Abstract. The γ-coloured Lithium Fluoride crystals visible fluorescence at room temperature was measured under excitation with ultra-violet continuous wave light at $\lambda = 244\,\mathrm{nm}$ from a frequency-doubled Ar^+-ion laser. The excitation wavelength lies close to the peak of the F-centre absorption band for LiF, and the visible luminescence spectrum consists of two broad emission bands, assigned to F_3^+ and F_2 centres in LiF. The luminescence spectrum changes in a short time after start of illumination. The main effect of 244 nm laser illumination is the observed change in the F_2 emission band intensity, which is strongly and permanently bleached by the CW UV light at 244 nm as soon as it illuminates the crystal. The time evolution of the fluorescence signals can be measured, and the corresponding curves were fitted by a single exponential decay, resulting in characteristic time constants of $\approx 9\,\mathrm{s}$ and $\approx 30\,\mathrm{s}$ for the F_2 and F_3^+ bands, respectively. The reported effects support previous work about UV laser illumination in the F-band absorption peak as an effective method to control the ratio of F_2 to F_3^+ centres in LiF, due to permanent bleaching of the former ones.

1. Introduction

Lithium Fluoride presents a few characteristics that differ from other alkali halides, since, e.g., it is almost non hygroscopic and crystals are very hard to cleave. Additive colouration of LiF crystals is not effective; however, it is possible to produce colour centres by irradiation with X-rays, $\gamma-$rays, electrons or ionized particles beams and even by means of high power laser pulses. Many colour centre absorption bands have been observed in coloured LiF crystals, and several of them were associated to F-centres and F-aggregate centres [1].

The F-centre in LiF presents an absorption band in the ultra violet (UV) spectral region, centred at 248 nm (5.0 eV) when measured at room temperature (RT). One of the most intriguing optical characteristics of coloured LiF is that the F-centre emission band has not yet been assigned. From Mollwo-Ivey Law, the emission band should lie at around ($\approx 900 - 910\,\mathrm{nm}$) [2]. Some experiments, which used direct pulsed laser pumping in the F-band at 248 nm, found an

Journal of Physics: Conference Series **249** (2010) 012002 doi:10.1088/1742-6596/249/1/012002

emission band, centred at 910 nm [3], which is superimposed to that assigned to the F_2^+ centre emission.

Photoluminescence in LiF has been studied by exciting through conventional broad-band c.w. (continuous wave) sources, followed by filters or monochromators, and the results usually show the interdependence of F_2 and F_3^+ luminescence bands. Some experiments with UV light (248, 315 nm) pointed to F_2 and F_3^+ emission bands. However, such first experiments were performed with low fluence light, and their results mainly show differences in emission spectra related to the crystal growth and irradiation techniques [4]. In any case, no evidence of the permanent bleaching of these centres has been observed. In fact, the regenerative optical pumping cycle of F_2 and F_3^+ centres is important for their use in colour centre lasers.

Recently,due to studies of waveguides and optical gain in LiF crystals and films [5, 6], there has benn a renewed interest in the optical emission properties of LiF. Grating assisted laser emission using femtosecond pulses to record the grating structures in LiF crystals were also reported [7, 8]. Low power UV c.w. laser light was also used to record submicrometric patterns on γ- and soft X-rays coloured crystals and films[9, 10, 11, 12]. In those papers, the authors also observed a permanent bleaching of F_2 centre emission in contrast to that of F_3^+ centres as a consequence of the UV illumination, which can lead to laser action in the green region of the spectrum.

2. Experimental details

The LiF crystals used in the experiments were grown and coloured by γ-rays at the Institute of Molecular and Atomic Physics, Minsk (Belarus); after irradiation, their colour ranges from light yellow to dark red, depending on the dose received ($\approx 10^6 - 10^7$ R).

Photoluminescence experiments were performed at RT using an Ar^+ ion laser (Spectra Physics) with an intracavity frequency doubler based on BBO crystal (GWU Lasertechnik) operating at 244 nm, which is normally used to record fibre Bragg gratings at the Physics Department, University of Aveiro (Portugal). Profiting from the recording set-up, prior to the sample the laser beam is deflected by a 460 nm pitch phase mask (Ibsen Photonics), so that only UV light impinges on the crystal. The laser power is measured with a thermal detector placed in front of the phase mask. Light emitted from the crystal is collected by a 400 μm diameter fibre optic placed close to the illuminated spot. Photoluminescence spectra and time evolution of the emission signals are acquired by using an Ocean Optics Mini-Spectrometer (USB4000) and their specific SpectraSuite software package in fast acquisition mode.

Due to the transient characteristics of the photoluminescence, the exciting and collecting geometry is first aligned by using one spot in the crystal surface. The sample is then moved with a micrometric translation stage, with the laser beam stopped, so that another spot can be illuminated to acquire the initial emission spectrum as soon as illumination begins. A final photoluminescence spectrum, after the established illumination period, is also measured. After that, again with the beam stopped, the coloured crystal is moved to another position in order to obtain the fluorescence time evolution. The total time from the opening of the exciting laser beam to the final photoluminescence spectrum acquisition is in the order of a few seconds, ($\approx 2 - 5$) s.

Optical transmittance spectra of the used crystals were measured at RT by using a Lambda 950 Perkin-Elmer spectrophotometer in the NUV and VIS spectral range, extending from 290 to 800 nm at a wavelength resolution of 2 nm. Photoluminescence spectra were also performed at RT with the same spectral resolution with an ISA-SPEX Jobin Yvon Fluorolog-3 spectrofluorometer equipped with a 450 W Xenon lamp and a solid-sample holder in front face detecting geometry, where the emitted signal is collected at an angle of 20° with respect to the normal direction of the sample surface, coincident with the exciting beam direction. A proper mask was used to select illuminated or non illuminated areas of the crystal.

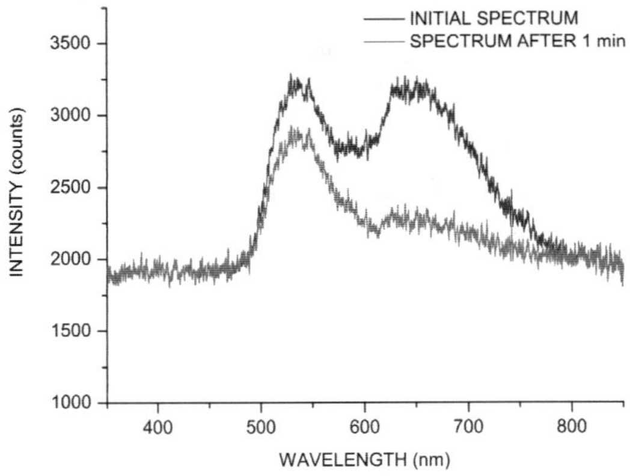

Figure 1. Luminescence spectra of a LiF sample taken just after the start of illumination and after 1 min, integration time 300 ms.

3. Results and Discussion

Figure 1 shows the comparison between the photoluminescence spectra of a LiF crystal under c.w. UV excitation, acquired just after the UV illumination started and about one minute later. The γ-ray irradiation dose of that sample was estimated to be about 8.4×10^4 Gy. Using the Smakula's the formula, the F centre concentration is determined to be $\approx 1.33 \times 10^{18}$ cm^{-3}. It is clear from the spectra in Figure 1 that the F_2 centre emission is strongly reduced after the UV illumination, whereas the F_3^+ photoemission shows a remarkable lower depletion. The effect is permanent (at least for a period of a few months), thus suggesting that the UV laser light at 244 nm really depletes the F_2 centre population, due to an unknown non-radiative recombination mechanism.

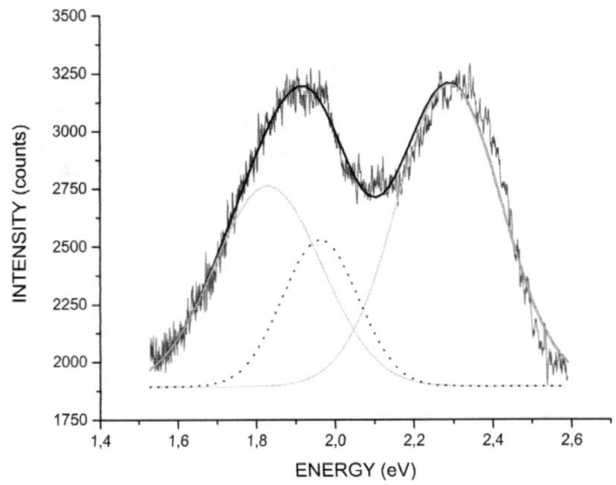

Figure 2. Fitting of the initial luminescence spectrum using three Gaussian functions with fixed peak energy parameter. Scattered line: experimental data, ——: resulting best fit, ······: individual components of the fit.

The emission spectra can be fitted as a sum of Gaussian bands, whose main peaks are centred at 1.83 eV and 2.29 eV. Within the experimental errors, these values are compatible with previous measurements of F_3^+ and F_2 defects photoemission bands in LiF, respectively. They are generally observed under light excitation at around 450 nm, close to the absorption peak of the M-band, which consists of the superposition of F_3^+ and F_2 centres fundamental absorption bands [1]. However, the quality of the fitting can be improved when a third band, peaked at 1.96 eV, is included, as shown in the graph of Figure 2 (notice the X-axis in energy units). This band was reported by several authors and tentatively assigned to perturbed F_2 centres [13].

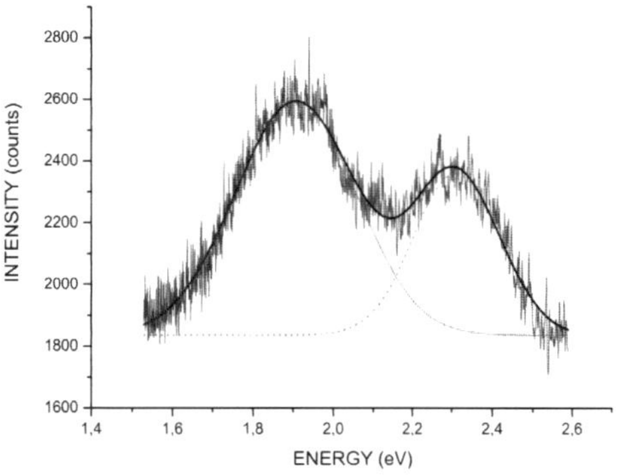

Figure 3. Fitting of the initial luminescence spectrum using two Gaussian functions with free peak energy parameter. Scattered line: experimental data, ——: resulting best fit, ······: individual components of the fit.

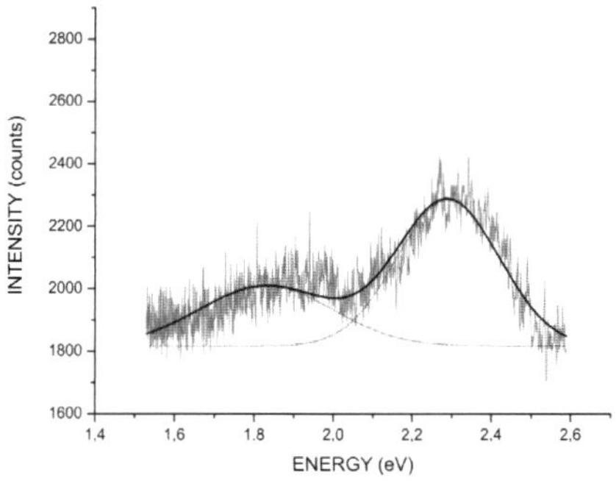

Figure 4. Fitting of the final luminescence spectrum using two fixed peak energy Gaussian functions. Scattered line: experimental data, ——: resulting best fit, ······: individual components of the fit

Figures 3 and 4 show the spectra obtained in similar conditions for another LiF sample (X-axis in energy), irradiation dose $\approx 8.4 \times 10^4$ Gy, together with the corresponding best fit by using two Gaussian bands. Even though the irradiation dose is similar to the previous sample, the F centre concentration, as calculated from Smakula's formula, is lower for the second sample ($\approx 5.01 \times 10^{18}\,\text{cm}^{-3}$). The graph in Figure 3 is obtained with the peak energies as a free parameter in the fit, whereas the second fit, Figure 4, fixes the peak energy of each component to the known F_3^+ and F_2 emission bands. In both cases, the resulting best fits are of the same quality, so the usual F_3^+ and F_2 centre structure is assumed.

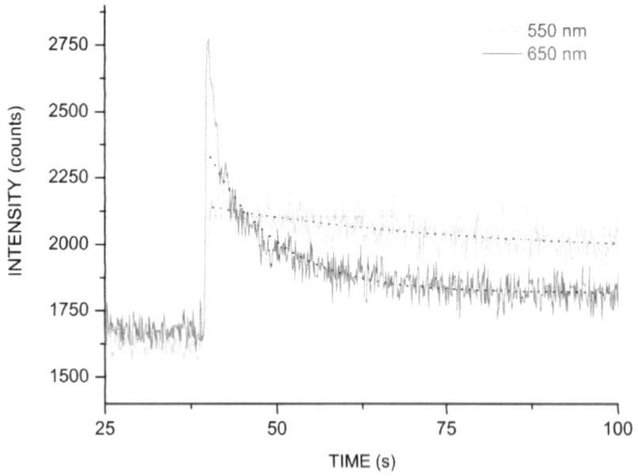

Figure 5. Time decay of the photoluminescence signals of a coloured LiF crystal under c.w. excitation at 244 nm, integration time 100 ms. The dotted lines show the best-fit of the experimental data with a single exponential decay.

From the spectra presented it is clear that the F_3^+ photoluminescence band intensity variation is much lower than that of the F_2 band, which is almost completely bleached after a short period of illumination, even by using low power c.w. light at 244 nm. This behaviour differs from the known F_3^+ and F_2 emission intensity evolution when coloured LiF is illuminated by c.w. laser wavelengths at 458 nm, in the M-band absorption band [14].

Figures 5 and 6 show, respectively, the decay of the fluorescence signals obtained for the two previously mentioned samples. The general behaviour is the faster decay of the red component as compared to the green one. The decay data are fitted with a single exponential decay law to obtain the characteristics times.

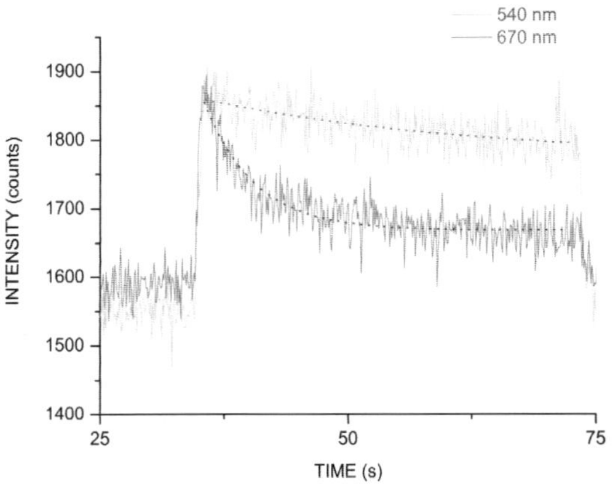

Figure 6. Time decay of the photoluminescence signals of the coloured LiF crystal (see text for details) under c.w. excitation at 244 nm, integration time 100 ms; notice that the monitored emission wavelengths are different from the previous figure. The dotted lines show the best-fit of the experimental data with a single exponential decay.

For the first sample the resulting characteristic time constants are 9.9 s for the emission wavelength of 650 nm and 36.2 s for the 550 nm intensity; they are obtained with a fluence of about $0.12\,W/cm^2$. For the second sample, the corresponding time constants are 5.0 s and 27.1 s at the observed wavelengths of 670 nm and 540 nm, respectively.

4. Conclusion

From the measurements of the RT photoluminescence of coloured LiF crystals under c.w. laser illumination at 244 nm, it was derived that the emission bands associated to F_2 and F_3^+ centres possess different time behaviour. Their intensities decay are fitted with a single exponential law, with a faster decay rate for the F_2 emission band. The bleaching process is non-reversible, to the scale of several months. No other well-assigned emission bands were observed in the initial and final emission spectra, investigated in the $300 - 900$ nm wavelength range.

The reported changes in the fluorescence spectra under c.w. laser illumination at 244 nm and the associated time constants are compatible with our previous work about UV laser illumination in the F-band absorption peak as an effective method to control the ratio of F_2 to F_3^+ centres in LiF [9, 10] due to permanent bleaching of the former defects. This characteristic can be useful for obtaining LiF laser action in the green region of the spectrum.

Both the time constants and the relative spectral intensities are strongly dependent on the initial F-centre concentration in the sample, as well as the UV laser power density. Further experiments are in progress to clarify these points.

Acknowledgments

The authors thank Dr. Vladimir Kalinov, Institute of Molecular and Atomic Physics, Minsk (Belarus) for kindly supplying the samples used. Prof. Kalinowski thanks ENEA for a fellowship during which part of this work has been done. This work has been partially funded by CAPES, CNPq, Fundação Araucária and FINEP (Brazil) and by FCT (Portugal).

References

[1] Nahum J and Wiegand D A 1967 Optical properties of some F-aggregate centers in LiF *Phys. Rev.* **154** 817
[2] see, e.g., W. Beall Fowler (ed.), Physics of Color Centers, Academic Press (1968)
[3] Baldacchinni M, Cremona M, Montereali R M, Grassano U M and Kalinov V 1992 Emission properties of γ irradiated LiF crystals excited in the F absorption band by an excimer laser *Proc. XII International Conference on Defects in Insulating Materials (Schloss Nordkirchen)* (Singapore: World Scientific) p 1103
[4] Görlich P, Karras H and Kötitz G 1963 On the fluorescence of sythetic LiF crystals *Phys. stat. sol.* **3** 1803
[5] Montereali R M, Piccinini M and Burattini E 2001 Amplified spontaneous emission in active channel waveguides produced by electron-beam lithography in LiF crystals *Appl. Phys. Lett.* **78** 4082
[6] Montereali R M and Piccinini M 2002 Optical gain of F_2 colour centres in LiF confining structures realised by electron-beam lithography *Opt. Commun.* **209** 201-208
[7] Kurubori T, Kawamura K, Hirano M and Hosono H 2003 Simultaneous fabrication of laser-active colour centres and permanent microgratings in lithium fluoride by a single femtosecond pulse *J. Phys.: Condens. Matter* **15** L399-L405
[8] Kurubori T, Wang Z, Hirose Y and Sakai T 2008 Femtosecond-laser-encoded dual-beam distributed-feedback colour centre laser in bulk LiF *Proc. International Conference on Luminescence - ICL 08 (Lyon)*
[9] Vincenti M A, Almaviva A, Montereali R M, Kalinowski H J and Nogueira RN 2006 Permanent luminescent micropatterns photoinduced by low-power ultraviolet irradiation in lithium fluoride *Appl. Phys. Lett.* **89** 241125
[10] Vincenti M A, Kalinowski H J and Montereali R M 2007 Permanent bleaching of F_2 color centers in lithium fluoride under UV light illumination *phys. stat. sol.(c)* **4** 1139
[11] Bonfigli F, Vincenti M A, Almaviva S, Montereali R M, Nichelatti E, Kalinowski H J and Nogueira R N 2008 *Proc. RIAO/Optilas 2007 (Campinas)* vol CP992 (Melville: American Institute of Physics) p 710
[12] Bonfigli F, Vincenti M A, Almaviva S, Montereali R M, Nichelatti E, Nogueira R N and Kalinowski H J 2009 *Appl. Optics* **48(31)** G38-G43
[13] Schepina L I, Lobanov N T, Maksimova V M and Kostyukov 1984 Perturbed F_2 centers in LiF crystals *Opt. Spectroscopy* **57** 817
[14] Baldacchini G, Cremona M, d'Auria G, Montereali R M and Kalinov V 1996 Radiative and nonradiative processes in the optical cycle of the F_3^+ center in LiF *Phys. Rev.* B **54** 17508

F and F-aggregates colour centres in lithium fluoride for high spatial resolution x-ray imaging

R. M. Montereali[1*], S. Almaviva[1], F. Bonfigli[1], I. Franzini[1], D. Pelliccia[2], A. Cedola[3] and S. Lagomarsino[3]

[1]ENEA, Physical Technologies and New Materials Dep., C.R. Frascati, V. E. Fermi 45, 00044 Frascati (Rome), Italy

[2] Institut für Synchrotronstrahlung – ANKA, Forschungszentrum Karlsruhe Herman-von-Helmholtz-Platz 1, 76344 Eggenstein-Leopoldshafen, Germany
[3]Istituto Fotonica e Nanotecnologie, CNR, V. Cineto Romano 42, 00156 Rome, Italy

Abstract. Stable formation of primary F and aggregate F_2 and F_3^+ colour centres in LiF crystals irradiated with a X-ray table-top source was investigated by optical absorption and photoluminescence measurements as a function of the irradiation dose. The F absorption band intensity follows a power law behaviour as a function of dose with a single exponent equal to 0.77, a value in excellent agreement with the predictions of a unified model for colour centres formation. The aggregate defects photoluminescence response appears suitable for applications in high dynamic range imaging plate X-ray detectors based on photoluminescence of coloured LiF. The choice of polycrystalline LiF thin films grown on different substrates allows one to obtain high resolved images even when more penetrating X-rays are used, as the colour centres formation is limited in depth by the film total thickness.

1. Introduction

A novel soft X-ray imaging detector based on the photoluminescence (PL) of active colour centres in Lithium Fluoride, LiF, thin layers was lately proposed [1,2] and is currently under further development [3] for investigation of biological specimens, characterization of materials and devices, as well as transfer and test of light-emitting micro and nano-patterns. A relevant advantage of this type of LiF-based detector is that it can be used both in the form of crystals and of films deposited onto different substrates, like silicon or glass [4]. Large area homogeneous thin films can be obtained by simple and well assessed physical deposition methods.

Recently, this X-ray imaging detector based on optical reading of colour centres PL was proposed and successfully tested also in the hard X-ray region [5] using a table-top X-ray micro-source. The high spatial resolution on a large field of view, a wide dynamic range, versatility and simplicity of use of the LiF-based imaging detector allow to obtain absorption radiographies of geometrical microstructures and biological objects, placed in close contact with its surface during X-ray irradiation. A spatial resolution of about 250 nm was demonstrated in the optical images obtained through confocal laser scanning fluorescence microscopy [5]; these micro-radiographies were read by simultaneously collecting the PL of F_2 and F_3^+ aggregate electronic point defects (two electrons

*Rosa Maria Montereali, montereali@frascati.enea.it; phone +39 06 94005296; fax +39 06 94005334

bounded to two and three anion vacancies, respectively). Their broad visible PL bands, peaked at ~ 678 an ~ 541 nm, respectively, are both simultaneously excited with the 458 nm line of an Argon laser, located in the M band spectral region [6], where their almost overlapping absorption bands are present [7].

In this work a careful spectroscopic investigation of LiF crystals irradiated in air by a laboratory X-ray tube with a Cu anode was carried out through systematic optical absorption and PL measurements performed at room temperature (RT) as a function of the irradiation dose. An example of X-ray micro-radiography of a biological object on a LiF imaging detector consisting of a LiF film grown on silicon is reported.

2. Experimental

Commercially available LiF crystals of diameter 25 mm and thickness 2 mm, polished on both sides, were irradiated in air with an X-ray micro-source (NOVA 600, Oxford Instruments) of nominal source size 15 μm at a 20 W power. This laboratory X-ray tube was equipped with a Cu anode emitting quasi-monochromatic radiation, strongly peaked at about 8 keV ($K\alpha_1$ and $K\alpha_2$ lines) with a minor contribution at 8.9 keV ($K\beta_1$ line). The emitted radiation is transmitted through a 245 μm thick Be window. The distance between the sample and the Cu target was 3.5 cm and several areas of typical dimensions (5x5) mm^2 were exposed for irradiation times from 1 to 90 minutes (fluencies in the range 68-6100 mJ/cm^2) by moving the LiF crystals behind an open aperture in a thick Pb sheet. Optical transmittance spectra were measured at RT by using a Lambda 19 Perkin-Elmer spectrophotometer in the near ultraviolet and visible spectral range, extending from 200 to 800 nm at a wavelength resolution of 2 nm. Visible PL spectral measurements of the irradiated LiF samples were performed under quasi-continuous optical excitation with an Argon laser line at 458 nm. The filtered PL signal was focused in a 40 cm focal length monochromator equipped with a photomultiplier detector connected to a lock-in amplifier.

3. Results and discussions

In figure 1 the RT optical absorption spectra of coloured LiF crystals irradiated for several times, corresponding to different doses, were reported. The main spectral features were the F and M absorption bands, located at ~ 245 nm [8] and at ~ 450 nm [6], respectively.

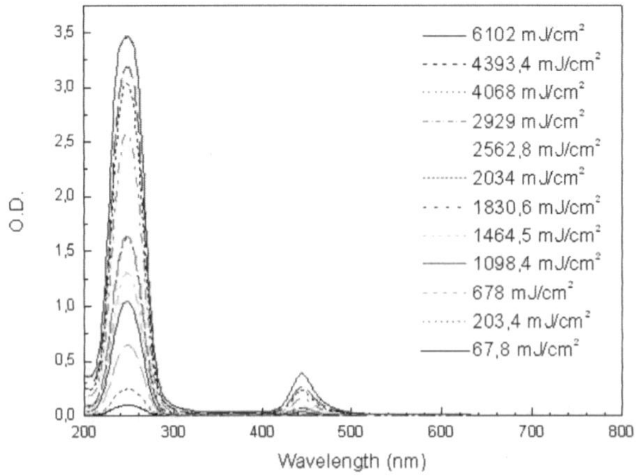

Figure 1. RT optical absorption spectra of X-ray coloured LiF crystals irradiated at several doses. The blank crystal contribution has been subtracted.

Their intensities grew with the increasing of dose. At the highest dose, the F band reached the instrumental saturation value. These optical absorption spectra were used in the quantitative evaluation of the amount of electronic defects inside the samples. In fact, the absorption band area is directly connected to the colour centres concentration *N* via the following Smakula formula [9], applied for bands of known shape,

$$N(cm^{-3}) = A\frac{n}{(n^2+2)^2}\frac{W(eV)}{f}\alpha_{\max}(cm^{-1})$$

where the band full-width at half-maximum, *W,* and the maximum value of the absorption coefficient, α_{max}, appear; *n* is the refractive index of the crystal at the wavelength corresponding to the absorption band peak and *f* the oscillator strength of the transition responsible for that absorption. *A* is a calculated constant of value 0.87×10^{17} for Gaussian bands.

Figure 2 reported the peak intensities of the F absorption bands, in O.D. , for LiF crystals irradiated at different X-ray doses (J/cm^2). They were derived from the X-ray source technical characteristics, taking into account the attenuation of the Be window and of the propagation in air. The X-ray doses were also reported in the top axis of figure 2 in keV/cm^3. These values were calculated taking into account the limited penetration of this kind of radiation in LiF, which was assumed equal to the attenuation length of X-rays at the energy of the Cu-Kα transition, which corresponds to ~ 330 μm. Obviously the same value was utilised to derive α_{max}. According to the Smakula formula, assuming *f* equal to 0.56 [10] and n = 1.4203 [11], the F centre densities for different X-ray doses were calculated and reported on the right axis of figure 2. The primary F defects concentration reached values higher than 10^{19} cm^{-3}.

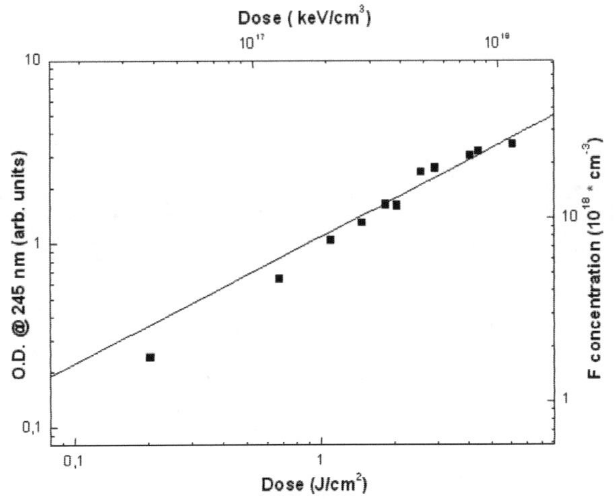

Figure 2. Optical density at 245 nm of LiF crystals irradiated at different X-ray doses (left axis). The F colour centre concentrations are also reported (right axis). The solid curve is a best fit of the experimental data (see text for details).

The best fit of the experimental points in figure 2, based on least squares minimisation, indicated that the F absorption band intensity follows a power law behaviour as a function of dose with a single exponent equal to 0.77. This result is in excellent agreement with the predictions of a unified model [12], suitable for the description of the general behaviour of colour centres formation in several alkali halide crystals under irradiation, which provided a value of 0.75 for LiF. The same value was found in stage III of the coloration curve for LiF crystals irradiated at RT by 3 keV electrons [13].

The energy needed to produce an F centre, from which the coloration efficiency can be estimated, is calculated between 0.24 and 0.61 keV, which appears quite low and it corresponds to more than 13 F centres/photon. Since approximately 50% of the absorbed X-ray radiation above the fluorine edge in LiF is re-emitted as X-ray fluorescence [14] this efficiency corresponds roughly to 0.5 – 1 keV per F centres. So again, several F defects were created for each X-ray photon absorbed.

The same procedure was applied to the M absorption band, which was decomposed as the sum of two Gaussian bands, according with the F_2 and F_3^+ bands' parameters reported in [7]. These aggregate defects reached maximum concentrations of the order of $7x10^{16}$ cm^{-3}, but it was not possible to describe their evolution as a function of dose with a similar power law. As a matter of fact, the ratio between the intensities of F to M band in the absorption spectra of figure 1 is quite high, of the order of 10 at the highest doses. This behaviour seems to indicate that the primary F centres concentration initial increases [14] and successively it is followed by the aggregation of larger complex defects at highest doses, in the used conditions with a dose-rate of 1.13 mW/cm^2.

In order to better investigate the F_2 and F_3^+ aggregate colour centre behaviour as a function of irradiation dose, laser induced visible PL spectra under laser excitation at 458 nm were also measured at RT in the same experimental conditions for all the samples. As example, in figure 3, the PL spectrum of a LiF crystal irradiated at a high dose is reported, together with the best fit obtained as the sum of two Gaussian bands, whose main parameters, i.e. peak energies and width, are consistent with the values reported in [7] for F_2 and F_3^+ red and green PL bands, respectively.

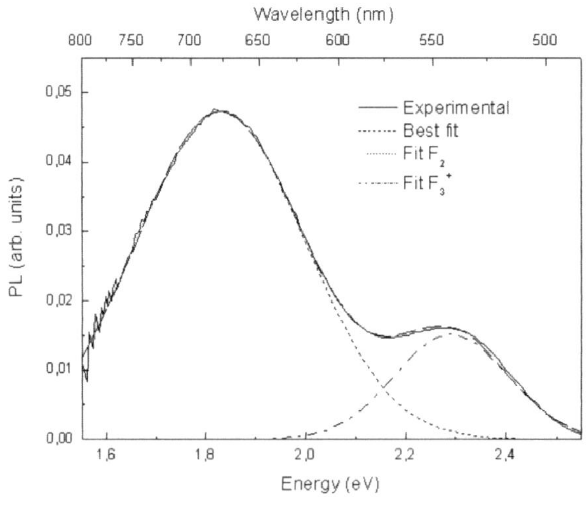

Figure 3. RT photoluminescence spectrum (solid) of a X-ray coloured LiF crystals irradiated at a dose of 4068 mJ/cm^2. The best fit (dotted) is obtained as the sum of two Gaussian bands, attributed to F_2 (dashed) and F_3^+ (dash-dotted) centres.

In figure 4 the PL intensities of the F_2 and F_3^+ PL bands in X-ray irradiated LiF crystals as a function of irradiation time are reported. Their emission signals cover two decades in the investigated dose range, a behavior which looks like suitable for practical use in X-ray imaging detectors based on the PL of these active defects.

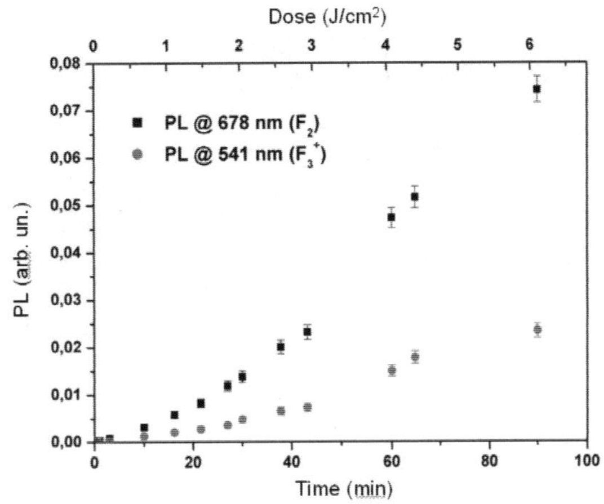

Figure 4. RT photoluminescence intensities of F_2 and F_3^+ aggregate colour centre in LiF crystals irradiated at different X-ray doses under laser excitation at 458 nm.

This imaging-plate X-ray LiF-based detector possesses intrinsic high spatial resolution on a large field of view, has a wide dynamic range and is easy to use, as it is insensitive to visible light and it does not need any development after irradiation. In order to fully exploit its capabilities, the use of LiF polycrystalline thin films thermally evaporated on glass or silicon substrates appears more suitable. Typical thicknesses between 0.5 and 2 μm for these films can be selected [4]; as a consequence they are entirely colored by high penetrating hard X-rays. However a large fraction of energy is deposited in the substrate and the effective PL signal is lowered with respect to thick crystals. Further investigations are in progress to test and optimize their response, in order to reduce the irradiation times. The wide dynamic range of the LiF film detector, already exploited for soft X-rays [3], has been confirmed by microradiographs of biological objects. As example, a human hair bulb X-ray radiography obtained on a LiF film, reported in figure 5, was obtained in the same experimental set-up with an irradiation time of 10 minutes.

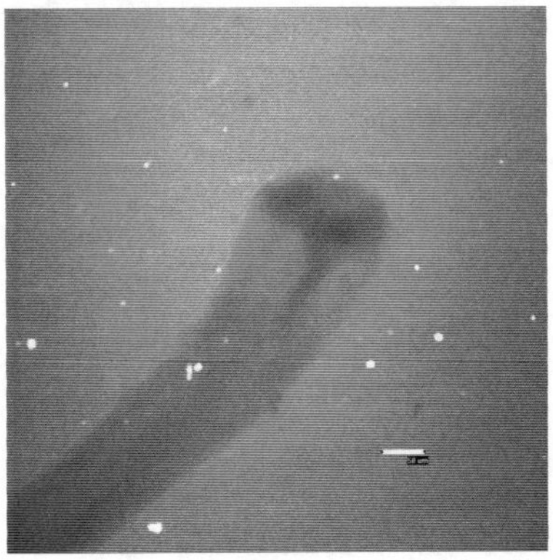

Figure 5. Confocal fluorescence image of a man-hair bulb X-ray microradiograph stored on a 1.4 μm thick LiF film grown on a silicon substrate, under optical excitation at 458 nm. The reference line is 50 μm.

The brightest areas correspond to the more irradiated ones, i.e. the less absorbing features within the sample. The internal structures are clearly distinguished. Several bright spots are visible in the figure; they are attributed to light scattered from dust particles and/or from film mechanical damage in the optical reading process. A better control of the handling procedures in all the experimental phases can drastically reduce such artifacts.

In conclusions, stable formation of primary F and aggregate F_2 and F_3^+ colour centres in LiF crystals irradiated with an X-ray table-top source equipped with a Cu anode was investigated as a function of irradiation dose. The formation efficiency for F defects was derived from careful optical absorption measurements and it follows a power law behaviour as function of the dose. The PL spectra intensities look suitable for the development of high dynamic response X-ray imaging detectors based on LiF thin films. Further experiments are in progress to better investigate the absorption and photoluminescence properties of F-aggregate colour centres in LiF crystals and thin films.

Acknowledgments

The authors would like to thank G. Baldacchini, A. Lai, E. Nichelatti and M.A. Vincenti for useful discussions and suggestions.

References

[1] ENEA Patent TO2002A000575, Metodo di rivelazione di immagini micrometriche e submicrometriche mediante irraggiamento di una maschera o di un campione biologico con radiazioni ionizzanti, G. Baldacchini, F. Bonfigli, A. Faenov, F. Flora, R. M. Montereali, A. Pace, T. Pikuz, L. Reale (International N°: WO 2004/005906 A1).

[2] Baldacchini G, Bonfigli F, Faenov A, Flora F, Montereali R M, Pace A, Pikuz T, and Reale L 2003 *J. Nanoscience and Nanotechnology* **3** 6 483

[3] Baldacchini G, Bollanti S, Bonfigli F, Flora F, Di Lazzaro P, Lai A, Marolo T, Montereali R M, Murra D, Faenov A, Pikuz T, Nichelatti E, Tomassetti G, Reale A, Reale L, Ritucci A, Limongi T, Palladino L, Francucci M, Martellucci S and Petrocelli G 2005 *Review Scientific Instruments* **76** 113104 (Also in the *Virtual Journal of Biological Physics Research* **10**(11), December 1, 2005)

[4] Montereali R M 2002 Point Defects in Thin Insulating Films of Lithium Fluoride for Optical Microsystems *Handbook of Thin Film Materials (Ferroelectric and Dielectric Thin Films* Vol.3) ed H S Nalwa (Academic Press) chapter 7, pp 399-431

[5] Almaviva S, Bonfigli F, Franzini I, Lai A, Montereali R M, Pelliccia D, Cedola A, Lagomarsino S 2006 Appl. Phys. Lett. **89** 54102

[6] Nahum J and Wiegand D A 1967 Phys. Rev. **154** 817

[7] Baldacchini G, De Nicola E, Montereali R M, Scacco A, Kalinov V 2001 *J. Phys. Chem. Solids* **61** 21

[8] Hughes E, Pooley D, Rahaman H U, Runciman W A 1967 *Harwell Atomic Research Establishment* R 5604

[9] Fowler W B 1968 *Physics of Color Centers* ed W B Fowler (New York and London: Academic Press)

[10] Perez A, Balanzat E, Dural J 1990 *Phys. Rev. B* **41** 7

[11] Palik E D 1997-98 *Handbook of Optical Constants of Solids* ed E D Palik (New York: Academic Press)

[12] Aguilar M, Agullo-Lopez F and Jacque F 1982 *Rad. Eff.* **61** 215

[13] Baldacchini G, d'Auria G, Montereali R M, Scacco A 1998 *J. Phys. C: Condensed Matter* **10** 857

[14] Sever B R, Kristianpoller N and Brown F C 1986 *Phys. Rev. B* **34** 2 1257

Journal of Physics: Conference Series **249** (2010) 012004 doi:10.1088/1742-6596/249/1/012004

Colour-centre concentrations in broad-band light-emitting strip waveguides in Lithium Fluoride

F. Bonfigli[1], M. Montecchi[2], R.M. Montereali[1], E. Nichelatti[3], M. Piccinini[4], F. Somma[5*]

[1]ENEA, Dip. Tecnologie Fisiche e Nuovi Materiali, C.R. Frascati, V. E. Fermi 45, 00044 Frascati (Roma), Italy

[2]ENEA, Dip. Tecnologie per l'Energia, Fonti Rinnovabili e Risparmio Energetico, C.R. Casaccia, V. Anguillarese 301, 00123 S. Maria di Galeria, Roma (Italy)

[3]ENEA, Dip. Tecnologie Fisiche e Nuovi Materiali, C.R. Casaccia, V. Anguillarese 301, 00123 S. Maria di Galeria, Roma (Italy)

[4]INFN-LNF, Via E. Fermi 40, 00044 Frascati, Roma (Italy)

[5]Physics Dep., University of Roma Tre, V. della Vasca Navale 84, 00146 Roma (Italy)

Abstract. The concentrations of primary F centres and F_2 and F_3^+ aggregate defects were evaluated as a function of the irradiation dose in broad-band light-emitting single-mode strip waveguides directly written with 12 keV electrons at the surface of a LiF crystal. From a careful best-fit procedure of the optical transmission spectra, measured on microstructures only few tens of micrometers wide, the spectral contributions of several types of colour centres were identified and quantified. This characterization provided useful information for the optimization of these confining microstructures and for the development of waveguide colour centre lasers.

1. Introduction

Lithium Fluoride (LiF) crystals and films have shown high potential for the realization of optoelectronic devices thanks to the very interesting optical properties of colour centres (CCs), which can be generated by low energy electrons [1,2] or by exposure to energetic photons, such as soft [3] and hard [4] X-rays or intense ultra-short laser pulses in the infrared [5,6]. In particular, the laser-active F_2 and F_3^+ aggregate CCs (two electrons bound to two and three anion vacancies, respectively) show broad and intense photoluminescence in their emission bands peaked at 670 and 535 nm, under simultaneous optical excitation in their almost overlapping absorption bands at about 450 nm, even at room temperature (RT). Nowadays, one of the most promising fields of application of CCs in LiF is in the development of active optical micro-devices, among them broad-band visible-emitting colour-centre waveguide lasers [6,7], based on the peculiar spectroscopic features of active electronic defects in this promising material.

* Fabrizia Somma, somma@fis.uniroma3.it; phone +39 06 57337006; fax +39 06 94005579303

Despite of the advantages the waveguide configuration provides for laser action and the successful results assessed in the case of low energy electrons [7,8] and ion-beam irradiation [9,10], the full spectral characterization, hence the optimization, of LiF-based single-mode active waveguides still remains a very difficult task.

The stable formation of high concentrations of primary and aggregate defects within a thin surface layer of the crystal locally induces a spectral-dependent modification in the refractive index of the irradiated volume with respect to the surrounding materials. In the case of single-mode strip waveguides induced in LiF crystals by 12 keV electron irradiation, an increase of the order of 10^{-3} in the real part of the index of refraction in the green-red wavelength interval [11] is sufficient to confine the laser-active F_2 and F_3^+ photoluminescence under optical pumping in the blue spectral range [7,8].

In this paper CC volume concentrations are evaluated as a function of the irradiation dose for three active micro-strips produced by 12 keV electrons on the same LiF crystal. The obtained results are discussed in connection with the modifications of the complex refractive index in the irradiated volume and the waveguide amplifying properties.

2. Experimental

Low-energy electron beams of energy 12 keV were used to produce three optically-active coloured strips, 3 mm long and about 150 μm wide, at the surface of a polished LiF single crystal of size (20x15x2.5) mm^3. Irradiations were performed at RT in a scanning electron microscope, at a fixed current of 10 nA, for increasing exposure times of 12, 24 and 48 minutes. These active strip waveguides were successfully characterized with confocal fluorescence microscopy [12] and optical transmittance measurements were taken in a highly versatile spectrophotometer [13], which can measure microstructures few tens of micrometer wide in the spectral range from 220 to 800 nm.

3. Results and discussions

The light-emitting strip waveguides have comparable width of ~145 μm, as deduced from the fluorescence image reported in figure 1 under blue illumination. The bright areas correspond to the irradiated channels, which emit green and red photoluminescence due to the stable formation of F_3^+ and F_2 aggregate defects, respectively. The fluorescence emission is clearly visible even by naked eyes under Argon laser excitation at the wavelength of 458 nm, as shown in figure 2.

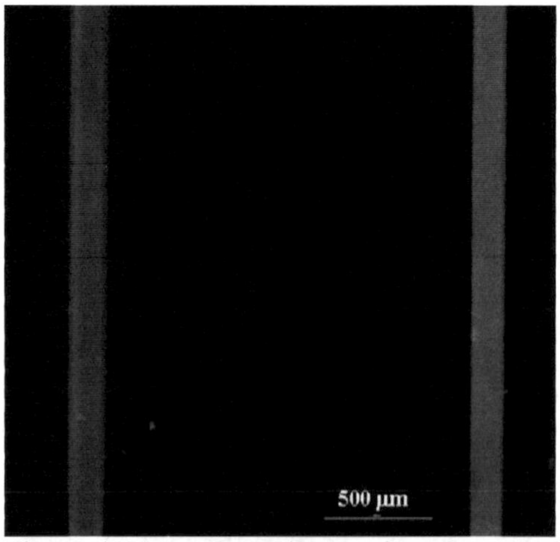

Figure 1. Fluorescence image of a coloured strip induced on a LiF crystal by irradiation with 12 keV electrons under blue illumination.

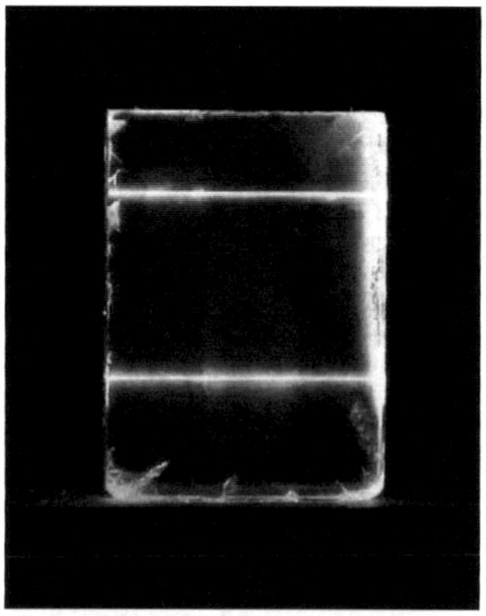

Figure 2. Photograph of coloured strips induced on a LiF crystal by irradiation with 12 keV electrons under Argon laser illumination.

The strips appear uniformly coloured, with quite well defined boundaries and constant width. In figure 2 the optical transmittance spectra measured outside (blank crystal) and inside the three coloured strips induced with 12 keV electrons at increasing doses are reported. The minima of the transmittance spectra at ≈ 245 nm and at ≈ 445 nm are clearly observed. They are ascribed to the primary F centre (an anion vacancy occupied by an electron) absorption band, and to the M one, due to the overlapping of F_2 and F_3^+ absorption bands [14]. The transmittance minima values at the peak wavelength of the M absorption bands decrease monotonically with the increase of irradiation dose.

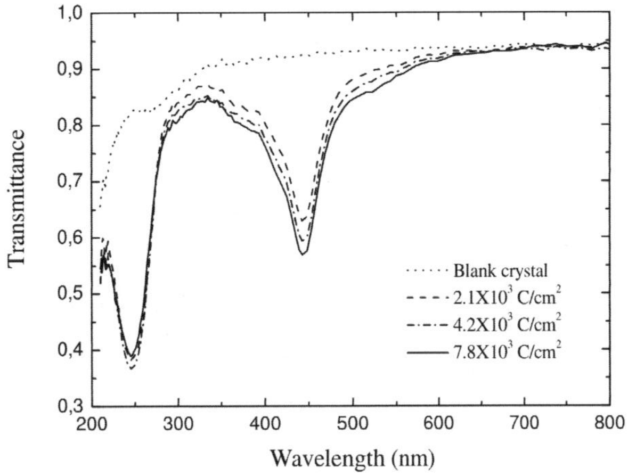

Figure 3. Optical transmittance spectra of three coloured strips induced on a LiF crystal by irradiation with 12 keV electrons at increasing doses. The measured spectrum of the blank crystal is also reported for comparison.

A crucial problem concerning heavily irradiated LiF crystals is the coexistence of several kinds of aggregate defects with often overlapping absorption bands, which makes it difficult to clearly evaluate the single contributions due to individual centres. Other spectroscopic features, ascribed to several kinds of larger aggregate defects, can be recognized [15]: e.g., the absorption features located at 315 and 380 nm are attributed to the R_1 and R_2 transitions of the F_3 defect, consisting of three adjacent F centres. The 515 and 545 nm absorption bands are attributed to the N_1 and N_2 transitions of the F_4 defects (four associated F centers). The M-band spectral range, where the F_2 and F_3^+ absorption bands overlap, is further complicated by the presence of another broad absorption band, which appears as a shoulder of the M band on the short wavelength side and it is associated to the presence of Li colloids [15,10]. A semi-classical dipole-electromagnetic field interaction model, which consider CCs in LiF as elementary two-level quantum systems interacting with the electromagnetic field by means of a dipole-type Hamiltonian, later generalized to include the contribution of inhomogeneous Gaussian absorption bands [16], was successfully applied to evaluate how CCs modify the complex refractive index [17]. The complex permittivity of the irradiated surface layer can be modeled as the sum of distinct absorption-band contributions, each one due to different types of CCs; it is dependent on the photon energy, on the amplitudes of the dielectric susceptibilities at the resonance absorption energies of CCs for Lorentzian and Gaussian bands, and on their half widths at half maximum (HWHMs). The dielectric susceptibilities, χ, are also dependent on the penetration depth of low-energy electrons, due to the proportionality of these quantities to the concentration of CCs along the depth [12], i.e. the dipole densities of harmonic oscillators, which were assumed as flat along the depth, fixed equal to 1.3 μm according to Monte Carlo simulations [2,12]. The fit parameters during data reduction of the experimental spectra were the resonance energies, the HWHMs, and the dielectric-susceptibility

Journal of Physics: Conference Series **249** (2010) 012004 doi:10.1088/1742-6596/249/1/012004

amplitudes of the bands associated with each type of CC, from which the contributions to the absorption coefficient were derived. The obtained best-fit results are summarized in table 1 for the strip irradiated with 12 keV electrons at the highest dose of 7.8x10^{-3} C/cm^2.

Table 1. Band parameters resulting from the best fit of the transmittance spectrum for the strip irradiated with 12 keV electrons at the highest dose of 7.8x10^{-3} C/cm^2. The confidence intervals were evaluated with a Montecarlo algorithm (40 statistical counts).

Centre type	Band type*	*E* (eV)	HWHM (eV)	$\langle \max[\mathrm{Im}(\chi)] \rangle$
F	G	5.038±0.003	0.363 ± 0.005	0.0369 ± 0.003
$F_3(R_1)$	L	3.99±0.05	0.14 ± 0.11	0.0017 ± 0.008
$F_3(R_2)$	L	3.31±0.12	0.39 ± 0.14	0.0068 ± 0.0009
Li colloids	G	2.98±0.02	0.144 ± 0.042	0.0072 ± 0.0023
F_2	L	2.79**	0.103	0.0162 ± 0.0044
F_3^+	L	2.77**	0.139	0.0213 ± 0.0034
$F_4(N_1)$	L	2.42±0.03	0.09 ± 0.19	0.0014 ± 0.0040
$F_4(N_2)$	L	2.49±0.13	0.17 ± 0.09	0.0036 ± 0.0033

* Gaussian (G) or Lorentzian (L).
** Kept fixed in the best fit.

The introduction of Gaussian bands, especially for the F one, improved significantly the best-fit of the measured transmittance spectra. All the contributions of larger aggregate defects listed in table 1 were necessary, as one could expect because of the high doses involved. The band located at 416 nm was also provisionally fitted with a Mie-based model, but the Gaussian band shape improved the best fit results, suggesting that other spectral contributions from not yet well identified defects could be present. By using Smakula's formula, the volume concentrations in the irradiated strips were derived for the primary F and aggregate F_2 and F_3^+ CCs and are reported in figure 4 as functions of the irradiation dose.

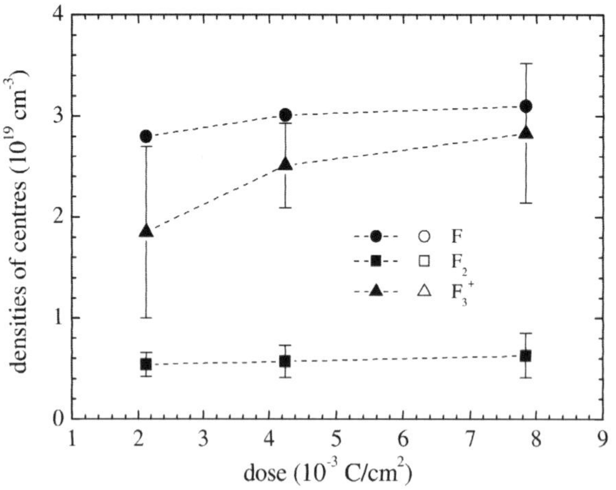

Figure 4. Volume concentrations of the primary F and aggregate F_2 and F_3^+ CCs in three coloured strips induced on a LiF crystal by irradiation with 12 keV electrons at increasing doses.

In the most coloured strip, a maximum F-centre concentration of 3.3x10^{19} cm^{-3} was estimated by assuming the oscillator strength *f* equal to 0.56 [18]. The F-centre density increases with the irradiation dose in the investigated range, even if the values seem to be close to saturation. As far as the M band is concerned, a preferential production of F_3^+ defects with respect to the F_2 ones was obtained, which could be ascribed to the nature of the bombarding charged particles, electrons, even for irradiation performed at RT. For the F_2 centres a value of *f* equal to 0.28 [18] was utilized, while for F_3^+ centres an absorption cross section $\sigma_{F^+_3} = (7.1 \pm 0.1) \times 10^{-17}$ cm^2 was adopted [19]. In the present case, the maximum of F_2 and F_3^+ densities were estimated as 7x10^{18} and 3x10^{19} cm^{-3}, respectively. The F_3^+ concentration progressively increases with the irradiation dose, while the F_2 appears to remain constant. It should be noticed that the *relative* errors of the calculated concentrations are similar for these two kinds of CCs.

In conclusions, spectral contributions of several types of CCs were identified and quantified vs. the irradiation dose in broad-band light-emitting single-mode strip waveguides directly written with 12 keV electrons at the surface of a LiF crystal. At highest doses, the increase in the primary F-centre concentration improve the confining properties of the F_2 and F_3^+ CCs photoluminescence [17]; a preferential formation of charged F_3^+ defects, which are laser-active in the green wavelength interval, was identified. However, the presence of larger aggregate defects, like F_4 centres, which absorb light in the same spectral range where the F_3^+ photoemission is located, increases the absorption losses and prevents amplification in the green [7]. In the case of red-emitting F_2 laser-active defects, the improved optical confinement properties makes possible to achieve positive net gain coefficients [7], which increase with irradiation doses.

References

[1] Nunes R A, Kalinowski H J, Paciornik S, De Souza A M and Scavarda do Carmo L C 1988 Electron-beam production of colour centres on alkali halide crystals and films *Nucl. Instr. Meth. Phys. Res. B* **32** 222

[2] Montereali R M 2002 Point defects in thin insulating films of lithium fluoride for optical microsystems *Handbook of Thin Film Materials (Ferroelectric and Dielectric Thin Films* Vol.3) ed H S Nalwa (Academic Press) chapter 7, pp 399-431

[3] Baldacchini G, Bonfigli F, Flora F, Montereali R M, Murra D, Nichelatti E, Faenov A, Pikuz T 2002 High-contrast photoluminescent patterns in lithium fluoride crystals produced by soft x-rays from a laser plasma source *Appl. Phys. Lett.* **80** 25

[4] Almaviva S, Bonfigli F, Franzini I, Lai A, Montereali R M, Pelliccia D, Cedola A, Lagomarsino S 2006 Hard x-ray contact microscopy with 250 nm spatial resolution using a LiF Film detector and a tabletop microsource *Appl. Phys. Lett.* **89** 54102

[5] Courrol L C, Samad R E, Gomes L, Ranieri I M, Baldochi S L, Zanardi de Freitas and Vieira Jr N D, 2004 Color center production by femtosecond pulse laser irradiation in LiF crystals *Opt. Express* **12** 288

[6] Kawamura K, Hirano H, Kurobori T, Takamizu D, Kamiya T and Hosono H 2004 Femto-second-laser-encoded distributed-feedback color center laser in lithium fluoride single crystals *Appl. Phys. Lett.* **84** 311

[7] Montereali R M, Piccinini M, Burattini E 2001 Amplified spontaneous emission in active channel waveguides produced by electron beam lithography in LiF crystals *Appl. Phys. Lett.***78** 26 4084

[8] Montereali R M, Mancini A, Righini G and Pelli S 1998 Active stripe waveguides produced by electron beam lithography in LiF crystals *Opt. Commun.* **153** 223

[9] Cremona M, Pereira J A M, Pelli S and Righini G C 2002 Optical waveguides produced in LiF by MeV ion beam bombardment *Appl. Phys. Lett.* **81** 4103

[10] Mussi V, Somma F, Moretti P, Mugnier J, Jacquier B, Montereali R M and Nichelatti E 2003 Mode analysis in He^+ implanted lithium fluoride planar waveguides *Appl. Phys. Lett.* **82** 388

[11] Rocchetti A, Assanto G, Montereali R M, Nichelatti E, Somma F 2006 Color-center waveguides in low-energy electron-bombarded lithium fluoride *Appl. Phys. Lett.* **88** 261111
[12] Montereali R M, Bigotta S, Piccinini M, Giammatteo M, Picozzi P, Santucci S 2000 Broad-band active channels induced by electron beam lithography in LiF films for waveguiding devices *Nucl. Instr. Meth. Phys. Res. B* **166-167** 764
[13] Montecchi M, Baccaro S, Dafinei I, Diemoz M, Montereali R M, Somma F 2004 Lumen: a highly versatile spectrophotometer for measuring the transmittance throughout very long samples as well microstructures *Rev. Sci. Instrum.* **75** (11) 4636
[14] Nahum J and Wiegand D A 1967 Optical properties of some F-aggregate centers in LiF *Phys. Rev.* **154** 817
[15] Montecchi M, Nichelatti E, Mancini A and Montereali R M 1999 Optical characterization of low-energy electron-beam-colored LiF crystals by spectral trasmittance measurements *J. Appl. Phys.***86** (6) 3745
[16] Montecchi M, Nichelatti E, Montereali R M, Piccinini M, Somma F 2004 Increase of refractive index induced by absorbing centres: homogeneous and inhomogeneous line-broadening cases *Optical and Quantum Electronics* **36** 43
[17] Montereali R M, Bonfigli F, Montecchi M, Nichelatti E, Piccinini M 2008 Advanced optical characterization of broad-band light-emitting color-center strip waveguides in lithium fluoride *Silicon Photonics and Photonic Integrated Circuits*, ed G C Righini, S K Honkanen, L Pavesi, L Vivien (SPIE Proceedings vol 6996) 69960J
[18] Perez A, Balanzat E, Dural J 1990 *Phys. Rev. B* **41** 7
[19] Baldacchini G, Cremona M, d'Auria G, Montereali R M and Kalinov V 1996 Radiative and nonradiative processes in the optical cycle of the F_3^+ center in LiF *Phys. Rev. B* **54** (24) 17508

Optical and Transport Properties of Rare-earth Trivalent Ions Located at Different Sites in Sol-gel SnO_2

E A Morais[1, 2], L V A Scalvi [1,3,*], L P Ravaro [1,3], M. Siu Li[4], E A Floriano[1,3]

[1] Physics Dept.- FC, UNESP – São Paulo State University, Bauru SP Brazil.
[2]Physics Dept.- UFMG - Minas Gerais Federal University, Belo Horizonte MG, Brazil
[3] Advanced Materials Group (MAV) and POSMAT- FC, UNESP, Brazil
[4] Physics Institute of São Carlos – USP, C. P.369, 13560-970 São Carlos SP Brazil

scalvi@fc.unesp.br

Abstract. Photoluminescence and photo-excited conductivity data as well as structural analysis are presented for sol-gel SnO_2 thin films doped with rare earth ions Eu^{3+} and Er^{3+}, deposited by sol-gel-dip-coating technique. Photoluminescence spectra are obtained under excitation with various types of monochromatic light sources, such as Kr^+, Ar^+ and Nd:YAG lasers, besides a Xe lamp plus a selective monochromator with UV grating. The luminescence fine structure is rather different depending on the location of the rare-earth doping, at lattice symmetric sites or segregated at the asymmetric grain boundary layer sites. The decay of photo-excited conductivity also shows different trapping rate depending on the rare-earth concentration. For Er-doped films, above the saturation limit, the evaluated capture energy is higher than for films with concentration below the limit, in good agreement with the different behaviour obtained from luminescence data. For Eu-doped films, the difference in the capture energy is not so evident in these materials with nanoscocopic crystallites, even though the luminescence spectra are rather distinct. It seems that grain boundary scattering plays a major role in Eu-doped SnO_2 films. Structural evaluation helps to interpret the electro-optical data.

1. Introduction

Tin dioxide (SnO_2) is a wide bandgap material (about 3.6eV), [1,2] which presents high transparency in the visible (90%) and high reflectivity in the infrared range [3]. In the undoped form, this compound exhibits n-type conductivity, due to oxygen vacancies and interstitial tin atoms, generated during sample processing, which acts as donors in the matrix. The wide bandgap makes SnO_2 an attractive oxide for electrically activated emission [4], since wide bandgap semiconductors have higher excitonic ionization energy. Besides, the wide gap decreases the luminescence quenching effects [5]. One of the most promising electroluminescent display technologies is the thin films electroluminescence (TFEL). Among the TFEL devices, luminescent properties are related to intra 4f rare-earth transition [6], mainly the ions Eu^{3+} and Er^{3+}, because their transitions take place on a wide energy range, with emissions from the visible to the infrared, allowing application in several optical communication devices.

We have investigated optical properties and optically induced transport properties of Eu and Er doped SnO_2 thin films, through measurements of photoluminescence and decay of photo-induced

* Corresponding author: scalvi@fc.unesp.br

conductivity. The acceptor nature of ions Er^{3+} and Eu^{3+} [7,8] creates a high charge compensation degree in the material, which is naturally of n-type, leading to a high resistivity compensated semiconductor. When produced by the sol-gel process, SnO_2 thin film presents nanoscopic crystallites (about 5-20nm). This nanostructured material exhibits a high concentration of grain interfaces by unit area, which favors the impact ionization of luminescent centers, because hot carriers are promoted by the interface active layer/dielectric in a TFEL device layout [9]. Then, the electrical properties are strongly influenced by grain boundary scattering. It is also important to mention that optical and transport properties of rare-earth doped SnO_2 is strongly influenced by the location of the doping ion, at lattice site, substitutional to Sn^{4+} or at grain boundary [8,10]

In this work, the luminescence spectra, under separated excitation with various distinct light sources, such as Kr^+, Ar^+ and Nd:YAG lasers, have been measured. Emission from these rare-earth ions confirms the different shapes of luminescence spectra whether the dominant luminescent center presents high symmetry, at lattice sites, substitutional to Sn^{4+}, or is located at asymmetric sites, segregated at boundary layer. Transport properties are also influenced by rare-earth ion location, because thin films excitation with the fourth harmonic of an Nd:YAG laser (4.65eV) at low temperature, yields distinct types of conductivity decay when the illumination is removed, which is related with the capture by rare-earth related defects located at different sites. Based on the model of trapping defects with thermally activated cross section [8,11], E_{cap} (capture energy) is evaluated for the trapping centers substitutional at Sn^{4+} lattice sites and located at grain boundary interface.

It is important to evoke that the thin film matrix presents a very useful shape for technological application, since it allows optical integration and the application of electric fields to operate electroluminescent devices.

2. Experimental

Colloidal suspensions of SnO_2 undoped and doped with Er^{3+} or Eu^{3+} have been prepared through the sol-gel route. Firstly, it has been prepared an aqueous solution of $SnCl_4.5H_2O$ (0.5 M). The incorporation of Eu has been provided by dissolution of Eu_2O_3 in HCl, being the desired amount added to the starting solution. In the case of Er doping, the reagent used was $ErCl_3.6H_2O$. Under stirring with a magnetic bar, NH_4OH was then added until the pH reaches 11, leading to hydrolysis and condensation reactions. The remaining precipitated was submitted to dialysis against distilled water for about 10 days, in order to eliminate Cl^- and NH_4^+ ions. This procedure leads to a stable suspensions (sol) of SnO_2:Eu and SnO_2:Er. Xerogels (powders) were obtained by drying the suspension at room conditions and treated at 1000°C during 6 hours. For photoluminescence (PL) measurements at low temperature, the xerogel (powder) was treated at 1000°C during 6 hours and compressed down with a load of $5x10^3$kgf/cm^2 to the shape of a pellet. Sample cooling was done in a He closed-cycle cryostat from Janis.

X-ray diffraction data of films were obtained with a a RIGAKU diffractometer, model D/MAX-2100/PC, with a scanning rate of 1°/min in range of 20–80 degrees.

For luminescence data presented here a few experimental setups were used: 1) for low temperature measurements, an Ar^+ laser from Spectra Physics model 2017 for sample excitation and signal recorded by a germanium detector from Edinburgh Instruments. 2) a xenon lamp of 450W, a fluorimeter SPEX F212I and a germanium detector North Coast Scient. Corporation model EO-8171. 3) excitation with a Kr^+ laser, tuned on an ultraviolet multi-line mode (3.67eV, 3.53eV and 3.47eV), 4) excitation with a Kr^+ laser, tuned on 2.57eV. A photomultiplier tube (PMT) R446 from Hamamatsu was used for signal measuring and recording in the experimental setup 3 and 4. 5) The fourth harmonic of a Nd:YAG laser, 4.65eV. The signal was detected by a CCD detector coupled with a Jobin Yvon monochromator, model T64000.

For the decay of photo-excited conductivity, it was used the fourth harmonic of a pulsed Nd:YAG laser (4.65eV), with 15Hz of frequency, and a Keithley 6517A for recording electrical characteristics. The AFM micrograph were obtained using a Nanoscope III-a from Digital Instruments and samples of imaging areas as large as 1 μm^2 were taken.

Journal of Physics: Conference Series **249** (2010) 012005 doi:10.1088/1742-6596/249/1/012005

3. Results and discussion

X-ray diffraction data are shown in figure 1 for undoped SnO_2 and doped with some concentrations of Er^{3+} or Eu^{3+}. Generally speaking, the definition of the obtained peaks is lower when compared to xerogels with the same composition [12]. It may be related to the influence of substrate, which decreases the signal magnitude obtained when the measurement is done under a shallow angle, 1.5°. Besides the thermal annealing temperature (550°C) does not allow a large crystallite growth, leading to broaden peaks in the X-ray pattern, typical of nanocrystalline domain. Comparing these results with rutile structure cassiterita phase (JCPDF, 41-1445) [13], there is good agreement as indicated by the main crystal directions shown in figure 1. The crystallite size, evaluate through the Scherrer equation [14], decreases for the increasing of rare-earth doping. For instance, taking into account the (101) direction, the crystallite size is 7.4, 5.2 and 3.5nm, for samples doped with 0.0, 0.1 and 4.0at% of Er respectively, whereas for Eu-doped SnO_2 films the evaluated value is about 7 nm for all the concentrations. The increase of rare-earth concentration inhibits the crystallite growth, which is similar to what happens in xerogels, because the excess of doping impurities segregates to the particles surface. Rietveld refinement data for rare-earth doped SnO_2 xerogels [12] allows evaluating the distortions for the rutile structure, caused by Er^{3+} and Eu^{3+} incorporation, as well as the distortion generated by the oxygen vacancies formation during its processing [15]. Comparing the undoped powder with Eu-doped samples, it is noticed that the structural disorder decreases with the introduction of up to 0.5% of Eu, and the increase of concentration to 1% Eu increases the disorder degree to a value close to the undoped xerogel. Then, the low doping leads to stabilization of cassiterite structure of SnO_2. Concerning the Er doping, there is an increase in the disorder for the sample doped with 0.1% of Er compared to the undoped sample. The disorder also increases when the doping level reaches 4%, similar to the observed behavior of samples highly doped with Eu. The excess of doping decreases the material crystallinity, because it generates clusters of a secondary phase, involving the SnO_2 matrix and inhibits the crystallite growth during the crystallization process. A secondary phase $Eu_2Sn_2O_7$ has been reported for Eu-doped SnO_2 with concentrations higher than 0.1at% of Eu [16], but this phase was not found in the present work.

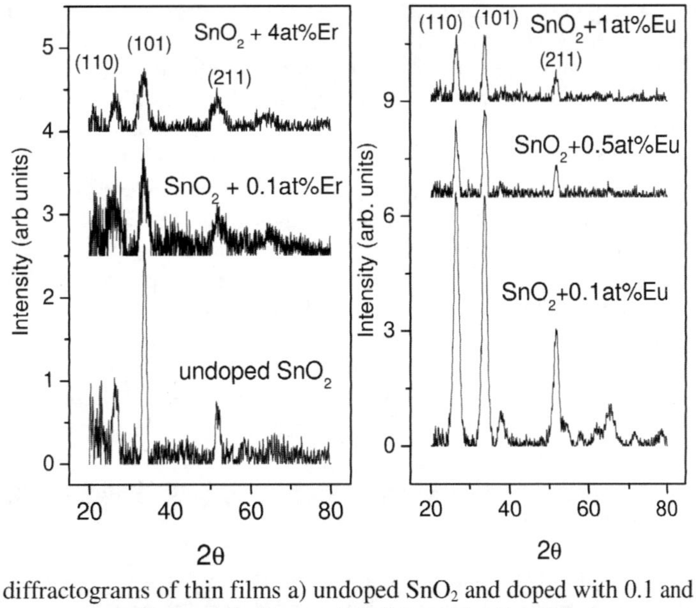

Figure1. X ray diffractograms of thin films a) undoped SnO_2 and doped with 0.1 and 4at% of Er, b) SnO_2 doped with com 0.1, 0.5 and 1at% of Eu.

Figure 2 (a) shows temperature dependence of photoluminescence (PL) spectra measured at temperatures 50 K and 280 K for a 2at% Er doped SnO_2 xerogel sample. Sample was excited with an Ar^+ laser, operating with 2.53eV of excitation energy, according to setup 1. The Er^{3+} transition $^4I_{13/2} \rightarrow {}^4I_{15/2}$ around 0.8eV is clearly evidenced by the presence of several peaks. The excitation energy (2.53eV) corresponds to the transition from ground state to $^4F_{7/2}$. The inset of Figure 2(a) shows PL spectra for tin dioxide xerogel codoped with 2at% of Er and Yb, excited by a Xenon lamp plus a monochromator (setup 2). It can be noticed in figure 2 that the cooling down leads to PL intensity increase, since the it decreases the lattice vibration of SnO_2, and then phonon assisted nonradiative processes, which enhance light emission from Er^{3+} core transition. It allows the observation of spectra details at low temperature, not observed at higher temperatures. In the measurement carried out at 50 K, it is clearly seen the presence of others peaks not observed at 280 K. This vanishing is probably related to lattice vibration. All the observed peaks are related to the decay of excited state $^4I_{13/2}$ to the ground state $^4I_{15/2}$. The presence of several peaks can be attributed to the splitting of $^4I_{13/2}$ and $^4I_{15/2}$ levels due to the matrix crystalline field [17]. The Er^{3+} $^4I_{13/2}$ manifold has upper states of the excited degenerate $^4I_{13/2}$ level populated due to local laser heating [17], as evidenced by the fact that such a line splitting does not takes place when the sample is irradiated with monochromatic light coming from a xenon lamp (inset of fig. 2(a)). In the inset of figure 2(a), two different excitation energies were used: 3.77eV and 2.35eV, according to the experimental setup 2. The codoping with Yb^{3+} allows verifying the energy transfer process, by exciting with 1.26eV, corresponding to transition $^2F_{7/2} \rightarrow {}^2F_{5/2}$ of Yb^{3+} and an energy transfer to transition $^2F_{7/2} \rightarrow {}^4I_{11/2}$ of Er^{+3}. However the verification of this energy transfer process is not the scope of this paper, and has been done elsewhere [18]. As can be seen in the inset of figure 2(a), the spectra present rather distinct shape depending on the energy excitation. They are related to two families of Er sites: substitutional to Sn^{4+} and segregated at boundary layer. The second family is related to the solubility limit of rare-earth (RE) ions in tin dioxide, which is about 0.05at%, and the doping excess become located at boundary layer [19]. For the high doping concentration (2%) of the samples reported here, it is expected very many Er complexes segregated at boundary layer. The inset of figure 2(a) shows the emission of Er^{+3} at room temperature, excited directly. The spectra observed by excitation with 2.35eV, is similar to the observed by excitation with 1.26eV [18], which is effective only for Er^{3+} complexes located at particles surface. On the other hand, the rare-earth ions located at Sn^{4+} sites are not substantially excited by this procedure. Substitutional sites located Er^{3+} ions are excited via band-to-band excitation of SnO_2 [18] and the spectra is as shown in the inset of figure 2(a) by 3.77eV excitation.

Figure 2(b) shows the green emission from an Er doped (4%) SnO_2 thin film, excited according to the experimental setup 3, and a SnO_2:2%Er pellet, excited by the experimental setup 4. The energy range in both curves of figure 2(b) are in good agreement and assures that the observed wide emission, with a peak about 2.33eV, includes the $^2H_{11/2} \rightarrow {}^4I_{15/2}$ radiative transition of Er^{+3} ion. The broad shape is probably related to random neighborhood of Er^{3+} in the SnO_2 lattice, which could be surrounded by a varying number of oxygen vacancies, interstitial tin atoms and doping ions (the doping concentration is rather high). Besides, it also could be affected by the distance form the grain boundary, since we are dealing with nanoscopic particles, in the range 5-20nm. Besides, the excess of Er is located at grain boundary, as discussed for figure 2(a), which are low symmetry sites and could contribute to the broad band. Luminescence from the matrix of SnO_2 itself has been reported recently [20], with very similar shape to the curves of figure 2(b), but shifted to UV-violet, presenting a peak about 3.10eV, which was attributed to oxygen vacancies. On the other hand, the observed broad band green emission centered about 2.33eV could be associated to the transition from the excited state $^2H_{11/2}$ to the $^4I_{15/2}$ ground state, as already mentioned. In the spectrum of figure 2(b), film is excited with the multilines UV from a Kr^+ laser (most intense line at 3.53eV). Figure 2(b) also shows photoluminescence (PL) emission from a pellet (powder compressed) of SnO_2:2%Er, with excitation with 2.57eV, which is coincident with $^4I_{15/2} \rightarrow {}^4F_{7/2}$ transition. Although the laser line at 2.57eV is not enough to excite the green transition in the SnO_2:4at%Er film, it is accountable for the excitation of observed transition in SnO_2:2%Er pellet. It means that to obtain efficiently the transition, the number of exciting centers in the beam light path becomes very relevant. This path is rather long for the pellets but much shorter for the film. However our results in the visible range make sure that luminescence is possible also for the film, which justifies a deeper investigation of film emission in the infrared. Crystallite nanoscopic dimensions could play a relevant role in the emission spectra, due to the possibility of quantum confinement effects [21,22].

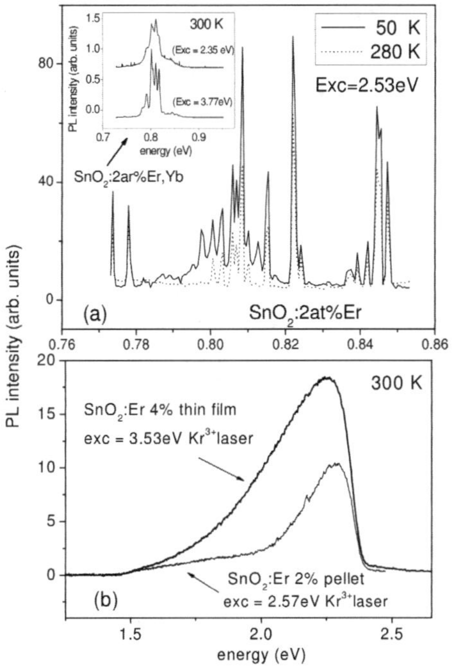

Figure 2. (a) Photoluminescence (PL) spectra for 2at% Er doped SnO_2 xerogel at 50 K and 280K. Sample was excited with an Ar^+ laser, operating with 2.53eV of excitation energy. Inset - PL of SnO_2:Er,Yb (2% at.) under excitation at: 3.77eV and 2.35eV (c). PL spectra of SnO_2:4% thin film, excited with the multi-lines ultraviolet (3.53eV) of a Kr^+ laser, and of SnO_2:2%Er pellet, excited with 2.57eV (482nm) from the same Kr^+ laser.

Figure 3 shows PL spectra obtained under excitation with 4.65eV (266nm) laser line (experimental setup 5) for Eu-doped SnO_2 xerogels with 1.0 and 0.5 at% Eu concentrations. The inset of Figure 3 shows PL spectra of 0.5%Eu doped SnO_2 xerogel, under excitation with 4.65eV and 2.53eV (488nm). The energy of 4.65eV is related to matrix band-to-band excitation, followed by energy transfer to Eu^{3+} ions. It is important to point out the nature of the transitions $^5D_0 \rightarrow ^7F_2$ (2.02eV) and $^5D_0 \rightarrow ^7F_1$ (2.10eV), labeled in figure 3. The latter one is a magnetic dipole transition and its intensity is not affected by structural changes of the nearest neighborhood, whereas the transition $^5D_0 \rightarrow ^7F_2$ is ruled by electric dipoles, being hypersensitive to the local crystalline field [23]. The ratio between the areas of these emission lines, i.e., ($^5D_0 \rightarrow ^7F_2$) / ($^5D_0 \rightarrow ^7F_1$) is called asymmetry ratio. An increase of asymmetry ratio from 0.01 to 5.2 is evaluated for samples with 0.5 to 1at% Eu, respectively, which means that for increasing Eu^{3+} concentration, energy transfer is more efficient to ions located far from the symmetry center, for example, ions located close to the particles surface. For SnO_2:0.5%Eu^{3+} samples the PL line shape suggests an asymmetric ratio very close to zero, indicating that the dominant Eu^{3+} ion is actually located at lattice sites, substitutional to Sn^{4+}. The xerogel SnO_2:1at%Eu presents PL spectra with the electric dipole $^5D_0 \rightarrow ^7F_2$ transition as dominant, suggesting that the Eu^{3+} ion is located close to the particles surface, with low symmetry. The excess of doping must cause the energy transfer to the Eu^{3+} sites belonging to the SnO_2 lattice to be less effective, when compared to luminescence from other centers with lesser symmetry. This behavior may be also related with decrease of particle size with increasing of Eu concentration. There is a decrease of the particle size as the Eu concentration increases. Excitation with 4.65eV leads to a very well defined PL spectra, with peaks centered at 2.11, 2.09 and 2.07eV, corresponding to the three components of $^5D_o \rightarrow ^7F_1$, as clearly observed in figure 3. In this case, the observed emission at 2.02eV presents low intensity. Then the asymmetric ratio is very close to zero, indicating that Eu^{3+} is actually located at lattice sites, substitutional to Sn^{4+} in the SnO_2 matrix [24]. On the other hand, directly excitation of Eu^{3+} ion with 2.53eV leads to an increase of the asymmetric ratio, as can bee seen in the inset of figure 3, for SnO_2:0.5at%Eu. The solubility limit of Eu in SnO_2 is about 0.05 to 0.06at% [16]. Then the excess of Eu segregates to particle surface, giving rise to a non-homogeneous impurity distribution. The observed broadening in the PL emission may be related to Eu^{3+} ions segregated at particle surface with random distribution of first neighbors. Then, we may conclude that the emission of Eu^{3+} substitutional in the SnO_2 lattice is more evident under excitation at 4.65eV, suggesting a high energy transfer from the matrix due to band-to band excitation. It is also possible to verify that the emission from Eu^{3+} ions located close to particles surface is more clearly observed by direct excitation of $^7F_2 \rightarrow ^5D_2$ transition with the 2.53eV laser line.

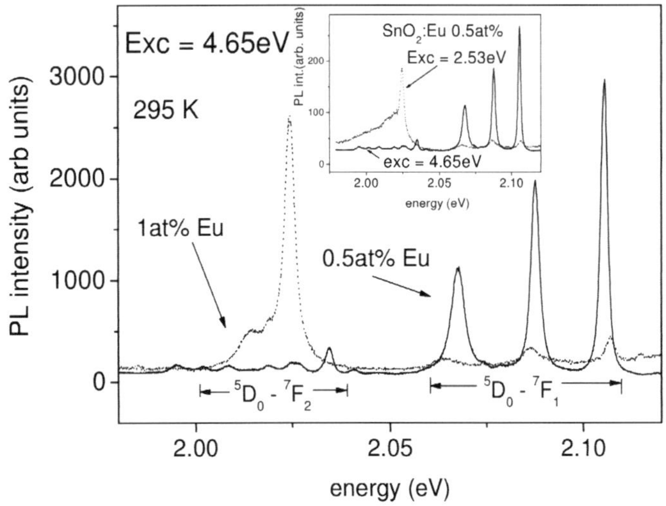

Figure 3. PL spectra for SnO_2:0.5at%Eu and SnO_2:1.0at%Eu xerogels at 295K, under excitation with the fourth harmonic of a Nd:YAG laser (4.65eV). Inset - PL for SnO_2:0.5at%Eu under excitation with the same 4.65eV laser line and with 2.53eV line of an Ar^+ laser.

Figures 4(a) brings results of decay of photo-excited conductivity at several temperatures, for thin films of SnO_2 doped with 0.1at% and 1at% of Eu^{3+}, respectively. In this measurement, films were irradiated with the fourth harmonic of a Nd:YAG laser (4.65eV), promoting electron-hole pair excitation, increasing the number of charge carriers and thus, the material conductivity. After 10 min of light irradiation, time enough to the excitation promotes a carrier saturation, laser illumination was removed and the decay of current was measured as function of time. Fig. 4(b) shows resistivity as function of temperature for SnO_2 thin films doped with 0.1%Er and 0.1%Eu, measured in the dark and also measured a long time after being submitted to the laser irradiation, corresponding to the measured decay presented in figure 4(a). As can be seen in figure 4(b), the Eu^{3+} doped sample presents a higher resistivity when compared to Er^{+3} doped films, which is probably related to the larger size of Eu^{3+} (0.947 Å) compared to Er^{3+} (0.890 Å) [25]. Besides the figure 4(b) also allows verifying the long lasting persistent photoconductivity phenomena in rare-earth doped SnO_2 films [26]. As the Er-doped sample remains a longer time under low pressure, its resistivity becomes higher than the Eu–doped film, due to slow oxygen adsorption. Figure 4 allows verifying that there is a concentration dependent decay of photoexcited current as well as a temperature dependency. Increasing the temperature and Eu concentration, the capture rate also increases, leading to a faster decay. According with the electron capture model presented previously [8,11], the slope of this curve allows evaluating the quantity (E_{cap} - ϕ), where E_{cap} is the trapping energy barrier, related to a relaxation process and ϕ is the grain boundary potential barrier. In this model, we consider the electron scattering at grain boundary as the dominant mechanism for the conductivity [8]. Average values where obtained for these films, fig. 4(a), leading to 113meV and 116meV, for 0.1% and 1at% doped films, respectively.

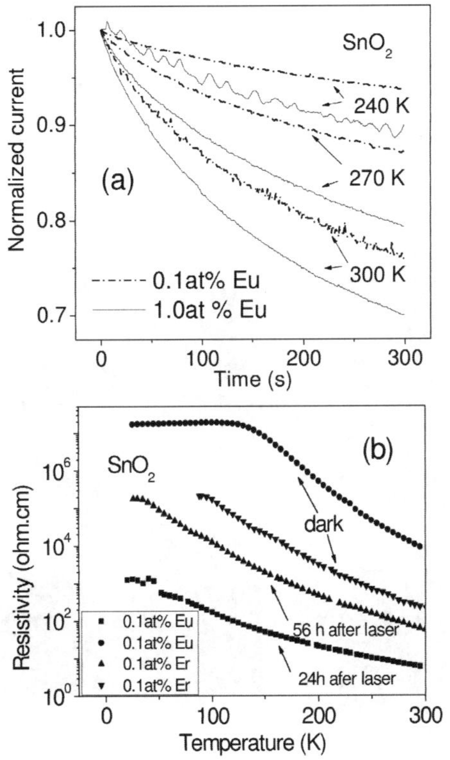

Figure 4. (a) Decay of photo-excited conductivity at several temperatures for SnO_2:0.1at%Eu and SnO_2:1at%Eu. (b) Resistivity as function of temperature for SnO_2:0.1at%Eu and SnO_2:0.1at%Er in the dark and long times after submission to the fourth harmonic of a Nd:YAG laser (4.65eV).

Recently, we have found the existence of different centers in thin films of SnO_2:Er, through measurements of decay of photoexcited conductivity [8,11]. The electron trapping energy by defects are related to oxygen vacancies as well as rare earth centers. Both types of defects are active in the capture process, but they present distinct capture cross section and trapping times. There is a strong indication that the trapping by Er^{3+} centers is dominant for short times, and for longer times the capture by oxygen vacancies becomes a competitive process. A larger variation in the quantity (E_{cap} - ϕ) was found for Er^{3+}-doped SnO_2 thin films [8,11], when the doping concentration is greatly varied, being found energies of 78, 70 and 118meV for 0.05, 0.1 and 4at% de Er^{3+}, respectively. Increasing the concentration from 0.05at% to 0.1at%, we observed a lower capture barrier, which is consistent with a faster decay detected in these measurements. On the other hand, when the Er^{3+} concentration increases to a high value (4at%), the value of (E_{cap} - ϕ) increases, which has been interpreted as due to a distinct location of Er^{3+}-related trap, located in the particles surface, in opposition to the lower doping case, where the dominant trapping center is located in the lattice, substitutional to Sn^{4+}. The increase in the value of (E_{cap} - ϕ) may be related to the presence of oxygen vacancies at grain boundary, which is reinforced by EXAFS measurements at K edge of Sb^{+5} in Sb-doped SnO_2 thin films [15], where it was verified that an increase in the doping concentration decreases the barrier height between grains. Then, if ϕ decreases, (E_{cap} – ϕ) may increase. Comparing (E_{cap} – ϕ) for SnO_2: 0.1at% Er^{3+} and SnO_2: 0.1at% Eu^{3+}, assuming that the intergrain potential barrier (ϕ) has the same height for both samples, the capture barrier (E_{cap}) is higher for Eu doping, which is consistent with a higher disorder obtained by the calculation of structural refinement of these samples [12].

Figure 5 shows atomic force microscopy data for undoped SnO_2 and 4%Er-doped. As can be seen, the undoped sample presents metallurgical grains of about 20-50nm, whereas the doped sample has its granular surface structure destroyed. Instead of grains, some spikes are observed. This figure may help the interpretation that in doped samples the crystallite are indeed the lowest structure, what does not happen in well formed films, where crystallites become part of a larger domain, composed of metallurgical grains. The spikes observed for the 4% doped sample may actually be related to the evaluated crystallite size, which is under investigation.

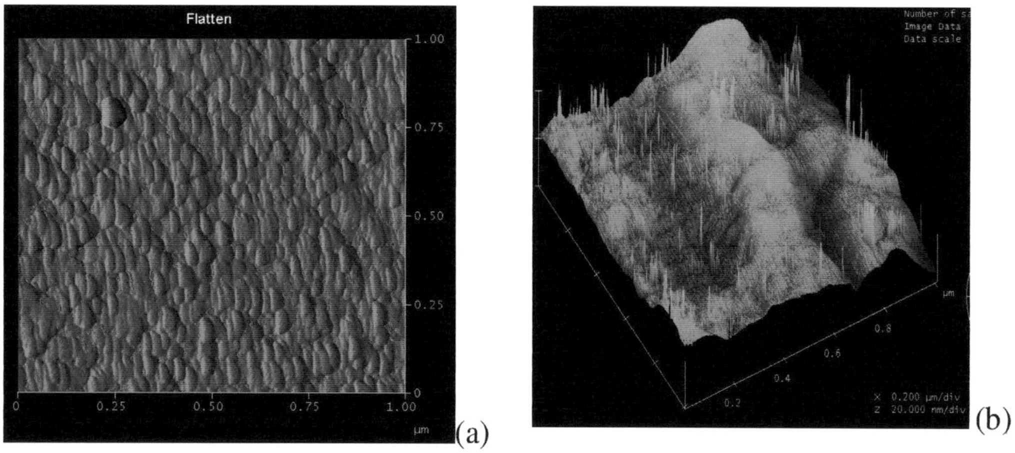

Fig 5 – AFM of a) undoped SnO_2, b) 4at% Er^{3+}-doped SnO_2

4. Conclusions

Temperature dependent excitation of 2at%Er in SnO_2 with 2.53eV promoted the excitation of ions Er^{3+} to its $^4F_{7/2}$ level and the probable decay to nonradiative level $^4I_{13/2}$, with subsequent emission to the ground state $^4I_{15/2}$, which means emission in the technologically interesting value of 0.8eV. Broad visible emission from Er-doped thin film has been observed with peak about 2.33eV (green). It corresponds to $^2H_{11/2} \rightarrow {}^4I_{15/2}$ radiative transition of Er^{3+} ion. The broad band profile may be related to the random distribution of oxygen vacancies, interstitial tin atoms and doping in the neighborhood of the Er^{3+} lattice sites, subtitutional to Sn^{4+} or segregated at grain boundary, which are low symmetry sites.

Eu-doped SnO_2 xerogels emission using 4.65eV and 2.53eV laser lines as excitation sources, allows distinguishing two Eu^{3+} families in SnO_2: one located in the lattice sites, substitutional to Sn^{4+}, and another one, where rare-earth impurity gets preferentially at particles surface due to segregation caused by excess of doping. Excitation with 4.65eV promotes matrix band-to-band excitation and an efficient energy transfer to Eu^{3+} substitutional to Sn^{4+}. Above the doping saturation limit, energy transfer is no longer efficient to Eu^{3+} at lattice sites and the energy transfer to ions located close to the particles surface, with lower symmetry, becomes operative.

The difference of capture barrier by the defects and intergrain potential barrier can be evaluated by a trapping model of photo-induced carriers as published elsewhere [8]. Its application for the thin films studied here, leads to higher capture energy for samples doped with Eu^{3+} compared to Er^{3+}, which may be an indication that the lattice relaxation is higher for the Eu-centers, even though the lattice relaxation mechanism in these films is still unknown.

The results shown in this paper are very promising under a technological point-of-view, since it reveals the possibility of controlling the emission from Er and Eu-doped SnO_2 concerning luminescence spectra structure through rare earth concentration and excitation procedure. Besides, we report emission from thin film, which is a very useful layout for this material, since allows the integration in optical systems and makes the electroluminescence phenomena in this material a reality in a very short time.

Acknowledgments

The authors wish to express their gratitude to Drs. A. Tabata, M. R Fernandes, and S. J L Ribeiro for their help with PL measurements in many ways. Authors also thank Brazilian financial agencies: CAPES, CNPq and FAPESP. Luis V A Scalvi thanks FAPESP for participation at ICDIM 2008.

References

[1] Cetinorgu E, Goldsmith S and Boxman R L 2006 *Semicond. Sci. Technol* **21** 364.
[2] Mulvaney P, Grieser F and Meisel D 1990 *Langmuir* **6** 567.
[3] Batzill M and Diebold U 2005 *Progress in Surf. Sci.* **79** 47.
[4] Kenyon A J, 2002 *Progress in Quantum Eletron* **26** 225.
[5] Favennec P N 1989 *Eletron. Lett.* **25** 718.
[6] Coffa S , Franzo G, Priolo F, Polman A and Serna R 1994 **Phys. Rev. B** *49* 16313.
[7] Morais E A, Scalvi L V A, Ribeiro S J L and Geraldo V 2005 *phys. stat. sol. (a)* **202** 301.
[8] Morais E A and Scalvi L V A 2007 *J. Eur. Cer. Soc.* **27** 3803.
[9] Perea-Lopes N, Gonzales-Ortega J A, Hirata G A 2006 *Opt. Mat.* **29** 43.
[10] Morais E A, Scalvi, L V A, Tabata A, Oliveira J B B and Ribeiro S J L 2008 *J Mat Sci* **43** 345
[11] Morais E A and Scalvi L V A 2007 *J. Mater. Sci.* **42** 2216.
[12] Morais E A, Scalvi L V A, Cavalheiro A A , Tabata A and Oliveira.J B B 2008 *J. Non-Cryst. Solids* **345** 4840
[13] Powder Diffraction File, Inorganic v. 21, Publ. By the JCPDS, Swathmore, 1983
[14] Cullity B D, *Elements of X-ray Diffraction* (Addison Wesley Publ. Comp., Reading Mass. 1978)
[15] Geraldo V, Briois V, Scalvi L V A and Santilli C V 2007 *J. Eur. Cer. Soc.* **27** 4265.
[16] Matsuoka T, Tohda T and Nitta T J.1983 *J. Eletrochemical Soc.* **130** 417
[17] Culp T D, Cederberg J G, Bieg B, Kuech T F, Bray K L, Pfeiffer D and Winter H C 1998 *J. Appl. Phys.* **83** 4918
[18] Morais E A, Ribeiro S J L, Scalvi L V A, Santilli C V, Ruggiero L O, Pulcinelli S H and Messaddeq Y 2002 *J. Alloys Compd.* **344** 217
[19] Matsuoka T, Kasahara Y, Tsuchiya M, Nitta T and Hagakawa S 1978 *J. Eletrochem. Soc.* **125** 102
[20] Wang Y, Ma J, Ji F,Yu X and Ma H 2005 *J Lumin.* **114** 71
[21] Yanes A C, Castillo J D, Torres M and Peraza J 2004 *Appl. Phys. Lett.* **85** 2342
[22] Dutta K and De S K 2007 *Mater. Lett.* **61** 4967
[23] Yu L, Song H, Lu S, Liu Z, Yang L, Wang T, and Kong X 2004 *Mat. Res. Bull.* **39** 2083
[24] Brito G E S, Ribeiro S J L, Briois V, Dexpert-Ghys J, Santilli C V and Pulcinelli S H, 1997 *J. Sol-Gel Sci. Technol.* **8** 261

[25] Shannon R D 1976 *Acta Crystallogr., Sect. A: Found. Crystallogr.* **32** 751.
[26] Morais E A, Scalvi L V A, Geraldo V, Scalvi R M F, Ribeiro S.J.L., Santilli C V and Pulcinelli S H 2004 *J Eur. Cer. Soc.* **24** 1857.

Model Pseudopotentials for Color Centers

P. Gash[a], R. H. Bartram[b] and T. J. Gryk[b,*]

[a]Physics Department, California State University, Chico, CA 95929-0202, USA
[b]Department of Physics, University of Connecticut, Storrs, CT 06269-3046, USA
*Pesent Address: Pratt & Whitney Division of UTC, 400 Main Street, East Hartford, CT 06108, USA

phil.gash@gmail.com

Abstract. The energy-dependent pseudopotentials for electron-excess color centers in ionic crystals adopted by Bartram, Stoneham and Gash (BSG) were determined by a smoothness criterion because the trial wave functions employed are slowly varying on the scale of the ion cores. Energy-independent norm-conserving pseudopotentials were introduced subsequently in the context of molecular-orbital calculations to ensure that the outer parts of valence pseudo-orbitals coincide with those of true valence orbitals. In the present investigation, whole-ion norm-conserving pseudopotentials calculated from numerical wave functions are employed in *F*-center calculations for comparison with BSG calculations. It is concluded that the smoothness criterion should take precedence in that application.

1. Introduction

Energy-dependent whole-ion pseudopotentials for electron-excess color centers were introduced by Bartram, Stoneham and Gash (BSG) [1] to address ion-size effects in alkali–halide *F*-band energies observed by Buchenauer and Fitchen (BF) [2] as systematic departures from the Ivey law [3]. These pseudopotentials were predicated on a smoothness criterion since the trial wave functions employed for electron-excess color centers are slowly varying on the scale of the ion cores. Energy-independent norm-conserving pseudopotentials were introduced subsequently in the context of molecular-orbital calculations on covalent complexes to ensure that the outer parts of valence pseudo-orbitals coincide with those of true valence orbitals [4], [5]. In the present investigation, energy-independent whole-ion norm-conserving pseudopotentials calculated from numerical wave functions [6], stabilized for anions by Watson wells [7], are employed in *F*-center calculations for comparison with corresponding BSG calculations [1], with the point-ion (PI) calculations of Gourary and Adrian [8] and with the BF experimental values [2].

2. BSG Pseudopotentials

The energy-dependent pseudopotentials introduced by BSG [1] were expressed in the form

$$V_p = V_{PI} + \sum\nolimits_{\gamma} C_{\gamma}\delta(\vec{r} - \vec{r}_{\gamma}), \tag{1}$$

where V_{PI} is the point-ion potential [8] and the pseudopotential coefficient C_{γ} for ion γ is given by

$$C_\gamma = A_\gamma + (\bar{V} - U_\gamma)B_\gamma, \tag{2a}$$

$$A_\gamma = \int (1-P_\gamma)(V_\gamma - V_{PI\gamma})d\tau - \int P_\gamma V_{PI\gamma} d\tau, \tag{2b}$$

$$B_\gamma = \int P_\gamma d\tau, \tag{2c}$$

$$\bar{V} = (\phi, V_p \phi)/(\phi,\phi), \tag{2d}$$

$$U_\gamma = \sum\nolimits_{\gamma' \neq \gamma} V_{PI\gamma}, \tag{2e}$$

where P_γ is a projection operator for the core orbitals of ion γ. The average value of the pseudopotential $\bar{V}$ was replaced by the total energy E in subsequent applications [9]; they are equivalent in the limit of smooth trial functions.

3. Norm-Conserving Pseudopotentials

The relevant equations for norm-conserving whole-ion pseudopotentials for monovalent ions are:

$$U_l(r) = \varepsilon_l - \frac{l(l+1)}{2r^2} + \frac{Z}{r} + \frac{\phi_l''(r)}{2\phi_l(r)} - V_c(r), \tag{3}$$

$$V_c(r) = -\frac{Q_{well}}{r_{well}} + Q_{well}\left(\frac{1}{r_{well}} - \frac{1}{r}\right)\theta(r - r_{well}) \text{ (anions)}, \tag{4a}$$

$$V_c(r) = 0 \text{ (cations)}, \tag{4b}$$

$$\phi_l(r) = \sum_{n=0}^{4} a_n^{(l)} r^{l+n+3}, \quad r < r_c, \tag{5a}$$

$$\phi_l(r) = \psi_l(r), \quad r > r_c. \tag{5b}$$

The matching conditions at $r = r_c \leq r_{well}$ can be expressed in the form

$$\sum_{n=0}^{4} A_n^{(l)} = \psi_l(r_c), \tag{6a}$$

$$\sum_{n=0}^{4} (l+n+3)A_n^{(l)} = r_c \psi_l'(r_c), \tag{6b}$$

$$\sum_{n=0}^{4} (l+n+3)(l+n+2)A_n^{(l)} = r_c^2 \psi_l''(r_c), \tag{6c}$$

$$\sum_{n=0}^{4}(l+n+3)(l+n+2)(l+n+1)A_n^{(l)} = r_c^3\psi_l'''(r_c), \tag{6d}$$

where $A_n^{(l)}$ is defined by

$$A_n^{(l)} \equiv a_n^{(l)} r_c^{l+n+3}. \tag{7}$$

Thus Equations (6) comprise four linear algebraic equations in the five unknown quantities $A_n^{(l)}$; by matrix inversion, one obtains solutions in the form

$$A_n^{(l)} = W_n^{(l)} + \beta_n^{(l)} A_0^{(l)}, \quad 0 < n \le 4. \tag{8}$$

A fifth relation is provided by the normalization condition,

$$N^{(l)} \equiv 1 - \int_{r_c}^{\infty} \psi_l(r)^2 dr = \int_0^{r_c} \phi_l(r)^2 dr = r_c \sum_{m=0}^{4}\sum_{n=0}^{4} \frac{A_m^{(l)} A_n^{(l)}}{m+n+7+2l}. \tag{9}$$

An example of a numerical wave function for $l=0$ together with the corresponding norm-conserving pseudowavefunction is presented in Figure 1 for Cl^{2-}. In each case, the matching radius r_c was chosen to coincide with the maximum of the outer lobe of $\psi_0(r)$. For anions, the radius r_{well} was specified as ~1.75 times the ionic radius and the charge Q_{well} was chosen to make the well depth 1.0 Hartree in each case.

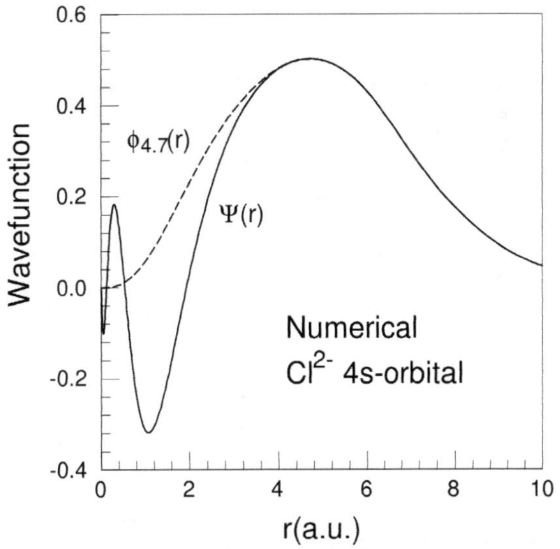

Figure 1. Numerical wave function $\psi(r)$ for Cl^{2-} in a Watson well, together with the corresponding norm-conserving pseudowavefunction $\phi(r)$.

Finally, the energy-independent pseudopotential coefficient C_γ is given by

Journal of Physics: Conference Series **249** (2010) 012006 doi:10.1088/1742-6596/249/1/012006

$$C_\gamma = 4\pi\int_0^\infty \left[U_0(r) - \frac{Z}{r} + \frac{Q_{ion}}{r}\right] r^2 dr = [\varepsilon_0 - V_c(r)]\frac{4\pi r_c^3}{3} + 2\pi Q_{ion} r_c^2$$
$$+ 2\pi\int_0^{r_c}\left[\frac{\sum_{n=0}^{4}(n+3)(n+2)a_n^{(0)}r^n}{\sum_{n=0}^{4}a_n^{(0)}r^n}\right]dr + 4\pi\int_{r_c}^\infty \frac{\psi_0''(r)}{2\psi_0(r)} r^2 dr. \qquad (10)$$

The calculated norm-conserving pseudopotentials for monovalent anions and cations are plotted in Figures 2 and 3, respectively.

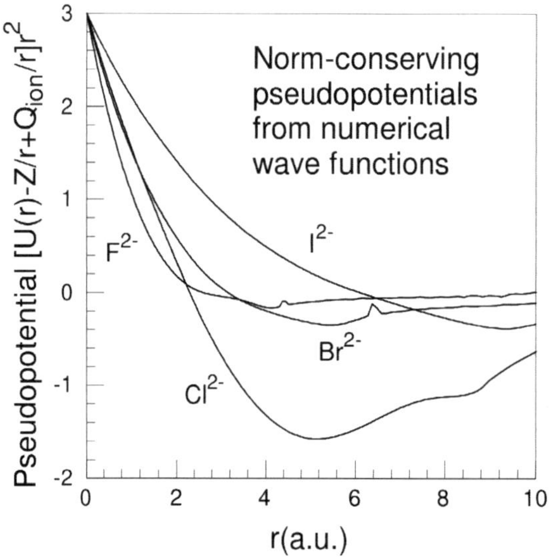

Figure 2. The integrand of pseudopotential coefficient C_γ as a function of r for halogen ions.

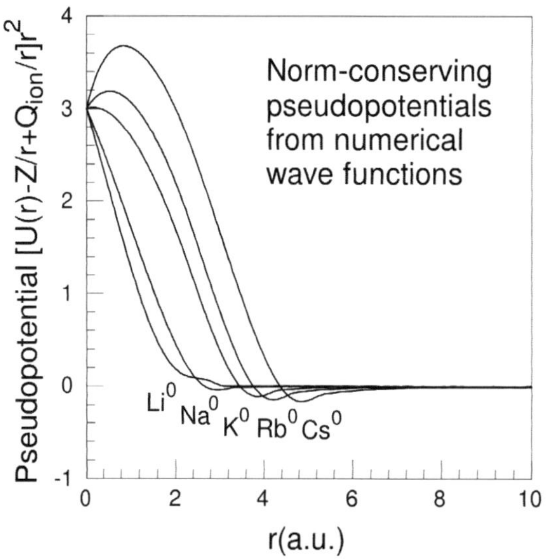

Figure 3. The integrand of pseudopotential coefficient C_γ as a function of r for alkali ions.

4. *F*-Center Calculations

F-center transition energies incorporating ion-size corrections based on norm-conserving pseudopotentials (NCP) were calculated for the seventeen alkali halides with rock-salt structure. The required input data are listed in Table I.

Table I. Whole-ion norm-conserving pseudopotential coefficients for alkali halides from Equation (10) and input for F-center calculations

Ion	C_γ (Hartree a_0^3)	A_H+A_X (Rydberg a_0^3)	B (a_0^3)
F^-	26.45	52.90	0.0
Cl^-	-50.42	-100.84	0.0
Br^-	26.07	52.14	0.0
I^-	34.61	69.22	0.0
Li^+	36.40	72.80	0.0
Na^+	44.76	89.52	0.0
K^+	77.85	155.70	0.0
Rb^+	94.05	188.10	0.0
Cs^+	127.08	254.16	0.0

Calculations were performed by the method of BSG [1] which proceeds from the PI method [8] augmented by corrections for ion-size effects, polarization and distortion. The ion-size correction employed by BSG was diminished by an adjustable parameter α. Ion-size corrections based on the NCP coefficients listed in Table I, similarly diminished, were employed in the present work. Calculated *F*-band energies, normalized to the Ivey law [3], are compared in Figure 4 with the experimental results of BF [2].

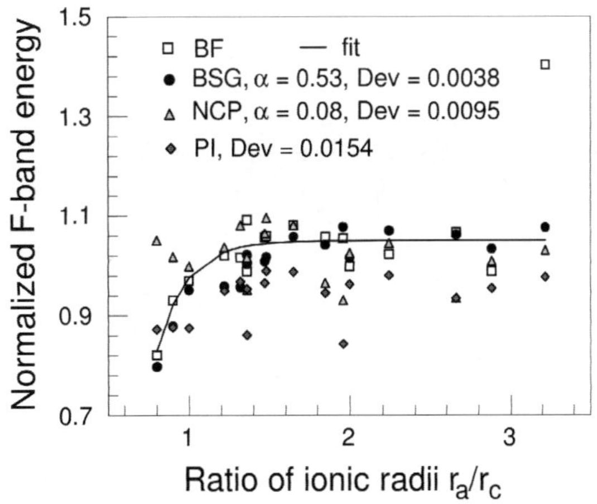

Figure 4. *F*-band energies normalized to the Ivey law [3]. White squares denote the experimental values of BF [2], the continuous curve is a plot of Equation (11), red diamonds represent the PI approximation [8] with corrections for polarization and distortion, black circles include the BSG [1] ion-size correction as well and green triangles include instead an ion-size correction based on the NCP coefficients C_γ listed in Table 1. Ion-size corrections are diminished by an adjustable parameter α and *Dev* is the sum of squares of energy differences, in Rydbergs, between theory and experiment.

The continuous curve is a saturating exponential fitted to the experimental data, given by

$$\frac{\Delta E(a,R)}{\Delta E_{Ivey}(a)} = 1.05 - 18.0\exp(-5.5R)\,, \tag{11}$$

$$\Delta E_{Ivey}(a) = 4.18a^{-1.84}, \quad (12)$$

where a is the nearest-neighbor distance in atomic units and R is the ratio of ionic radii,

$$a \equiv r_{anion} + r_{cation}, \quad (13)$$

$$R \equiv r_{anion} / r_{cation}. \quad (14)$$

5. Discussion

Energy-independent norm-conserving pseudopotentials were introduced as a formal device for application of the frozen-core approximation in molecular calculations and are still extensively employed for that purpose. However, it is evident from Figure 4 that whole-ion NCP's are inferior to the BSG pseudopotentials for calculating ion-size corrections to F-band energies, since the required reduction factor α is much smaller and their deviation from experiment is significantly greater. In particular, the calculation fails to reproduce the observed dependence of ion-size effects on the ratio of ionic radii [2]. Figure 1-3 illustrate the fact that norm conservation can be achieved only by imposition of a pseudopotential that is highly repulsive at the ion core and is thus incompatible with a smooth trial function. On the other hand, norm-conservation is essentially irrelevant for electron-excess color centers where covalent bonding is not involved.

References

[1] R. H. Bartram, A. M. Stoneham and P. Gash, Phys. Rev. 176, 1014-1024 (1968)
[2] C. J. Buchenauer and D. B. Fitchen, Phys. Rev. **167**, 846 (1968)
[3] H. F. Ivey, Phys. Rev. **72**, 341 (1947)
[4] P. A. Christiansen, Y. S. Lee and K. S. Pitzer, J. Chem. Phys. **71**, 4445 (1979)
[5] D. R. Hamann, M. Schlüter and C. Chang, Phys. Rev. Letters **43**, 1494 (1979)
[6] C. Froese-Fischer, T. Brage and P. Jönsson, *Computational Atomic Structure: An MCHF Approach* (Institute of Publishing, Bristol and Philadelphia, 2000)
[7] R. E. Watson, Phys. Rev. **111**, 1108 (1958)
[8] B. S. Gourary and F. J. Adrian, Phys. Rev. **105**, 1180 (1957)
[9] R. C. Alig, Phys. Rev. B2, 2108 (1970)

Optical Investigation of Sm doped ZrO_2

S. Lange[1], I. Sildos[1], M. Hartmanova[2], V. Kiisk[1], E. E. Lomonova[3], and M. Kirm[1]

[1]Institute of Physics, University of Tartu, Riia 142, 51014 Tartu, Estonia
[2]Institute of Physics, Slovak Academy of Sciences, Dubravska cesta 9, 845 11 Bratislava 45, Slovak Republic
[3]Institute of General Physics, Russian Academy of Sciences, 117942, Moscow, Russia

E-mail: sven.lange@mail.ee

Abstract. A samarium doped polycrystalline ZrO_2 bulk sample was investigated for its phase composition as well as optical properties within the temperature range 6–300 K. From micro-Raman measurements the existence of tetragonal as well as monoclinic phase was confirmed. The emission spectrum of Sm^{3+} reveals drastic changes under excitation near the region of fundamental absorption. Combined excitation-emission spectroscopy (CEES) measurements reveal three different sites for the dopant ions to be present in the sample. Nature and excitation scheme of the three sites are discussed.

1. Introduction

Together with several other transition metal oxides zirconium oxide (ZrO_2) has been under investigation in the search for novel luminescent materials for some time now. Wide band-gap (>5 eV) and low phonon energies (<650 cm^{-1}) make it a promising host for doping with a wide range of luminescent impurities for operation at elevated temperatures. In combination with the high refractive index and high optical damage threshold this could lead to new luminescent materials of superior performance for applications like phosphor-converted LEDs, integrated optical amplifiers, laser hosts and other photonic materials.

Doping of ZrO_2 with various rare earth (RE) ions and the creation and study of intrinsic defects in the pure crystal have been reported by several authors already [[1,2,3,4,5]]. Introduction of luminescent impurities, especially at high concentrations, required for several applications, markedly influences the phase content of zirconia. In addition, the nature of the host-sensitized excitation mechanisms leading to the emission of RE ions and intrinsic excitations is still a topic of ongoing debates [[5,6]]. This is mainly because of the rather complicated electronic and phase structure caused by the dopants being able to stabilize zirconia in tetragonal and cubic phases in addition to the only naturally stable monoclinic phase at room temperature. In relation to the above in the current work we aim to investigate the phase structure, the distribution of the luminescent impurities and the luminescent properties of Sm-doped ZrO_2.

2. Experimental

The crystalline ZrO_2 containing 4 mol% of Sm_2O_3 were grown by the directional solidification of the melt using the direct microwave heating in a water-cooled crucible (SMT). The growth rate of the crystal was 10 mm/h. The cooldown of the crystal from 2000°C took place in steps with decreasing temperature gradients from 200–30°C/min. More detailed explanation of the technique can be found elsewhere [7]. XRD study indicates the presence of nearly equal amounts of monoclinic and tetragonal phase (see Ref. [8] for in-depth description). In addition to standard XRD analysis the phase content and inhomogenicity of the sample was investigated via micro-Raman measurements. The measurements were conducted with a dedicated Renishaw micro-Raman setup with 514 nm laser excitations.

Combined excitation-emission spectroscopic (CEES) studies (eg. [9]) were applied to reveal different types of Sm centres in the sample. Individual PL spectra of CEES spectrum were measured using an Andor SR303i spectrograph equipped with an Andor ICCD detector and a Hamamatsu photomultiplier tube. A tunable multi-stage OPO (type NT 342/1/UVE) was used as the excitation source. PL spectra were corrected to the variable intensity of excitation source by recording the excitation spectrum of 623 nm line with a similar experimental setup having a continuous mode 150 W xenon lamp dispersed via MDR-23 monochromator as the excitation source. For low temperature measurements an UTREKS-type helium-bath cryostat was used.

3. Results

Micro-Raman measurements (spatial spot diameter ~1 μm) indicate a large inhomogenity of the sample. Depending on the detection location on the sample we can see a variation of the Raman lines intensities corresponding to monoclinic (M) or tetragonal (T) phases (Fig. 1a,b). In addition, some photoluminescence (PL) of Sm^{3+} ions is excited. The PL line is situated at around ~400 cm^{-1} (~525 nm) to the red from the excitation wavelength of 514 nm.

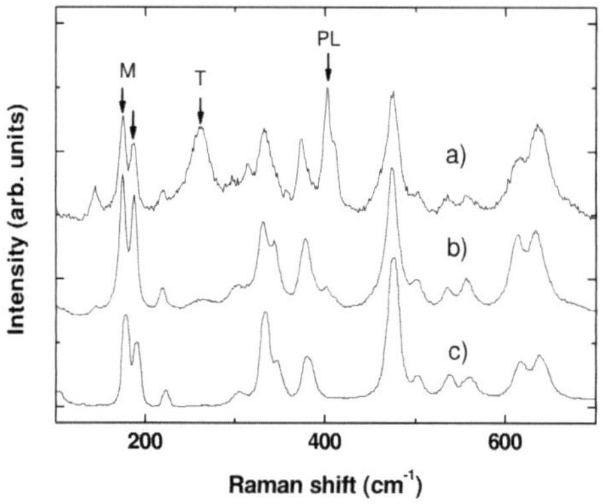

Figure 1. The micro-Raman spectrum of studied $ZrO_2{:}Sm^{3+}$ sample on two different locations (a) and (b) on sample (laser spot ~1 μm). A reference spectrum (c) of monoclinic ZrO_2 from reference [10]. Characteristic Raman lines of monoclinic (M) and tetragonal (T) phases as well as a photoluminescence (PL) band of Sm^{3+} are indicated.

For the measurement of the CEES spectra laser excitation in the interval of 210 to 480 nm was used. Chosen excitation range covers both indirect (210-300 nm)excitation near the fundamental absorption edge as well as direct (300-470 nm) excitation of Sm^{3+} through its f-f transitions. Luminescence spectra were registered in the interval of 550–750 nm, which corresponds to $^4G_{5/2}$– $^6H_{(J)}$ manifold of f-f transitions of Sm^{3+} ions. General look of the results obtained at 6 K and room temperature (RT) are depicted in figure 2.

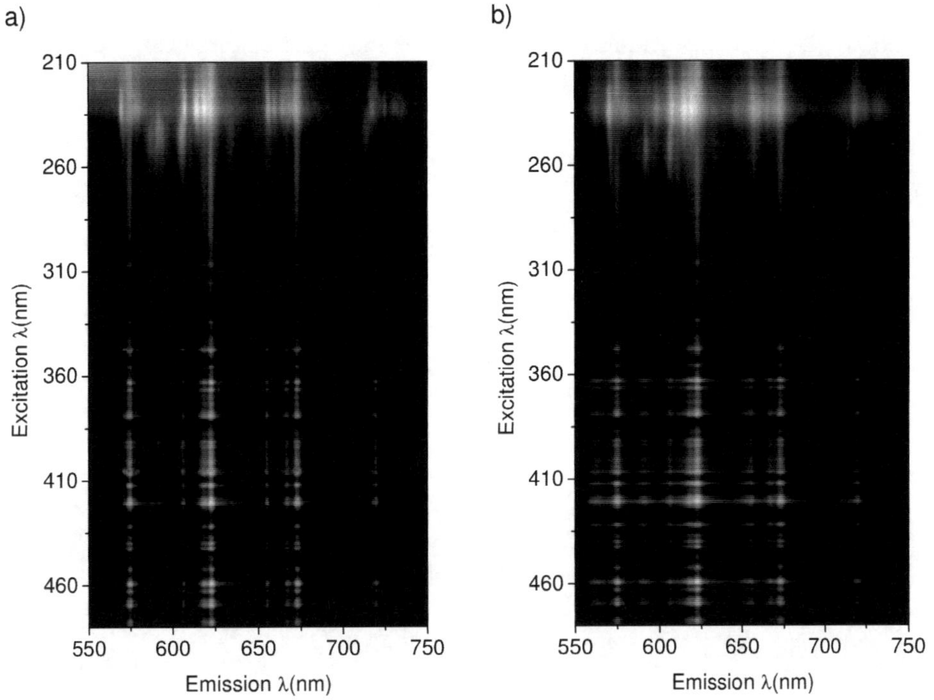

Figure 2. The combined excitation-emission spectra of Sm-doped ZrO_2 measured at 6 K (a) and RT (b). For coverage of the whole range of emission intensities a logarithmic scale is used.

From CEES spectra two distinct regions of Sm^{3+} excitation are revealed. At excitation wavelengths from 470 to 300 nm the directly excited PL emission characteristic to f-f transitions of Sm^{3+} ions is highly fluctuating in intensity. Below 300 nm the spectra represent a more uniform excitation region for the Sm^{3+} PL emission via indirect host mediated excitation of the ions.

The Stark splitting of the $^4G_{5/2}$ to ^6H_i (i = 11/2, 9/2, 7/2 and 5/2) emission lines of Sm^{3+} ion in tetragonal and monoclinic crystalline phases have been reported previously in [[3,11]]. By comparing low-temperature PL spectra derived from CEES spectrum at different excitation wavelengths we can suggest that in our sample Sm^{3+} ions both in the tetragonal and monoclinic crystal surrounding sites are indeed present (Fig. 3). Tetragonal phase is excited with varying intensity over the whole measured excitation wavelengths range from 210 to 480 nm (Fig. 4). Monoclinic sites are excited prevalently in noticeably smaller band centered at ~235 nm (Fig. 4). A third, yet unknown crystalline surrounding is presented by the additional PL lines at excitation near 240 nm. The latter maximum shifts to 250 nm at RT (Fig. 4b).

At 6 K a broad-band emission in the blue region of the PL spectra appears at excitation wavelength close to ZrO_2 fundamental absorption edge at 210 nm (Fig. 3a). This defect related emission is totally quenched at RT [[12]].

a) b)

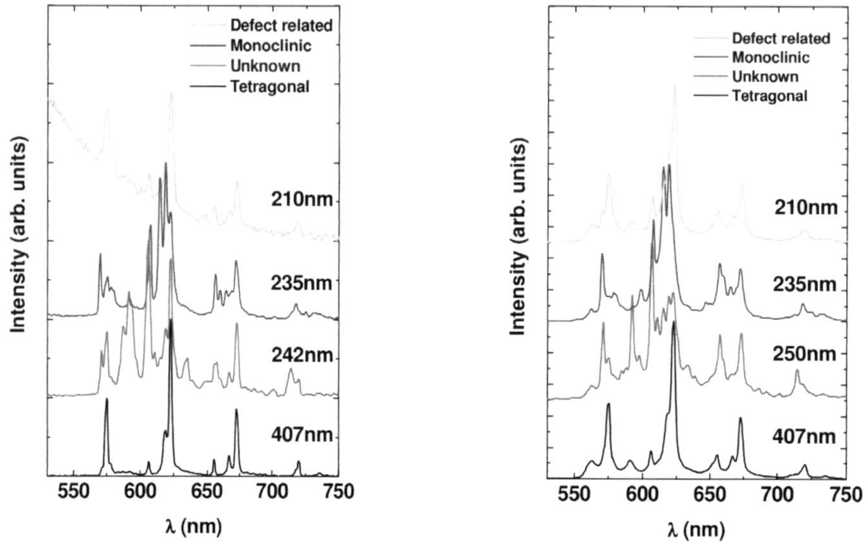

Figure 3. Sm^{3+} emission spectra at tetragonal, monoclinic and unknown sites of ZrO_2 at different excitation wavelengthts. Measured at 6 K (a) and RT (b), respectively.

a) b)

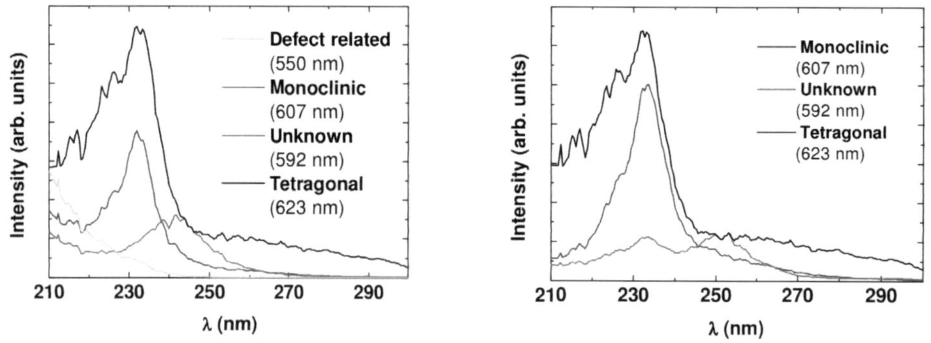

Figure 4. Excitation spectra of the Sm^{3+} emission in tetragonal, monoclinic and unknown phases of ZrO_2 at 6 K(a) and RT (b).

The PL decay curves of the three sites of Sm^{3+} have different behaviours (Fig. 5a). In tetragonal site the decay is nearly independent of temperature. Non-exponential behavior can be modelled airly well by assuming an energy transfer to randomly distributed acceptors via multipolar interaction [[13]]. According to this model the PL decay can be described by the following equation:

$$I(t) = A\exp\left[-\frac{t}{\tau_{\mathrm{rad}}} - \Gamma\left(1-\frac{3}{s}\right)\frac{C}{C_0}\left(\frac{t}{\tau_{\mathrm{rad}}}\right)^{\frac{3}{s}}\right], \qquad (1)$$

where τ_{rad} represents the radiative lifetime, C is the concentration of the acceptor species, C_0 is a critical concentration corresponding to the rate of energy transfer equal to the radiative decay rate, and s=6, 8 or 10 depending on the type of interaction. Best fit (with τ_{rad} ~ 3.5 ms) was achieved for s=8 characteristic of dipole-quadrupole interaction.

At 6 K the Sm^{3+} emission decay in monoclinic site is essentially non-exponential and has a prolonged tail. The decay profile for the unknown site occurred to be quite strongly temperature dependent. The decay curve is evidently non-exponential at RT but leads to the appearance of a well pronounced exponential part at 6 K (Fig. 5b).

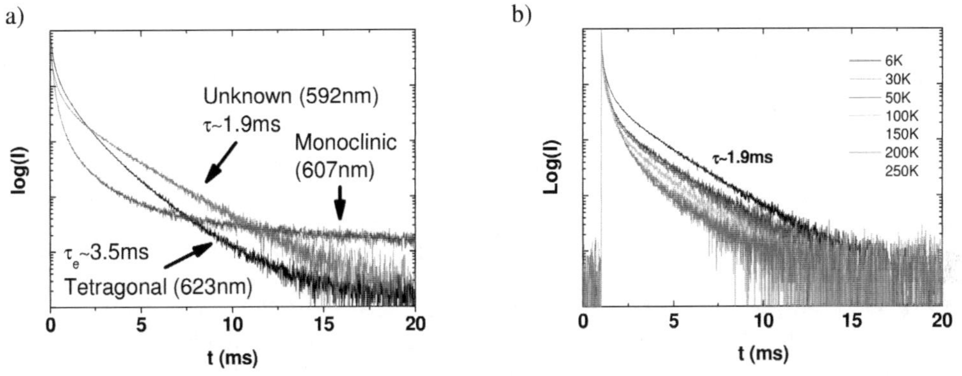

Figure 5. PL decay curves of Sm^{3+} ion emission in different surroundings of ZrO_2 matrix, measured at 6 K (a); temperature dependence of the PL decay curves for the of Sm^{3+} in the unknown site of ZrO_2 (b).

4. Discussion

According to Garvie and Nicolholson [14] the relative amount of monoclinic phase in mixed monoclinic/tetragonal ZrO_2 can be derived from the following empirical equation:

$$C_m = \frac{I_m^{179} + I_m^{190}}{I_T^{264} + I_m^{179} + I_m^{190}} \tag{2}$$

Here the different terms denote integral intensities of the Raman lines at indicated Raman wavenumbers. According to eq. (2) the Raman signal (Fig. 1) suggests that the phase-mixed sample has different surroundings present ranging from mixture of tetragonal (C_t=52%) and monoclinic (C_m=48%) phases to a lesser amount of almost totally monoclinic locations (C_m=98.6%).

The presence of comparable amounts of monoclinic and tetragonal phases in many parts of the crystal leads to a large phase boundary area being present. The latter can be the cause for the emission from the unknown site in the CEES spectrum.

The difference of surroundings for the dopant ions cause three different routes of excitation for Sm^{3+} ions in the respective sites. The most prominent is the presence of the tetragonal phase over the whole excitation range from 210 to 480 nm. Above 250 nm where no other sites are excited the excitation is achieved dominantly via direct absorption of photons by Sm^{3+} ion 4f shell electronic transitions. The luminescence decay here is temperature independent and is shortened probably by a cross-relaxation between neighbouring Sm^{3+} ions due to their relatively high concentration (4 mol%).

The luminescence characteristic of the monoclinic phase is present only when excited into a band peaking at 235 nm (at both RT and 6 K) coinciding with the exciton absorption [5]. The long-lasting decay of the PL indicates a migration of the host electronic excitations prior to the energy transfer to the PL centre [15]. Similar behavior was noticed for Sm^{3+} in TiO_2 matrix.

We propose that in the phase boundary regions in the crystal where the third unknown sites of Sm^{3+} are present yet another route of excitation is prevailing. At 6 K the excitation is taking place probably from a defect donor level slightly lower in energy than the excitonic states. Due to the exponential nature of the decay we can conclude that the excited charge carriers are transferring the excitation to the Sm^{3+} ions relatively fast and no long-lasting migration is taking place (compared to monoclinic phase). At RT thermally induced relaxation processes cause the decay to become non-exponential and shortened and also lead to the appearance of another excitation peak close to the exciton absorption band (~235 nm). We suggest that it could be an indication that a donor level from the monoclinic phase has been opened for Sm^{3+} ion excitation via temperature assisted multiphonon relaxation. The existence of such possibility also suggests that the unknown phase must spatially be placed close to monoclinic phase where noticeable charged carrier migration was detected in PL decay curve. Additional spatially resolved Raman measurements are needed to confirm this.

5. Conclusion

Based on micro-Raman study it is inferred that the highly Sm-doped polycrystalline zirconia produced by the scull-melting technique contains nearly equal amounts of micrometer-scale distributed monoclinic and tetragonal phases of ZrO_2. However, from the luminescence measurements it is concluded that the essential part of Sm^{3+} emission originates from the tetragonal phase with a clear indication of cross-relaxation in the decay kinetics. A lesser amount of active ions are situated in the monoclinic phase and appear to be preferentially host-sensitized with the excitation mediated by excitons. The third site revealed by CEES might be related to the boundary region of the two phases and can at RT partly be excited by charged carriers from monoclinic phase.

6. Acknowledgements

The work was partially supported by Estonian Science Foundation (Grants No's 6999 and 6658).

References

[1] E. De la Rosa-Cruz, L.A. Diaz-Torres, P. Salas, R.A. Rodriguez, and G.A. Kumar 2003 J. Appl. Phys. 94 3509-3515
[2] R. Reisfeld, M. Zelner, and A. Patra 2000 J. Alloys Comp. 300 147-151
[3] V. Kiisk, I. Sildos, S. Lange, V. Reedo, T. Tätte, M. Kirm and J. Aarik 2005 Appl. Surface Science 247 412-417
[4] V.R. PaiVerneker, A.N. Petelin, F.J. Crowne, D.C. Nagle 1989 Phys. Rev. B 40 12
[5] M. Kirm, J. Aarik, M. Jürgens, I. Sildos 2005 Nuclear Instruments and Methods in Physics Research A 537 251-255
[6] N. G. Petrik, D.P. Taylor, T. M. Orlando 1999 J. of Appl. Phys. 85 9
[7] M. Hartmanova, J. Schnaider, V. Navratil, F. Kundracik, H. Schulz, E.E. Lomonova 2000 Solid State Ionics 107 136-137
[8] D. Miller, L. Grigorjeva, W. Lojkowski, A. Opalinska 2005 Solid State Phenomena 106 103-107
[9] Z. Fleishman, P. S. Dafon, V. Dierolf, C. Munasinghe, A. J. Steckl, 2007 Phys. Stat. Sol. (c) 4 No. 3 834-837
[10] S. N. Tkachev, M.H. Manghnani, A. Niilisk, J.Aarik, H. Mändar, J of Material Science 2005 40 4293-4298
[11] E. De la Rosa-Cruz, L.A. Diaz-Torres, P. Salas, R.A. Rodriguez, and C. Angeles 2004 Proc. Of SPIE Vol. 5510 57-67
[12] S. Lange, I. Sildos, V. Kiisk, J. Aarik, and M. Kirm 2005 phys. stat. sol. (c) 2 No. 1 326–329
[13] M. Inokuti, F. Hirayama 1965 Journal of Chemical Physics 43 1978-1989
[14] D. R. Clarke and F. Adar 1982 Journal of American Cheramic Society 65 284
[15] S. W. S. McKeever 1985 Thermoluminescence of Solids Cambridge University Press

Different incorporation of Cu^+ and Cu^{2+} in lithium tetraborate single crystals

G Corradi[1], V Nagirnyi[2], A Watterich[1], A Kotlov[2] and K Polgár[1]

[1]Research Institute for Solid State Physics and Optics, Hungarian Academy of Sciences, Budapest, P.O.Box 49, H-1525, Hungary
[2]Institute of Physics, University of Tartu, 142 Riia Str., 51014 Tartu, Estonia

corradi@szfki.hu

Abstract. The incorporation site of Cu in the scintillator and thermoluminescent dosimeter material lithium tetraborate $Li_2B_4O_7$:Cu has been shown to depend on the charge state of the dopant. As confirmed by a refined analysis, the EPR spectra of an unpaired hole can be ascribed to Cu^{2+} situated near the Li lattice site with C_1 symmetry, in contrast to the higher (C_2) overall symmetry reported earlier for Cu^+ in its relaxed excited state, the latter suggesting interstitial incorporation for the monovalent dopant. Consequences for charge transfer processes involving copper ions are discussed.

1. Introduction

The acousto- and optoelectronic material lithium tetraborate ($Li_2B_4O_7$) owes its remarkable properties to its low-symmetry crystal structure (space group $I4_1cd$ belonging to the point group C_{4v}) and the related large polarizability of its lattice. The Li^+ ions are accommodated at general positions with no symmetry (C_1) and stabilize the =B-O-B≡ covalent network consisting of a frame of alternating oxygen bonded BO_4 and BO_3 oxyanions [1] (see figure 1). The basic structural unit of the network is a cradle-like B_4O_7 group. Similarly to other borates $Li_2B_4O_7$ is transparent also in the ultraviolet (UV) up to 160 nm, which is useful for non-linear optical applications based on higher harmonics. Appropriately doped $Li_2B_4O_7$ is also a scintillator suited for neutron detection; stable intrinsic defects introduced by neutron irradiation have been characterised by EPR [2, 3].

If activated by small amounts of Cu, $Li_2B_4O_7$ can be used as a tissue-equivalent thermoluminescent (TL) dosimeter; outstanding sensitivities exceeding that of the well-known LiF:Mg,Ti phosphor (TLD-100) can be achieved if Cu is used together with co-dopants like Ag and P [4, 5]. Recent studies on $Li_2B_4O_7$:Cu single crystals indicate that Cu is present in two charge states. Cu^{2+}, traceable by EPR, is incorporated at a site with C_1 symmetry which may be a strongly relaxed Li site [6], while the non-paramagnetic Cu^+ charge state, detectable due to its UV-induced luminescence, displays higher, apparently C_2 overall symmetry, in its dynamic excited state [7]. As further EPR studies, showing also regular saturation behaviour, exclude a similar dynamic configuration for Cu^{2+}, further attempts to understand the relation between both charge states and their incorporation in the lattice have been undertaken. The results have recently been summarized in Ref. [8]. Here the main points stressing the difference between the sites in question will be discussed concerning also the consequences for charge trapping at copper ions, which in turn may lead to a better understanding of the TL mechanisms.

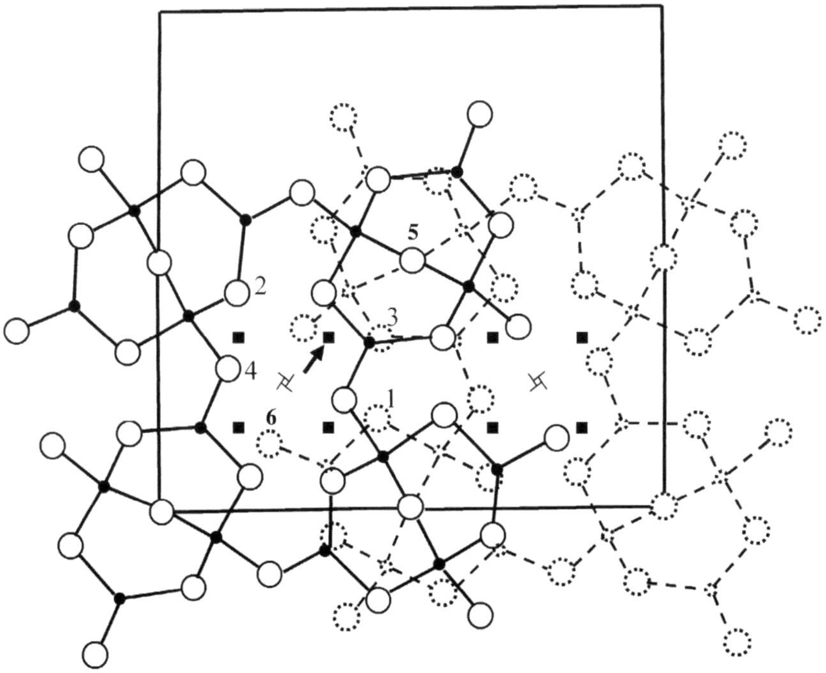

Figure 1. Structure of the $Li_2B_4O_7$ lattice [1] shown as projection of part of the tetragonal elementary cell along the [001] or C_2-direction, showing the =B—O—B≡ covalent network. Some positions of the vertical fourfold screw axes are also indicated. The nearest oxygen neighbours of a chosen Li site (indicated by an arrow) are numbered 1-6.

2. Ground state properties of Cu^{2+} derived from EPR measurements

Our earlier EPR investigations on copper doped lithium tetraborate [6, 8] showed that the principal directions of the **g**-tensor (see table 1) and the Cu^{63} hyperfine tensor practically coincide as expected for a well-localized $3d^9$ single-hole type state of Cu^{2+} ions. All principal axes (shown in figure 2) form large angles with the single C_2 symmetry axis of the lattice. The main (third) principal axis corresponding to the smallest g-value (g_3) and the largest HF splitting is nearly parallel to a [110] type direction of the lattice. As the minimal g-value tends to discern a direction where a pair of nearest oxygen neighbours with Cu-O distances near 2 Å are located (compare Cu^{2+} and isoelectronic Ni^+ ions in other oxides like $LiNbO_3$ [9] or $PbWO_4$ [10]), the only straightforward choice for the dopant's location in its divalent state is a more or less displaced Li site. As shown by a calculation assuming a displacement of only the dopant ion, a relaxation (indicated in figure 2) of the Cu^{2+} ion from the Li site roughly towards the midpoint of the nearest oxygen neighbours 1 and 2 in fact may result in a configuration explaining all main features of the spin Hamiltonian tensors [8](for more detail see table 2). It should be noted that such a relaxation qualitatively improves the agreement: while the g_3 direction is essentially given by the line connecting oxygen neighbours 1 and 2, the principal direction corresponding to the second g-eigenvalue (g_2) tends to be defined by the next nearest oxygen neighbour which is no more oxygen No. 3 but No. 6 (see figure 2). The directions in question show an agreement (compare tables 1 and 2) which can be considered as remarkable taking into account that no local adjustment of the covalently bound =B-O-B≡ frame has been assumed. No other site for the

Journal of Physics: Conference Series **249** (2010) 012008 doi:10.1088/1742-6596/249/1/012008

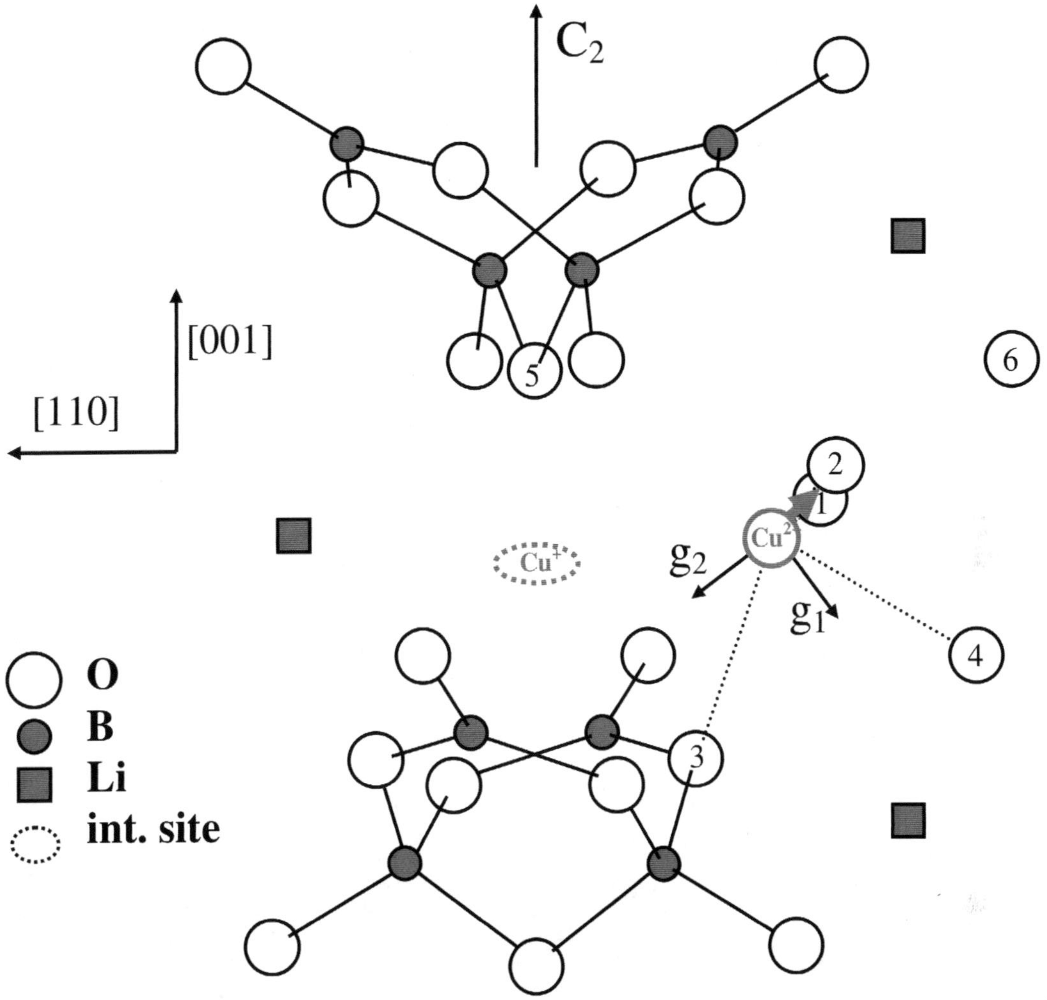

Figure 2. Projection along a [110] type axis, showing Cu^+ at an interstitial position and Cu^{2+} substituted at a Li site with possible relaxation indicated by an arrow. The nearest oxygen neighbours of the chosen Li site are numbered 1-6; the principal directions of the Cu^{2+} **g** tensor given in Table 1 are also indicated. Note that the main principal axis is nearly perpendicular to the plane of the projection.

Cu^{2+} ion, and in particular no other pair of potential first-neighbour oxygens, with the properties required for reproducing the EPR results even approximately, can be found in the $Li_2B_4O_7$ lattice.

It should be pointed out that on the basis of the EPR data presently no conclusion about the presence and location of local charge compensator(s) can be made, the simplest assumption being a Li vacancy at one of the two nearest neighbour Li sites (appearing as the closest Li neighbours also on the projection shown in figure 2). However, the EPR spectra belong to a well-defined species, either locally charge-compensated, or not.

Table 1. Principal values and eigenvector coordinates of the **g** tensor [8].

g tensor		
2.515 ± 0.013	2.220 ± 0.010	2.0265 ± 0.0010
0.423	0.551	-0.720
0.389	0.606	0.693
0.818	-0.573	0.042

Table 2. Distances and position vectors of the six nearest oxygen neighbours with respect to a Li^+ site [1]. The chosen Li^+ site is shown in figures 1 and 2 having the cell coordinates x/a=y/a=0.342 and z/c=0.36, where a=9.47 Å and c=10.26 Å are the lattice constants. Also included are a possible relaxation vector **S** of the Cu^{2+} ion substituted at this Li^+ site and the changed position vectors of its oxygen neighbours.

Position vectors	$\mathbf{O}^1$	$\mathbf{O}^2$	$\mathbf{O}^3$	$\mathbf{O}^4$	$\mathbf{O}^5$	$\mathbf{O}^6$
r (Å)	1.815	2.08	2.15	2.28	2.55	2.76
x/r	0.475	-0.842	0.432	-0.859	0.585	-0.429
y/r	-0.851	0.446	0.004	-0.245	0.585	-0.710
z/r	0.226	0.301	-0.902	-0.450	0.562	0.558

	$\mathbf{S} = \frac{1}{2}(\mathbf{O}^1+\mathbf{O}^2)$	$\mathbf{O}^2-\mathbf{S} = \frac{1}{2}(\mathbf{O}^2-\mathbf{O}^1)$	$\mathbf{O}^3-\mathbf{S}$	$\mathbf{O}^4-\mathbf{S}$	$\mathbf{O}^5-\mathbf{S}$	$\mathbf{O}^6-\mathbf{S}$
r (Å)	0.75	1.81	2.83	2.18	2.80	2.08
x/r	-0.594	-0.722	0.485	-0.696	0.692	-0.355
y/r	-0.411	0.683	0.112	-0.115	0.643	-0.795
z/r	0.692	0.113	-0.867	-0.709	0.327	0.491

3. Discussion of the excited state properties of Cu^+

As reported in refs. [7, 8], the optical absorption and luminescence properties of $Li_2B_4O_7$:Cu single crystals, observed for excitations up to 20 eV, also reflect the low-symmetry properties of the host. The only Cu-related emission band at 3.35 eV is due to the Cu^+ charge state and can be ascribed to the transition from the lowest excited triplet state of its $3d^94s$ term to the $3d^{10}$ ground state [8,11,12], a transition which is normally parity- and spin-forbidden. This band corresponds to the spectral properties of the first prominent TL peak observed near 350 K in $Li_2B_4O_7$ powders co-doped by Cu, Ag, and P [5]. The corresponding $3d^{10} \rightarrow 3d^94s$ transitions in the excitation and absorption spectra near 5 eV have substantial relative intensities compared to the allowed $3d^{10} \rightarrow 3d^94p$ transitions near 7 eV. Along with the observed polarisation anisotropy of all spectra this indicates some low-symmetry position for the Cu^+ charge state [7, 8].

More stringent requirements on the local symmetry of Cu^+ in its relaxed excited $3d^94s$ type state are imposed by the anomalous decay kinetics of the emission [7, 8]. In the temperature region 6-295 K the decay kinetics is represented by a single exponential decay component. As shown by the two-plateau type temperature dependence of its time constant, for temperatures near 6 K only the lowest metastable singlet level of the excited triplet is populated. However, below this temperature a second, faster exponential decay component shows up, that can only be explained by assuming a tunnel splitting of the singlet level. This can occur only for a site with at least twofold symmetry having at the same time a vibronic instability, leading to a double or multi-well potential at low temperatures [13]. Similar phenomena have been observed for the Cu^+ ground state in KCl and most other alkali halides with a cation-anion distance exceeding ~3 Å (for a summary see [14]). Positions like the relaxed Li site derived above for Cu^{2+} in $Li_2B_4O_7$ may have some approximate local symmetry due to a nearly linear O-Cu-O configuration, but as all further neighbours occupy strongly asymmetric positions (see table 2), tunnelling phenomena in such a configuration are rather improbable. For the excited Cu^+ ion therefore an alternative interstitial site on the C_2 symmetry axis has to be considered (which is the only nontrivial local symmetry element of the lattice space group $I4_1cd$). Accordingly the Cu^+ ion has to be assumed at some medium position between two B_4O_9 groups having the same C_2 axis (see figure 2). A Cu^+ ion at ~1.7 Å 'below' the on-axis oxygen (5), while clearly restricted to move towards these B_4O_9 groups, would have no oxygen neighbours within ~3.1 Å in near-'horizontal' directions. Such a site is clearly characterized by a double-well potential: in fact, the exact on-centre position may be less advantageous for Cu^+ mainly because of the presence of the on-axis oxygen and two boron neighbours at 2.3 Å, but there would be enough space for off-centre displacements. The site in question is an interstitial site where the Cu^+ ion may require some charge compensation. However, any local charge compensators, if present, also have to comply with C_2 symmetry, leaving as the only reasonable local possibility a pair of nearest neighbour Li vacancies (see the Li sites at both sides of the interstitial site shown in figure 2). These, being actually an overcompensation of the surplus charge, would provide additional freedom of motion. It should be pointed out that the luminescence spectra observed originate from a well-defined Cu^+ species, either locally charge-(over)compensated, or not, similarly to the statement made for Cu^{2+} in the previous section.

4. Discussion of Cu sites and charge transfer processes

The above-described results on the copper ions in $Li_2B_4O_7$ clearly indicate the difference between the incorporation sites of both charge states. While for the Cu^{2+} ion a stable low-symmetry position more than 2 Å away from the C_2 axis has been found, the Cu^+ ion should be located in a double-well potential preferably centred on the C_2 axis. Though this configuration for Cu^+ has been derived for its relaxed excited state, it can be assumed that the $3d^{10}$-type ground state also possesses tunnelling characteristics at low temperatures, given the smaller radius of the Cu^+ ion in the ground state and the similar properties observed for this state in the alkali halides with sufficiently large lattice constant [14].

The difference between Cu^{2+} and Cu^+ incorporation shows how sensitively the $Li_2B_4O_7$ lattice reacts to an additional charge of the dopant. This is clearly facilitated by the properties of the Li sublattice, where the non-covalently bound Li^+ ions occupy weakly defined positions with relatively large distances to most of their oxygen neighbours (see table 2), which should be compared to the tetrahedrally coordinated Li^+ ion in Li_2O with a Li-O distance of 1.963 Å [15]. Accordingly, in addition to possible Li vacancies, local relaxation of the Li sublattice has to be a major ingredient of charge compensation for both copper ions.

Given the substantially different sites, anomalies may be expected for charge trapping at a copper site leading to $Cu^+ \leftrightarrow Cu^{2+}$ transfer in either direction, as such a process may involve major displacements of the Cu ion, possibly with a symmetry change of the Cu defect site between C_1 and C_2, and/or the appearance of other, less stable Cu species. This would tend to reduce the effective trapping cross sections. The difference between the sites may be further stressed if the charge states found stable have different nearby charge-compensating defects. In the latter case charge trapping probably *would* lead to the appearance of some further, differently compensated Cu species. Preliminary luminescence measurements made on X-irradiated $Li_2B_4O_7$:Cu crystals, however, failed to demonstrate the presence of a new Cu^+ species. This finding does not contradict the predictions about the differently incorporated and possibly differently charge-compensated Cu^{2+} and Cu^+ ions. In addition to further experimental work the described properties and open alternatives for both Cu defects should be verified by detailed calculations of the electronic structure.

5. Summary

As shown by the analysis of the experimental results obtained recently by the complementary methods of EPR and time resolved luminescence, the markedly different but well-defined substitution sites of the Cu charge states are a puzzling property of the $Li_2B_4O_7$:Cu system. While Cu^{2+} has a frozen-in substitutional position near to a Li site with strong additional distortion of its surroundings, Cu^{+} has a dynamic excited state with an overall C_2 symmetry suggesting interstitial incorporation of the monovalent dopant ion. The different character and the large distance between these states should result in large relaxational movements during charge trapping processes. Evidently, the properties of the different Cu sites are related to the low symmetry of the matrix and the soft character of its Li sublattice being also at the origin of the large ionic polarizability of the $Li_2B_4O_7$ lattice.

Acknowledgements

Grants from the European Community, the Hungarian Scientific Research Fund (OTKA K60086) and the Estonian Science Foundation (grant 7274) are acknowledged.

References

[1] Krogh-Moe J 1962 *Acta Cryst.* **15** 190-3
[2] Malovichko G I, Grachev V G and Matkovskii A O 1991 *Fiz. Tverd. Tela* **33** 1966-72; Engl. transl. *Sov. Phys. Solid State* **33** 1107-11
[3] Malovichko G I, Vitruk L E, Yurchenko N Yu, Burak Ya V, Grachev V G, Matkovskii A O and Sugak D Yu 1992 *Fiz. Tverd. Tela* **34** 509-12; Engl. transl. *Sov. Phys. Solid State* **34** 272-4
[4] Prokic M 2002 *Radiat. Prot. Dosim.* **100** 265-8
[5] Can N, Karali T, Townsend P D and Yildiz F 2006 *J. Phys. D: Appl Phys* **39** 2038-43
[6] Corradi G, Watterich A, Polgár K, Nagirnyi V, Hofstaetter A, Rakitina LG and Meyer M 2007 *phys. stat. sol. (c)* **4** 1276-9
[7] Nagirnyi V, Kotlov A, Corradi G, Watterich A and Kirm M 2007 *phys. stat. sol. (c)* **4** 885-8
[8] Corradi G, Nagirnyi V, Kotlov A, Watterich A, Kirm M, Polgár K, Hofstaetter A and Meyer M 2008 *J. Phys.: Cond.. Matter* **20** 025216/1-9
[9] Corradi G, Polgár K, Bugaii A A., Zaritskii I M, Rakitina L G, Grachev V G and Derjugina N I 1986 *Fiz. Tverd. Tela* **28** 739-48 Engl. transl. *Sov. Phys. Solid State* **28** 412-7
[10] Hofstaetter A, Laguta V V, Meyer B K, Nikl M, Rosa J and Zhu R Y 2004 *Radiat. Meas.* **38** 703-6
[11] Ignatovich M, Holovey V, Watterich A, Vidóczy T, Baranyai P, Kelemen A, Ogenko V and Chuiko O 2003 *Radiat.Phys. Chem.* **67** 587-91
[12] Ignatovich M, Holovey V, Watterich A, Vidóczy T, Baranyai P, Kelemen A and Chuiko O 2004 *Radiat. Meas.* **38** 567-70
[13] Liidja G, Nagirnyi V, Soovik T and Zazubovich S 1989 *Phys. Stat. Sol. (c)* **152** 563-75
[14] Payne S A 1987 *Phys. Rev.* B **36** 6125-31
[15] Rodeja J G, Meyer M, Hayoun M 2001 *Model. Simulat. Mater. Sci. Eng.* **9** 81; http://www.cond-mat.physik.uni-mainz.de/~horbach/cecam2005/cecam_hayoun.ppt#1

Fluorescence Properties of Colour Centres Produced by Ultrashort Laser Irradiation in LiF Crystals

R E Samad [1], L C Courrol [2], L Gomes [1], I M Ranieri [1], S L Baldochi [1], A Z de Freitas [1] and N D Vieira Jr [1]

[1] Centro de Lasers e Aplicações, IPEN–CNEN/SP, São Paulo, SP - Brazil, 05508-000
[2] Departamento de Ciências Exatas e da Terra, UNIFESP, Diadema, SP - Brazil, 09972-270

Corresponding author e-mail: resamad@gmail.com

Abstract. LiF is a nonhygroscopic alkali halide crystal which possesses good optical and physical properties, and can host different species of colour centres at room temperature. Colour centres in LiF present broad absorption and emission bands in the near UV, visible and near IR regions of the spectrum. In this paper we study spectroscopic properties of colour centres produced in LiF by ultrashort laser pulses. The absorption and emission properties of these materials were measured showing that during the irradiation F, F_2, F_2^+ , F_3^+ and F_2^- colour centres were created in the crystals. A colour centres formation dose-like curve as a function of the ultrashort pulse energy was determined using fluorescence spectroscopy.

1. Introduction

Colour centres in ionic crystals present very interesting optical properties such as optical transitions sensitive to the surrounding lattice, broad absorption and emission bands in the near UV, visible and near IR regions of the spectrum [1]. Some of them present a four level optical cycle suitable for laser action, and are stable at room temperature[2, 3].

LiF occupies a special place among the alkali halide crystals because it is nonhygroscopic , possesses good physical and optical qualities, and hosts different species of colour centres at room temperature[4]. The formation of optically active colour centres in LiF by ionizing radiation such as charged particles (electrons and ions) as well as by gamma and X-rays is well-known.

It was reported that colour centre formation in LiF increases its refractive index, and active channel waveguides were successfully fabricated by electron-beam lithography [5-13]. Recently, it was shown that is possible to create colour centres in LiF using ultrashort laser pulses [14-17]. This method allows the production of defects with dimensional control by focusing the high-intensity ultrashort laser pulses inside the material, and optical waveguides and microgratings were produced deep inside LiF crystal.

In this work, we study the fluorescence optical properties of colour centres produced in LiF crystals by ultrashort laser pulses and determine the F_2, F_3^+, and F_2^- formation dose-like curve in LiF crystals as a function of the ultrashort pulse energy.

2. Experimental Setup

The samples of ultra pure LiF single crystals used in this work were grown in our crystal growth facility by the Czochralski technique under Argon atmosphere[18]. All samples used in the reported experiments were cleaved.

For creating the colour centres, 60 fs pulses from a Ti:Sapphire CPA laser system (Quantronix Odin) operating at 830 nm, at 1 kHz repetition rate, in a beam with a M^2=1.6, were used. The pulses maximum energy was 275 μJ, and the beam was focused inside the samples by an 83 mm converging lens to a radius of 12 μm, in such a way that there was no ablation at the sample surface. All irradiations were done at room temperature.

For the absorption and emission spectra measurements, a 40 × 15 × 5 mm^3 sample was irradiated with the laser impinging on one of the 40 × 5 mm^2 face (*xz* surface in Figure 1), and propagating inside the sample in the *y* direction. The sample was placed in a computer controlled translation table, and was independently moved in the *x* and *z* axes in sinusoidal oscillations with different frequencies during 5 minutes, in order that a volumetric region with colour centres was formed. The amplitudes of the movements in the *x* and *z* directions were 3 mm and 0.5 mm, respectively. The sample was irradiated, at different positions, with 80, 96, 126, 152, 194, 232 and 275 μJ pulses, during 5 minutes for each irradiation, forming regions with different colour centres densities, as can be seen in Figure 1. Two days after irradiation, the absorption spectra were measured, through the *z* direction, in the range 200 nm-800 nm by a dual beam Spectrophotometer (Varian Cary 17D). For obtaining a colour centre formation dose curve, each colour centre region was excited by 440 nm laser light from the second harmonic from a Femtolasers Rainbow oscillator, and its emission spectrum was measured by a Newport OSM-400 VIS-IR fibre spectrometer.

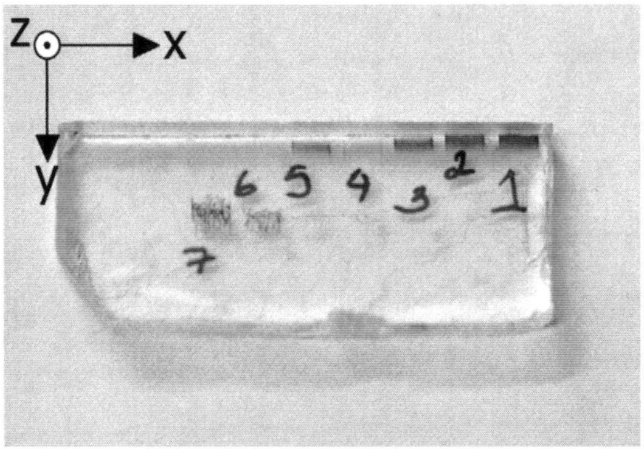

Figure 1. Regions of colour centres produced by ultrashort pulses; the regions marked as 1, 2, 3, 4, 5, 6 and 7 correspond to colour centres produced by pulses with energies of 275, 232, 194, 126, 152, 96 and 80 μJ, respectively.

3. Results

Figure 2a shows the absorption spectra of the colour centres created in the LiF crystal under three different ultrashort pulses energies. A strong absorption band around 250 nm, corresponding to F centres, could not be resolved due to the thickness of the centres region, and for this reason is not shown. The shoulder observed in the region between 315 nm and 375 nm is due to F_3 centres (three associated F centres) absorption. F_3^+ and F_2 centres are responsibly by the almost overlapping absorption band at 444 nm (F_2 or two associated F centres) and 448 nm (F_3^+ or ionized F_3 centres); the 520 nm and 545 nm bands are attributed to F_4 centres (four associated F centres). The band due to the

Journal of Physics: Conference Series **249** (2010) 012009 doi:10.1088/1742-6596/249/1/012009

F_2^+ centre (ionized F_2 centre) at 632 nm is not observed. The presence of these aggregates indicates that a high density of F centres is formed by ultrashort pulses during the irradiation.

From the absorption spectra shown in figure 2a it is possible to estimate the F_2 and F^+_3 centres concentrations in each sample using the Smakula formula [19] explicitly written in terms of the sample optical density (OD) and length (l):

$$Nf = 1.29\times 10^{17}\left(\frac{n}{n^2+2}\right)^2 \frac{2.303\times \mathrm{OD}}{l} W \quad (1)$$

For these calculations, whose results are shown in Table 1, the refractive index value used was 1.39, the width of the absorption bands at half-maximum, W, at room temperature, was taken from the literature[20] (0.16 eV for F_2 and 0.29 eV for F^+_3 centres); the optical density for each band was obtained by a Gaussian decomposition of the sample spectrum, show in Figure 2b only for the most intense spectrum for clarity. An oscillator strength of $f = 0.5$ was assumed, and the colour centres' region thickness was considered to be $l = 0.5$ mm since the samples were moved by this amount during the irradiation.

Table 1. Concentrations of F_2 and F_3^+ colour centres produced with different ultrashort pulse energies 232, 126 and 96 μJ.

Pulse Energy (μJ)	Centres Concentration (10^{17} cm^{-3}) F_2	F_3^+
232	(OD = 1.00) 2.376	(OD = 0.37) 1.593
126	(OD = 0.60) 1.426	(OD = 0.18) 0.775
96	(OD = 0.31) 0.737	(OD = 0.11) 0.474

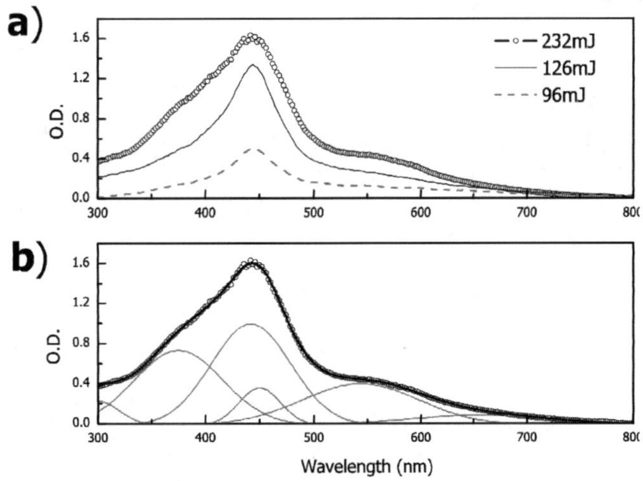

Figure 2. a) Absorption spectra of the colour centres produced in LiF by ultrashort pulses irradiation for three different energies: 232, 126 and 96 μJ. b) Gaussian decomposition of the spectrum correspondent to the 232 μJ irradiation along with the experimental data (circles); the gray lines are the individual Gaussians and the black line is its sum.

Figure 3 shows the emission spectrum of colour centres produced by the 232 μJ ultrashort pulses, exciting the sample with 440 nm laser light propagating in the y direction and detected by a fibre spectrometer looking at the z direction. A 3-Gaussians fit was done to the experimental data to determine the curves for the F_3^+, F_2 and F_2^- centres emissions. The emission of the F_2^+ centre is centred on 1.4 eV (900 nm) and could not be fitted due to the 2nd order of the laser scattering that peaks at 880 nm. In this spectrum we observe the typical broad emission bands related to transitions of the F_3^+ and F_2 centres at 2.35 eV and 1.85 eV (528 nm and 670 nm), respectively, shown in table 2, and also an intense emission band at 1.17 eV (1060 nm) due to the F_2^- emission.

Table 2. Optical characteristics of colour centres produced in LiF crystals.

Centre	Absorption		Emission	
	Energy (eV)	Wavelength (nm)	Energy (eV)	Wavelength (nm)
F	5.00	248	-	-
F_3	3.92, 3.32	316, 374	-	-
F_2	2.79	444	1.85	670
F_3^+	2.77	448	2.35	528
F_2^-	1.29	960	1.17	1060

The peak intensities and band areas of F_3^+ ,F_2, and F_2^- colour centres emissions, obtained from the 3-Gaussians fit, were plotted as a function of the pulses energies to obtain the dose-like curves shown in figures 4a and 4b. The growth of the emission intensities and areas is clearly observed with the increase in the pulse energy, evidencing that a greater density of centres is created at higher intensities, corroborating the concentration growth calculated using Smakula formula, shown in table 1. It can be seen that, under the irradiation condition used in this work, the density of the centres created by ultrashort pulses seems to approach saturation, but this behaviour has to be investigated at higher pulse energies to be confirmed.

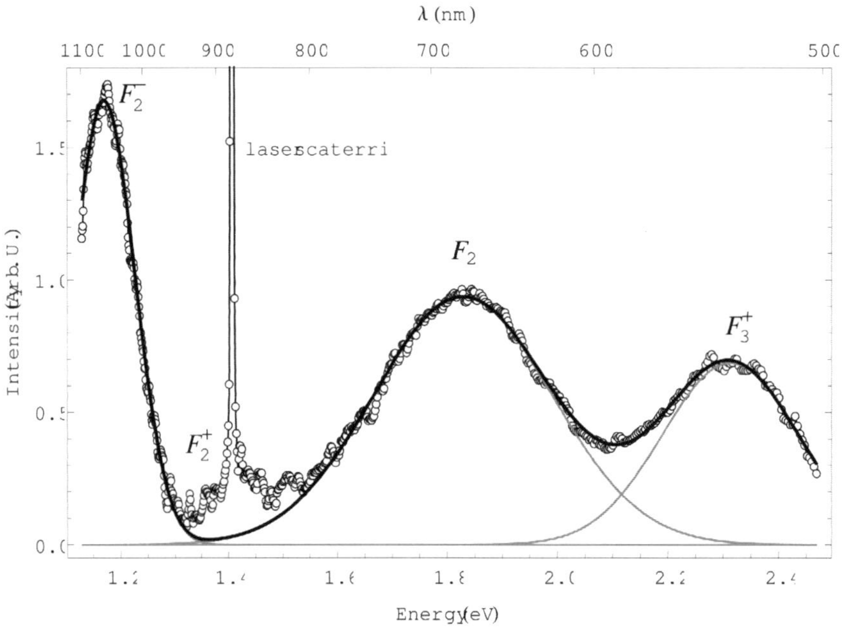

Figure 3. Emission spectra obtained under excitation at 440 nm for centres produced by 232 μJ ultrashort laser pulse energies, showing a 3-Gaussians fit to the centres emission peaks. The laser scattering peak is the grating second order of the 440 nm emission.

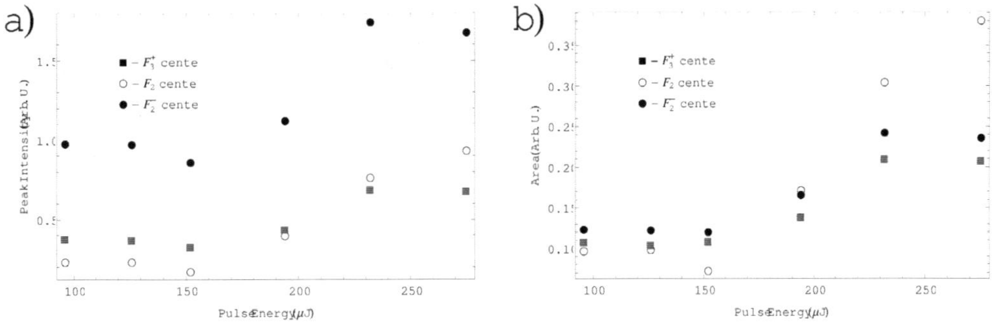

Figure 4. Colour centres dose-like curves obtained as a function of the ultrashort pulse energy, calculated from the 3-Gaussians fit; a) emission peak intensity b) emission band area.

4. Conclusions

After ultrashort pulse laser irradiation, various stable colour centres and colour centres aggregates were formed in LiF crystals along the beam path. The centres created are stable at room temperature. We proposed that the colour centres creation was initiated by multiphoton excitation, followed by vacancy creation by pushing neutral atoms away from its equilibrium position in the lattice by the electrons quivering motion in the laser electric field. By measuring absorption and emission spectra, the colour centre were determined to be F, F_2, F_3^+ and F_2^- centres. There is a high density of secondary centres (aggregates) formed during the irradiation, and a colour centre production dose-like curve was obtained. This dose curve suggests that the density of the centres created by ultrashort pulses reaches an intensity saturation, but this behaviour has to be investigated at higher pulse energies.

5. References

[1] Gellermann W 1991 Color Center Lasers *J. Phys. Chem. Sol.* **52** 249-97

[2] TerMikirtychev V V and Tsuboi T 1996 Stable room-temperature tunable color center lasers and passive Q-switchers *Prog. Quant. Electron.* **20** 219-68

[3] Baldacchini G and Montereali R M 2001 New perspectives of coloured LiF for optoelectronic devices *Opt. Mat.* **16** 53-61

[4] Baldacchini G 2002 Colored LiF: an optical material for all seasons *J. Lumin.* **100** 333-43

[5] Montereali R M and Piccinini M 2002 Optical gain of F_2^- colour centres in LiF confining structures realised by electron-beam lithography *Opt. Commun.* **209** 201-8

[6] Almaviva S, Baldacchini G, Piccinini M and Montereali R M 2007 Confocal laser scanning microscopy of active color centers based strips induced on LiF by low energy electron beams *J. Non-Cryst. Sol.* **353** 687-91

[7] Fornarini L, Martelli S, Montereali R M, Righini G C, Pelli S and Cremona M 2000 Realisation and characterisation of LiF/NaF thin film planar waveguides *Thin Solid Films* **358** 191-5

[8] Montecchi M, Nichelatti E, Montereali R M, Piccinini M and Somma F 2004 Increase of refractive index induced by absorbing centres *Opt. Quant. Electron.* **36** 43-55

[9] Montereali R M 1997 Prospects for active waveguides in alkali fluoride films *Proceedings of the 13th International Conference on Defects in Insulating Materials - ICDIM 96* **239** (Zurich-Uetikon, ed G E Matthews and R T Williams 711-6

[10] Montereali R M, Mancini A, Righini G C and Pelli S 1998 Active stripe waveguides produced by electron beam lithography in LiF single crystals *Opt. Commun.* **153** 223-5

[11] Montereali R M, Piccinini M and Burattini E 2001 Amplified spontaneous emission in active channel waveguides produced by electron-beam lithography in LiF crystals *Appl. Phys. Lett.* **78** 4082-4

[12] Montereali R M 2002 Miniaturized structures based on color centers in lithium fluoride: Optical properties and applications *Radiat. Eff. Defect. S.* **157** 545-51
[13] Montereali R M, Mancini A, Martelli S, Menchini F and Picozzi P 2001 Micro and sub-micro fabrication techniques for optically confined structures based on lithium fluoride *Appl. Organomet. Chem.* **15** 407-13
[14] Courrol L C, Ranieri I M, Baldochi S L, Samad R E, de Freitas A Z, Gomes L and Vieira N D 2007 Study of color centers produced in thulium doped YLF crystals irradiated by electron beam and femtosecond laser pulses *Opt. Commun.* **270** 340-6
[15] Courrol L C, Samad R E, Gomes L, Ranieri I M, Baldochi S L, de Freitas A Z and Vieira N D 2004 Color center production by femtosecond pulse laser irradiation in LiF crystals *Opt. Expr.* **12** 288-93
[16] Courrol L C, Samad R E, Gomes L, Ranieri I M, Baldochi S L, Freitas A Z, Morato S P and Vieira N D 2006 Color center production by femtosecond-pulse laser irradiation in fluoride crystals *Las. Phys.* **16** 331-5
[17] Courrol L C, Samad R E, Ranieri I M, Gomes L, Baldochi S L, de Freitas A Z and Vieira N D 2005 Production of stabilized color centers in $YLiF_4$ crystals by high-intensity ultrashort laser pulses *J. Opt. Soc. Am. B* **22** 2560-3
[18] Baldochi S L and Ranieri I M 2001 Alkali Halide Crystals: Growth in *Encyclopedia of Materials: Science and Technology* vol. 1, ed K H J Buschow, R W Cahn, M C Flemings, B Ilschner, E J Kramer, S Mahajan, and P Veyssière (Amsterdam: Pergamon, Elsevier Science) pp 74-8.
[19] Izerrouken M, Meftah A and Nekkab M 2007 Color centers in neutron-irradiated $Y_3Al_5O_{12}$, CAF_2 and LiF single crystals *J. Lumin.* **127** 696-702
[20] Baldacchini G, De Nicola E, Montereali R M, Scacco A and Kalinov V 2000 Optical bands of F_2 and F_3^+ centers in LiF *J. Phys. Chem. Sol.* **61** 21-6

Journal of Physics: Conference Series **249** (2010) 012010 doi:10.1088/1742-6596/249/1/012010

Energy transfer rates of KY_3F_{10}:Yb:Nd:Tm crystals

L C Courrol [1], H M S M D Linhares [2], A F H Librantz [2,3], S L Baldochi [2], L Gomes [2], I M Ranieri [2]

[1] Departamento de Ciências Exatas e da Terra, UNIFESP, Diadema, SP, Brazil, 09972-270
[2] Centro de Lasers e Aplicações ,IPEN–CNEN/SP, São Paulo – SP - Brazil, 05508-000
[3] Exact Sciences Department, UNINOVE, São Paulo, SP 02111-030, Brazil

Corresponding author e-mail: lcourrol@gmail.com

Abstract. In this work we present the spectroscopic properties of KY3F10 (KY3F) single crystals doped with thulium and also co-doped with ytterbium and/or neodymium, KY3F:Yb:Nd:Tm and KY3F:Nd:Tm. The most important processes that lead to the thulium up conversion emissions were identified. The absorption spectra of the samples were measured at room temperature in the range of 200 nm-1200 nm. The emission spectra were obtained by exciting the samples with a 797 nm laser diode and were analyzed using a lock-in amplifier technique. A time-resolved luminescence spectroscopy technique was employed to measure the luminescence decays and to determine the mechanisms involved in the energy transfer up-conversion processes. Analysis of the energy transfer processes dynamics in KY3F:Yb:Tm:Nd crystal, using selective pulsed-laser excitations, shows that the energy transfer from Nd3+ to Yb3+ ions is the mechanism responsible for the enhancement of the blue up-conversion efficiency when compared with the Yb:Tm system. In the case of KY3F:Nd:Tm it is observed emissions at 350, 355 and 452 nm excited by an additional Yb:Tm step cross relaxation, Yb (2F5/2) : Tm (1G4) that populates the 1D2 (Tm3+) excited level. A study of the energy transfer processes in KY3F:Yb:Tm:Nd crystal showed that the 1G4 excited level is mainly populated by the sequence of two nonradiative energy processes that starts well after the Nd3+ and Tm3+ excitation at 797nm: Nd (4F3/2) → Yb (2F7/2) followed by Yb (2F5/2) → Tm (3H4) → Tm (1G4).

1. Introduction

The study of solids doped with thulium ions has received great interest in the last decades due to the numerous applications of these materials. Concerning to the thulium emission wavelength, these solids can be used as laser materials [[10]] for different applications in life sciences [[11]], LIDAR [[9]] and industry or can be applied as the S-band Tm-doped fiber amplifier (TDFA) in wavelength-division-multiplexing telecommunication [[6]], or for optical devices [[12]], color displays [[8]] and optical memories [[7]].

Due to the attractive thermo-mechanical properties, wide transparency and high optical damage threshold, KY3F has been studied as laser materials when activated by several RE ions, which can easily substitute for Y^{3+} ions in a non-center-symmetrical site [[1] [2-5]].

Journal of Physics: Conference Series **249** (2010) 012010 doi:10.1088/1742-6596/249/1/012010

KY3F has relatively low phonon energy (~500 cm^{-1}) that is an important point in avoiding energy loss by non-radiative relaxation. KY3F is the only compound in the KF–YF_3 system that melts congruently without any phase transition. It crystallizes in the cubic fluorite-type structure (Fm3m) with lattice parameter a = 11.54 Å [[5]], and the site symmetry of the Yttrium is tetragonal (C_{4v}) and is isotropic crystals.

In this paper, KY_3F_{10} (KY3F) doped with Thulium (Tm) ions that are sensitized by Neodymium (Nd) and Ytterbium (Yb) were studied, and the multiples processes of energy transfer that occurs when these materials are excited around 800 nm were inspected.

2. Experimental Setup

The rare earth fluorides were prepared from pure oxide powders (Alpha-Johnson Matthey, 99.99%) by hydrofluorination at high temperature in HF atmosphere. The powder was contained in a cylindrical platinum boat, which was inserted in a sealed platinum tube. The LiF-LnF_3 (Ln=Y, Yb, Nd, and Tm) mixture was melted using an open platinum boat in the same atmosphere, with a composition of 1.02 LiF: 1 LnF_3. LiF powder (Alpha-Johnson Matthey, 99.9%) was zone-refined before it was added to the mixture. The KY3F samples were obtained in the same system by slow cooling. Mixtures of KF-LnF_3 were contained in a cylindrical platinum boat and melted around 990°C and cooled with a rate of ~10°C/h. The synthesized material had always transparent crystalline regions from which the samples were extracted.

The rare earths concentrations were obtained by ICP-OES analysis and yielded the results are shown in the table 1.

Table 1. The rare earths concentrations obtained by ICP-OES analysis.

Nominal concentration (mol%)	Measured concentrations (mol%)
KY3F:Yb (20.00): Nd (1.30): Tm (0.50)	KY3F:Yb (18.50): Nd (0.91): Tm (0.48)
KY3F:Yb (10.00): Nd (1.30): Tm (0.50)	KY3F:Yb (9.25): Nd (0.87): Tm (0.49)
KY3F: Nd (1.30): Tm (0.50)	KY3F: Nd (1.25): Tm (0.47)
KY3F: Yb (10.00): Tm (0.50)	KY3F: Yb (9.50): Tm (0.48)

The absorption spectra of all samples were measured in the range 700nm-1000nm at room temperature using a Varian Cary 17D/OLIS Spectrometer. The emission spectra were obtained by exciting the samples that had the same optical thickness, with a SDL diode laser at 797 nm. The emissions of the samples were analyzed with a 0.5 m monochromator (Spex) and a PMT detector. The signal was amplified with an EG&G 7220 lock-in and processed by a computer. The relative errors in the emission measurements are estimated to be < 5%. In the lifetime measurements, the samples were excited by pulsed laser radiation generated by a tunable OPO-IR pumped by 2W of Nd:YAG Q-switch laser of 4 ns and 10 HZ. Luminescence signals were analyzed by the 0.25 m Kratos monochromator, detected by the EMI S-20 PMT or InSb 77K infrared detector (Judson) and resolved by the EGG Boxcar Processor. Detector response time was of approximately 0.5 μs. This detector might be pretty slow which could distort the decay curve measurements even in the microsecond time scale and the evaluated decay times might be distorted due to such reason as well.

3. Results and discussion

Optical absorption spectra of KY3F doped Yb:Tm:Nd crystals are shown in Figure 1. This crystal has two main absorptions in the near infrared around 960 nm (Yb^{3+}) and around 800 nm due to Nd^{3+} and Tm^{3+} ions. The most intense absorption is near 960 nm due to high concentration of ytterbium in the samples.

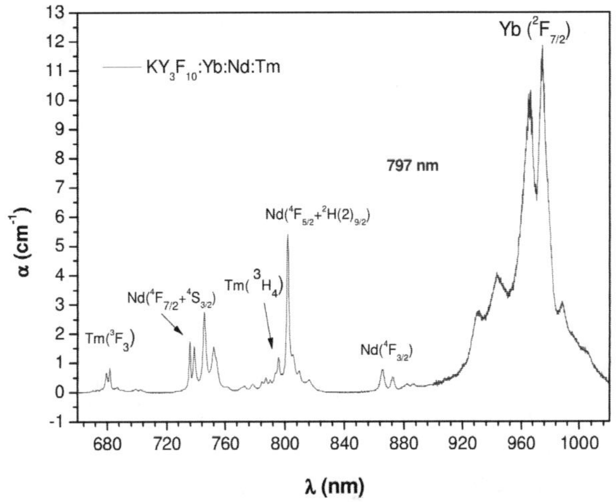

Figure 1. Optical absorption of KY3F crystal doped with neodymium, thulium and ytterbium.

When KY3F crystal containing Tm^{3+} co-doped with Yb^{3+} or Yb^{3+} and Nd^{3+} is excited at 797 nm, ultraviolet (350 nm) and blue (~450 and 480 nm) Tm^{3+} emissions are observed as it can be seen in the Figure 2.

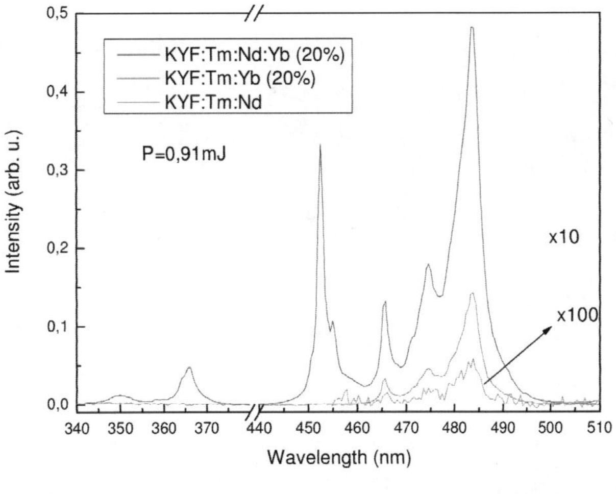

Figure 2. Ultraviolet and blue emission of KY3F:Yb:Tm:Nd, KY3F:Tm:Yb and KY3F:Tm:Nd crystals excited at 797 nm.

As observed in the Figure 2, the Nd (~1 mol%) effect in the KY3F:Yb:Tm:Nd crystal causes an enhancement of the Tm^{3+} blue emission indicating that Nd^{3+} ions significantly contribute to the population of the $^1G_4$ excited level that emits around 480nm. The population of the $^1G_4$ excited level

Journal of Physics: Conference Series **249** (2010) 012010 doi:10.1088/1742-6596/249/1/012010

consequently interacts with $^2F_{5/2}$ (Yb^{3+}) exciting the $^1D_2$ (Tm^{3+}) level, which emits near 360 nm and 450 nm.

Figure 3 shows the schematic energy diagram levels of Yb/Tm/Nd system [13] (most of indicated processes will be discussed and proved to be essential for the blue emission up-conversion by the luminescence dynamics analysis in the sequence).

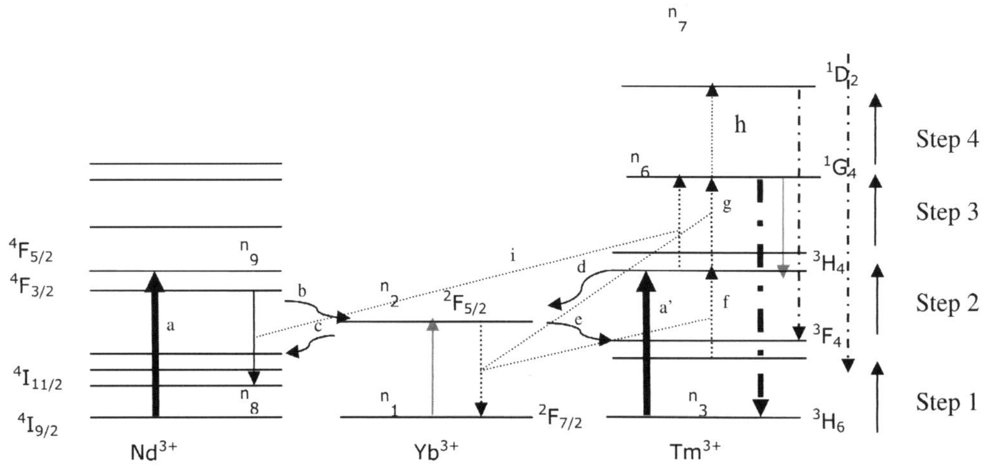

Figure 3. Energy levels scheme and energy transfer mechanisms of Yb/Tm/Nd system. Continuous line (up) 797 nm excitation. Dash dot line (down) thulium emissions. Dot lines (up and down) Yb emission and cross relaxation processes.

When the Yb/Tm/Nd sample is excited at 797 nm the following processes are observed to occur:

(a) Ground state absorption of Nd ($^4I_{9/2}$) $\rightarrow$ Nd($^4F_{5/2}$)

(a') Ground state absorption of Tm ($^3H_6$) $\rightarrow$ Tm ($^3H_4$)

(b) Energy transfer Nd - Yb:

Nd($^4F_{3/2}$) + Yb ($^2F_{7/2}$) $\rightarrow$ Nd($^4I_{11/2}$) + Yb ($^2F_{5/2}$)

(c) Back transfer Yb-Nd

Yb ($^2F_{5/2}$) + Nd ($^4I_{9/2}$) $\rightarrow$ Yb ($^2F_{7/2}$) + Nd($^4I_{15/2}$)

(d) Back transfer Tm-Yb:

Tm ($^3H_4$) + Yb ($^2F_{7/2}$) $\rightarrow$ Tm ($^3H_6$) + Yb($^2F_{5/2}$)

(e) Cross-relaxation Yb x Tm:

Yb($^2F_{5/2}$) + Tm ($^3H_6$) $\rightarrow$ Yb ($^2F_{7/2}$) + Tm ($^3F_4$)

(f) Cross-relaxation Yb x Tm:

Yb ($^2F_{5/2}$) + Tm ($^3F_4$) $\rightarrow$ Yb ($^2F_{7/2}$) + Tm ($^3H_4$)

(g) Cross-relaxation Yb x Tm:

Yb($^2F_{5/2}$) + Tm ($^3H_4$) $\rightarrow$ Yb ($^2F_{7/2}$) + Tm ($^1G_4$)

(h)) Cross-relaxation Yb x Tm:

$Yb(^2F_{5/2}) + Tm\ (^1G_4) \rightarrow Yb\ (^2F_{7/2}) + Tm\ (^1D_2)$

(i) Cross-relaxation Nd x Tm:

$Nd(^4F_{3/2}) + Tm\ (^3H_4) \rightarrow Nd\ (^4I_{11/2}) + Tm\ (^1G_4)$

(p) Energy transfer Tm-Nd:

$Tm\ (^3F_4) + Nd\ (^4I_{9/2}) \rightarrow Tm\ (^3H_6) + Nd\ (^4I_{15/2})$

(q) Energy transfer Tm-Nd:

$Tm\ (^3H_4) + Nd\ (^4I_{9/2}) \rightarrow Tm\ (^3H_6) + Nd\ (^4F_{5/2})$

(r) Cross-relaxation Tm x Tm:

$Tm\ (^3H_4) + Tm\ (^3H_6) \rightarrow Tm\ (^3H_5) + Tm\ (^3F_4)$

(s) Cross-relaxation Tm x Tm:

$Tm\ (^1G_4) + Tm\ (^3H_6) \rightarrow Tm\ (^3H_5) + Tm\ (^3H_4)$

3.1. Nd-Yb interaction:

Figure 4 shows the decay time transient of $^4F_{3/2}$ level of Nd^{3+} measured at 1052 nm and the $^2F_{5/2}$ luminescence decay of Yb^{3+} (measured at 1000 nm) for Yb(10):Tm(0.5):Nd(1.3):KY_3F_{10} crystal under 797 nm (a and b) and 960 nm pulsed laser excitations (c).

Best fit of luminescence decay of $^4F_{3/2}$ level (solid line of Figure 4(a)) was done using Inokuti-Hirayama model from where best fitting parameters were $\gamma = 504\ s^{-1}$ and $\tau_{R9} = 334\ \mu s$ (integrated decay time was $\tau_{d9} = 6.5\ \mu s$). Transfer rate of *process b* was obtained from the relation $b = 1/\tau_{d9} - 1/\tau_{R9}$, which gave $b = 1.15x10^5\ s^{-1}$. Figure 4(b) shows that Yb^{3+} luminescence transient measured at 1000 nm exhibits a raise time of 11 μs consistent with the decay time of Nd^{3+} excited ions (797 nm). Figure 4(c) shows that the decay time constant of $^2F_{5/2}$ (Yb^{3+}) level is 753 μs (τ_{d2}), which is shorter than the radiative lifetime (1.65 ms). Similarly, the transfer rate of *process c* was obtained from the relation $c = 1/\tau_{d2} - 1/\tau_{R2}$, which gave $c = 714\ s^{-1}$.

3.2. Yb-Tm interaction (step 1):

Figure 5 shows the decay time of $^3F_4$ luminescence transient of Tm^{3+} measured at 1900nm.

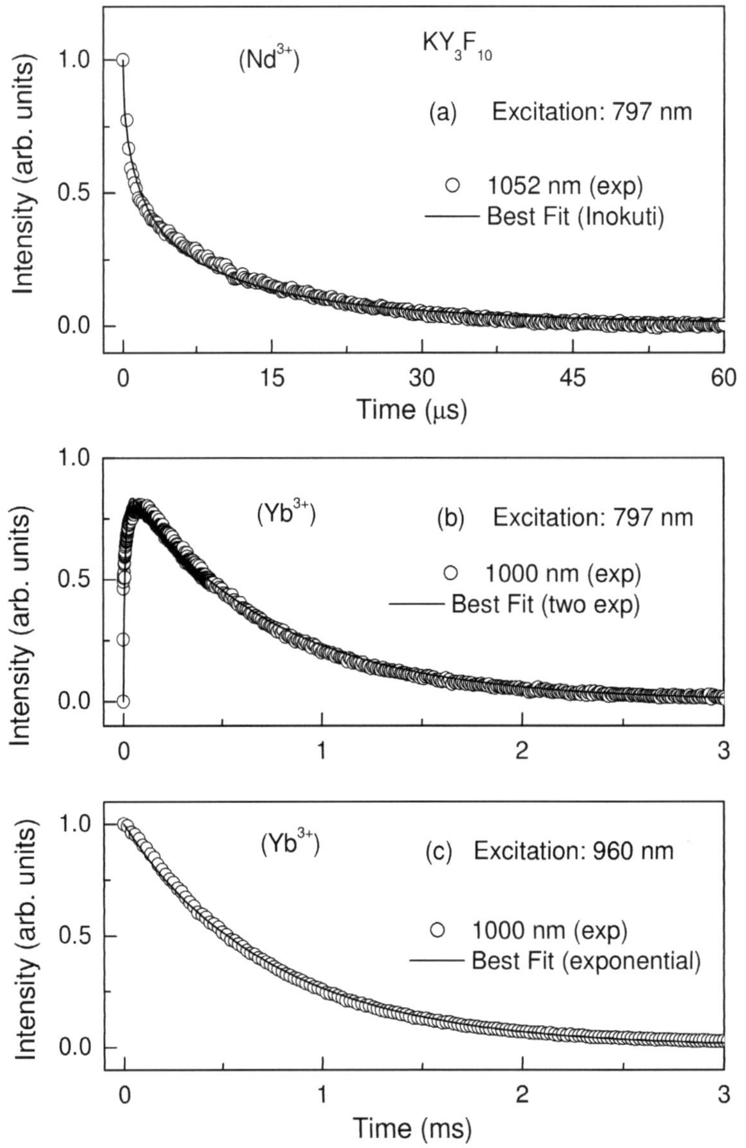

Figure 4. Decay time transient of $^4F_{3/2}$ level of Nd^{3+} measured at 1052 nm and the $^2F_{5/2}$ luminescence decay of Yb^{3+} (measured at 1000 nm) for Yb(10):Tm(0.5):Nd(1):KY_3F_{10} crystal under 797 nm (a and b) and 960 nm pulsed laser excitations (c).

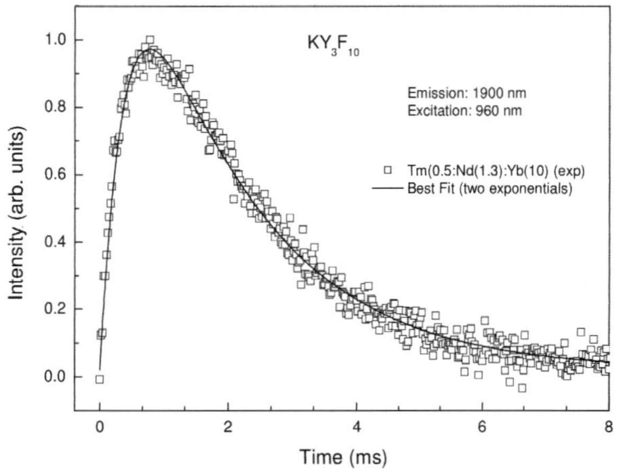

Figure 5. Decay time of $^3F_4$ luminescence transient of Tm^{3+} measured at 1900 nm for Yb(10):Tm(0.5):Nd(1):KY_3F_{10} crystal excited by laser pulse at 960 nm.

Best fit of luminescence curve (solid line of Figure 5) was done using two exponentials function $F_1(t) = A\ (\exp(-t/\tau_{d4}) - \exp(-t/\tau_{s4}))$, where A is a constant, τ_{s4} is the luminescence rise time and τ_{d4} is the decay time constants. Best fitting parameters found were $\tau_{s4} = 326$ μs and $\tau_{d4} = 2.2$ ms. Transfer rate constant of *process e* was obtained from relation $e = 1/\tau_{s4} - 1/\tau_{d2}$. Using $\tau_{d2} = 753$ μs one gets $e = 1739\ s^{-1}$. Transfer rate of *process p* was obtained using the relation $p = 1/\tau_{d4} - 1/\tau_{R4} = 348\ s^{-1}$, where $\tau_{R4} = 9.4$ ms (radiative lifetime of $^3F_4$ level) .

3.3. Yb-Tm interaction (step 2):

The $^3H_4$ luminescence decays of Tm^{3+} observed at 800 nm were measured after pulsed laser excitations (960 and 783 nm) and showed in the figure 6.

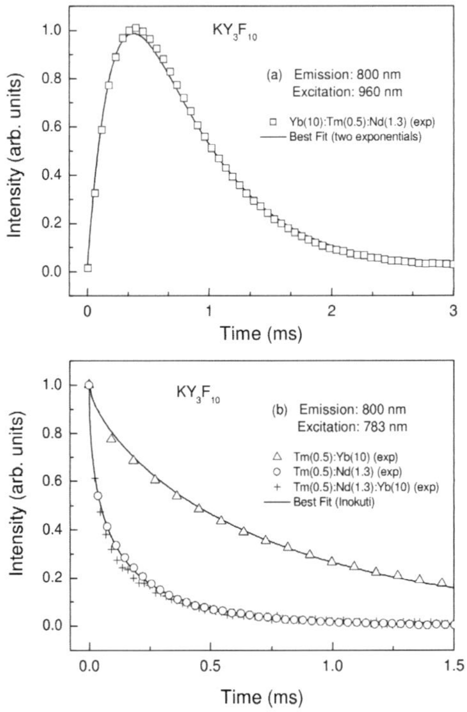

Figure 6. The $^3H_4$ luminescence decays of Tm^{3+} observed at 800 nm were measured after pulsed laser excitations in Yb(10):Tm(0.5):Nd(1):KY_3F_{10} crystal at 960 nm (a) and in Tm(0.5):Yb(10) and Tm(0.5):Nd(1) excited at 783 nm (b).

For this case, the best fit of luminescence curve exhibited in Figure 6(a) was done using the fitting function $F_2(t) = A\ (\exp(-t/\tau_{d5})-\exp(-t/\tau_{s5}))$, where τ_{s5} is the luminescence rising time and τ_{d5} the decay time constants of $^3H_4$ level. Best fitting parameters obtained were τ_{s5} = 293 μs and τ_{d5} = 476 μs. One may observe that τ_{s5} is longer than the decay time constant verified for $^3H_4$ level in Tm:Yb:Nd system direct excited at 783 nm giving τ_{d5} equal to 143 μs (see Figure 6(b)). However, τ_{s5} value (293 μs) is consistent with the expected rise time of a *donor like state* obtained by multiplication of $^2F_{5/2}$-luminescence decay (Yb^{3+}) and the $^3F_4$-luminescence transient (Tm^{3+}) (called here $\tau_s^{(donor)}$). Once, we are working with exponentials (see $F_1(t)$), one can deduce that $\tau_s^{(donor)}$ is given by

$$\tau_{s(24)}^{(donor)} = [1/\tau_{d2} + 1/\tau_{s4}]^{-1} \qquad (1)$$

Using τ_{d2} = 753 μs and τ_{s4} = 326 μs one gets $\tau_{s(24)}^{(donor)}$ = 228 μs that is consistent with the experimental value τ_{s5} = 293 μs. By the same argument, one can get the decay time constant of the *donor like state* using that

$$\tau_{d(24)}^{(donor)} = [1/\tau_{d2} + 1/\tau_{d4}]^{-1} \quad (2)$$

Using τ_{d2} = 753 μs and τ_{d4} = 2.2 ms (both measured in this work) we get $\tau_{d(24)}^{(donor)}$ = 561 μs, which is longer than the experimental value τ_{d5} = 476 μs obtained from best fitting in Figure 6(a). The transfer rate of *process f* can now be calculated using the relation

$$f = 1/\tau_{d5} - 1/\tau_{d(24)}^{(donor)} \approx 318\ s^{-1} \quad (3)$$

Best fit parameters of $^3H_4$ luminescence decay of Tm:Yb system were γ = 15.5 $s^{-1/2}$ and τ_{R5} = 1.14 ms given an integrated lifetime $\tau_5^{(integrated)}$ = 830 μs. Best fit parameters of $^3H_4$ luminescence decay of Tm:Nd system were γ = 96 $s^{-1/2}$ and τ_{R5} = 1.14 ms that gives τ_5 = 143 μs (integrated).

The transfer rate of *process d* can be calculated using the relation

$$d = 1/\tau_5^{(integrated)} - 1/\tau_{R5} \sim 328\ s^{-1} \quad (4)$$

where $\tau_5^{(integrated)}$ = 830 μs (Tm:Nd) and τ_{R5} = 1.14 ms.

Results presented in Figure 6(b) show that the $^3H_4$ (Tm^{3+}) excited level is strongly deactivated by Nd^{3+} ions in Tm:Nd:Yb (*process q*) similar to the case of Tm:Nd system. Its rate constant can be calculated from relation

$$q = 1/\tau_5 - 1/\tau_{R5} \quad (5)$$

Using that τ_5 =143 μs and τ_{R5} = 1.14 ms (experimental measures) one gets q = 6.12 x10^3 s^{-1}. Tm($^3H_4$):Tm($^3H_6$) cross-relaxation rate (*process r*) was calculated using the relation

$$r = 1/\tau_5 - 1/\tau_{R5} \quad (6)$$

Using τ_5 =830 μs and the radiative lifetime of $^3H_4$ level (τ_{R5} = 1.14 ms) one gets r = 328 s^{-1}.

3.4. Yb-Tm interaction (step 3):

The luminescence transient of $^1G_4$ level of Tm^{3+} was measured for 797 nm laser pulsed excitation and showed in the Figure 7.

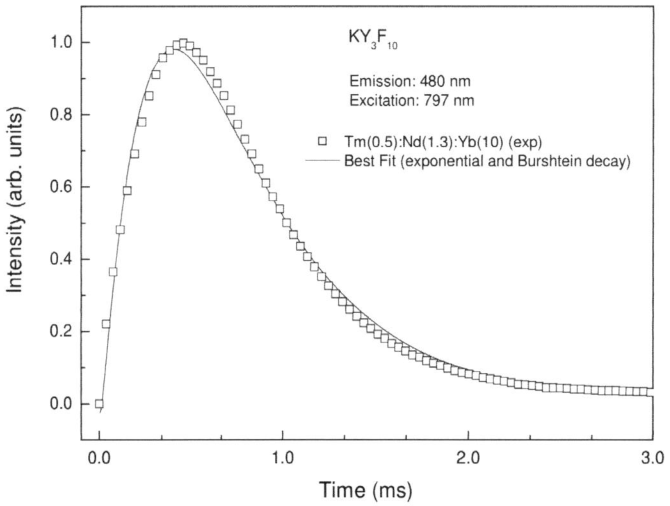

Figure 7. The decay time of $^1G_4$ luminescence transient of Tm^{3+} excited by pulsed laser at 797 nm was measured for Yb(10):Tm(0.5):Nd(1):KY_3F_{10} crystal.

In this case, best fit of luminescence curve (solid line of Figure 7) was done using a fitting function given by

$$F_3(t) = A\,[\exp(-t/\tau_{d6} - \gamma\sqrt{t} - wt) - \exp(-t/\tau_{s6})], \qquad (7)$$

where A is a constant, τ_{s6} is the rise-time constant of luminescence and τ_{d6} is the decay-time constant of $^1G_4$ level (the Burshtein function was used in $F_3(t)$ because the nature of the decay of $^1G_4$ level that includes a cross relaxation *process s*). Best fitting parameters obtained were τ_{s6} = 283 μs for the luminescence rising and γ = 6.13 s^{-1}, w = 538 s^{-1} and τ_{R6} = 649 μs for the luminescence decay - integrated decay time constant was $\tau_{d6}^{(integrated)}$ = 428 μs. It is noticeable that the $^1G_4$ luminescence generated at expenses of cross interaction between $^2F_{7/2}$ (Yb^{3+}) and $^3H_4$ (Tm^{3+}) excited levels directly excited by 797 nm pulsed laser excitation can not be produced by Tm^{3+} close by Nd^{3+} ions, because in this case, the decay time of the *donor like state* obtained is too short, i.e., $\tau_{s(25)}^{(donor)} = [1/\tau_{d2} + 1/\tau_5]^{-1}$ ~120 μs compared with the $^1G_4$ luminescence rise time (τ_{s6} = 283 μs).

On the other hand, the Tm^{3+} ion close to Yb^{3+} but isolated from Nd^{3+} has a longer lifetime for the $^3H_4$ level equal to 830 μs . In this case, the *donor like state* may have a lifetime of $\tau_{s(25)}^{(donor)-2}$ ~395 μs, which is longer than the rise time constant of $^1G_4$ luminescence transient (~283 μs). The transfer rate constant of *process g* was obtained from relation

$$g = 1/\tau_{s6} - 1/\tau_{s(25)}^{(donor)-2} \qquad (8)$$

where τ_{s6} = 283 μs and $\tau_{s(25)}^{(donor)-2}$ = 395 μs measured for Tm:Yb system, which gave g = 1000 s^{-1}.

Tm($^1G_4$):Tm($^3H_6$) cross-relaxation rate (*process s*) was calculated using the decay constant of $^1G_4$ level measured for Tm:Nd:Yb system, where $s = \tau_{d6}^{(integrated)} - 1/\tau_{R6}$. Using the measured $\tau_{d6}^{(integrated)}$ = 428 μs and the radiative lifetime τ_{R6} = 649 μs, we get s = 796 s^{-1}.

3.5. Yb-Tm interaction (step 4):
The luminescence transient of $^1D_2$ level of Tm^{3+} was measured for 797 nm laser pulsed excitation and showed in the Figure 8.

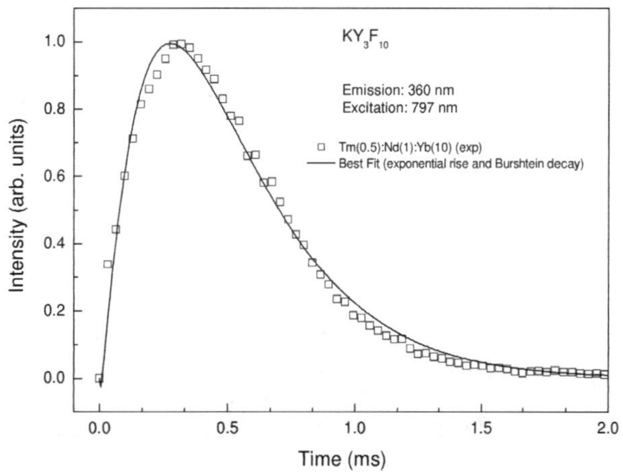

Figure 8. Decay-time transient of $^1D_2$ luminescence from Tm^{3+} measured in Yb(10):Tm(0.5):Nd(1.3):KY_3F_{10} crystal after 797 nm pulsed laser excitation.

Best fit of luminescence curve (solid line of Figure 8) was done using $F_4(t) = A\,[\exp(-t/\tau_d - \gamma\sqrt{t} - wt) - \exp(-t/\tau_{s7})]$ where A is a constant, τ_{s7} is the rise-time constant and τ_d the decay-time of the luminescence transient (the Burshtein function was used in $F_4(t)$ because the nature of the decay of $^1G_4$ level that composes the *donor like state* ($^2F_{5/2}(Yb^{3+})$ x $^1G_4(Tm^{3+})$). In this case, the radiative lifetime of $^1D_2$ level is too short comparable to the time constant observed in the luminescence transient (τ_{R7} = 72 μs). Best fit parameters obtained were τ_{s7} = 224 μs (raise-time) and τ_d = 302 μs, γ = 2.8 s^{-1}, w = 216 s^{-1} that gives an integrated decay-time constant equal to τ_{d7} = 271 μs. These time constants values (measured here) are much closer to the values estimated for the *donor like state* obtained by the ($^2F_{5/2}(Yb^{3+})$ x $^1G_4(Tm^{3+})$) product that gives $\tau_{s(26)}^{(donor)}$ = 205 μs (raise-time) and

$\tau_{d(26)}^{donor}$ = 272 μs (decay-time). As a consequence, the transfer rate constant of *process h* could not be obtained from the luminescence transient exhibited in Figure 9.

All the energy transfer parameters or rates (s^{-1}) involved in the Yb/Tm/Nd system, which were obtained in this work, are given in Table 1. *Process i* could not be observed using direct excitation of Nd^{3+} and Tm^{3+} ions with 797 nm laser excitation because the $^4F_{3/2}$ (Nd^{3+}) excited level rapidly transfer its population to the Yb^{3+} (~6.5 μs) so triggering process h – process I will be considered negligible in Tm:Nd:Yb:KY3F crystal.

Table 2. Energy transfer parameters (or rates) of the KY_3F:Yb:Tm:Nd crystal

Ions Interaction	Parameter	Rate (s^{-1})	Ions Interaction	Parameter	Rate (s^{-1})
Nd-Yb	b	$1.15x10^5$	Tm-Nd	p	348
Yb-Nd	c	714	Tm-Nd	q	6120
Tm-Yb	d	-	Tm-Tm	r	328
Yb-Tm	e	1739	Tm-Tm	s	796
Yb-Tm	f	318			
Yb-Tm	g	1000			
Yb-Tm	h	-			

4. Conclusions

Studying the optical properties of KY3F doped with Yb/Tm/Nd it can be concluded that**:** the crystal is efficient and generate ultraviolet blue emission by three and two photon processes mechanisms arising from the 797 nm excitation, respectively. The KY3F:Yb:Tm:Nd has the maximum absorption peak at 802 nm.

A very important energy transfer mechanism between Nd^{3+} and Yb^{3+} ions was noticed considering that Nd^{3+} emission is very shortened exhibiting a lifetime of 6.5 μs. The cross-relaxation Yb($^2F_{5/2}$) x Tm($^3H_4$) leads to the $^1G_4$ population grown by two-photon process, while in the case of Yb:Tm system, a Tm($^3H_4$) → Yb($^2F_{7/2}$) energy transfer is required first to excite an Yb^{3+} ion with a lower transfer rate constant of ~328 s^{-1}.

In order to verify the possibility of Yb:Nd:Tm system (KY3F) to present population inversion (Δn) for the $^1G_4 \rightarrow {}^3H_6$ emission of Tm^{3+} (~483 nm) under continuous pumping, we are solving the rate equations system. Numeric solutions are being used to simulate Δn as a function of the pump intensity at 797 nm.

Acknowledgements
The authors would like to thank FAPESP and CNPq for financial support.

References

[1] Barnes N P 2007 *Ieee J Sel Top Quant* 13 435-47
[2] Boulma E, Diaf M, Jouart J P, Bouffard M, Doualan J L and Moncorge R 2006 J Phys-Condens Matr 18 6721-7
[3] Boulma E, Jouart J P, Bouffard M, Diaf M, Doualan J L and Moncorge R 2008 Opt Mat 30 1028-32
[4] Ito M, Hraiech S, Goutaudier C, Lebbou K and Boulon G 2008 J Cryst Growth 310 140-4
[5] Kim K J, Jouini A, Yoshikawa A, Simura R, Boulon G and Fukuda T 2007 J Cryst Growth 299 171-7
[6] Kishi Y and Tanabe S 2006 J Am Ceram Soc 89 236-40
[7] Miura K, Kawamoto H, Kubota Y, Nishimura N and Kita Y Up-conversion laser material used in, e.g., optical memories|comprises fluoride, chloride, and bromide, glass and single crystal of

fluoride, chloride, bromide and iodide. Central Glass Co Ltd)
[8] Rao R P 2005 J Lumin 113 271-8
[9] Scholle K, Heumann E and Huber G 2004 Laser Phys Let 1 285-90
[10] Tang Y L, Yang Y, Cheng X J and Xu J Q 2008 Chinese Opt Let 6 44-6
[11] Tollefson M K, Gettman M T and Frank I 2008 J Urology 179 365-
[12] Tsonev L 2008 Opt Mat 30 892-9

Site Selective Spectroscopy on Erbium Ions in Stoichiometric Lithium Tantalate

K. Miyahara, A. Toulouse, N. Woodward, P. Capek, and V. Dierolf

Physics Dept. Lehigh University, Bethlehem, PA 18015,USA

vod2@lehigh.edu

Abstract. Ferroelectric materials such as lithium niobate ($LiNbO_3$) and the isostructural lithium tantalate ($LiTaO_3$) play an important role in integrated optics since they allow the possibility to combine their favourable electro-optical, acousto-optical, and nonlinear properties with the ability to add additional functional groups by doping. Examples are rare earth ions that act as active centres for laser and optical amplifier applications. We present our sites-selective spectroscopic studies on Er^{3+} doped nearly stoichiometric $LiTaO_3$ that include results about the assignment of excitation and emission peaks to different sites, symmetry properties of these sites, energy transfer among major sites, and up-conversion efficiencies. We compare the results in $LiTiO_3$ with the corresponding ones in the much better studied $LiNbO_3$ host and find that the type of centres and their spectral feature are very similar.

1. Introduction The first section in your paper

Ferroelectric materials such as lithium niobate (LiNbO3) and the isostructural lithium tantalate ($LiTaO_3$) play an important role in integrated optics since they allow the possibility to combine their favorable electro-optical. acousto-optical, and nonlinear properties with the ability to add additional functional groups by doping. Examples are rare earth ions that act as active centres for laser and optical amplifier applications. For this reason, extensive studies of the spectroscopic properties of many rare earth ions have been performed for the $LiNbO_3$ host material [1-5].

However, much less is known for lithium tantalite ($LiTaO_3$), although it offers better resistance against optical damage making it a more favourable host for high-power lasers in the visible spectral region. We present our site selective spectroscopic studies on Er^{3+} doped, nearly stoichiometric $LiTaO_3$ samples. While the Er^{3+} ion is well known for its highly efficient emission in the telecommunication window around 1.5 µm, it offers also emission in the visible at around 550nm and 650nm that can be excited through up-conversion with widely available powerful semiconductor lasers around 980nm. To this end, we focus on the excitation around 980nm and address the following questions:

- What are the Er^{3+} incorporation sites and how do they compare with the ones found in $LiNbO_3$?
- How do the different sites participate in the up-conversion processes?
- How does the defect distribution change under thermal annealing?

2. Experimental Technique

We address these questions using combined excitation emission spectroscopy in which we record a large number of emission spectra while continuously varying the excitation wavelength. Details of this technique and its potential are described in Ref [3, 6]. In our study, we use a tunable semiconductor laser with an external cavity that can be tuned smoothly from 965nm to 990nm. This allows us to excite the $^4I_{15/2}$ to $^4I_{11/2}$ transition and observe the emission at 1.5µm ($^4I_{13/2}$ to $^4I_{11/2}$). Moreover, the laser has sufficient power (approx 10mW) that allows the observation of emission at 550nm that is excited through up-conversion processes. The relevant transitions for this work are indicated with the expected symmetry properties for C_{3V} symmetry in Fig. 1. The emission spectra are recorded in reflection mode such that the propagation of excitation and emission is counter-propagating and parallel to the normal of the selected sample surface. Hence, for z-cut samples, the polarization of the emission and excitation are perpendicular to the c-axis of $LiTaO_3$ allowing only σ-type transitions for Er^{3+} ions in C_{3v} symmetry.

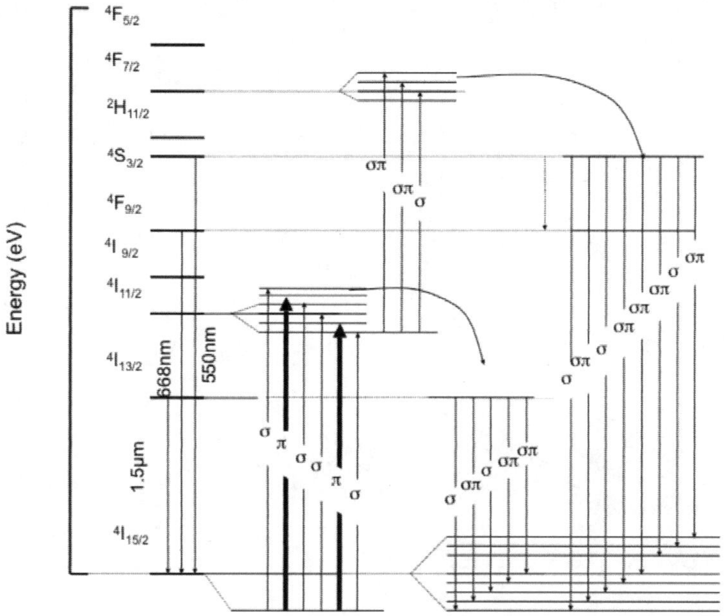

Figure 1. Energy level scheme of Er^{3+} in $LiTaO_3$. The transitions relevant for this work are indicated including the expected polarization properties for C_{3v} point symmetry. In particular, two excitation transition are highlighted that are not expected to appear for our reflection mode detection for a z-cut sample

3. Results and Discussion

3.1. Identification of Sites

For identification of the energy levels belonging to different incorporation environments in the crystal, we use 980nm excitation and detect the emission at around 1.5µm. From an inspection of the energy level scheme, we see that 6 excitation transitions and 8 emission transitions are expected. Although we limit our emission detection range to 5 transitions, we are still dealing with 30 transition pairs for a single site. In identifying these pairs, we make use of the CEES data (shown in a combined contour/image plot in Fig. 2 for 4 excitation/emission transition pairs) and exploit the property that the same emission transitions are expected for all seven excitation transitions and vice versa. As a consequence, pairs of excitation/emission transitions must lie on corners of square such as those drawn in Fig. 2. Through careful inspection of the whole data-set we are able to identify at least 7 different sites labeled (A…G). The corresponding transition energies are determined and listed in Table 1. Using these transitions, the crystal field splitting of the ground state $^4I_{15/2}$ and the excited $^4I_{11/2}$ state can be determined by simple subtraction of the transitions energies between the two respective lowest levels in each multiplet. The results are listed in Table 2. The ground state splitting can be calculated from all measured emission transitions and the same results have to be obtained. For our data this is fulfilled within ±0.1meV.

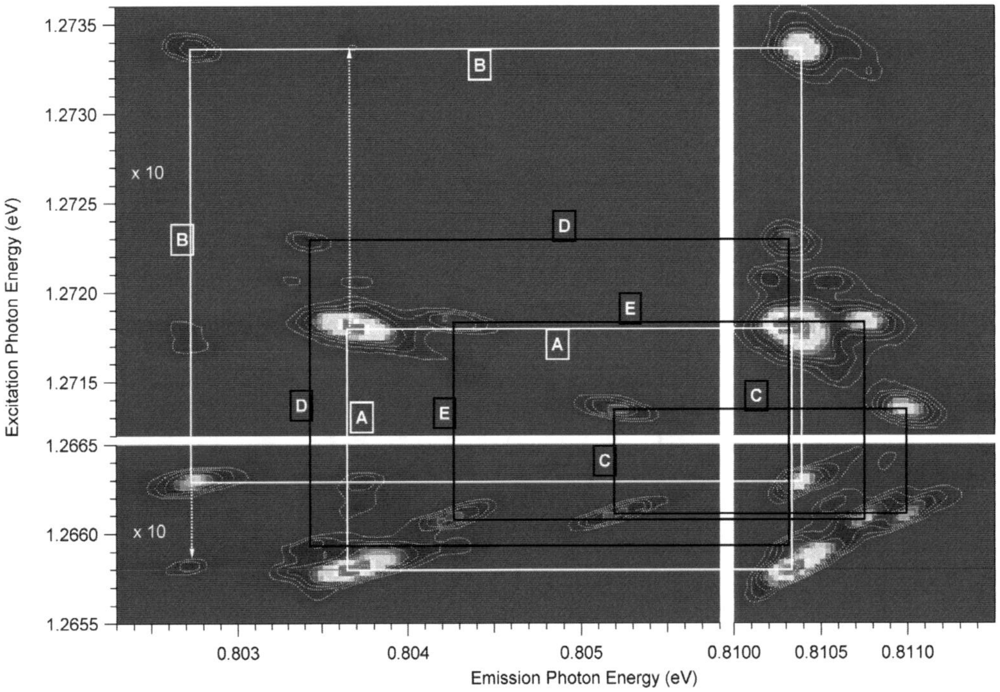

Figure 2. Image plot of CEES data for two excitation (#1, #3) and two emission (#1, #2) transitions. Excitation/Emission peaks that belong to the same sites are identified using squares and are labeled.

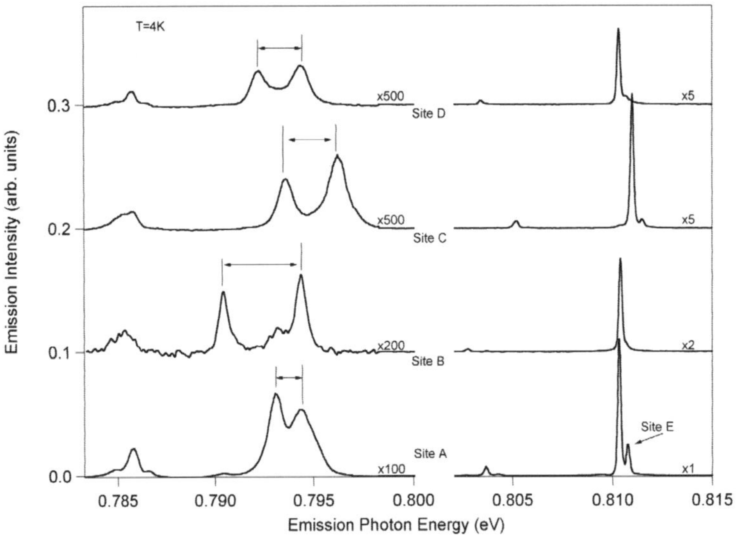

Figure 3. Selectively excited emission spectra. The indicated splitting of the levels #3 and #4 are used to compare the sites with the corresponding ones in $LiNbO_3$.

Table 1: Emission and Excitation energies of Er^{3+} in $LiTaO_3$ for transitions in the IR and green. Several different sites are listed and the labels related to the ones used in publications on $LiNbO_3$

Site	**A**	**B**	**C**	**D**	**E**	**H1**	**H2**
Label In Ref [1]	5	9	2			7	
Label in Ref[2]	A1	A3	B0			A2	
	1.2658	1.2663	1.2661	1.2659	1.2661		1.2662
	1.2706	1.2717	1.2705	1.2709	1.2708	1.2709	1.2708
Excitation	1.2718	1.2735	1.2713	1.27225	1.2718	1.27208	1.27205
Energies for $^4I_{15/2}$	1.2761	1.2783	1.2750	1.2768	1.2758	1.27648	1.27640
to $^4I_{11/2}$ transition	1.278	1.2799	1.2770				
	1.2792	1.2818	1.2787		1.2794		
Up conversion only	1.2811	1.2837	1.2805				
	0.8104	0.8103	0.811	0.81031	0.8107	0.81018	0.8105
Emission energy	0.8037	0.8027	0.8052	0.8034	0.8043	0.80331	0.8037
for $^4I_{13/2}$ to $^4I_{15/2}$	0.7943	0.7943	0.7962	0.7945	0.79503	0.79446	0.79446
transition	0.7931	0.7903	0.79363	0.7923	0.7932	0.79251	0.79251
	0.7858	0.7853	0.7855	0.7858	0.7856	0.78579	0.78585
	2.2653	2.2662	2.266	2.266	2.659		
	2.2586	2.2585	2.2602	2.26056	2.593		
Emission energy	2.2493	2.2501	2.2512				
for $^4S_{3/2}$ to $^4I_{15/2}$	2.2482	2.2461	2.2484				
transition	2.2408	2.2411	2.2405				
	2.2261	2.224	2.229				
	2.2216	2.2194	2.224				
	2.2153	2.2128	2.2163				

In the studies on $LiNbO_3$, it was noted that the splitting of the emission peaks to the 3rd and 4th ground state level (which corresponds to the splitting between them) is a good signature for the grouping of the sites into type 0,1,2,3, see Ref. [3]. In Fig. 3, we depict the emission spectra excited selectively for different sites. We note that again the splitting is characteristically different for the different sites. Through comparison of these splittings, we connect the sites A...G identified for $LiTaO_3$ with those in $LiNbO_3$. The assignments are included in Table 1. We find good correspondance and similarities in the relative strength and hence abundance amongst corresponding sites. However, as apposed to $LiNbO_3$, the relative strengths of the emission lines from different sites remain virtually unchanged after 5h annealing at 200°C. Only a narrowing of the emission lines can be observed indicating that just the intrinsic defects are ordering under these conditions while the Er^{3+}-related defects remain stable.

Table 2: Crystal field splitting of the $^4I_{11/2}$ and $^4I_{15/2}$ state.

Site	A	B	C	D	E	F	G
	0	0	0	0	0	*	0
	4.8	5.4	4.4	5	4.7	0	4.6
Splitting	6	7.2	5.2	6.35	5.7	1.18	5.85
of ^{4}I1$_{11/2}$	10.3	12	8.9	10.9	9.7	5.58	10.2
state in meV	12.2	13.6	10.9				
	13.4	15.5	12.6		13.3		
	0	0	0	0	0	0	0
	6.7	7.7	5.8	6.9	6.4	6.9	6.8
Splitting	16	16.1	14.8	15.8	15.7	15.7	16.0
of $^4I_{15/2}$	17.2	20.1	17.5	18.0	17.5	17.7	18.0
state [meV]	24.5	25.1	25.5	24.5	25.1	24.4	24.7
	39.2	42.2	37				
	43.7	46.8	42				
	50.0	53.4	49.7				

* lowest level unknown; splitting calculated relative 2nd level

3.2. Local symmetry of defects

To further compare the sites between the two host materials, we compare their emission properties in terms of local point symmetry. In our observation geometry, we do not expect to see π-type transition for a z-cut sample while these transitions should appear for x-cut. For the latter crystal cut, both σ and π type transitions are expected depending on the polarization of the pump laser and the emission detection. In Fig. 4, we compare excitation spectra for the 3 major sites for excitation onto the z and x-surfaces. These excitation spectra can conveniently by extracted from our CEES data by taking vertical cross sections at emission wavelengths that are characteristic for a particular site.

We find for site A that the 2nd and 5th emission transition are essentially absent for the z-cut sample, indicating that we are dealing with a site with C_{3V} symmetry. Site B and C show all six excitation transitions and most hence be connected with incorporation with lower site symmetry suggesting charge compensating defects that are off the c-axis. The situation was analogous for $LiNbO_3$.

3.3. Energy transfer

Close inspection of the CEES data reveal that emissions for Site A can always be observed for any excitation energy of site B. The relative strength of site A to site B is 20%. We can interpret this behavior by assuming an energy transfer from Site B to Site A. The branching ratio of such a transfer vs. emission within site B would be approximately 1:5 in our sample. The reverse transfer can be observed as well with a branching ration of 1:20. It is reasonable that the energy transfer occurs while the ions are in the lowest level of the excited $^4I_{13/2}$ state, which has a long radiative lifetime (approx ms). As can be seen by the emission data, the energy levels of the two sites are almost identical making energy transfer likely in both directions. The asymmetry in the branching ratio may be attributed to the difference in the relative abundance of site A (majority site) and site B.

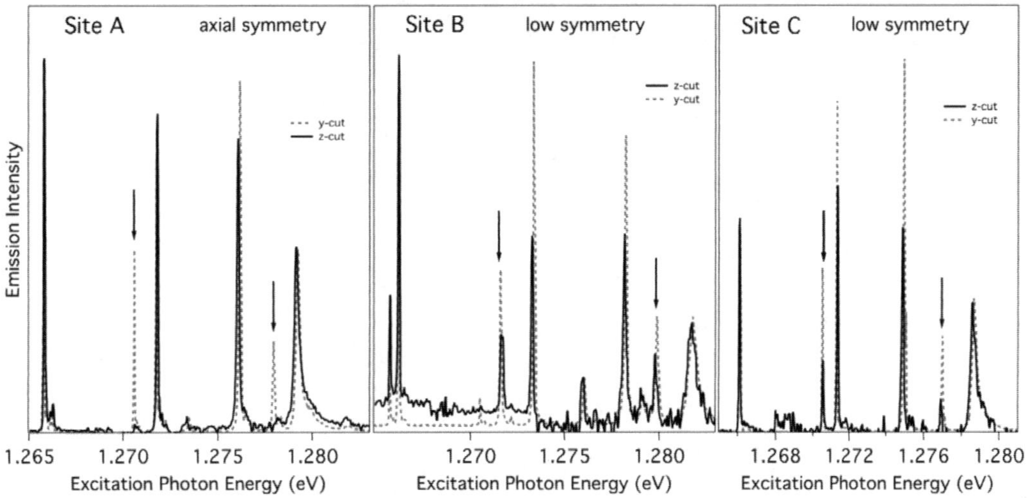

Figure 4. Selected Excitation spectra for the three major sites obtained by exciting along the z- and x axis. The data were obtained as vertical cross sections of the CEES data

3.4. Up-conversion

The presence of a rather efficient energy transfer among the different sites suggest that up-conversion through energy transfer is possible by which an already excited ion receives energy from a near-by ion that in turn relaxes to the ground state. Unlike the energy transfer considered above, this process mostly occurs in the $^4I_{11/2}$ state, which still has a reasonably long lifetime. In Fig. 5, we present the CEES data that were obtained under 980nm-range excitation but emission detection this time at 550nm. The data reveals again excitation peaks that can be associated with the major sites that we identified above. This is indicated for site A, B, and C. However, a large number of additional sites with quite different energy levels can be found as well. These sites that are absent in the direct excitation scheme are most likely associated with rare earth ions clusters.

Turning the attention back to the up-conversion behavior of the majority sites we depict the excitation spectrum for the up-conversion emission (Fig. 6) and we notice that up-conversion of site A and B can also be achieved through excitation of the respective other site in the initial step. Apparently energy transfer can occur in the $^4I_{11/2}$ state as well, despite the smaller lifetime and the somewhat larger energy mismatch. Moreover, efficient up-conversion can be achieved even for excitation energies that do not correspond to levels of the $^4I_{13/2}$ state. In this case, the first step of the up-conversion excitation will invoke transitions that involve the weakly coupled phonons or local vibrations. For site B the energy difference (17.5meV or 141cm^{-1}) from the lowest lying level to the phonon assisted excitation peak corresponds closely to the lowest energy E(TO) or E(LO) mode [7]. For site A, however, the corresponding energy difference is smaller (15.3meV or 123cm^{-1}) and does coincide with a known phonon mode. We suspect that this difference corresponds to localized mode caused by the Er^{3+} dopant itself.

Overall, the behavior of Er^{3+} ions in $LiTaO_3$ in terms of up-conversion resembles again very closely the results for $LiNbO_3$[3]. This result does not come unexpected due to the similarities in the incorporation of the ion, the energy levels, and the phonon spectrum of the two host materials.

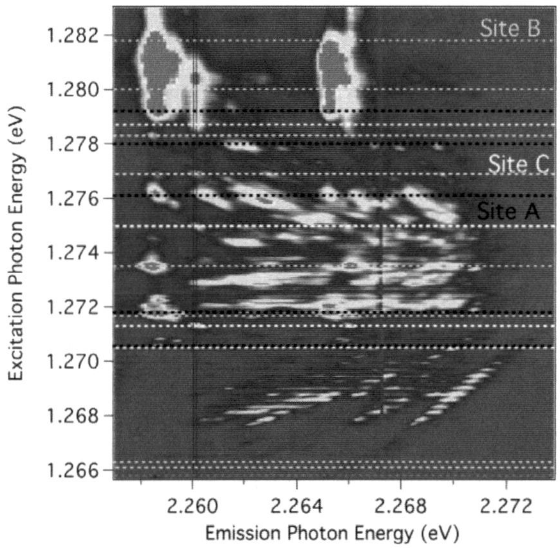

Figure 5. Image plot of CEES data obtained under excitation around 980nm and emission detection around 550nm. The dotted lines indicate the excitation position of the major sites. T=10K

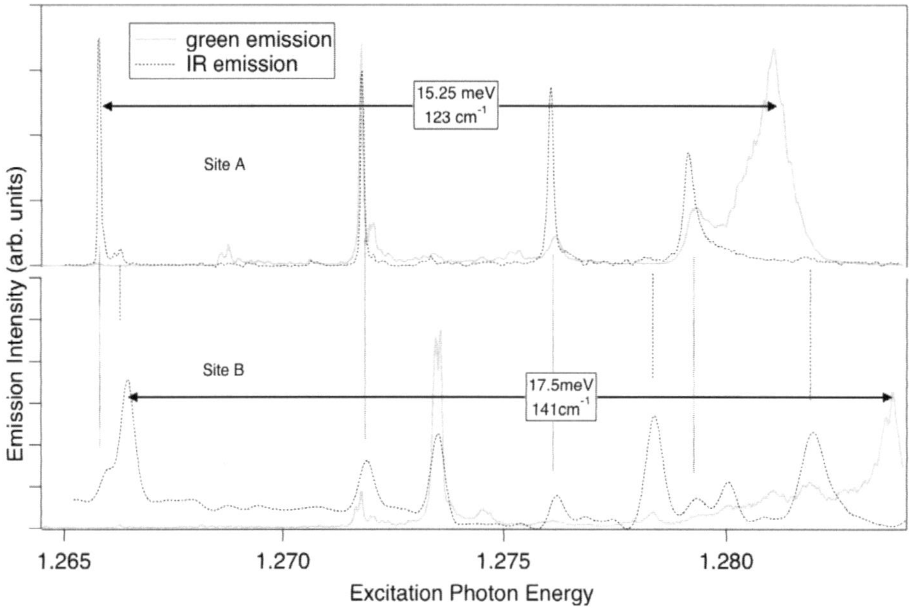

Figure 5. Excitation spectrum for up-conversion (green) and regular IR emission. The spectra are obtained from the data in Fig. 2 and Fig 5 as vertical cross sections. The dotted vertical lines indicate the spectral positions for which energy transfer from A to B and from B to A take place. The energy difference between lowest and highest energy band is indicated.

4. Summary and Conclusions

Using the technique of combined excitation emission spectroscopy, we determined the energy levels of a large variety of sites. We find a strong resemblance with the sites in $LiNbO_3$ and hence assume that their nature is very similar. The large number of sites is due to the required charge compensation that lead to sites with local point symmetry lower than C_{3V}. The majority site in the host has almost perfect C_{3V} symmetry and may be attributed to rare earth ions without a local compensation or with compensation that is arranged along the c-axis. Based on the analogy with $LiNbO_3$ for which the lattice location has been determined experimentally [8,9] we find no evidence for defect pairs with Er^{3+} ions on adjacent Li and Nb sites such as proposed for $LiNbO_3$ by Jackson et al[10]. In such a pair, strong energy transfer would be expected. Site A and B that are exhibiting a weak energy transfer have different local symmetries and hence cannot be such a pair for which C_{3V} symmetry is expected for both ions.

Energy up-conversion takes place for modest excitation levels in ways that resemble $LiNbO_3$, including the involvement of local modes and bulk phonons.

Overall, the Er^{3+} ion in $LiTaO_3$ shows a very similar behavior to Er^{3+} in $LiNbO_3$. From this result we suspect that the ion can be used as an efficient probe for perturbation and variation in materials such as it has been demonstrated in $LiNbO_3$ for the imaging of waveguides [11] and domain walls [12].

References

[1] D. M. Gill, L. McCaughan, and J. C. Wright, Physical Review B (Condensed Matter) **53**, 2334 (1996).
[2] O. Witte, H. Stolz, and W. von der Osten, Journal of Physics D (Applied Physics) **29**, 561 (1996).
[3] V. Dierolf and M. Koerdt, Phys. Rev. B **61**, 8043 (2000).
[4] W. Jia, K. Lim, H.Lui, Y. Wang, J. Ju, S. Yun, F. Fernandez, And W. Yen, J. of Lumin. 66-67, 190 (1995).
[5] L. Nunez, B. Herreros, R. Duchowicz, G. Lifante, J. Tocho, F. Cusso, J. of Lumin. 60-61, 81 (1994).
[6] V. Dierolf and Chr. Sandmann, Journal of Lumin 125 (2007) 67–79.
[7] X. Yang, G. Lan, B. Li, and H. Wang, phys. stat. sol. (b) **141**, 287 (1987).
[8] T. Gog, M. Griebenow, and G. Materlik, Physics Letters A **181**, 417 (1993).
[9] T. Gog, P. Schotters, J. Falta, G. Materlik, and M. Grodzicki, Journal of Physics: Condensed Matter **7**, 6971 (1995).
[10] R. M. Araujo, K. Lengyel, R. Jackson, L. Kovács and Mário E G Valerio, . *Phys.: Condens. Matter* **19,** 046211 (2007).
[11] V. Dierolf, C. Sandmann, .J. Lumen **102-103** 201, (2003)
[12] V. Dierolf, C. Sandmann, V. Gopalan, S. Kim, and K. Polgar, J. Appl. Phys. **93**, 2295 (2003).

Effect of Zn^{2+}, Fe^{3+} and Cr^{3+} addition to hydroxyapatite for its application as an active constituent of sunscreens

T S de Araujo[1], S O de Souza[1], E M B de Sousa[2]
[1]Physics Department, Federal University of Sergipe, São Cristóvão, SE 49100-000, Brazil
[2]Center of Development of the Nuclear Technology – CDTN/CNEN, Belo Horizonte, MG 30123-970, Brazil

E-mail: tatiana.araujo@cefetse.edu.br

Abstract. Biocompatible phosphate materials are used in different applications like bone and dental implants, drug delivery systems and others, but could also be applied in inorganic sunscreens. Using sunscreens is extremely necessary, because long time exposure to sun can cause skin cancer. In this work chemical precipitation method has been used to produce hydroxyapatite. Cr^{3+}, Zn^{2+} and Fe^{3+} doped samples were characterized using powder X-Ray Diffraction (XRD) and Optical Absorption techniques. X-ray diffraction measurements confirmed the materials were in the expected crystalline structures. The crystallite size as measured from the X-ray pattern was 23–27 nm (±1).The absorption spectra in the ultraviolet and visible ranges indicate that appropriately doped and sized hydroxyapatite particles may have potential applications as active constituents of sunscreens.

1. Introduction

The sun emits a wide spectrum of electromagnetic waves and ultraviolet light (UV) is the most aggressive to human tissues, mainly the skin. Large amounts of UVB and UVC are absorbed by earth's atmosphere. Solar UV radiation that reaches the earth as well as our skin, is composed by 5–10% energetic UVB (290–320 nm) and 90–95% UVA (320–400 nm) which is less energetic, but penetrates deeper the skin due to its longer wavelength [1]. Daily use of sunscreens became imperative in certain geographical regions to prevent damages caused to skin. The number of new cases of skin cancer is growing continually in the world, especially in Brazil; where this number is around 120 thousand a year [2]. Sunscreens might be capable to absorb or to reflect the incident UV radiation protecting the individuals [3].

The sunscreens are divided into two major classes: i- the so-called chemical filters, made of substances capable of absorbing solar UV converting it to less energetic photons that are emitted; and ii- the physical filters, composed by physical barriers, on the skin, reflecting radiation. The well known physical filters are zinc oxide and titanium dioxide [4].

In the case of long-term and frequent usage, particular attention must be paid to efficiency and safety. Many organic sunscreens penetrate into the skin causing photoallergies, phototoxic reactions and skin irritations [5]. Therefore, there is an urgent need for the development of a safer sunscreen system. This can be achieved by formulations with little penetration into the skin and using biocompatible materials. For these very reasons inorganic sunscreens represent one of the best means to protect the skin and are recommended for children and people with sensitive skins [6]. Titanium dioxide and zinc oxide are popular inorganic sunscreen agents. Their high refractive indices, however, can make the skin look unnaturally white when incorporated into the products. Additionally, their high photocatalytic activity facilitates the generation of reactive oxygen species, which can oxidize and degrade other ingredients in the formulation, raising safety concerns [7-9].

For solving problems described with titanium dioxide and zinc oxide, it is interesting to develop calcium phosphate that absorbs in the ultraviolet region. Calcium phosphates were already investigated and showed many possible applications in medicine [10,11], and, due to its chemical similarity with the mineral phase of the bones and teeth, they showed no toxicity [12-15]. Hydroxyapatite, ($Ca_{10}(PO_4)_6(OH)_2$) actually used for highly stable and inert bone and teeth transplants, is also a promising candidate for advanced sunscreens. Furthermore, its UV absorption limit can be tuned to absorb in the desired range from UVB (290–320 nm) to UVA (320–400 nm) by simply introducing a doping.

Inorganic sunscreens must be constituted preferably with particles smaller than 400 nm, avoiding Rayleigh scattering, because the intensity of the scattering depends on the wavelength and it is proportional to the particle radius. Some sunscreens versions involve particles of 70–200 nm [2].
The objective of this study is to develop nanocrystalline HAP powders with appropriate absorption in the UV region, suitable as active components of inorganic sunscreens.

2. Experimental procedure

Hydroxyapatite (HAP) was produced using a precipitation route of $Ca(NO_3)_2.4H_2O$ with $(NH_4)_2HPO_4$ [14,15].
The HAP reaction formation is expressed as follows:

$$10Ca(NO_3)_2.4H_2O + 6(NH_4)_2HPO_4 + 8NH_4OH \rightarrow Ca_{10}(PO_4)_6(OH)_2 + 20NH_4NO_3 + 6H_2O \qquad (1)$$

The samples were prepared mixing a solution of $Ca(NO_3)_2.4H_2O$ to a $(NH_4)_2HPO_4$ solution at a rate of 1.5 ml/min, keeping the pH constant at 10.4. The pH was adjusted during the mixing using a solution of NH_4OH. The mixture was then kept at room temperature for 50 hours. This period is called "maturation time". The resulting suspension was filtered, repeatedly washed with distilled water and dried and calcined at 500 °C for 1 hour. At the end of the calcinations a ceramic in powder form is obtained, constituted by particles agglomerated due to weak bonds acting between them. After a light grinding in agate mortar these bonds are broken and we have hydroxyapatite in the form of a fine powder.

For the Cr^{3+}, Zn^{2+} and Fe^{3+} -doped samples, a 0.01 M solution of $Cr(NO_3)_3.9H_2O$, $Zn(NO_3)_2.6H_2O$ and $Fe(NO_3)_3.9H_2O$, respectively, were added to $Ca(NO_3)_2.4H_2O$ solution.
The precipitated HAP powders was chemically and physically characterized using powder X-ray diffraction (XRD) and UV-vis diffuse reflectance. For characterization of the samples, the powders were drizzled and selected those between 75mm/μm and 150 mm/μm.
Powder XRD measurements were performed at room temperature in a Rigaku DMAX2000 difractometer using CuKα radiation at 40 kV/40 mA with a spinning sample holder and a monochromator detector.
The UV–vis diffuse reflectance spectrum was measured on a UV–vis recording spectrophotometer (UV-2401, Shimadzu). Barium sulphate was used as reference to calibrate the spectrophotometer.

3. Results and discussions

X-ray powder diffraction (XRD) technique was used to study the effect of dopants on the phase identification of the samples. The typical XRD patterns of the pure and Zn^{2+}, Fe^{3+}, Cr^{3+} -doped HAP samples, which perfectly match with the JCPDS pattern 9-432 for HAP, are presented in figure 1. The incorporation of Cr^{3+} and Fe^{3+} into the material apparently leads to the formation of less crystalline HAP. This effect seems natural as the dopants assumed to substitute at Ca^{2+} sites (ionic radius 0.99 Å) have larger charge and smaller ionic radius (Cr^{3+} 0.63 Å, Fe^{3+} 0.64 Å). Several ions, usually with an ionic radius smaller than that of Ca^{2+} are known to inhibit the formation of HAP [17].

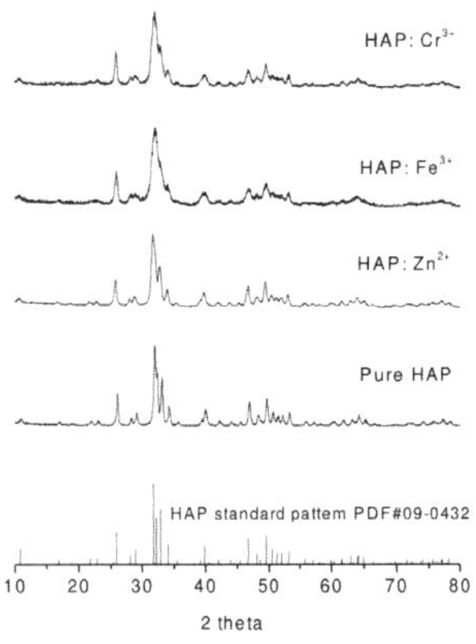

Figure 1. X-ray diffraction pattern of the pure HAP and Cr^{3+}, Fe^{3+} and Zn^{2+}-doped HAP compared to the HAP standard pattern PDF # 9-432.

Contradictory data are reported about zinc (ionic radius 0.74 Å) incorporation into HAP structure. According to some authors [18-20] Zn^{2+}-doped HAP can be obtained without producing significant alterations in the structure of HAP. On the other hand, Le Geros et al [21] reported that incorporation of reduced amounts of Zn^{2+} into HAP structure induces an evident reduction of the degree of crystallinity of the HAP phase. In this work, HAP-Zn^{2+} exhibited high crystalline peak at about 33,11°; almost identical patterns were recorded for pure and doped powders.

The relatively broadening of the XRD peaks for Cr^{3+}, Fe^{3+}-doped HAP indicates that the sizes of the crystallite grains of those materials are smaller than the pure HAP and Zn^{2+}-doped HAP. Through Scherrer's equation, within the JADE Rigaku computer code, the medium crystallite sizes were estimated. The results are shown in table 1. Note that the effect of Zn^{2+} on the growth of HAP particles is more remarkable than those of Cr^{3+} and Fe^{3+}, although the reason for this difference remains unclear at the present. For Harry [22] the particle size of a powder for application as sunscreen should obey, in beginning, the size of the passage of a sieve 200 mesh. At the present work, the low values of crystallite sizes (<30 nm) indicates the formation of nanoparticles.

Table 1. Medium size crystallite obtained via XRD measurements using Scherrer's equation to HAP prepared with different dopant at room temperature.

HAP	Crystallite size ± 1 (nm)
Pure	27
Zn^{2+}	25
Fe^{3+}	23
Cr^{3+}	23

An inorganic sunscreen which functions well when absorbing or reflecting the UV light, but, tends to be opaque and white on the skin is unacceptable for cosmetic use. The higher refractive index of titanium dioxide compared with zinc oxide (2.6 vs 1.9) [23] explains its whiter appearance. The still lower refractive index of HAP (around 1.6 [24, 25]) leads to a transparent appearance and a higher cosmetic acceptability.

For a composition to be an efficient sunscreen two main characteristics are necessary: (I) absorption in the UV region (II) to dissipate the energy absorbed with the minimum impact to skin, impeding, therefore, the formation of reactive species. The particle size is of great importance for the effectiveness of the sunscreen, nanoparticles improve the absorption in the UV region and increase the absorption intensity [26]. Production of doped HAP in nano scale has been developed, exploring its absorption capability and its intense scattering of the solar radiation in the UV region. To evaluate the photoprotection action of that material the technique of diffuse reflectance was used.

For the sake of comparison diffuse reflectance measurements were performed in titanium dioxide (TiO_2) commonly used in sunscreens formulations in cosmetics industry. As can be seen in figure 2, TiO_2 showed one large absorption around 268-419 nm.

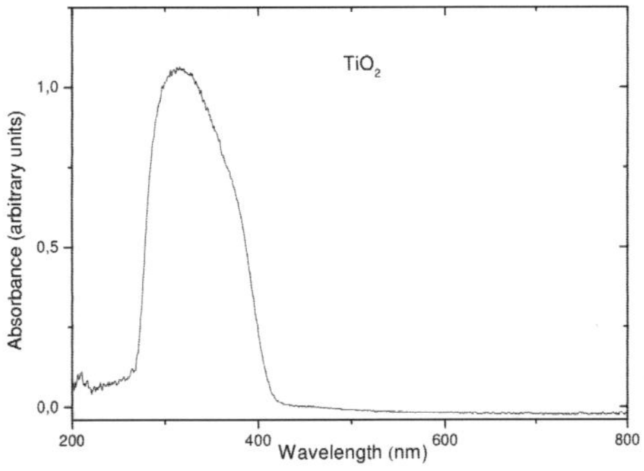

Figure 2. UV–Vis absorption spectrum of TiO_2.

Even if they have a good performance in optical absorption, TiO_2 and ZnO catalyze the formation of superoxide and hydroxyl radicals which can initiate oxidations, putting at risk people's health. TiO_2 is a semiconductor that can absorb light and, under certain conditions, generates free radicals. The band gap of 3 eV for TiO_2 is a measure of the minimum energy required to promote an electron from its valence band to its conduction band [27]. A compound with this kind of band gap can be excited by radiation at wavelengths below ~380 nm. Thus, TiO_2 may be susceptible to excitation by UVB and UVA in sunlight. Photoexcitation of TiO_2 could promote a single electron from the valence band to the conduction band, leaving a positively charged space (a hole) behind. Usually, the electron recombines with the hole, but sometimes the hole migrates to the particle surface, where it can react with absorbed species. In an aqueous environment it can react with water or hydroxyl ions, forming hydroxyl radicals [28]. Such processes are well known for aqueous preparations of TiO_2, exposed to either artificial UV light or natural sunlight. In this capacity, the photocatalytic potential of TiO_2 has been used experimentally to degrade suspensions of organic materials and purify drinking water.

In literature it was found that zinc oxide blocks mostly UVA radiation and has an attenuation maximum around 380 nm, but this can vary slightly with particle size [21, 29]. Its photocatalytic mechanism has been proven to be similar to that of TiO_2 [30].

In 1993, FDA (Food and Drug Administration) defined that a sunscreen should retain UVA around 360 nm to allege to be anti-UVA and the material used in the formulation should be able to present absorption in this spectrum region.

In figure 3 is presented the UV–Vis absorption spectra of pure and Cr^{3+}, Zn^{2+}, Fe^{3+}-doped HAP produced in this work and the maximum absorption wavelengths for TiO_2, pure and doped HAP are shown in the table 2. Pure HAP presents optical absorption in UV region (200-340 nm), with a strong band below 247 nm. Similar results for pure HAP were found by Nishikawa [31] when producing HAP by an aqueous precipitation procedure from $Ca(OH)_2$ and H_3PO_4 at temperature 1150 ^{0}C.

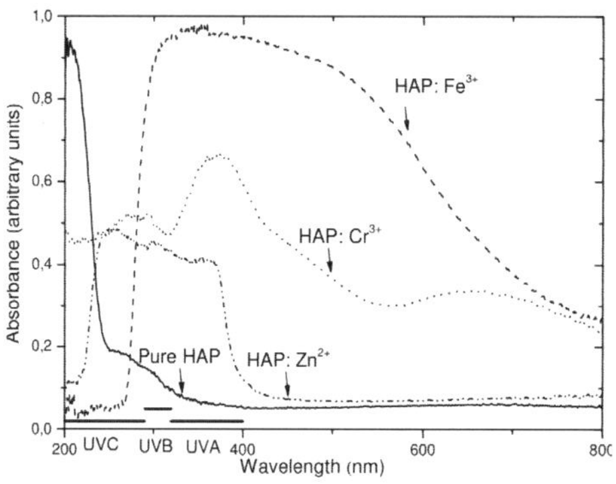

Figure 3. UV–Vis absorption spectra of Zn^{2+}, Fe^{3+}, Cr^{3+} -doped HAP.

Table 2. Maximum absorption wavelengths of the powders.

Powders	Maximum Absorption (nm)
Pure HAP	207
Zn^{2+}	240-372
Fe^{3+}	310-427
Cr^{3+}	368
TiO_2	316

As can be seen in figure 3, the optical absorption of HAP had suffered significant changes with the introduction of Cr^{3+}, Fe^{3+} and Zn^{2+}. Cr^{3+}-doped HAP generates absorption bands from UV to the visible (231-315, 318-542, 579-800 nm). Fe^{3+}-doped HAP showed a wide absorption from UV to the visible (267-800 nm) regions. The absorption bands in the visible region can induce the possibility of an undesirable visual color effect in sunscreens. During sunscreen production care has to be taken to avoid the agglomeration of particles, because the final effect of the product will decrease and the bad dispersion will generate a colored film. This problem can be healed if there is a good dispersion of the sunscreen on the skin an thus, a colored sunscreen will have transparent appearance when correctly applied on the skin [2].

Even if the Fe^{3+} and Cr^{3+}-doped HAP had good results in their absorption spectrum, the Zn^{2+}-doped HAP is the highlighted sample, because it showed absorption features in the spectral interval of 213-420 nm, which is similar to that obtained for TiO_2. Moreover, it is known that an excellent sunscreen should have maximum absorption capacity in relationship the UV radiations in wavelengths between 296.7 and 320 nm [22]. With this aim, the most promising sample produced in this work is the Zn^{2+} -doped HAP.

4. Conclusions

Sunscreen efficiency is directly related to the physical and chemical properties of the ingredients, in particular their absorption and/or dispersion capabilities with respect to solar radiation. The present work contributes information on the production and optical absorption of Zn^{2+}, Fe^{3+} or Cr^{3+}-doped HAP as possible active components of sunscreens.

A simple and low cost technique to produce doped HAP has been worked out, resulting in single-phase samples as shown by XRD powder analysis. The doped HAP samples display absorption bands in the ultraviolet region as required, similarly to TiO_2 powders frequently used in sunscreen formulations. The observed absorption bands in the visible region may result in some visual color of the products, however, no damage to the success of the application is expected if the HAP component is appropriately incorporated in the sunscreen formula leading to its sufficient dispersion on the skin. Due to the excellent biocompatibility of HAP combined to the low toxicity of the Zn^{2+} ions, the Zn^{2+} -doped HAP has a promising potential to be applied as a sunscreen. However, the long road from the sunscreen's elaboration to its commercialization necessitates several further studies by specialists of different areas in order to guarantee the protection against the undesirable effects of the ultraviolet radiation.

5. References

[1] Verschooten L, Claerhout S, Laethem A V 2006 *Photochemistry and Photobiology* **82** 1016-1023
[2] Flor J, Davolos M R 2007 *Quím. Nova* **30** 1
[3] Nohynek G J, Schaefer H 2001 *Regul. Toxicol. Pharm* **33** 285-299
[4] Shaath M A, Pathak M A 1997 *Sunscreens development, evaluation, and regulatory aspects* (New York : Marcel Dekker) 589-600
[5] Schulz J et al. 2002 *Advanced Drug Delivery Reviews* **54** 157-163
[6] Serpone N, Dondi D, Albini A 2007 *Inorganica Chimica Acta* **360** 794–802
[7] Masui T, Hirai H, Imanaka N 2006 *Journal of Alloys and Compounds* **408–412** 1141–1144
[8] Lili W U, Youshi W U, Yuanchang S H I, Huiying W E I 2006 *Rare Metals* **25** 68-73
[9] Yabea S, Satob T 2003 *Journal of Solid State Chemistry* **171** 7–11
[10] Herzog B, Quass K, Schmidt E, Müller S, Luther H. 2004 *Journal of Colloid and Interface Science* **276** 354–363
[11] Lynch S E et al. 1999 *Tissue Engineering* (New York: quintessence books)
[12] Rodriguez L M, Vallet R J M F 2001 *Biomaterials* **22** 583-588
[13] Springsteen A, Yurek R, Frazier M, Carr K F 1999 *Analytica Chimica Acta* 380 155-164
[14] Araujo T S, Macedo Z S, Oliveira P A S C, Valerio M E G 2007 *Journal of Materials Science* **42** 7 2236-2243
[15] Araujo T S, Lima T A R M, Valerio M E G 2007 *Scientia Plena* **3** 7 285-291
[16] Lin K, Chang J, Lu J, Wu W, Zeng Y 2007 *Ceramics International* **33** 979-985
[17] Miyaji F, Kono Y, Suyama Y 2005 *Materials Research Bulletin* **40** 209-220
[18] Rao S V C 1976 *J. Indian Chem. Soc.* 347
[19] Tripathy N K, Patel P N, Panda A 1989 J. *Solid State Chem.* **80** 1
[20] Panda A, Patel P N, Tripathy N K 1990 *Indian J. Chem.* **29A** 70
[21] LeGeros R Z, Taheri M M, Quirologico G B, Le Geros J P 1989 *Proc. 2nd Int. Phosphorus.* (Boston: Imphos) 89-103
[22] Harry R G 1973 *Harry's cosmeticology* 6nd edition Leonard Hill Books (London)
[23] Hernández J R V, Goymann C C M 2006 *International Journal of Pharmaceutics* **322** 161–170
[24] Smiddy M A, Martin J E G H, Kelly A L, Kruif C G, Huppertz T 2006 *J. Dairy Sci.* **89** 1906-1914
[25] Pardo J A, Peña J I, Merino R I, Cases R, Larrea A, Orera V M 2002 *Journal of Non-Crystalline Solids* **298** 23–31
[26] Araujo T S, Souza S O, 2009 *Scientia Plena* **4** 11 1-7
[27] Gasparro F P, Mitchnick M, Nash J F 1998 *Photochemistry and Photobiology* **68** 3 243-256
[28] Brezova V, Stasko A 1994 *J. Catal.* **147** 156-162
[29] Pinnell et al. 2008 *Dermatol Surg.* **26** 4 309-314
[30] Dindar B, Icli S 2001 *J. Photochem. Photobiol. A. Chem.* **140** 263–268
[31] Nishikawa H. 2001 *Materials Letters.* **50** 364 –370

Acknowledgments
The authors wish to thank the Brazilian funding agencies CNPq and CAPES for financial support.

Thermoluminescent properties studies of spodumene lilac sample to dosimetric applications

L L Lima[1], R A P Oliveira[1], H R B R Lima[1], H N Santos[1], J O Santos[2], A F Lima[1] and S O Souza[1, 3]

[1] Physics Department, Federal University of Sergipe, P. O. Box 353, 49100-000 São Cristóvão, SE, Brazil.
[2] Federal Center of Technology Education of Sergipe, UNDED Lagarto, Road Lourival Batista, Carro Quebrado, 49400-000, Lagarto, SE, Brazil.

Abstract. This work investigates the thermoluminescent (TL) dosimetric properties in natural spodumene, $LiAlSi_2O_6$, called kunzite, from Minas Gerais State, Brazil. The mineralogical and chemical composition of this material was identified by means X-ray fluorescence and X-ray diffraction. Some dosimetric properties were studied, such as thermoluminescent emission curves as function of gamma dose. The glow curves of annealed kunzite presented two very intense TL peaks at 215 °C (peak II) and 350 °C (peak III), after gamma irradiation, being both of first kinetic order. These two most prominent peaks analyzed do not presented a linear growth in the range of 50 to 5000 Gy in the range of doses studied. The peak II also presented a very short calculated lifetime, which means it is hardly can be used in dosimetry, while the peak III has a longer lifetime and could be used in some applications for high doses dosimetry.

1. Introduction

Spodumene ($LiAlSi_2O_6$) is a natural lithium aluminosilicates mineral of significant technological interest. Depending on temperature and pressure, it has been observed in either C2/c (α-spodumene) or $P2_1/c$ (β-spodumene) symmetry [1]. There are three varieties (thiphane, kunzite and hiddenite) of beautiful coloration that are considered as semi-precious gemstones. Besides, the natural crystal is strongly luminescent under UV, electron beam or X-ray excitation. These optical characteristics are caused by lattice defects which exist in natural samples. The most frequent are exchange of Si^{4+} and Al^{3+} ions and the presence of different impurities on substitution and interstitial sites [2-5].

On exciting the samples by ionizing radiation, electrons and holes can be trapped by these sites. They can be thermally released from traps causing a radioactive recombination and thermoluminescent peaks. The characterization of the thermoluminescence properties in aluminosilicates is of special interest for solid state dosimetry applied to dose reconstruction and dating [6-7]. Its striking luminescence turns the spodumene an interesting material to be applied as dosimeter and/or scintillator. However, there is no much information about this potentiality in literature.

---

[3] Corresponding author; e-mail: sosouza@fisica.ufs.br

The use of thermoluminescence (TL) in radiation dosimetry requires knowledge of the TL properties of the material. Besides the necessary care during the experiments, it is vitally important the knowledge of the curve shape of the emission and the answer depending on dose, which can be presented as a theoretical model able to predict results. They also provide explanations of the material in terms of its internal structure and the lifetime of the peak. This lifetime is indirectly calculated using the kinetics activation energy (E), the frequency factor (s) and kinetic order (b) [8].

In this work we report some thermoluminescent properties of a natural kunzite (or lilac α-spodumene) from Minas Gerais State, Brazil. In order to derive reliable values of the dosimetric properties two different methods (peak shape and curve-fitting) of TL glow peaks analysis were studied, as well some calibration curves showing the dose range for spodumene TL response.

2. Experimental Details

The natural lilac spodumene sample was obtained from Minas Gerais State, Brazil. The mineralogical composition of this material was identified by means of X-ray diffraction and its impurity content was analyzed with X-ray Fluorescence (XRF). According to previous paper, the lilac color of kunzite is correlated with the presence of Mn and with a low Fe/Mn ratio [9]. The lilac color occurs whenever the concentration ratio of Mn/Fe is larger than 1. The sample here investigated presented the Fe/Mn ratio of 1.11, showing a lilac color in the *c*-axis direction and a pale yellow in the other directions.

The crystal was ground into powders and sieved to retain the grains between 0.080 to 0.180 mm. Gamma irradiation was carried out at room temperature (RT) using a ^{60}Co source with a dose rate of 7.5 $kGy.h^{-1}$. For pre-irradiation heating process, the samples were annealed in normal air atmosphere at 600 °C for one hour and cooled to RT suddenly.

The glow curves were obtained from a Daybreak 1100 series automated TL reader system equipped with a photomultiplier EMI 9635B and they were recorded from RT up to 500 °C, with a heating rate of 4 $°C.s^{-1}$ in a nitrogen atmosphere. The TL spectra were recorded using a homemade TL reader equipped with an Unicom 100-FUNBEC monochromator with an accuracy of 5.0 nm, the light being detected by a Hamamatsu 551S photomultiplier.

3. Methods

In the TL models, the probability per unit time for any electron leaving the trap is given by

$$p=s\exp(-E/kT)$$

where s is the frequency factor, E the activation energy, k the Boltzmann constant and T the temperature in which occur thermal liberation.

The lifetime of TL peaks is nothing but the inverse of p:

$$\tau=s^{-1}\exp(E/kT)$$

As theoretical methods to calculate the kinetic parameters, E and s, it was employed the Peak Shape and the Curve Fitting Methods. Both are based on the glow curve shape [8,10].

Curve fitting (CF)

The Curve Fitting Method assumes, for a first order kinetics, that the intensity of the glow peak is given by

$$I(T) = n_0 s \exp\left(-E/kT\right)$$
$$\exp\left[-(s/\beta)\int_{T_0}^{T} \exp(-E/kT)dT\right]$$

This method consists in to solve numerically the previous integral. The computed curve is then compared with the actual experimental curve and a root-mean-square (RMS) deviation between the both is calculated. The procedure continues by sequentially changing the E and s values until a minimum value of the RMS deviation is obtained.

Journal of Physics: Conference Series **249** (2010) 012013 doi:10.1088/1742-6596/249/1/012013

Peak shape (PS)
The method based on the TL Peak Shape utilizes just two or three points from the glow curve. Usually, theses are the temperature at the peak maximum T_m and either the low- and high-temperature half-heights at T_1 and T_2. The activation energy can be calculated through the Chen's formula

$$E = c_y (kT_m^2 / y) - b_y (2kT_m)$$

where γ is $\omega = T_2 - T_1$, $\tau = T_m - T_1$ *or* $\delta = T_2 - T_m$.

The constants c_γ and b_γ for the 1st kinetic order are given in table 1.

Table 1. The constants c_γ and b_γ for the 1st kinetic order

	ω	τ	δ
c_γ	2,52	1,51	0,976
b_γ	1,0	1,58	0

4. Results and discussion

The figure 1 shows X-ray diffractogram of the samples not annealed and annealed at 600 °C during 1 h. In both cases the crystalline structures correspond to the pattern of the natural α-spodumene that has the main planes appointed in the figure.

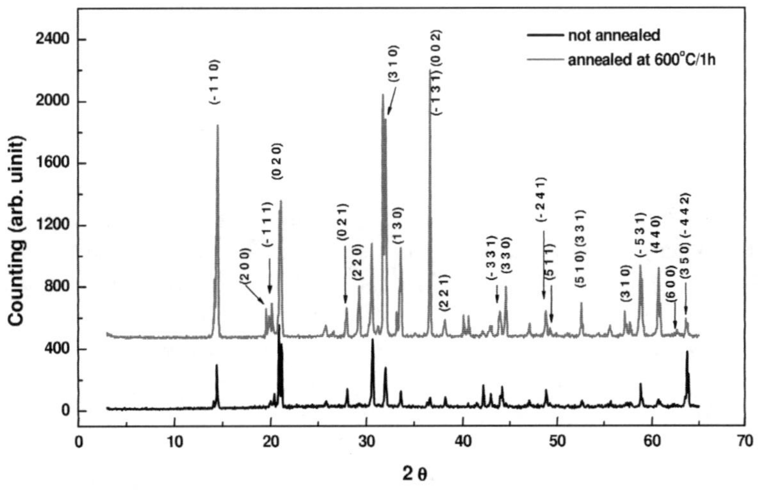

Figure 1. X-ray diffractogram of the not annealed and annealed at 600 °C during 1 h samples. The numbers between brackets are the reflection planes from the α-spodumene pattern.

Some glow curves of the not annealed sample irradiated with different gamma doses are shown in figure 2. There are two most prominent TL peaks around 230 and 430 °C, called peaks II and IV, respectively. This was previous reported by Souza et al. [11]. The peak at 230 °C is more highlighted with respect to the dose increment, unlike of the peak at 430 °C which looks like to disappear with higher doses. The peaks do not change their temperature position as dose increases that lead one to conclude they are of first kinetic order.

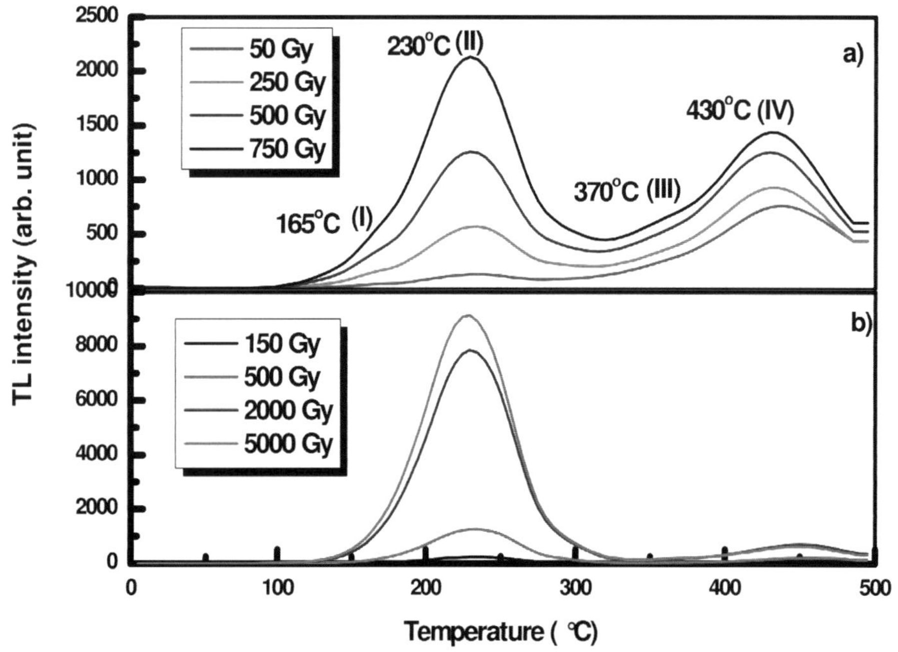

Figure 2. Glow curves of the not annealed sample irradiated in different gamma doses range, from 50 to 750 Gy (a) and from 150 to 5000 Gy (b).

As the main objective was to characterize the spodumene for applications in dosimetry, the focus of this work was given for samples annealed at 600 °C for 1 h and subjected to varied gamma radiation doses. In Figure 3, in contrast to the not annealed, the annealed sample do not present the higher temperature peaks up to 50 Gy, but only one emission around 215 °C, which turns very prominent with increasing radiation dose. However, it is approaching to saturation at about 500 Gy. Unlike, the higher temperature peaks from 500 Gy become easily seen, being the most intense one around 350 °C. However, it is clear that there is more than one peak for temperatures above 300 °C. There must be at least 3 of them at 350, 370 and 460 °C. Moreover, one can perceive the presence of a "shoulder" on the peak of 215 °C, around 145 °C. As the latter must have a very low stability at RT, it decays rapidly after the irradiation and before the TL reading, and thus it is very difficult to follow its actual behavior. It must be done in lower temperatures. Here again one can conclude that most prominent peaks are of first kinetic order, because they do not change their temperature position as dose increases.

Comparing figure 2 and 3, it is noticeable that the more prominent peak in both glow curves have different characteristics. Besides having been moved to a slightly lower temperature (from 230 to 215 °C), it is more intense in the annealed sample. It is also quite clear that the annealing favored the appearance of peaks at higher temperature with higher doses, suggesting that heating causes an increase in the number of deep traps, reducing the number of shallower traps. Another important observation is that there is no change in the position of the peak with the radiation dose increment, which is an indication that it is of first kinetics order and favors its use in dosimetry. This change in the shape of the curve, especially with the appearance of the peaks at higher temperatures, can be used to the identification of the thermal history of the sample, which is important in geology, for example. After all, if the crystal suffered a annealing at 600 °C it will show only the 215 - 230 °C peak with to doses up to 150 Gy, while the not annealed sample will display principally the higher peaks at 370 and 430 °C.

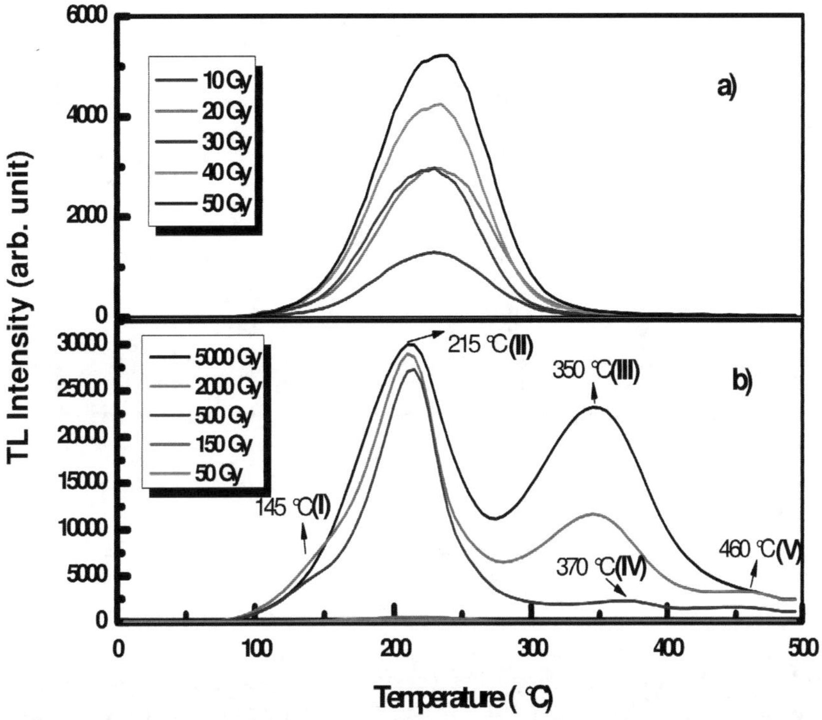

Figure 3. TL glow curves of the annealed samples at 600 °C during 1 h with additional gamma irradiated dose: (a) low dose and (b) high doses. Heating rate is 4 °C/s.

To assess more precisely the behavior of the TL intensity of each peak observed it was plotted their TL intensity versus gamma dose. For not annealed sample, figure 4, it appears that the all the TL peaks have a similar behavior, with an initially sublinear growth, followed by supralinearly region up to about 1500 Gy, entering in saturation around 2 kGy. For annealed sample, figure 5, their behavior is slightly different. Initially the present a sublinear growth, followed by supralinearly region up to about 500 Gy, entering in saturation at about 1 kGy.

An explanation of the supralinearity of the TL response is given based on a model with two traps and a recombination center. Empty traps are filled, initially, linearly with radiation dose. One of the filled traps results in the TL peak in the studied temperature region while the other may be a thermally disconnected deep trap, i.e., a TL peak resulting in an upper temperature region. Both traps compete to capture electrons from the conduction band during irradiation. In this context, the supralinearity occurs when the deep trap saturates before the trap shallower, thus leaving to compete for the capture of electrons. In this case, the shallower trap is then filled with a higher rate. If the resulting rate during the transition between the two regimes is more than linear, it is then called supralinear. In spodumene, there is the possibility that traps corresponding to peaks at higher temperature than the here studied are responsible for supralinear behavior of the TL peaks between RT and 500 °C. This behavior does not preclude the use of these peaks in thermoluminescent dosimetry, only complicates their use due to the fact that the growth curve must be well established for evaluating the dose received by a spodumene dosimeter in future.

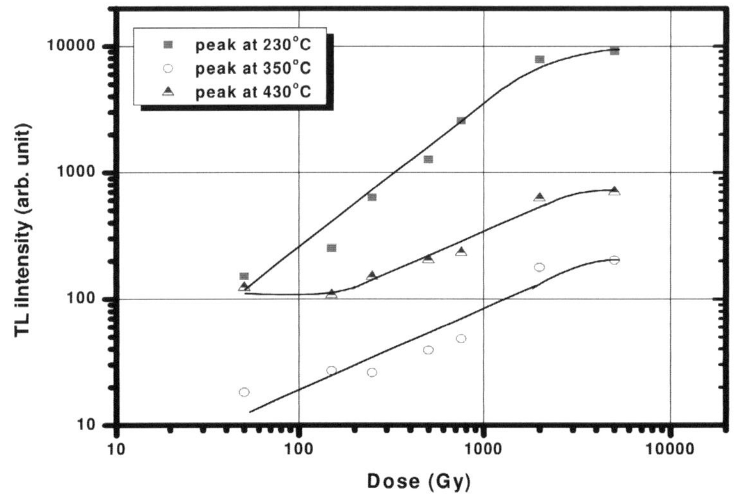

Figure 4. Behavior of the three main TL peaks as function of the gamma dose increment from the non annealed sample

It was calculated the frequency factor, activation energy and the lifetime to the most prominent glow peaks of pre-annealed sample using curve fitting and peak shape methods, being the results shown on the table 2. The different methods applied produced small differences in the parameters. As can be noted, the lifetime of these TL glow peak is slight sensitive to the calculated method, however they are in the same order of magnitude, being the peak 215 °C of about 60 days and the peak 350 °C of about a hundred days.

Table 2. Kinetic parameters resulting from the analyses of the TL main peaks of lilac spodumene. The peaks were considered of 1st order kinetics. (Methods: PS - Peak Shape and CF - Curve Fitting)

Peak temperature	Activation energy (eV)	Frequency factor (s^{-1})	Method of analysis	Lifetime at 15 °C
215 ± 1	0,87± 0,04	1,6 ± 0,4)	PS	67,0 ± 0,6 d
215 ± 1	0,82 ± 0,01	(0,4 ± 0,5)	CF	58,0 ± 0,4 d
350 ± 1	0,91 ± 0,10	(0,08 ± 0,60)	PS	106,0 ± 1,2 y
350 ± 1	0,971 ± 0,005	(0,24± 0,30)	CF	330,0 ± 0,8 y

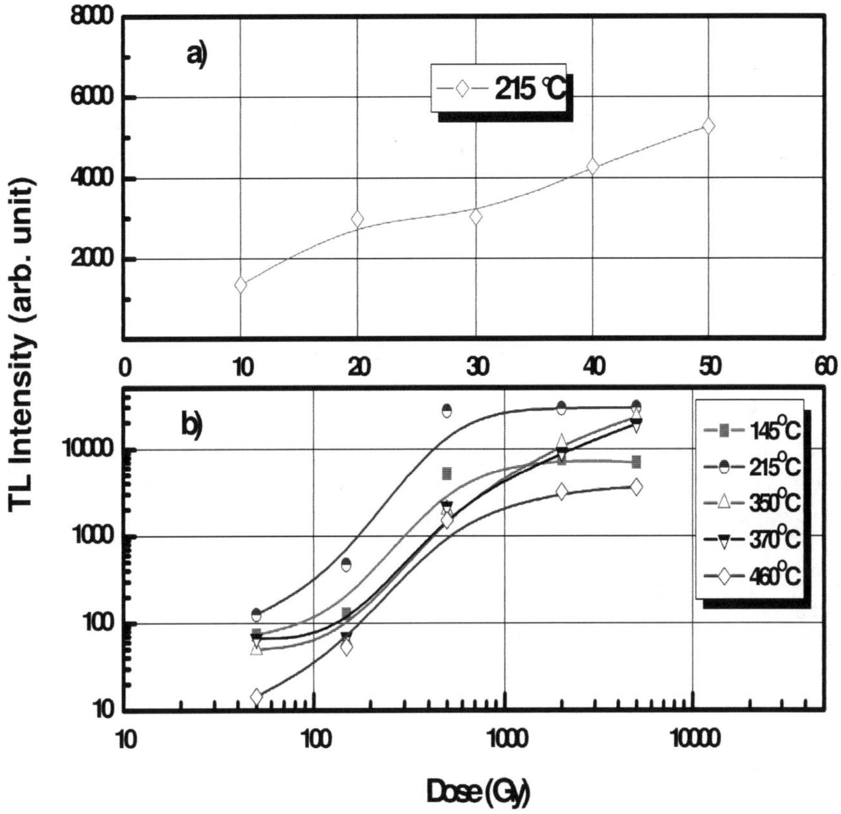

Figure 5. Behavior of the TL peaks of the annealed sample at 600 °C during 1 h as function of the increasing of the gamma dose. a) only the principal TL peak (215 °C) with doses between 10 and 50 Gy b) all identified peaks with doses of 50 to 5000 Gy.

Although the simulation has resulted in a slightly better fit for general order kinetics to 215 °C peak, the difference in settings was not very significant and the order of kinetics was found b = 1.22 ± 0.05, which is also very close first-order. Therefore, the results seem to indicate that the peak is that of the first order, confirming the preliminary assessment. This approach was also found to the 350 °C peak, confirming a good approximation to the possibility of first-order kinetics.

5. Conclusion

The present work was focused on studies of the dosimetric properties of the lilac α-spodumene. The behavior of the TL intensity of the peaks was analyzed as a function of increasing dose, and the main peaks did not show linear behavior, which is desired in a good dosimeter. However, the steps in doses increments were high, and there is the possibility to have a linear behavior in lower dose ranges, mostly between 100 and 1000 Gy, which should be investigated.

The lifetime was calculated using the temperature of 15 °C, as is usually done in literature. The values found were about 2 months to peak at 215 °C, which is relatively short, but indicates that the spodumene may be used for retrospective dosimetry in cases of radioactive accident if the elapsed time of analysis is less than 2 months. The lifetime of the peak at 350 °C is about a few hundred years, which is short for geological dating, but could be used for dosimetry. However, this peak is not well defined as at 215 °C which makes it difficult to use. It is necessary to apply more methods and techniques to confirm their possible applications.

Acknowledgements

The authors gratefully acknowledge the CNPq, FAPITEC and Posgrap–UFS for partial financial support. We are also thankful to IPEN for samples irradiation and to DMP-IG/USP for the X-ray diffractogram.

References

[1] Arlt T and Angel R J 2000 *Phys. Chem. Miner.* **27** 719
[2] Claffy E W 1953 *Am. Mineral* **38** 919
[3] Fujii A and Isotani S 1988 *An. Acad. Cie.* **60** 127
[4] Holuj F 1968 *Canad. J. Phys.* **46** 287
[5] Walker G, Eljaer A, Sherloc R, Glynn T J, Czaja M and Mazurac Z 1997 *J. Lumin.* **278** 72
[6] Sanchez-Muñoz L, Correcher V, Garcia-Guinea J and Delgado A 1999 *Rad. Prot. Dos.* **84** 543
[7] Gómez-ros J M, Correcher V, Garcia-Guinea J and Delgado A 2002 *Rad. Prot. Dos.* **100** 399
[8] Chen R and McKeever SWS *Theory of thermoluminescence and related phenomena* (Singapore:World Scientific)
[9] Ito A S and Isotani S 1991 *Rad. Eff. def. Solids* **116** 307
[10] McKeever S W S 1985 *Thermoluminescence of solids* (Cambridge: Cambridge University Press)
[11] Souza S O, Watanabe S, Lima A F and Lalic M V 2007 *Acta Phys. Polonica* A **112** 1007

Concentration Dependence of Afterglow Suppression in CsI:Tl,Sm

L. A. Kappers[a], R. H. Bartram[a], D. S. Hamilton[a], A. Lempicki[b], C. Brecher[b], V. Gaysinskiy[c], E. E. Ovechkina[c] and V. V. Nagarkar[c]

[a]Department of Physics, University of Connecticut, Storrs, CT 06269-3046
[b]ALEM Associates, 303A Commonwealth Avenue, Boston, MA 02115, USA
[c]Radiation Monitoring Devices (RMD) Inc., 44 Hunt St., Watertown, MA 02472, USA

lawrence.kappers@uconn.edu

Abstract. Combined radioluminescence, afterglow and thermoluminescence experiments on single-crystal samples of co-doped CsI:Tl,Sm suggest that deeper samarium electron traps scavenge electrons from shallower thallium traps and that electrons subsequently released by samarium recombine non-radiatively with holes trapped as $V_{KA}(Tl^+)$ centers, thus providing a mechanism for suppression of trapped-charge accumulation in repetitive applications. In the present investigation, experiments performed on two single-crystal samples of CsI:Tl,Sm with nominal concentrations of 0.11% Tl^+ and of 0.2% and 0.05% Sm^{2+}, respectively, support the inference that electrons tunnel freely between samarium ions and are trapped preferentially near $V_{KA}(Tl^+)$ centers where non-radiative recombination is the rate-limiting step.

1. Introduction

The feasibility of substantially diminishing afterglow in CsI:Tl scintillator material by co-doping with Sm^{2+} has been demonstrated. [1] Rate equations informed by experiment predict that deeper samarium electron traps, involving diffuse 6*s* orbitals of Sm^+, scavenge electrons from shallower thallium traps, and combined radioluminescence, afterglow and thermoluminescence experiments on single-crystal samples of CsI:Tl,Sm suggest that most electrons released by Sm^+ recombine non-radiatively with holes trapped as $V_{KA}(Tl^+)$ centers, thus providing a mechanism for suppression of trapped-charge accumulation in repetitive applications. [2] A linear-coupling model in the harmonic approximation, based on quantum-chemistry calculations with selective lattice relaxation, suggests further that non-radiative charge transfer is enabled by low-energy excited states of Sm^{2+} within the ground configuration and is mediated by spin-orbit interaction, as illustrated in Figure 1. Although recombination is primarily non-radiative, tracking of the process with increased sensitivity of the optical detection system is enabled both by radiative recombination following thermal ionization and by a small radiative component of the charge-transfer recombination associated with magnetic-dipole transitions from low-energy excited states of Sm^{2+}. In the present investigation, combined scintillation and afterglow experiments were performed on two single-crystal samples of CsI:Tl,Sm with nominal concentrations of 0.11% Tl^+ and of 0.2% and 0.05% Sm^{2+}, respectively, in order to determine the concentration dependence of the rate of non-radiative charge transfer and thus to construct a more nuanced model of the non-radiative process.

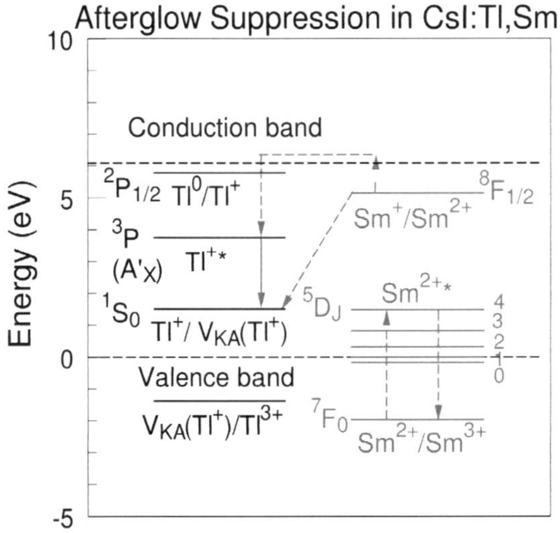

Figure 1. Energy-level diagram illustrating the mechanism for afterglow suppression in CsI:Tl,Sm by non-radiative charge-transfer transitions.

2. Experiment

An electron Van de Graaff accelerator operated at a beam voltage of 1.0 MeV was employed as the primary radiation source with the 1.0 μA electron beam stopped by a thin copper target that served as a point source of 0.5 MeV gamma rays. The sample was mounted on a heated pedestal and cooled by flowing nitrogen gas, the sample temperature was monitored by a thermocouple and luminescence was conducted to a photomultiplier by a shielded optical fiber. Immediately following four-minute irradiations at each of several temperatures, the gain was increased by three orders of magnitude to monitor the decay of the afterglow at constant temperature. Light output data were recorded at four-second intervals. The afterglow, normalized to the scintillation light output, is plotted as a function of time after irradiation in Figures 2 and 3 for the samples with 0.2% and 0.05% Sm^{2+} concentration, respectively.

3. Rate Equations

Complete rate equations for each phase of the process were presented in [2]. Rate equations for the irradiation phase in terms of the normalized trapped-hole concentration $\tilde{n}_h$ and normalized light output $\tilde{I}$ are

$$\frac{d\tilde{n}_h}{dt} \cong \frac{\tilde{f}_r(1-\tilde{n}_h) - \tilde{n}_h^2 p_e^{(Sm)}}{1+\tilde{n}_h} - \alpha\tilde{n}_h^2 , \tag{1}$$

$$\tilde{I} \cong \tilde{f}_x + \tilde{f}_r - \left(\frac{d\tilde{n}_h}{dt} + (1-r)\alpha\tilde{n}_h^2 \right), \tag{2}$$

$$\alpha \cong s_A \exp(-E_A / k_B T), \tag{3}$$

$$p_e^{(Sm)} \cong s^{(Sm)} \exp(-E^{(Sm)} / k_B T), \tag{4}$$

where $\tilde{f}_x$ and $\tilde{f}_r$ are, respectively, the normalized rates of production of excitons and of electron-hole pairs during irradiation. The parameters α and r respectively determine the rate of charge-transfer recombination and its radiative fraction, and the parameter $p_e^{(Sm)}$ determines the rate of thermal ionization. It is assumed that Tl^+ is too shallow to be an effective electron trap in the temperature range of interest. With the added assumption that Sm^{2+} does not trap electrons in the afterglow phase, thus reducing the kinetics to first order, rate equations for the afterglow phase are

$$\frac{d\tilde{n}_h}{dt} \cong -p_e^{(Sm)}\tilde{n}_h - \alpha\tilde{n}_h^2, \tag{5}$$

$$\tilde{I} \cong -\frac{d\tilde{n}_h}{dt} - (1-r)\alpha\tilde{n}_h^2 = r\alpha\tilde{n}_h^2 + p_e^{(Sm)}\tilde{n}_h, \tag{6}$$

with explicit solution as a function of afterglow time t:

$$\tilde{n}_h \cong \frac{\tilde{n}_{ho}\exp[-p_e^{(Sm)}t]}{1+[\alpha\tilde{n}_{h0}/p_e^{(Sm)}]\{1-\exp[-p_e^{(Sm)}t]\}}. \tag{7}$$

Subscript 0 refers to initial values immediately following irradiation.

These equations were fitted to afterglow data as shown in Figures 2 and 3. Values $\tilde{f}_r = 0.5\,\text{min}^{-1}$ and $\tilde{f}_x = 1.2\,\text{min}^{-1}$ are assumed as in [2] and an additive constant was included to accommodate PM dark current. The parameters α and $p_e^{(Sm)}$ are plotted as functions of temperature in Figure 4 and fitted to Equations (3) and (4) with the optimized parameters listed in Table 1. Optimized values of r are typically ~1%.

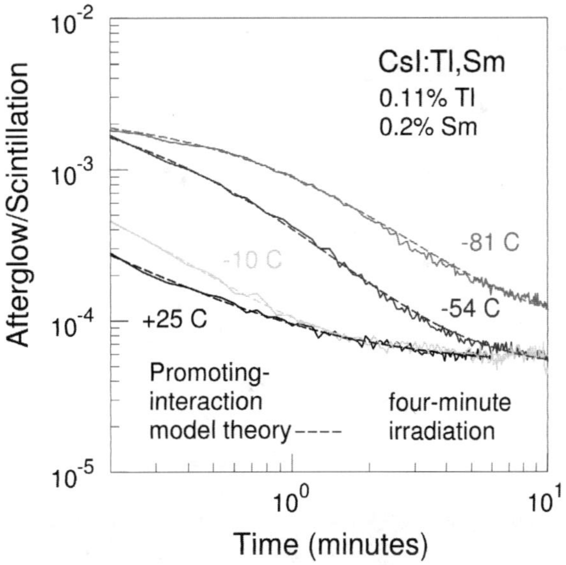

Figure 2. Recorded ratio of afterglow and scintillation for 0.2% Sm^{2+} as a function of time following four-minute irradiations at four temperatures (continuous curves), compared with fitted theory from Equations (1)-(7) (dashed curves).

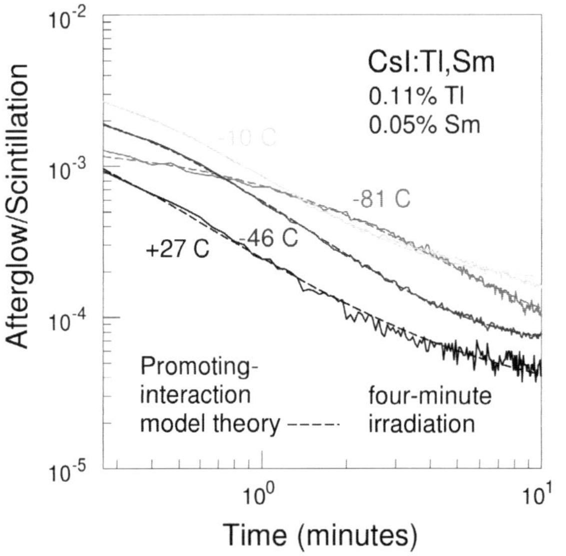

Figure 3. Recorded ratio of afterglow and scintillation for 0.05% Sm^{2+} as a function of time following four-minute irradiations at four temperatures (continuous curves), compared with fitted theory from Equations (1)-(7) (dashed curves).

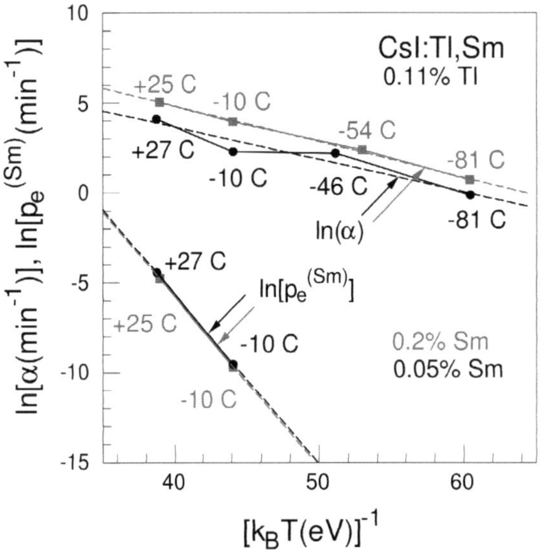

Figure 4. Comparison of ln(α) and $\ln(p_e^{(Sm)})$ as functions of $1/k_BT$ (continuous curves and solid circles) with Equations (3) and (4) (dashed curves) and the optimized parameters listed in Table 1. Curves in red are for 0.2% Sm^{2+} and curves in black are for 0.05% Sm^{2+}.

Table 1. Optimized parameters for thermally activated processes

%Sm^{2+}	E_A (eV)	$\ln(s_A)$	s_A (min^{-1})	$E^{(Sm)}$ (eV)	$\ln(s^{(Sm)})$	$s^{(Sm)}$ (min^{-1})
0.2%	0.199	12.81	3.66×10^5	0.946	32.069	8.46×10^{13}
0.05%	0.180	10.87	5.26×10^4	0.925	32.072	8.48×10^{13}

4. Promoting-Interaction Model

In the theoretical model of [2], the promoting interaction is proportional to the square of the matrix element of the spin-orbit interaction between initial and final electronic states. Among other factors, it is approximately proportional to the square of the value of the diffuse 6*s* orbital of Sm^+ at the site of the target $V_{KA}(Tl^+)$ center. In application of the theoretical model, we have assumed a fixed value of the promoting interaction for each value of the samarium concentration, implying a fixed separation of the initial and final sites. The model can be elaborated further by adopting a continuum approximation for CsI in which a substitutional Sm^{2+} ion is represented by a point charge +*e* embedded in a medium of dielectric constant $\kappa = 6.31$ [3] and effective mass ratio $m^*/m = 1.0$. In this approximation, the diffuse 6*s* orbital centered on the samarium ion at $\vec{r}_0$ is represented by the pseudowavefunction

$$\phi_{6s}(\Delta r) = \frac{\exp(-\Delta r / \kappa a_0)}{\pi^{1/2}(\kappa a_0)^{3/2}}, \quad \Delta r \equiv |\vec{r} - \vec{r}_0| . \tag{8}$$

We proceed by assuming that the ratio of separations of initial and final sites in the two crystal samples of CsI:Tl,Sm satisfies

$$\frac{\Delta r_2}{\Delta r_1} = \frac{\rho_1^{1/3}}{\rho_2^{1/3}} = 1.25, \tag{9}$$

where $\rho_1 = 0.0031$ and $\rho_2 = 0.0016$ are the relative concentrations of dopants, including both Tl^+ and Sm^{2+}. The ratio of promoting interactions is given by

$$\frac{s_{A1}}{s_{A2}} = \left[\frac{\phi_{6s}(\Delta r_1)}{\phi_{6s}(\Delta r_2)}\right]^2 = \exp[2(\Delta r_2 - \Delta r_1)/\kappa a_0]. \tag{10}$$

It follows from Equations (9) and (10) and Table 1 that $\Delta r_1 / a = 2.88$ and $\Delta r_2 / a = 3.59$, where $a = 4.56$ Å is the lattice parameter.

5. Tunneling Model

An earlier tunneling model proposed by Delbecq *et al.* [4] was based on somewhat different assumptions: A random distribution of trapped charges is accumulated during the radiation phase, and tunneling progresses in the order of increasing separation of initial and final ion sites. These authors derive a simple expression for the light output *I*(*t*) in terms of irradiation time t_0 and afterglow time *t*:

$$I(t) = \frac{C}{t_0}\ln\left(1 + \frac{t_0}{t}\right), \tag{11}$$

where *C* is an adjustable parameter. Unfortunately, Equation (11) provides an egregiously poor fit to the afterglow data for CsI:Tl,Sm, as evidenced by comparison of Figures 2 and 5; the best that can be said for it is that *I*(*t*) decreases monotonically. The first assumption of the tunneling model is transparently inappropriate for prolonged, intense radiation that culminates in a dynamic equilibrium between trapping and radiative recombination of electrons and holes. The second assumption of the tunneling model differs fundamentally from the premise of the promoting-interaction model, since the rate of tunneling is assumed to depend sensitively on the separation of initial and final sites.

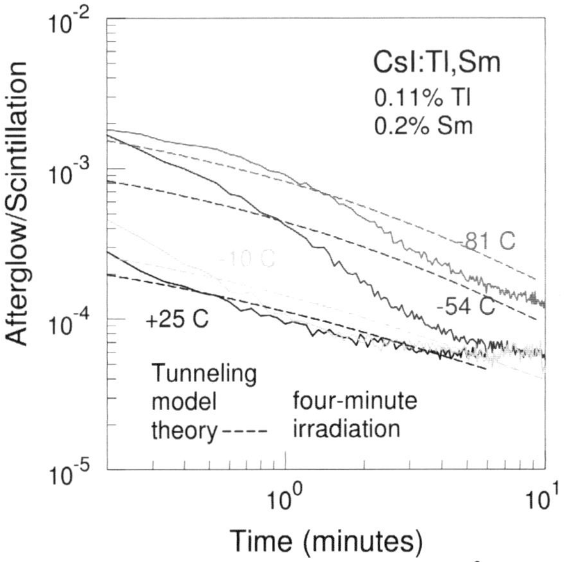

Figure 5. Recorded ratio of afterglow and scintillation for 0.2% Sm^{2+} as a function of time following four-minute irradiations at four temperatures (continuous curves), compared with fitted theory from Equation (11) (dashed curves).

6. Discussion

The calculated values for the separation of initial and final sites are plausible for the nearest dopant sites but not for a random distribution of trapped charges substantially less than the dopant concentration. However, the overlap of 6*s* orbitals on nearby samarium sites is substantial and the energy barrier for tunneling between them should be relatively small. Accordingly, the promoting-interaction model can be extended to incorporate the assumption that electrons tunnel freely between samarium ions and are trapped preferentially near $V_{KA}(Tl^{+})$ centers where non-radiative recombination is the rate-limiting step, thus explaining the single non-radiative transition rate for fixed temperature and dopant concentration.

7. Acknowledgement

We thank NIH and DOE for support of this work under Grants No. R43 RR022921-01 and DE FG02 06ER84434, respectively.

References

[1] V. V. Nagarkar, C. Brecher, E. E. Ovechkina, V. Gaysinskiy, S. R. Miller, S. Thacker, A. Lempicki and R. H. Bartram, "Scintillation properties of CsI:Tl crystals codoped with Sm^{2+}", *IEEE Transactions on Nuclear Science* **55**, 1270-1274 (2008)

[2] R. H. Bartram, L. A. Kappers, D. S. Hamilton, A. Lempicki, C. Brecher, V. Gaysinskiy, E. E. Ovechkina and V. V. Nagarkar, "Afterglow suppression and non-radiative charge-transfer in CsI:Tl,Sm", *IEEE Transactions on Nuclear Science* **55**, 1232-1236 (2008)

[3] *CRC Handbook of Chemistry and Physics, 77th Edition*, Ed. David R. Lide (CRC Press, Boca Raton, 1996), p.**12**-47

[4] C. J. Delbecq, Y. Toyozawa and P. H. Yuster, "Tunneling recombination of trapped electrons and holes in KCl:AgCl and KCl:TlCl", *Phys. Rev. B* **9**, 4497 (1974)

High pressure spectroscopy study of SCF $Tb_3Al_5O_{12}$:Mn

K Wiśniewski[1], Yu. Zorenko, V. Gorbenko, T. Zorenko[2], B. Kukliński[1], M. Grinberg[1*]

[1]Institute of Experimental Physics Gdansk University, 80-952 Gdansk, Poland
[2]Laboratory of Optoelectronic materials, Electronic Department of Ivan Franko National University of Lviv, 79017 Lviv, Ukraine

fizmgr@univ.gda.pl

Abstract. Single crystal film of $Tb_3Al_5O_{12}$:Mn, with concentration of Mn approximately equal to 0.07 at%, has been grown by liquid phase epitaxy on $Y_3Al_5O_{12}$ substrate. We have measured the absorption, luminescence, luminescence excitation spectra and luminescence kinetics of the film. To analyze the energetic structure and to estimate the crystal field strength we have measured the photoluminescence spectra under high hydrostatic pressure applied in diamond anvil cell up to 110 kbar. The Mn emission was consisted with the band peaked at 16800 cm^{-1} that shifts toward the lower energy with increasing pressure with the rate -9 cm^{-1}/kbar. To characterize the Mn charge state we performed calculations of Tanabe –Sugano diagrams for $3d^5$, $3d^4$ and $3d^3$ system. The negative pressure shift of the 16800 cm^{-1} band yields attribution of this band to the $^4T_1 \rightarrow {}^6A_1$ transition in dodecahedrally coordinated Mn^{2+} ion. We have estimated quantity of *Dq/B* equal to 1.8 ±0.2. The configurational coordinate diagram of the $Mn^{2+}(3d^5)$ has been calculated to analyze the nonradiative deexcitation processes $^4T_1 \rightarrow {}^2T_2 \rightarrow {}^6A_1$.

1. Introduction

The $Tb_3Al_5O_{12}$ (TbAG) garnet is considered as a promising host for luminescent materials, [1]. The advantage of TbAG results form effective energy transfer from Tb^{3+} cation sublattice to the luminescence centers of impurity ions. Up to present the TbAG doped Ce^{3+} ions [1-4] and simultaneously Ce3+ and Eu^{3+} ions [3, 4] were analyzed. In this work we focus on spectral investigations of TbAG:Mn single crystal film (SCF) deposited on deposited on $Y_3Al_5O_{12}$ (YAG) monocrystal.

Manganese is a well-known activator in garnet host used mainly for tunable solid-state laser media [5]. Since in Mn^{3+} bound 4 electrons at the unfilled 3d shell the activation energy for creation of Mn^{4+} ($3d^3$ system) and Mn^{2+} ($3d^5$ system) is small enough that in many materials the ground and the first excited states of the Mn^{3+}, Mn^{2+} and Mn^{4+} can be located in the bandgap of the host and can contribute to the luminescence. Actually the luminescence related to Mn ions in different charge states (2+, 3+ and 4+) in garnets was studied in several works [5-10]. Luminescence of Mn^{2+} consisting of broad band emission with maximum at 590 nm attributed to $^4T_1 \rightarrow {}^6A_1$ transition has been reported in $Y_3Sc_2Ga_3O_{12}$ and YAG, [7, 8, 10]. The Mn^{3+} intensive luminescence related to the $^5T_2 \rightarrow {}^5E$ and $^1T_2 \rightarrow {}^5E$ transitions in the range 650 nm and 700-900nm respectively have been observed in several garnets [6, 7, 10]. The relative intensity of $^5T_2 \rightarrow {}^5E$ and $^1T_2 \rightarrow {}^5E$ transitions strongly depends on the crystals field strength and temperature [6]. Luminescence of Mn^{4+} ions was investigated in details in YAG [7, 8] and in Ca-co-doped $Gd_3Ga_5O_{12}$ (GGG) crystals, [11]. The luminescence spectrum of the Mn^{4+} ion in octahedral coordination consists the sharp R-lines around 640-670 nm related $^2E \rightarrow {}^4A_2$ transition. It is considered that existence of Mn^{4+} in octahedral coordination in YAG is accompanied by Mn^{2+} ions occupying dodecahedrally coordinated Y^{3+} sites, [10].

The energetic structure of the transition metal ion system can be analysed in the strong crystal field approximation by means of Tanabe –Sugano diagrams [12, 13]. According to this model the energetic structure is determined by Racah parameters *B* and *C* that describe Coulomb and exchange interaction of electrons in the d shell and crystal field parameter *Dq* that describes the splitting of d level in the cubic crystal field. In ionic crystals the quantity of *Dq* is proportional to the -5 power of central ion ligands distance, *R* , and therefore can be significantly different in different lattices. On the other hand quantities *B* and *C* are diminished with respect to the free ion quantities by nephelauxetic effect related to mixing of 3d wave functions with valence ones and are less depend on the lattice host.

One of the most effective tool for characterisation of transition metal ion in crystal lattice is high pressure spectroscopy. High pressure compresses materials and causes the increase of crystal field strength *Dq*. As the result one observes the shift of the emission bands related to the transitions between the states belonging to different electronic configurations. High pressure spectroscopy has been successfully used to analyze materials doped with Cr^{3+}, Cr^{4+}, Ti^{3+} and Mn^{4+}. The broad reviews of this topic can be finding in the papers, [14, 15]. The main purpose of presented contribution is to characterise of the Mn ions that incorporated into TbAG SCF. We have applied the high-pressure luminescence spectroscopy for determination of the valence state of manganese ions. Analysis of luminescence lineshape and luminescence kinetics obtained at high hydrostatic pressure confirmed existence of Mn^{2+} in the investigated host.

2. Samples preparation and experimental technique

The TbAG:Mn films with thickness in the 19-64 μm range was crystallised by LPE onto YAG substrates from melt-solutions of Tb2O3 and Al2O3 in PbO–B2O3(12:1) . The growth temperature was varied between 970 and 1000 0C. Concentration of Mn ions varied from 0.019 to 0.07 at%. Details of the sample preparation are described in the paper [16].

Steady state luminescence has been excited by Ar laser with 457 nm and He-Cd laser with 325nm. The emission has been dispersed by PGS 2 spectrometer working as a monochromator and collected by an R943-02 Hamamatsu photomultiplier working in the photon counting regime.

The luminescence kinetics was measured at room temperature using a Hamamatsu C4334-01 model Streak Camera. The luminescence was spectrally separated with a Bruker Optics 2501S model monochromator. The optical parametric generator PG401/SH pumped by the tripled output of a Nd:YAG pulse laser has been used for excitation of luminescence.

High hydrostatic pressure has been applied in a diamond anvil cell (DAC) of the Merrill–Bassett system, [17]. Pressure appears when diamonds are moved and is transmitted to the fluid medium (dimethylsiloxane oil). A ruby crystal is used as the pressure detector.

3. Results and discussion

3.1. Luminescence spectra

Luminescence spectra of the TbAG:Mn (0.07 at. %) SCF on YAG obtained at different temperatures under excitation 457 nm are presented in Fig.1 a. The spectra consist of two broad bands peaked at 18700 cm^{-1} and 16800 cm^{-1} which have been attributed to d-f transition of Ce^{3+} in YAG [18, 19] and to Mn emission, respectively. The sharp lines around 14550cm^{-1} correspond to the $^2E \rightarrow {}^4A_2$ transition accompanied by phonons side-band in octahedrally Cr^{3+} ions in YAG [20, 21].

Luminescence spectra obtained at different temperatures under excitation at 325 nm are presented in Fig.1 b. One notices that the spectrum lineshape depends strongly on temperature. At 10K one observes the sharp lines at 18420 cm^{-1}, 17080 cm^{-1} and 16500 cm^{-1} related respectively to $^5D_4 \rightarrow {}^7F_6$, $^5D_4 \rightarrow {}^7F_5$ and $^5D_4 \rightarrow {}^7F_4$ transitions in Tb^{3+} ions and broad band with maximum at 16800 cm^{-1} attributed to the Mn ion emission. The Tb^{3+} emission is strongly quenched when temperature increases. Considering Mn luminescence by means of electron –lattice coupling one notices that the band consists with zero-phonon line (band) and one phonon repetition with phonon energy approximately equal to 600 cm $^{-1}$ and Huang –Rhys factor smaller than unity. The similar emission has been related to the $^4T_1 \rightarrow {}^6A_1$ transition of the Mn^{2+} ion, [6, 10].

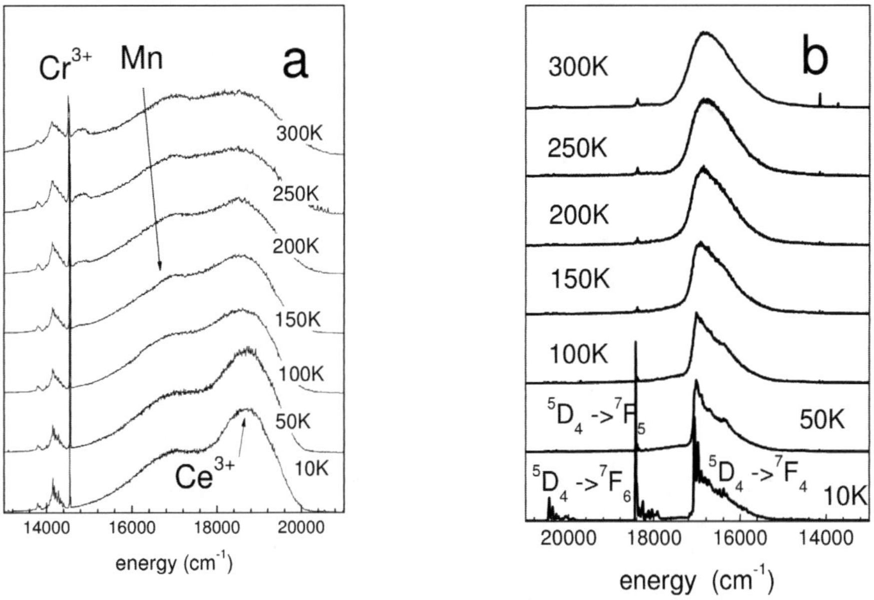

Figure 1. Steady state lluminescence spectra of TbAG:Mn (0.07 at. %) SCF obtained at different temperatures under excitation at 457 nm (a) and 325 nm (b)

3.2. High-pressure luminescence and luminescence kinetics

Ambient temperature luminescence excited with 325 nm, obtained at different pressures are presented in Fig. 2 a. In Fig. 2 b the energies of the emission peaks versus pressure are presented. One notices that increasing pressure causes the significant red shift of the emission. The shift rate of the maximum of the emission is constant and is equal to -9 cm^{-1}/kbar.

The luminescence decays obtained at various pressures are presented in Fig. 3 a. The emission decays almost exponentially for all pressures and emission lifetime decreases with pressure from 0.62 ms to 0.01 ms, for ambient pressure and 125 kbar, respectively. The luminescence decay times measured as the ratio $\tau = \dfrac{\int I(t)tdt}{\int I(t)dt}$ versus pressure are presented in Fig. 3 b. The decreasing of the luminescence lifetime can be attributed to pressure induced increase of the nonradiative internal-conversion process.

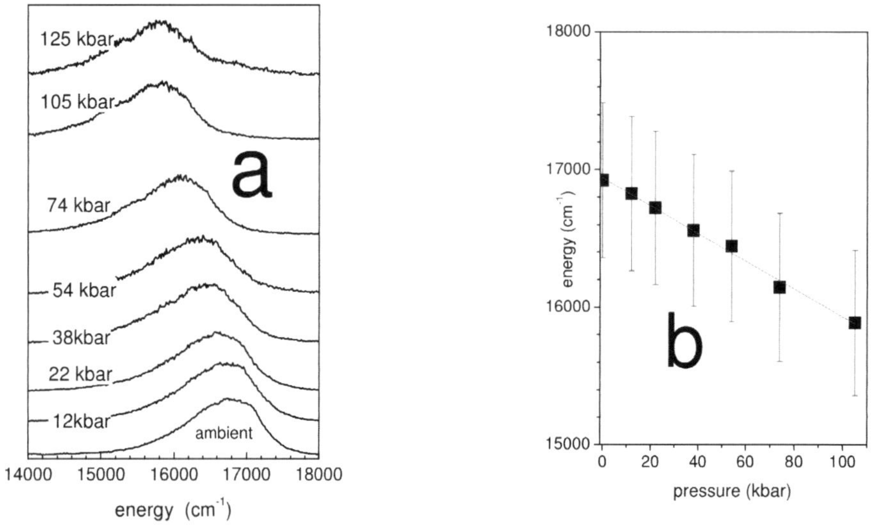

Figure 2. Steady state lluminescence spectra of TbAG:Mn (0.07 at. %) SCF obtained at different pressures at ambient temperature under excitation at 325 nm (a). Energy of the emission spectra versus pressure (b)

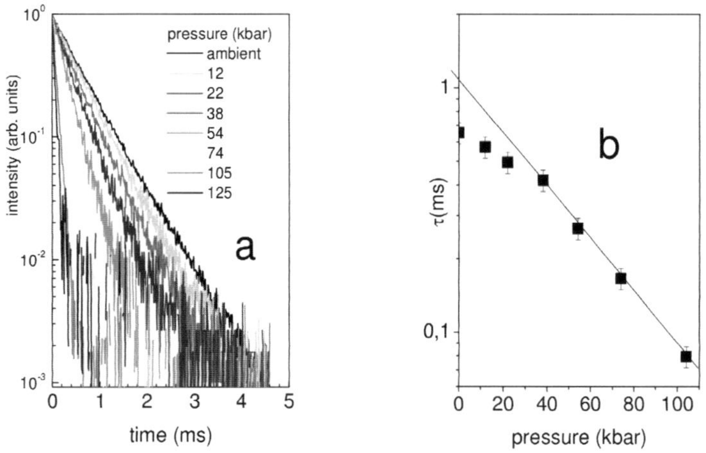

Figure 3. (a) Luminescence decay for ambient temperature obtained at different pressures. Luminescence has been excited with 325 nm laser pulse anfd monitored at 16800 cm^{-1}. (b) Luminescence decay times versus pressure.

3.3. Energetic structure of the Mn2+ , Mn3+ and Mn4+ in garnet lattice

We have considered the possibility of existence of Mn ions in the Mn^{2+}, Mn^{3+} and Mn^{4+} charge states and therefore have analyzed our experimental data by means of respective Tanabe –Sugano diagrams. The Tanabe –Sugano diagrams presenting the lowest states of octahedrally coordinated Mn^{3+} ($3d^4$) system, octahedrally (positive *Dq/B*) and tetrahedrally (negative *Dq/B*) coordinated Mn^{4+} ($3d^3$) system, and dodecahedrally coordinated Mn^{2+} ($3d^5$) system are presented in Fig. 4 a, b and c, respectively. The expected radiative transitions in all cases are represented by arrows. One notices that in the reasonable range of the quantity $1.8 < Dq/B < 2.8$ the ground state of Mn^{3+} is the ^5E state belonging to t_2^3e electronic configuration, whereas the luminescent states are $^5T_2$ from $t_2^2e^2$ electronic configuration and $^1T_2$ from the t_2^4 electronic configuration. It is expected that similarly like in YAG:Mn^{3+} the ground state is splitted by strong static Jahn –Teller effect of $E\otimes\epsilon$ type with stabilization energy $E_{st} = \frac{L_\epsilon^{\ 2}}{6\hbar\omega} = 1850\,\text{cm}^{-1}$ [10]. According to the diagram in Fig. 4 a only the spin forbidden $^1T_2 \rightarrow {^5E}$ transition can be characterized by negative pressure shift of the luminescence

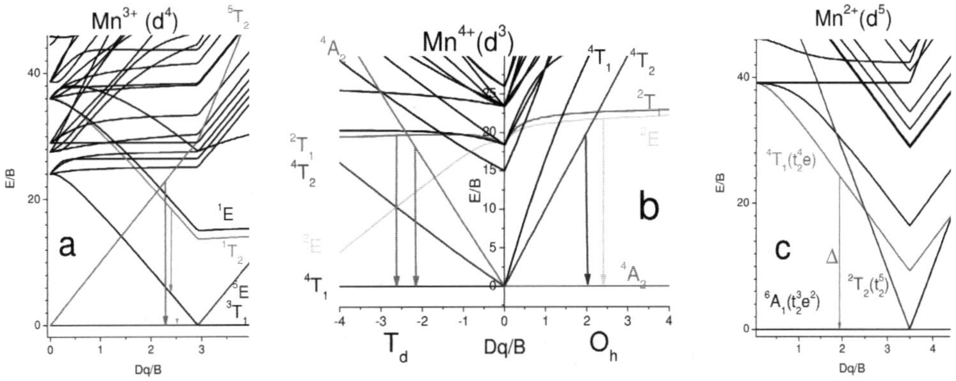

Figure 4. (a) Tanabe –Sugano diagram of Mn^{3+} ($3d^4$ system) in octahedral coordination. (b) Tanabe –Sugano diagram of Mn^{4+} ($3d^3$ system) in octahedral (positive *Dq/B*) and tetrahedral (negative *Dq/B*) coordination. (c) Tanabe –Sugano diagram of Mn^{2+} ($3d^5$ system) in dodecahedral coordination.

and long luminescence lifetime. The spin allowed $^5T_2 \rightarrow {^5E}$ transition yields luminescence that should shift toward the blue region with increasing pressure. If one would attribute the luminescence presented in Fig. 2 a to the Mn^{3+} to $^1T_2 \rightarrow {^5E}$ transition the crystal field parameter *Dq/B* should be greater than 2.5. In this case one expects the emission energy at ambient pressure to be smaller than 14000 cm^{-1}, whereas we have obtained this energy 16800 cm^{-1}. This discrepancy excludes the ascription of obtained emission to the $^1T_2 \rightarrow {^5E}$ transition in Mn^{3+}.

In the case of Mn^{4+} in octahedral coordination the ground state is the $^4A_2$ state of ground electronic configuration t_2^3, whereas the first excited state depends on the crystal field strength. For $Dq/B>2.4$ it is the ^2E state and for $Dq/B<2.4$ it is $^4T_2$ state of the first excited electronic configuration t_2^2e. One expects that the energy of the spin –allowed $^4T_2\rightarrow{^4A_2}$ emission would increase with increasing pressure, whereas the energy of the $^2E\rightarrow{^4A_2}$ spin forbidden transition is pressure independent. We have obtained the large negative pressure shift of the emission that is contradict to the assumption that we deal with Mn^{4+} in octahedral coordination. In the case of tetrahedrally coordinated Mn^{4+} one expects the spin allowed emission related to $^4A_2\rightarrow{^4T_1}$ and/or spin –forbidden emission related to $^2T_1\rightarrow{^4T_1}$ transitions. However it is very probable that this emission is quenched by non-radiative transitions to the $^4T_2$ and ^2E states.

In the case of Mn^{2+} presented in Fig. 4 c the ground state is $^6A_1$ form the $t_2^3e^2$ electronic configuration whereas the first excited state is $^4T_1$ form t_2^4e electronic configuration. The spin forbidden transition $^4T_1\rightarrow{^6A_1}$ can yield the luminescence with peak at 16800cm^{-1} that energy diminishes with increasing pressure.

The difference between energy of $^4T_1$ (t_2^4e) and $^6A_1$ ($t_3^3e^2$) states (labeled as Δ in Fig. 4c) can be used for calculation of Racah parameters and crystal field strength. For the case when $\frac{80}{9}Dq \gg B, C$ one can use following approximation:

$$\Delta \approx 10B + 6C - 10Dq - \frac{9B^2}{4(9B + C + \frac{80}{9}Dq)} - \frac{C^2}{\frac{160}{9}Dq} \quad (1)$$

The free Mn^{2+} ion Racah parameters are B= 860 cm^{-1} and C=3850 cm^{-1} [18]. With these values relation (1) yields Dq= 1600 cm^{-1}. Usually values B and C in crystals are smaller than the free ion ones and therefore quantity Dq is smaller than 1600 cm^{-1}. Considering the reasonable values B=600 cm^{-1} and C=3500 cm^{-1} one obtains Dq=1060 cm^{-1}. Diminishing of energy Δ with increasing pressure is caused mainly by increase of the value of Dq. Obtained quantity of pressure shift -9 cm^{-1}/ kbar is similar like in other transition metal ions, where one has observed the transitions between states from different electronic configurations [15].

The effect of electron - lattice interaction is responsible for existence of vibrational broadening of the emission band. In the case of Mn^{2+} one should consider coupling in the excited states $^4T_1$ and $^2T_2$ to the one dimensional breathing mode a_1, two dimensional mode "e" with coordinations Q_θ, Q_ε and three dimensional mode τ with coordinations $Q_\varsigma, Q_\xi, Q_\eta$. Hamiltonian of the electron –lattice interaction can be expressed as follows

$$H_{e-l} = \frac{1}{2}\hbar\omega\sum_\alpha Q_\alpha^2 + \sum_\alpha \frac{dV}{dQ_\alpha} Q_\alpha = \frac{1}{2}\hbar\omega\sum_\alpha Q_\alpha^2 + \sum_\alpha V_\alpha Q_\alpha \quad (2)$$

where V is the crystal field potential. As the result of electron –lattice interaction the system being in the "l" state reaches minimum energy after the shift of the ligands ions in the configurational space to the point defined by new minimum energy $(Q_1^l, Q_\theta^l, Q_\varepsilon^l Q_\varsigma^l, Q_\xi^l, Q_\eta^l)$.One can define the stabilization energy, E_{st}^l as the energy which is released in the relaxation process.

$$E_{st}^l = \frac{1}{2}\sum_\alpha (Q_\alpha^l)^2 \quad (3)$$

In the case of breathing mode stabilization energy is equal to $S\hbar\omega = \frac{V_1^2}{2\hbar\omega}$, where S is Huang –Rhys factor and $\hbar\omega$ is phonon energy, and in the case of Jahn –Teller effect stabilization energy is equal to $\frac{(V_e^l)^2}{6\hbar\omega}$ and $\frac{(V_\tau^l)^2}{9\hbar\omega}$ in the case of coupling with e and τ mode, respectively, [22]. In general case the stabilization energy is a more complicated function of parameters V_α^l. One can consider the pressure effect on lattice relaxation and stabilization energy; [15];

$$\frac{dE_{st}}{dp} = \frac{E_{st}}{3B_0}[2(n+1) - 6\gamma] \quad (4)$$

In relation (4) B_0 is a local compressibility (bulk modulus), γ is Grüneisen parameter and n exponent describing the dependence of crystal field on central ion–ligand distance. Parameter n is equal 5 for symmetrical a_1 mode (for Jahn -Teller mode n can be equal to 3). Typical values of Grüneisen parameter are between 1 and 2. Thus one notices that contribution related to increase the crystal field strength $2(n+1)$ and that related increase of material stiffness 6γ cancel each other. Therefore rather small changes in the electron-lattice coupling and Jahn-Teller effect with increasing pressure are expected. This prediction is confirmed by our experiments since pressure does not change the bandwidth of the spectrum (see Fig. 2 a).

The relatively long luminescence lifetime results form the fact that the $^4T_1 \rightarrow {}^6A_1$ transition is forbidden with respect to parity and spin. Decrease of the luminescence lifetime with pressure is accompanied by diminishing of the luminescence intensity. This correlation suggests that the increase of nonradiative internal conversion process is responsible for this effect. The luminescence lifetime in presence of nonradiative internal conversion is given by following relation

$$\tau = \left[\frac{1}{\tau_{rad}} + P_{nr} \cdot e^{-\frac{E_{nr}}{kT}} \right]^{-1} \tag{5}$$

where τ_{rad} is the radiative lifetime, P_{nr} is probability of the nonradiative internal conversion to the ground state $^6A_1$ and E_{nr} is activation energy of the nonradiative processes.

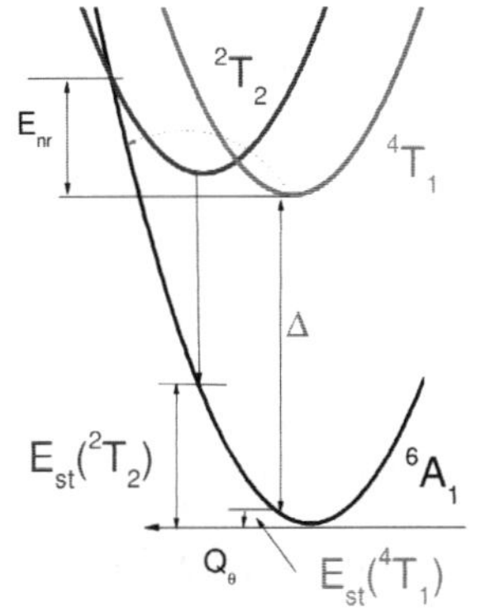

Figure 5. One dimensional configurational coordinate diagram representing the ground $^6A_1$ and the first excited $^4T_1$ and $^2T_2$ electronic manifolds of Mn^{2+}.

To analyze the nonradiative pathway one considers the diagram presented in Fig. 5, where the electronic manifolds representing the $^6A_1$ ($t_3^3e^2$), $^4T_1$ (t_2^4e) and $^2T_2$ (t_2^5) states are presented. Energy barrier for nonradiative processes is defined as a difference between energy of parabolas cross-over and minimum energy of the excited electronic manifold. Considering the $^6A_1$ ($t_3^3e^2$) and $^4T_1$ (t_2^4e) states one obtains quantity of energy barrier given by following relation:

$$E_{nr}^0 = \frac{\Delta^2}{4E_{st}} \tag{6}$$

One notices that stabilization energy in the $^4T_1$ (t_2^4e) state is small and therefore energy of crossing of the $^6A_1$ ($t_3^3e^2$) and $^4T_1$ (t_2^4e) parabolas obtained from (6) is very high (actually is not indicated in Fig. 5). On the other hand the electron-lattice interaction and shift in the configurational state is much larger in the $^2T_2(t_2^5)$ state since this quantity depends on difference in electronic configuration between the $^6A_1$ ($t_3^3e^2$) and $^2T_2(t_2^5)$. Thus one considers the nonradiative process that involves the transition of the system from $^4T_1$ (t_2^4e) to the $^2T_2$ (t_2^5) state and then to the ground state $^6A_1$ ($t_3^3e^2$). In Fig. 5 energy E_{nr} is the activation energy of this three step process.

Considering Tanabe –Sugano diagram presented in Fig. 4c one notices that energy of the $^2T_2$ (t_2^5) state should strongly diminish with pressure. Therefore pressure diminishes also the energy of the barrier E_{nr} and shortens the luminescence lifetime.

4. Conclusions

The effective energy transfer from TbAG host to Mn ions takes place in TbAG:Mn and results in the intensive orange Mn^{2+} luminescence. It has been found that the manganese in the TbAG:Mn film locate mainly in the Mn^{2+} state in the dodecahedral positions of the lattice. We have not observed luminescence related to Mn^{3+} in octahedral coordination. Also we have not observed Mn^{4+} in octahedral coordination, however we cannot exclude the existence of Mn^{4+} ions in tetrahedral coordination replacing Al^{3+} ions, that compensate Mn^{2+} replacing Tb^{3+}.

Considering the ionic radiuses of Mn^{4+}, which are 0.53 Å and 0.39 Å, [23] in octahedral and tetrahedral positions, respectively one notices that they are almost equal to the respective quantities of ionic radiuses of Al^{3+} which are equal to 0.535 Å and 0.39 Å. Thus Mn^{4+} can easy replace Al^{3+} in octahedral and tetrahedral sites as well. As it has been discussed in this contribution tetrahedrally coordinated Mn^{4+} ions are not seen in the luminescence due to the effective nonradiative processes involving the $^4T_2$ and ^2E states.

Acknowledgements.
The work was supported by Gdansk University, grant no 5200-5-0049-8.

References

[1] Kummer, F., Zwaschka, F., Ellens, A., Debray, A., Waitl, G., 2001. Luminous substance for a light source and light source associated therewith. Intern. Patent Appl. №: WO 01/08452.
[2] Y. Zorenko, T. Voznyak, V. Vistovsky, T. Zorenko, S. Nedilko, M. Batentschuk, A. Osvet, A. Winnacker, G. Zimmerer, V. Kolobanov, D. Spassky, Radiation Meas. **42** (2007) 648.
[3] M. Batentschuk, A. Osvet, G. Schierning, A. Klier, J. Schneider, A. Winnacker, Radiat. Meas. **38** (2004) 539.
[4] Yu. Zorenko, V. Gorbenko, T. Voznyak, M. Batenshchuk, A. Oswet, A.Winnacker, J. Luminescence **128** (2008) 652.
[5] Sh. Singh, P. Gavrilovic. US patent No. 22483, 1994.
[6] S. Kuck, S. Hartung, S. Hurling, K. Peterman, G. Huber, Phys. Rev. B **57** (1998) 2203.
[7] F. Donegan, T. J.Glynn, G. F. Imbusch, J. P. Remeika, Journal of Luminescence **36** (1986) 93.
[8] M. A. Noginov, G. B. Loutts. J. Opt. Soc. Am. **B 16** (1999) 3.
[9] Su-Hua Yang , Ching-Yi Lu, J. Electrochem. Soc., **154** (2007) J397.
[10] M. A. Noginov, G.B. Loutts, and M. Warren, J. Opt. Soc. Am. B 16 (1999). 475
[11] D. Galanciak, M Grinberg, W.Gryk, S.Kobyakov, A. Suchocki, G. Boulon, and A. Brenier, J. Phys.: Condens. Matter 17 (2005) 7185
[12] Y. Tanabe, S. Sugano, J. Phys. Soc. Jpn. 9 (1954) 753
[13] B. Henderson, and G. F. Imbusch, Optical Spectroscopy of Inorganic Solids, (Clarendon, Oxford, 1989)
[14] K.L. Bray, Topics Curr. Chem. 213 (2001) 1.
[15] M. Grinberg, Optical Materials 28, 26-34 (2006) and references therein
[16] Yu. Zorenko, V. Gorbenko, Radiation Measurements **42** (2007) 907.
[17] L. Merrill, W.A. Bassett, Rev. Sci. Instrum. 45 (1974) 290
[18] P. Dorenbos, J. Lumin. 91 (2000) 155 and references therein
[19] X. Liu, X. Wang, Z. Wang, Phys Rev. B 39,(1989) 10633
[20] Y.R. Shen, K.L. Bray , Phys. Rev. B 56 (1997) 10882
[21] Y.R. Shen, T. Riedener, K.L. Bray, Phys. Rev. B 61 (2000-I) 11460
[22] R. Englman, The Jahn-Teller Effect in Molecules and Crystals Wiley –Interscience London, New York Sydney Toronto 1972
[23] R. D. Shannon, Acta Crystallographica A32 (1976) 751

Deformation-induced phonon scattering by debris defects in alkali halides

H P Strunk and W Wasserbaech

Institute of Metal Science, University of Stuttgart, Heisenbergstr. 3, 70569 Stuttgart, Germany

E-mail: strunk@mf.mpg.de

Abstract. A number of investigations have shown that the thermal conductivity of alkali halide crystals is markedly affected by crystalline defects, chiefly in the low temperature region, say below around 10 K. The effects are very prominent after plastic deformation, where the conductivity is reduced by more than a factor of 10 after deformation by several percent already. Albeit a considerable amount of data and respective models have been elaborated, the correspondence between models and the body of experimental results still needs improvement, especially in the temperature range between around 1 and 5 K. As will be shown for the case of NaCl, where most of the experimental results exist, a satisfactory description of the thermal conductivity properties can be obtained when relying on, instead of dislocations and/or vacancy defects, a debris defect as suggested by transmission electron microscopy.

1. Introduction

Revisiting work on phonon scattering obtained at temperatures below approximately 25 K from plastically deformed alkali halide single crystals, one realizes a difficulty existing for decades in joining scattering and plasticity properties. Plastic deformation causes an increase in scattering by a factor of more than 15 (as compared to the undeformed state) in a temperature range between 1 and about 10 K [1,2]. The offered explanations are based, of course, on the defects introduced during deformation, i.e. dislocations and point defects. Unfortunately, however, all applied models, be it coherent or incoherent, dynamic or static scattering, have been insufficient in this temperature range, because of either wrong sizes of the computed scattering cross-sections or wrong temperature dependence of the thermal conductivity.

This paper will relieve this unsatisfactory situation based on a deformation-induced defect typical of deformed alkali halide crystals. Discussions concerning this defect date back to the so-called memory effect which describes the capability of an alkali halide crystal to 'remember' the passage of a dislocation through a volume even if this dislocation has left the crystal in the course of the deformation [3]. In metals for example this effect has not been observed. Models for the underlying defect structure, afterwards coined 'debris', have been suggested assuming edge dislocation dipoles [4] or distributed vacancies and vacancy clusters [5]. These models failed by not being able to explain the deformation-induced volume expansion [5]. With alkali halide crystal's defect structures becoming accessible by transmission electron microscopy, the nature of the debris defect could be modelled essentially as a microcrack [6] which model is applied here to the phonon scattering properties.

In the following, by focussing on NaCl single crystal properties, we first collect the appropriate phonon scattering results, then the relevant debris aspects. The discussion will bring together both fields.

2. The state of the research

2.1. Phonon scattering in NaCl single crystals

The typical result for the thermal conductivity of a variety of alkali halide and specifically of NaCl crystals is depicted in Fig. 1. The thermal conductivity of the undeformed specimen shows a strong increase with T^3 dependence – as expected from boundary scattering of phonons [7] - from very low temperatures, T, a maximum at around 10 K and a decay towards higher temperatures. Very important results have to be noted here. The defects introduced by plastic deformation (here to a length reduction of ε=4 %, i.e. plastic shear in the primary glide system γ = 8%) reduce the thermal conductivity by a factor of around 15 in the very low temperature range, say between approximately 1 and 5 K, whereby the T^3 dependence of the thermal conductivity is kept. It is also important for the discussion that this part of the curve corresponds to a defect-induced relaxation time (i.e. scattering cross-section) that is independent of the phonon frequency ($1/\tau = 1.1 \times 10^7$ sec^{-1}) [1]. Experiments to recover the thermal conductivity loss are shown in Fig. 2a, where the thermal resistivity values (normalized with the respective values of the undeformed state) as measured at 2, 3 and 4 K after isochronal anneals (15 minutes at each annealing temperature) are plotted. One recognizes complete recovery in an annealing range from 450 to 770 K. For comparison reasons to be used below we also have added sketches for the isochronal annealing of the dislocations (Fig. 2b) and of the deformation-induced volume expansion for NaCl crystals (Fig. 2c).

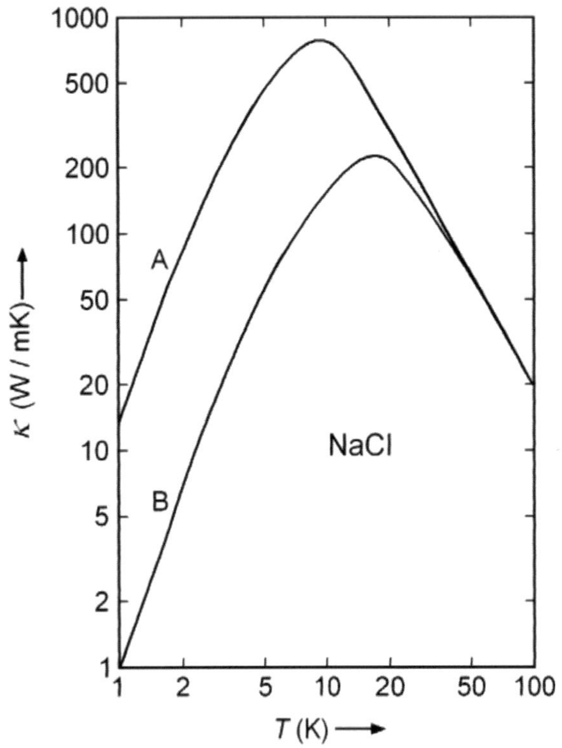

Fig. 1. Thermal conductivity, κ, as a function of the temperature for NaCl single crystals in annealed (graph A) and compressed state (4%) (graph B). Measurement and compression along <001>. After Taylor et al. [1].

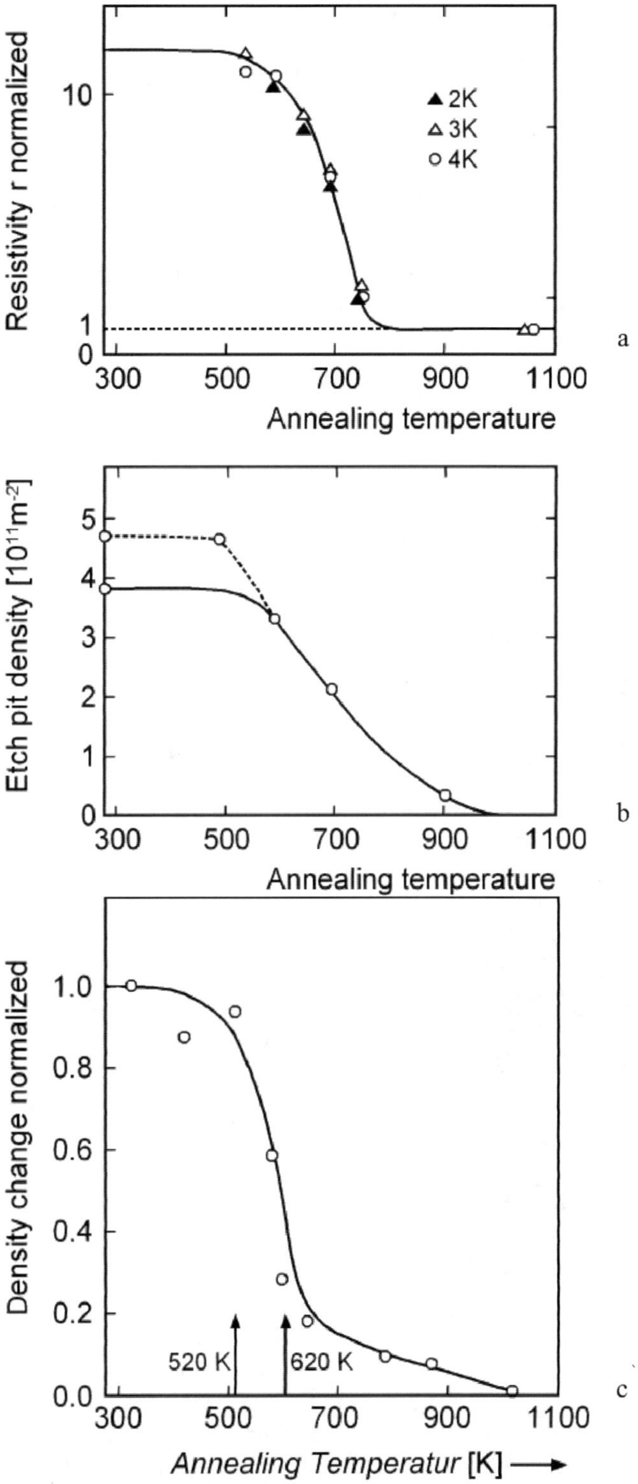

Fig. 2. Comparison of isochronal annealing behaviours.

a) Recovery of the thermal resistivity r (=inverse of conductivity) for the temperatures 2, 3 and 4 K, plotted as relative values normalized with the respective values of the undeformed crystal. Isochronal annealing steps 15 minutes each. After Taylor et al. [1].

b) recovery of the dislocation density as measured with etch pits. Dashed line describes the additional density by dislocation dipoles (interpretation of the authors, after Taylor et al. [1]). Annealing steps 15 minutes.

c) Recovery of the deformation-induced volume expansion, normalized to the initial volume expansion. Annealing steps 60 minutes. After Davidge and Pratt [2] and modified by Strunk [6].

2.2. The debris defect in plastically deformed NaCl single crystals

The transmission electron microscope (TEM) investigations of NaCl single crystals, although hampered initially by their irradiation sensitivity, have clarified to a large extent the deformation-induced defect structure. First, the total dislocation densities as determined by TEM corroborate those determined by etch pitting [8,9,10] and X-ray measurements [11] in all three work-hardening stages of the deformation along {001}, ref. 11 concerns deformation along {110}. The square root of the total dislocation density, $\sqrt{\rho}$, is proportional to the resolved shear stress τ (see also Fig. 4). Secondly, an additional class of deformation-induced defects has been detected. In the following we restrict ourselves on Stage I work hardening that is characterized by easy glide on one <-101>{101}glide system only. Figure 3 shows a respective micrograph in which, in addition to a few edge dislocation dipole segments, a high density of damage clusters is present. These clusters have formed due to the extensive electron irradiation during observation in the electron microscope. Careful inspection shows that the clusters are aligned parallel to these dislocation dipoles. Obviously linear defects not visible otherwise have been decorated by the irradiation damage clusters.

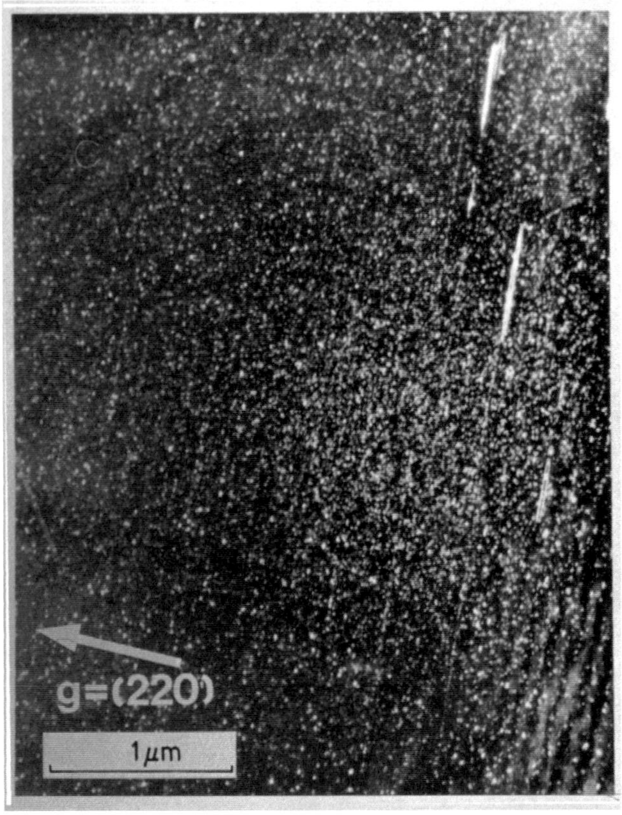

Fig. 3. Defect arrangement in the glide plane of a NaCl crystal deformed along <001>. Transmission electron micrograph, weak-beam dark field with g/-3g, g: diffraction vector. Visible are a few dislocation dipole segments (bright lines, partly both dislocations visible) and a high density of electron irradiation damage clusters that are aligned parallel to the dislocation dipoles (best seen at glancing angle). Deformation data:, resolved strain 1.7 MPa, resolved strain rate 3.7 x 10^{-4} s^{-1}, deformation temperature 220 K, debris density (=density of microcracks) ρ_{debris} = 3.2 x 10^{12} m^{-2}.

Before discussing the possible structure of the debris defect, we consider the dependencies of both the dislocation and debris densities on the resolved flow stress, τ. Figure 4 shows for three deformation temperatures that the square roots of the densities depend linearly on the flow stress. One notices one important difference (apart from the excessive densities of the debris at all deformation temperatures): The dislocation densities extrapolate to the origin of the coordinate system in agreement with earlier work [8,9,10], but the debris densities extrapolate to the respective critical resolved flow stress, τ_0. This fact indicates that debris form as a consequence of macroscopic slip. In addition we can infer from the similar slopes that the functional dependence of the debris densities on the flow stress is, in a first approximation, the same for all temperatures, at least for the range covered by the experiments, i.e. the density depends only on the work hardening, $\Delta\tau = \tau - \tau_0(T)$. Seen this way, since the flow stress squared is proportional to the work done to the crystal during deformation, the density of debris is proportional to the work done during macroscopic deformation.

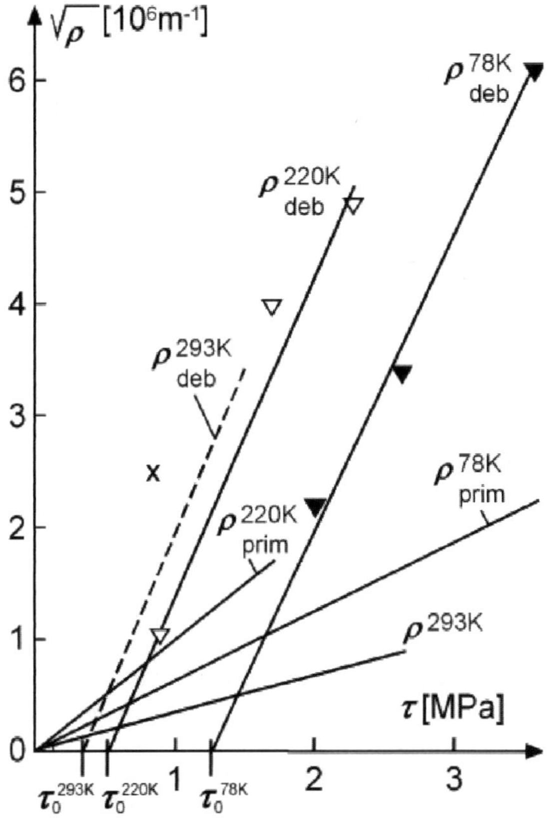

Fig. 4. Plot of the square root of the density, ρ, of both dislocations and debris for three different deformation temperatures T as a function of the flow stress, τ. Each debris defect is counted two dislocations (see model in Fig. 5). τ_0^T denotes the critical resolved flow stress at deformation temperature, T.

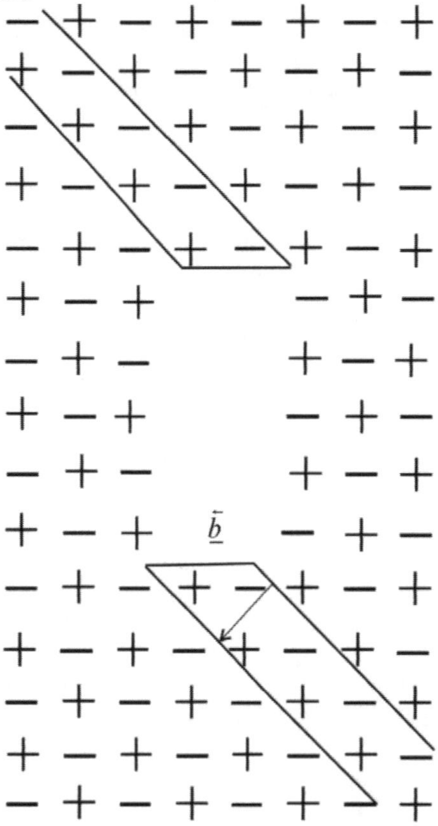

Fig. 5 Model of the debris defect. This defect forms from an edge dislocation dipole (see respective inserted half planes) by decohesion, i.e. cracking, along the plane of the dipole, which is a {100} cleavage plane. $\bar{b}$: Burgers vector.

For the consideration of the possible atomic structure of the debris defect, the observation is important that the decorated defects are aligned parallel to the dislocation dipoles. This observation suggests certain relatedness between both defect types. A respective debris model is sketched in Fig. 5 and describes a narrow edge dislocation dipole of vacancy type where the highly strained volume between both dislocation lines has been removed by decohesion along the cleavage plane of type {100} connecting both dislocation cores of the original dipole. This model fulfills all requirements [12] that follow from the debris properties, notably it produces a high lattice expansion and is of low energy (no dislocation strain field anymore and only low energy {001} internal surfaces). An energy balance between the elastic strain field energy of a dipole to the energy of the internal {100} surface shows that dislocation dipoles in NaCl with a dipole width (distance between the two dislocation cores) below 4.7 nm (equivalent to 16 elementary {001} lattice plane distances) are prone to the microcrack transformation [12]. We have to note here that interstitial type edge dislocation dipoles cannot convert by a similar process. At most they may break up into segments of dipoles, however the dislocations still exist. Thus they further-on can take part in the deformation process and have therefore a high chance, during glide, to encounter a dislocation suitable to form a vacancy type dipole.

The annealing properties of the debris and of the dislocations are different. As shown in Fig. 2, the dislocations, during isochronal annealing, anneal in a much broader and higher reaching temperature range [1], see Fig. 2b, than do the debris defects (as measured in density change) [5], e.g. Fig. 2c. TEM observations on annealed crystals, which had been deformed before at 78 K [13], corroborate this interpretation. Anneal at 520 K (see arrow in Fig. 2c) for one hour reduces the debris density to 95 % of the initial value and the dislocation dipole density to 40 %. After anneal at 620 K (see arrow in Fig 2c) for one hour, the experiments to decorate debris defects failed completely, which fact can be valued to indicate very low or even zero debris density left (or the remaining debris have changed into a structure that cannot be decorated anymore). Only 8 % of the dislocation dipole density has survived, the density of free dislocations is still 85 % at this annealing stage as compared to the pre-anneal value [13].

3. Discussion

In view of the quantification of debris density and the model of the debris defect as sort of a microcrack, both based on transmission electron microscope observations, two questions arise with regard to the phonon scattering discrepancy mentioned in the introduction. First, can the debris defect population explain the deformation-induced volume expansion in NaCl crystals? Secondly, how and to which degree can the high debris density contribute to the understanding of the phonon scattering dependency on temperature and defect density?

3.1. Debris defect

Concerning the deformation-induced volume expansion of NaCl, all respective attempts based on dislocation densities have failed. As these dislocation data have been confirmed by electron microscopy [9,10] all earlier arguments that rule out dislocations to produce the overall volume expansion (see e.g. [5]) still hold. Moreover, it should be mentioned here without further discussion that the energy stored during plastic deformation in the crystal is too small by a factor of 2 to 4 to explain the observed volume expansion on the basis of vacancies and small clusters thereof [14].

We shall now argue that the debris as discussed here can account for both volume expansion and stored energy. In correspondence to the work of Taylor et al. [1] we have to refer to a resolved shear strain of 8% (strain in compression 4%), where the volume expansion measured by Davidge and Pratt [5] corresponds to 28 g m^{-3}. The 8% strain is equivalent to a flow stress at room temperature of 1.1 MPa (see Fig. 3) and Fig. 4 yields a debris density of around 1.8×10^{12} m × m^{-3} (mind one debris is counted as two dislocations). Equating now the total debris volume (width × height × total length, i.e. density) times the density (2.1×10^6 g m^{-3}) to this volume expansion we obtain a debris height of almost 5 elementary {001}. This number is well below the calculated maximum height of 16 elementary {001} planes and shows that the present debris defect can certainly describe the observed volume expansion.

As regards the stored energy we consider the debris as an internal {001} surface. Unfortunately, however, experimental data on stored energy are sparse and a broad range of values for the {001} surface energy exist. Therefore we restrict ourselves here to show that the debris defects can principally explain the deformation-induced stored energy. Referring to Gurov et al.'s data [14], one finds for a compressive deformation rate (5×10^{-4} s^{-1}, similar to ours of $\sim 2 \times 10^{-4}$ s^{-1} [13]) for stage I deformation a stored energy of approximately 13×10^3 J/m^3 at an (extrapolated) 4% compressive strain. On the other hand, we have derived above for these deformation conditions a debris density of 1.8×10^{12} m × m^{-3}. The total surface energy contained in the debris can be calculated from the debris circumference of the cross-section (see fig. 5) × total debris length × surface energy. A very helpful collection of theoretical and experimental surface energy values can be found in ref. [15], according to which the energies range from 144 to 384 mJ/m^2 (9 to 24 meV/$Å^2$). (The discussion of the question whether free surface energy values can be used for the internal surfaces of the debris is beyond the scope of this paper.) Equating stored energy and debris surface energy yields for the debris height a span between 31 and 87 elementary {001} planes. This result is somewhat larger than the theoretical maximum debris height of 16 elementary {001} planes. In view of the many uncertainties of this assessment we may take this result as a principal proof for debris to account essentially for the stored energy. In addition we note that the contribution of dislocations and point defects to the stored energy need to be considered of course also.

3.2. Phonon scattering

At this point it appears reasonable to briefly describe the basis of the evaluation of phonon scattering due to crystals lattice defects. In the following, the phonon-dislocation interaction with different lattice defects will be considered, in order to investigate their influence on the lattice thermal conductivity:

i) dislocations
ii) point defects
iii) clusters and debris (as analyzed here)

Phonons are scattered by mobile dislocations and by the elastic distortion field of sessile dislocations (dynamic and static nature of scattering, resp.). From theoretical calculations [16] it is expected that the scattering of phonons by the static strain field of randomly distributed dislocations should produce a thermal resistivity W_{ds} proportional to T^{-2}, i.e. the thermal conductivity $\kappa_d = 1/W_{ds}$ varies with T^2, whereas in the experiment a thermal conductivity proportional to T^3 is found after plastic deformation [1].

Anderson and co-workers [17, 18] have established that the decrease in thermal conductivity after plastic deformation is primarily due to dynamic scattering of phonons by dislocations. The vibrating string model according to Granato and Lücke [19] has been rather successful in explaining the dynamic behaviour of dislocations as found in acoustic experiments. If phonons induce a dislocation to vibrate, the incident wave will be dissipated as the dislocation radiates elastic waves (reradiation scattering). Especially phonons with frequencies in the vicinity of the resonance frequency ω_0 are scattered strongly. In several alkali halides, Anderson and co-workers [17,18] found from thermal conductivity measurements a resonance temperature of about 0.2 K. In order to explain the temperature of the thermal conductivity minimum, it was necessary to assume an average dislocation length of several 10^{-8} m which is much less than has been determined ultrasonically [20].

Kneezel and Granato [21,22] improved this scattering model by taking into account several additional effects such as phonon focusing and resonance angle scattering and applied it to the scattering of phonons by edge dislocation dipoles. The dipoles were assumed to have a dipole spacing in the range $3\,b \leq d \leq 300\,b$ and a density which is about 100 times the monopole density. In this case high-frequency phonons will excite dipoles into the optical mode of vibration (i.e. the dislocations move in opposite directions). The optical mode resonant frequency is higher than that of independently vibrating dislocations, especially for narrow dipoles. Assuming an average dipole spacing of 60 b, a resonant frequency $\omega_0 \approx 10^{11}$ Hz was obtained, which corresponds well with the temperature ($T \approx 0.2$ K) at which the minimum of the thermal conductivity was observed in deformed LiF [23] and KCl [24] single crystals. Although the model describes the measured thermal conductivity rather good in the temperature range below 1 K, discrepancies become increasingly larger at higher temperatures. At high temperatures high-frequency phonons ($\omega >> \omega_0$) are excited which are out of resonance and hence not scattered. In addition, other scattering processes become effective.

The scattering of phonons by point defects (such as vacancies) increases rapidly with frequency (proportional to ω^4) and gives rise to a thermal resistivity proportional to the temperature T [16]. Hence the scattering is effective only in the temperature range where the maximum of the conductivity occurs and can be neglected at lower temperatures.

The scattering of elastic waves by inclusions (such as precipitates or clusters) embedded in an elastic medium has been investigated by Truell and co-workers [25,26,27]. In the low-frequency limit (i.e. long wavelengths) the scattering cross-section of spherical obstacles (including a spherical cavity) varies with ω^4. Thus the long-wavelength scattering cross-section is Rayleigh-like as in the case of point defect scattering. In the high-frequency limit (i.e. short wavelengths), one has a problem in geometrical optics [28]. A spherical precipitate presents a constant cross section to all incoming phonons, and the scattering cross section is independent of phonon frequency. If the precipitate is not spherical, then its effective size depends on the direction of the incident phonon. However, the cross-section must again be frequency-independent and must contain some average size of the precipitate [28]. The relaxation rate is then given by [28]

$$1/\tau_{pr} = N_{pr}\, v\, \sigma \qquad (1)$$

where N_{pr} is the concentration of the scatterers and v is the sound velocity. In the case of a frequency-independent scattering process the temperature dependence of the thermal conductivity is expected to be proportional to T^3 as in the case of boundary scattering. The transition from one scattering type to the other occurs for [28]

$$q\, d_{pr} > 1.5 \qquad (2)$$

where q is the magnitude of the wave vector ($q = 2\pi/\lambda$) and d_{pr} the diameter of the precipitates. The influence of precipitates and clusters on the thermal conductivity has been investigated for example in the case of quenched and annealed KCL and KBr crystals containing Ba^{++} and Sr^{++} [28], neutron irradiated magnesium oxide [29] and γ-irradiated LiF [30]. In all cases the thermal conductivity below the conductivity maximum was reduced and had a T^3 temperature dependence down to about 1 K, indicating extended defects. At the lowest temperatures, however, the corresponding thermal resistance fades away, suggesting that the scattering changed from a geometrical like to a Rayleigh like scattering process.

We now attribute the effect of the plastic deformation on the thermal conductivity of deformed NaCl crystals as observed by Taylor et al. [1] to the influence of extended defects in the shape of the present debris model. The dominant frequency of the phonons that contribute mainly to the thermal current is given by $\omega = 5 \times 10^{11}$ T [K^{-1} s^{-1}] [17,18]. Hence, we obtain in the temperature range above 1 K a dominant wavelength of the phonons smaller than about 40 nm. According to eqn. (2) the effective diameter of extended defects must be larger than about 9 nm and the corresponding scattering cross-section larger than 64 nm^2. As pointed out in section 3.1 the debris have a rectangular shape about 1 μm long and about 5 nm wide. Thus their cross-section is about 2000 nm^2 and hence large enough for geometrical scattering of the phonons. From eqn. (1) we obtain, with a debris density of about 1.8×10^{18} m^{-3}, a relaxation rate of about 10^7 s^{-1} similar to that derived for the deformed specimens [1]

4. Conclusions

In this paper we have elaborated for the case of plastically deformed NaCl single crystals that the models based on dislocation content, measured by etch pitting, and produced vacancy density, derived from the deformation-induced volume expansion, cannot satisfactorily describe the available results on the thermal conductivity at low temperatures, specifically in the temperature range between around 1 and around 5 K. Transmission electron microscopy suggested a different deformation-induced defect that in the simplest form can be classified as a conversion of a vacancy type dislocation dipole into an internal crack. This paper shows that this defect can account for all four criteria important to be explained. For the deformation related criteria, it accounts for the deformation-induced volume expansion and for the energy stored during deformation. For the thermal conductivity related criteria, it accounts for the T^3 dependence of the thermal conductivity and for the relaxation rate that is independent of the phonon frequency (geometrical scattering).

References

[1] Taylor A, Albers H R and Pohl R O 1965 *J. Appl. Phys.* **36** 2270

[2] Ishioka S and Suzuki H J 1963 *J. Phys. Soc. Japan* **18** Suppl. II 93
[3] Gilman J J and Johnston W G 1957 *Dislocations and Mechanical Properties of Crystals* eds J C Fisher et al (London: Chapman & Hall) p 116
[4] Gilman J J and Johnston W G 1960 *J. Appl. Phys.* **31** 632
[5] Davidge R W and Pratt P L 1964 *phys. stat. so.l* 6 759
[6] Strunk H P 1985 *Proc. Yamada Conf. IX on Dislocations in Solids* eds H Suzuki et al. (Tokyo: University of Tokyo Press) p 483
[7] Casimir J B 1938 *Physica* **5** 495
[8] Hesse J 1965 *phys. stat. sol.* **6** 209
[9] Hesse J and Hobbs L W 1972 *phys. stat. sol. (a)***14** 599
[10] Kemter L and Strunk H 1977 *phys. stat. sol. (a)* **40** 385
[11] Matucha K H, Franzbecker and Wilkens M 1969 *phys. stat. sol. b* **33** 493
[12] Strunk H P 1985 *Proc. Int. Symp. on Lattice Defects Related Properties of Dielectric Materials* (Turawa, Poland, 7-12 October 1985) ed B Macalik (Wroclaw: Institute for Low Temperature and Structure Research, Academy of Sciences) p 101
[13] Strecker H 1982 Doctoral thesis, University of Stuttgart
[14] Gurov A F, Nadgornyi E M, Polovov V M and Vilenskaya T M 1975 *phys. stat. sol. (a)* **28** 571
[15] Li B, Michaelidis A and Scheffler M 2007 *Phys. Rev. B* **76** 075401
[16] Klemens P G 1958 *Solid State Physics* **7** 1
[17] Anderson A C 1983 *Dislocations in Solids* ed F R N Nabarro (Amsterdam: North-Holland) p 235
[18] Anderson A C 1984 *Phonon Scattering in Condensed Matter* eds W Eisenmenger et al (Berlin: Springer) p 309
[19] Granato A V and Lücke K 1956 *J. Appl. Phys.* **27** 583
[20] Wire G 1972 Ph. D. thesis, University of Illinois
[21] Kneezel G A and Granato A V 1981 *Dislocation Modelling of Physical Systems* eds M F Ashby et al (Oxford: Pergamon Press) p 454
[22] Kneezel G A and Granato A V 1982 *Phys. Rev. B* **25** 2851
[23] Roth E P and Anderson A C 1979 *Phys. Rev. B* **20** 768
[24] In-Sang Yang and Anderson A C 1989 *Phys. Rev. B* **39** 13498
[25] Ying C F and Truell R J 1956 *Appl. Phys.* **27** 1086
[26] Einspruch N G and Truell R 1960 *J. Acoust. Soc. Am.* **32** 214
[27] Einspruch N G, Witterholt J and Truell R J 1960 *Appl. Phys.***31** 806
[28] Schwartz J W and Walker C T 1967 *Phys. Rev.* **155** 969
[29] Tsai C L, Moodenbaugh A R and Weinstock H 1981 *Thermal Conductivity 16* ed D Larsen New York: Plenum Press) p 139
[30] Ghantous M A, Guckelsberger K and Locatelli M 1980 *J. Phys. Chem. Sol.* **41** 473

Rare earth crystal field spectra as a probe of librational motions in BaY_2F_8 solid state laser crystals

R Capelletti[1], A Baraldi[1], E Buffagni[1], N Magnani[2] and M Mazzera[1]

[1] Physics Department, University of Parma, Viale G P Usberti 7A, Campus Universitario, 43100 Parma, Italy and Consorzio Nazionale Interuniversitario per le Scienze Fisiche della Materia (CNISM)

[2] European Commission, Joint Research Centre, Institute for Transuranium Elements, Postfach 2340, D-76125 Karlsruhe, Germany

E-mail: rosanna.capelletti@fis.unipr.it

Abstract. The fine structure (FS) accompanying a few lines, originated by crystal field (CF) transitions of rare earths (RE), as Er^{3+} and Tm^{3+}, in BaY_2F_8 single crystals, is analyzed as a function of the RE^{3+} concentration (0.5÷20 at%) and temperature (9-300 K), by using high resolution (as fine as 0.02 cm^{-1}) Fourier transform spectroscopy and linear dichroism measurements. The 9 K absorption spectra show that FS includes weak, narrow, and closely spaced (0.4÷0.8 cm^{-1}) lines, covering a few cm^{-1} range on both sides of the narrowest among the CF lines. The FS increases by increasing the RE^{3+} concentration and vanishes at rather low temperature (40 and 60 K for Er^{3+} and Tm^{3+}, respectively). The polarized light spectra confirm the association of a given set of FS lines to a specific CF line. The FS is ascribed to the simultaneous excitation of an electronic CF transition and of a local librational (or hindered rotation) mode of the RE^{3+}-F^- group. The attribution is supported 1) by specific features of the host matrix and guest rare earths, which allow some mobility of F^- ions, and 2) by the spacing of the FS lines, which is in excellent agreement with the calculated RE^{3+}-F^- group rotational constant.

1. Introduction

The spectroscopic evidence for molecular group rotations (both free and hindered) in condensed matter comes mostly from rotovibrational spectra where such modes are simultaneously excited together with the stretching ones. The pioneering works dealt, for example, with diatomic molecular groups, as cyanide and hydroxyl ions, substituting for the halogen in alkali halides [1-2] and more recently with HD molecular groups in silicon [3]. The stretching mode absorption line, falling in the mid-infrared region, is accompanied by side lines: their separation from the central line supplies information about the rotational (or librational) mode energy. At low temperatures barriers separating different energy minima restrict the molecular group motion allowing only librations (or hindered rotations): tunneling across the barriers may occur as well. By increasing the temperature, thermally activated free rotations may take place. IR spectroscopy allowed monitoring even rotations of polyatomic molecular groups in condensed matter, e.g. NO_2 and NO_3 in alkali halide crystals [4], OCS and SF_6 in liquid He clusters [5-6]. A number of narrow, rotationally resolved side lines, spaced only by ~0.02 cm^{-1} and grouped into P and R branches, were detected around the ν_3 vibrational absorption of the SF_6 group [6].

Much less is known about rotational or librational lines accompanying electronic transitions, very likely due to the broadness of lines arising from such transitions, which may hide the finely spaced rotational spectrum. However, a nice example is provided by optical measurements performed on benzoic acid doped with pentacene: the absorption and emission spectra show two line progressions, spaced by 12.2 and 16.7 cm^{-1}, which were attributed to the lowest frequency libration modes of pentacene in the electronic ground and excited state, respectively [7]. The width of the electronic absorption line (a few cm^{-1}) is much smaller than the libration mode frequency, thus allowing the detection of the side lines.

In the present work the remarkable sharpness of absorption lines (as narrow as 0.13 cm^{-1}) originated by electronic transitions of rare earths (RE), as Er^{3+} and Tm^{3+}, in monoclinic BaY_2F_8 single crystals [8-9], is exploited to probe possible librational or rotational modes. Such crystals are also of technological interest being good rare earth hosts for laser applications [10]. More in detail, the fine structure (FS) accompanying a few lines, induced by the f-f electronic transitions of both RE^{3+}, is analyzed as a function of the RE^{3+} concentration and temperature, by taking advantage of both high resolution Fourier transform spectroscopy and linear dichroism measurements. The attribution to fluorine libration with respect to the RE^{3+} is discussed on the basis of BaY_2F_8 crystal structure [11], rotational constants of RE^{3+}-F^- complexes, and spacing between the FS lines.

2. Experimental Details

The BaY_2F_8 single crystals were grown by a computer-controlled Czochralski technique under a purified argon atmosphere [12]. The investigated samples were doped with 0.5, 2, 12, and 20 Er^{3+} at% and with 1 Tm^{3+} at% [8]. A nominally pure sample was also analyzed as a reference.

The optical absorption spectra were monitored by means of a Fourier Transform (FT) spectrometer Bomem DA8, operating in the 2000-24000 cm^{-1} range and capable of a nonapodized resolution as fine as 0.01 cm^{-1}. The sample temperature was controlled in the 9-300 K range by means of a CTI Cryogenics model 22 Cryodine Refrigerator equipped with KRS5 and fused silica windows. Measurements with linearly polarized light in the IR region (4000-10000 cm^{-1}) were performed by using a gold grid polarizer deposited onto a KRS5 substrate.

3. Results and discussion

Figure 1 displays the fine structure, which accompanies the line peaking at 6530.2 cm^{-1} (A1 line), originated by the Er^{3+} zero phonon crystal field (CF) transition from the lowest sublevel of the $^4I_{15/2}$ ground manifold to the lowest sublevel of the $^4I_{13/2}$ excited manifold [8], in two BaY_2F_8 samples doped with different Er^{3+} amounts, i.e. 0.5 and 2 at% (see curves a and b, respectively). The absorption spectrum, measured at 9 K with an apodized resolution as fine as 0.04 cm^{-1}, shows on both sides of the stronger A1 line (indicated by an arrow) the FS, consisting in other weaker, narrow lines laying very close to the A1 and a progression (spanning over a few cm^{-1}) of much weaker ones. A magnification of the weakest lines is provided by the inset of Fig. 1, where the spectra in the 6524.1-6529.5 cm^{-1} range are compared for two Er^{3+} concentrations: the line pattern and position are the same for both samples, while their amplitude increases by increasing the Er^{3+} doping level. At the highest concentrations (12 and 20 at%), the FS grows accordingly (as the A1 line does), but it is partially blurred due to line inhomogeneous broadening and overlapping of absorptions induced by RE^{3+} clusters [8]. These results, joined to the absence of FS in pure samples, support that such a structure is closely related to the presence of Er^{3+}. FS lines have been detected to 'decorate' other among the narrowest lines induced by Er^{3+} CF transitions.

Similar FS patterns accompany the narrowest CF lines induced by Tm^{3+} doping (1 at%) in BaY_2F_8 as, for example, the A1 and A2 lines at 5681.85 and 5683.46 cm^{-1}, respectively, see figure 2(a). The two Tm^{3+} CF lines originate from two transitions reaching the same final state within the excited $^3F_4$ manifold and starting from the first two sublevels (separated only by 1.61 cm^{-1}) of the ground $^3H_6$

manifold [9]. The spectra were measured at 9 K by light propagating along the c axis [11] and with the electric field orthogonal and parallel to the a axis (curve a and b, respectively): they prove the strong optical anisotropy of the Tm^{3+} CF transitions. The A1 and A2 lines show an opposite dependence on the polarization angle. The weak FS lines, indicated by arrows in figure 2(a), exhibit the same polarization dependence as the CF A1 and A2 lines do (solid red and dashed blue arrows, respectively), supporting the hypothesis that a group of FS lines is associated with a well defined CF line. Detailed analysis of the FS and CF line amplitudes vs. the polarization angle for both Er^{3+} and Tm^{3+} transitions confirms such close association.

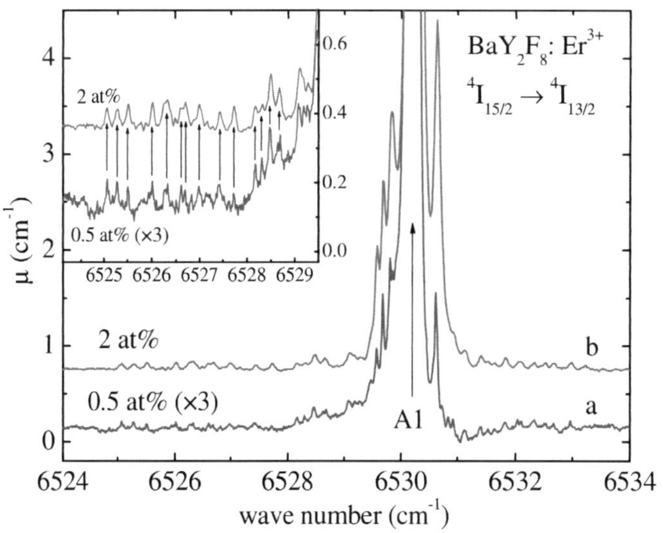

Figure 1. Optical absorption spectra of BaY_2F_8 samples doped with different Er^{3+} concentrations (curves a: 0.5 at%, res. 0.02 cm^{-1}; curve b: 2 at%, res. 0.04 cm^{-1}) measured at 9 K in the region of the $^4I_{15/2} \rightarrow {}^4I_{13/2}$ transition. The inset shows a magnification of the 6524.1-6529.5 cm^{-1} range. Arrows indicate some lines of the FS pattern.

The fine structure vanishes at rather low temperatures: the FS lines are no longer distinguishable from the CF ones they are associated with, already at 40 and 60 K for Er^{3+} and Tm^{3+} doping, respectively. Their amplitude shows a much faster thermally induced decrease than the CF lines due either to isolated RE^{3+} or to RE^{3+} clusters (the α lines in Ref. 8).

As shown by figures 1 and 2(a) the separation of the closest FS lines from the related CF one may be as small as ~0.4 cm^{-1} (i.e. ~5×10^{-5} eV) and the spacing between neighbor FS lines is in the ~0.2-0.8 cm^{-1} range.

In principle the FS lines might be attributed to 1) hyperfine splitting due to ^{167}Er (22.94% natural abundance) and ^{169}Tm (100% natural abundance) nuclear spins (I=7/2 and 1/2, respectively) interaction with the 4f-electron total angular moments; 2) isotopic effects due to the presence, in addition to ^{167}Er, of ^{166}Er, ^{168}Er, and ^{170}Er of comparable natural abundance (33.41, 27.07, and 14.88%, respectively); 3) vibronic transitions in which the incoming photon excites simultaneously a CF electronic transition and a BaY_2F_8 lattice mode; 4) simultaneous excitation of both an electronic transition and a low energy local mode (e.g. hindered rotations, librations) induced by the RE^{3+} doping.

The FS shows a large number of lines of different amplitudes, see figures 1 and 2(a), which cannot be accounted for by the 2I+1 lines of comparable amplitudes expected on the basis of a possible hyperfine interaction for Er^{3+} and Tm^{3+} (eight and two, respectively). Furthermore hyperfine interaction should be excluded, because once the value of the coupling constant for a given RE and the CF eigenfunctions of that RE in the host crystals [8] are known, its associated features and their energy splittings should follow determinate quantitative and qualitative patterns [13-14] which are not compatible with the present observations. Even isotopic effects should be ruled out in the case of Er^{3+} doped samples, because the number of lines expected is lower than that observed and the spacing among lines should be much smaller, i.e. ~1×10^{-3} cm^{-1}, as in the case of Er^{3+} doped $LiYF_4$ tetragonal single crystals [15]. The FS lines cannot be induced by vibronic transitions [16-17], because the separation of the strongest among the FS lines from the CF lines is extremely small (of the order of a few tenths of cm^{-1}, i.e. ~10^{-4} eV): the density of states for lattice modes of such low energies is negligible.

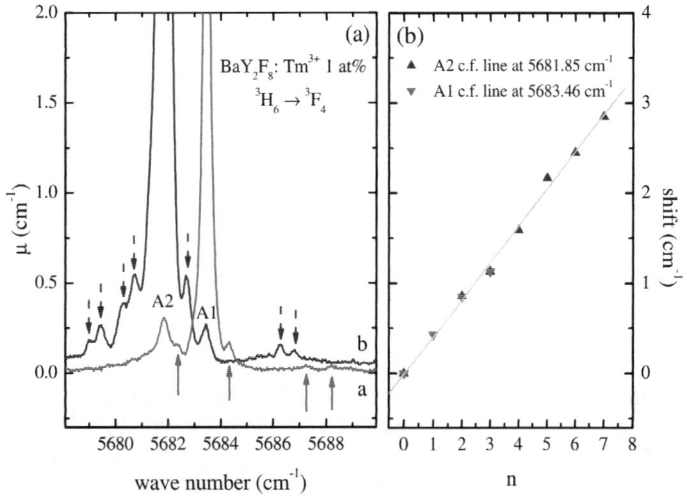

Figure 2. (a) Optical absorption spectra of a BaY_2F_8: Tm^{3+} 1 at% sample in the region of the $^3H_6 \rightarrow {}^3F_4$ transition measured at 9 K with linearly polarized light propagating along the c axis (res.=0.04 cm^{-1}). Curves a and b: electric field orthogonal and parallel to the a axis, respectively. Arrows indicate some of the FS lines associated with the A1 (solid red) and A2 (dashed blue) CF lines. (b) The FS line shifts from the related CF A1 (▼) and A2 (▲) lines as a function of a quantum number n. Solid line is the linear fit of the experimental data.

The last hypothesis, i.e. the simultaneous excitation of both electronic transition and local modes (e.g. hindered rotations, librations) is supported by 1) the BaY_2F_8 structure, 2) RE^{3+} size, and 3) rotational constant B of RE^{3+}-F^- groups ($B\equiv\hbar^2/2M_I$, where ħ is the reduced Planck's constant and M_I is the RE^{3+}-F^- group moment of inertia). In monoclinic BaY_2F_8, Y^{3+} ion is surrounded by eight F^- anions: four are classified as F1, two as F2, and two as F3 [11]. The rather large thermal parameter suggests that the F3 may be slightly out of the binary axis and may benefit of some mobility. Furthermore, Er^{3+} and Tm^{3+} ionic radii (1.14 and 1.13 Å, respectively, for fluorides and coordination number=8, as in BaY_2F_8 [18]) are smaller than that of Y^{3+} (1.16 Å), which they substitute for, thus providing additional room for possible F^- motion. From the separation F3-Y^{3+} (or F3-RE^{3+} in doped samples)=2.242 Å [11] and the Er^{3+}, Tm^{3+}, and F^- masses, 2B for the F3-Er^{3+} and F3-Tm^{3+}complexes can be easily calculated as $2B_{calc}$~0.39 cm^{-1}.

On the basis of the above premises the separations (shifts) of the FS lines from the related CF line were collected 1) for all the 'decorated' CF lines appearing in unpolarized spectra of Er^{3+} and Tm^{3+} doped samples and 2) for different Er^{3+} concentrations. The shifts of the lines, symmetrically located with respect to the related CF line, were selected and plotted vs. a quantum number n. The experimental data could be well fitted to straight lines, from whose slope the 2B parameter could be easily evaluated as $2B_{exp}$=(0.36±0.006) cm^{-1} for both Er^{3+} and Tm^{3+} FS line progressions. The average $2B_{exp}$ value, obtained for different transitions and concentrations, is in good agreement with the $2B_{calc}$ reported above for F3-RE^{3+} complexes. Slightly different $2B_{exp}$ values were obtained by considering, among the FS lines related to a given CF line, only the 'strongest' ones (i.e. those closest to the CF line), for example in the case of the FS accompanying A1 and A2 Tm^{3+} CF lines displayed in figure 2(a): the plot of the FS line shifts from the related CF A1 and A2 lines vs. quantum number n, portrayed in figure 2(b), supplies a value of $2B_{exp}$=(0.41±0.01) cm^{-1}, which is anyway in excellent agreement with $2B_{calc}$~0.39 cm^{-1}. Similar linear dependence of the lines belonging to P and R branches as a function of a rotational quantum number was found for OCS molecules embedded in liquid He droplets [5].

In spite of the overall self consistency of the description supplied, some criticisms and refinements may be put forth. 1) The description of F^--RE^{3+} complex as a free-rotator is naïve: more likely the F^--RE^{3+} complex mobility is restricted and characterized by hindered rotations and/or librations, and 2) a few of the FS lines actually monitored by the spectra do not fit in the linear plots. Beyeler [19] has calculated the vibration-rotation absorption spectra of 'rotator' hindered by fields of cubic symmetry for different energy barriers (separating potential minima) and temperatures. The simulation supplies line-rich spectra, which resemble closely those accompanying the RE^{3+} CF lines (originated by electronic transitions) in the present work. The lines are not equally spaced, unless the barrier vanishes and/or the temperature is high enough to allow 'free rotations'. At low temperatures the lines are fewer and are due to librations of the molecular group within the potential wells: in this case the librational absorption lines appear on the high energy side, well separated from the central line. For $kT \ll V_0$, where V_0 is the energy barrier, the molecular group occupies levels below the top of the barrier and the energy levels are those of a harmonic oscillator of frequency ν_{libr} which depends on V_0 and on the rotational constant B as

$$\nu_{libr}=2\,(V_0\,B)^{1/2} \tag{1}$$

where ν_{libr}, V_0, and B are expressed in cm^{-1} [4]. Tunnelling across the energy barrier is also expected.

In the case of BaY_2F_8: Tm^{3+}, in addition to the closely spaced FS lines overlapping the A2 and A1 CF lines at 5681.8 and 5683.45 cm^{-1}, respectively, there are two groups of much weaker FS lines on the higher energy side, separated from the A1 and A2 lines by about 4.5 and 4.7 cm^{-1}, respectively, see figure 2(a). The main lines of the two groups peak at 5686.27 and 5688.2 cm^{-1}, respectively, and exhibit the same polarization dependence as the A1 and A2 lines do, see figure 2(a): they may be regarded as librational lines associated to the A1 and A2 lines. Although Eq. (1) was written for a cubic potential, which is not necessarily that probed by the F^--Tm^{3+} complex in BaY_2F_8, however V_0 may be roughly estimated by assuming librational frequencies ν_{libr}=4.5 and 4.7 cm^{-1} and $2B_{exp}$=0.41 cm^{-1} (see above). The resulting values of V_0 are about 25 and 27 cm^{-1}, i.e. the energy barrier 'seen' by the librating F^--Tm^{3+} molecular group, when the Tm^{3+} electronic transition starts from the first excited sublevel and from the lowest one within the ground $^3H_6$ manifold, respectively, to reach the first sublevel of the excited $^3F_4$ manifold (A2 and A1 lines). The low values of V_0 (of the order of 3×10^{-3} eV) explain why the FS lines are no longer detectable yet at T>60 K (see above), due also to the thermally induced line broadening. At high temperature the F^--Tm^{3+} molecular group experiences a precession around a fixed axis: a thermally activated process makes it able to pass over the barrier which separates different angular positions, corresponding to the potential minima. However the typically free-rotator line spectrum cannot be detected due to the simultaneous line broadening. At low temperatures the barrier can be crossed only by tunneling and the F^--Tm^{3+} molecular group performs only small angular oscillations (librations) within the well.

In BaY_2F_8: Er^{3+} spectra it is difficult to identify unequivocally a librational line, from whose position V_0 might be, at least roughly, estimated according to Eq. 1. The Er^{3+} FS lines are no longer distinguishable at temperatures as low as 40 K (see above): this suggests that the energy barrier which separates different orientations of the F^--Er^{3+} complex is still less than that estimated for the F^--Tm^{3+} one. Thus 9 K can be regarded as a temperature high enough to forbid the librational line detection.

In conclusion, by taking advantage of high resolution spectroscopy applied at low temperatures and sharpness of CF lines, due to Er^{3+} and Tm^{3+}, it was possible to detect in BaY_2F_8 crystals a fine structure, composed by series of weak and closely spaced lines, accompanying the narrowest among the CF lines, which is ascribed to the simultaneous excitation of an electronic CF transition and of a local librational (or hindered rotation) mode of the RE^{3+}-F^- group. The attribution is supported by specific features of the host matrix and guest rare earths, which allow some mobility of F^- ions and by the spacing of the FS lines which is in excellent agreement with the calculated RE^{3+}-F^- group rotational constant. The energy barrier which separates different potential minima could be, at least roughly, evaluated as low as 3×10^{-3} eV in BaY_2F_8: Tm^{3+}. The FS lines can be regarded as a rot(electr)onic structure: its pattern resembles that displayed by the extensively studied rotovibrational spectra (anyway not well resolved in crystalline solids [1-2]), but, at variance of the latter, its detection is quite rare and difficult, especially in the case of low energy 'rotational' modes, as in the present work.

Acknowledgments
The Authors like to express their gratitude to Prof. Mauro Tonelli and co-workers (Physics Department, University of Pisa) for providing BaY_2F_8 crystals and to Carlo Mora (IMEM-CNR, Parma) for technical help.

References
[1] Narayanamurti V and Pohl R O 1970 *Rev. Mod. Phys.* **42** 201 and references therein
[2] Krantz M and Luty F 1988 *Phys. Rev.* B **37** 7038
[3] Shi G A, Stavola M, Fowler W B and Chen E E 2005 *Phys. Rev.* B **72** 85207
[4] Narayanamurti V, Seward W D and Pohl R O 1966 *Phys. Rev.* **148** 481
[5] Grebenev S, Hartmann M, Havenith M, Sartakov B, Toennies J P and Vilesov A F 2000 *J. Chem. Phys.* **112** 4485 and references therein
[6] Hartmann M, Miller R E, Toennies J P and Vilesov A 1995 *Phys. Rev. Lett.* **75** 1566
[7] Molenkamp L W and Wiersma D A 1984 *J. Chem. Phys.* **80** 3054
[8] Baraldi A, Capelletti R, Mazzera M, Ponzoni A, Amoretti G, Magnani N, Toncelli A and Tonelli M 2005 *Phys. Rev.* B **72** 075132
[9] Amoretti G, Baraldi A, Capelletti R, Magnani N, Mazzera M, Riolo P, Sani E, Toncelli A and Tonelli M 2005 *phys. stat. sol.* (c) **2** 248
[10] Kaminskii A A 2007 *Laser & Photon. Rev.* **1** 93
[11] Guilbert L H, Gesland J Y, Bulou A and Retoux R 1993 *Mat. Res. Bull.* **28** 923
[12] Toncelli A, Tonelli M, Cassanho A and Jenssen H P 1999 *J. Lumin.* **82** 291
[13] Agladze N I and Popova N M 1985 *Solid State Commun.* **55** 1097
[14] Baraldi A, Capelletti R, Magnani N, Mazzera M, Földvári I and Beregi E 2007 *Phys. Rev.* B **76** 165130
[15] Macfarlane R M, Cassanho A and Meltzer R S 1992 *Phys. Rev. Lett.* **69** 542
[16] Blasse G 1992 *Intern. Rev. Phys. Chem.* **11** 71
[17] Baraldi A, Capelletti R, Mazzera M, Ponzoni A, Sani E and Tonelli M 2003 *Rad. Effects and Defects in Solids* **158** 241
[18] Kaminskii A A 1990 *Laser Crystals* (Berlin: Springer-Verlag)
[19] Beyeler H U 1974 *J. Chem. Phys.* **60** 4123

TUNNELLING PROCESSES-DRIVEN RADIATIVE RECOMBINATION IN COMPLEX OXIDE SCINTILLATORS

M. Nikl[1], A. Vedda[2], G.P. Pazzi[3], E. Mihokova[1,2], M. Fasoli[2], J. Pejchal[3,4], P. Bohacek[1], A. Yoshikawa[4], G. Ren[5], K. Nejezchleb[6]

[1]Institute of Physics AS CR, Cukrovarnicka 10, 162 53 Prague, Czech Republic
[2]Dipartimento di Scienza dei Materiali dell' Università di Milano-Bicocca, Via Cozzi 53, 20 125 Milano, Italy
[3] Institute of Applied Physics "N.Carrara"(IFAC) of CNR, Via Madonna del Piano 10, 50019 Sesto Fiorentino (Firenze) Italy
[4]IMRAM, Tohoku University, 2-1-1 Katahira, Aoba-ku, Sendai , Japan
[5]Shanghai Institute of Ceramics, Chinese Academy of Sciences, Shanghai 200050, China
[6]Crytur Ltd., Palackeho 175, 511 19 Turnov, Czech Republic

nikl@fzu.cz

Abstract. Photo- and thermo-luminescence techniques are applied to several kinds of single crystal complex oxide scintillators to reveal tunnelling-driven radiative recombination processes. The tunnelling mechanism provides the physical ground to interpret the observed luminescence and scintillation characteristics which could not be explained by classical radiative recombination schemes.

1. Introduction

Scintillator materials should efficiently and quickly transform high-energy photons or accelerated particles into UV/visible photons [1]. A critical stage of the scintillation mechanism is the transport of free electrons and holes in the conduction and valence bands, respectively. The presence of traps in the forbidden gap can result in electron/hole temporary localization causing a considerable delay in their radiative recombination at the emission centers [2]. Consequently, slow components appear in the scintillation decay, that deteriorate the temporal performance and the overall figure-of-merit of a scintillator.

To evaluate the charge transport delay, it is usual to consider the thermally stimulated detrapping of electrons (or holes) into the conduction (or valence) band. In such a case the decay time τ can be considered to be proportional to a factor $\exp[E_T/kT]$, with E_T being the trap depth and kT the thermal energy at a given temperature T (usually room temperature (RT)). However, in the case of close proximity between electrons in traps and holes at the recombination site, the radiative recombination of charge carriers may occur also via a tunnelling process without their previous delocalization in the conduction or valence bands. The tunnelling probability decays exponentially with the electron-hole distance in a single pair. Due to spatial averaging over many pairs with various distances in a macroscopic volume, the radiative decay via tunnelling mechanism manifests a t^{-p} time dependence, with p=1-1.5 [3]. The tunnelling process can also be thermally assisted, at temperatures sufficiently high to populate excited states of the trap [4].

In this paper we present some examples of tunnelling-driven radiative recombination processes in well-known complex oxide crystalline scintillators such as Ce-doped aluminum garnets ($A_3Al_5O_{12}$, A=Y, Lu), perovskites ($YAlO_3$), orthosilicates ($(Lu\text{-}Y)_2SiO_5$), and $PbWO_4$. The tunnelling mechanism can satisfactorily account for the shape of the photoluminescence and scintillation time decays when shallow traps are present in the material; this approach can substitute the usually considered analysis of such decays by a sum of exponential functions, where the slower components would lack any physical interpretation. In thermally stimulated luminescence (TSL) measurements, thermally assisted tunnelling can justify the presence of a unique trap depth corresponding to various glow curve peaks in a wide temperature range. Moreover, the presence of tunnelling processes proves the spatial correlation of traps and recombination centers and poses a question if such correlation can be understood exclusively as a matter of statistics or if it reveals defect aggregation.

2. Experimental

Wavelength-resolved TSL measurements were performed in the 10-310 K range using a CTI closed-cycle refrigerator after x-ray irradiation at 10 K (with a Philips 2274 x-ray tube operated at 20 kV), with a home made apparatus allowing the detection of the TSL signal both as a function of temperature and wavelength in the 200-1100 nm interval. A heating rate of 0.1 K/s was adopted. Phosphorescence decays were measured by the same set-up after x-ray irradiation at fixed temperature in the 10-30 K interval.

Radioluminescence (RL) spectra and scintillation decays were measured at room temperature (RT) under X-ray tube and ^{22}Na radioisotope excitation (511 keV photons), respectively; in the latter case a modified spectrofluorometer (199S Edinburgh Instrument) was used.

Photoluminescence (PL) decays were measured after excitation by excimer laser (308 nm, XeCl line, 10 ns pulse width (FWHM), 5 mJ per pulse, 10 Hz repetition rate).The decays were detected by a fast photomultiplier (Hamamatsu R1398) in the current regime coupled with a digital scope (Tektronix TDS 680B). At each temperature several decays were recorded with different scales both in amplitude and time. By a sofware procedure these partial decays were „patched up“ to obtain an overall decay curve, covering a broad range in both time and dynamical scales.

Single crystals with typical dimensions ∅20x50 mm of Ce-doped (0.1-0.5mol% in the crystal) $Y_3Al_5O_{12}$ (YAG) and $YAlO_3$ (YAP) were grown by the Czochralski technique from 5N Y_2O_3, 4N Al_2O_3 and 5N Lu_2O_3 raw powders in a molybdenum crucible under reducing atmosphere and by μ-PD technique (∅3x20 mm, YAP:Ce). 0.5%Ce-doped $(Lu\text{-}Y)_2SiO_5$ orthosilicate and $PbWO_4$ single crystals were grown by Czochralski technique from iridium and platinum crucibles respectively, using 4N raw powders. Plates of about 7×7×1 mm^3 and cubes of about 1 cm^3 were cut and polished for TSL, RL, PL and scintillation decay experiments, respectively.

3. Experimental results and discussion

3.1. Ce-doped aluminum garnets

In Fig. 1 the TSL glow curve of YAG:Ce after x-irradiation at 10 K is shown. The rich TSL pattern below RT has been already shown and discussed in the literature [5,6], but the smooth and structureless signal below approximately 80 K was not discussed. In fact, it is qualitatively indentical to that of LuAG:Ce, which has been ascribed to an athermal electron tunnelling from an electron trap towards an adjacent hole localized at Ce^{3+} ion [7]. A further evidence of such a mechanism in LuAG:Ce was obtained by measuring the phosphorescence decay at constant temperature which at long times approaches a ~t^{-1} regime. The same measurement was performed within 10-30 K on YAG:Ce and the same decay characteristics were obtained (Fig. 2). Hence, also in YAG:Ce such a tunnelling-driven radiative recombination mechanism can explain the TSL glow curve pattern below 80 K. In LuAG:Ce a combination of an exponential and an inverse power law (t^{-1}) function was successfully used to fit the scintillation decay at RT and was interpreted in the way that the tunnelling recombination is responsible for the observed slow sub-microsecond decay components [7]. In YAG:Ce, however, the smaller amplitude of slower decay components, prevents any similar reliable fit [6].

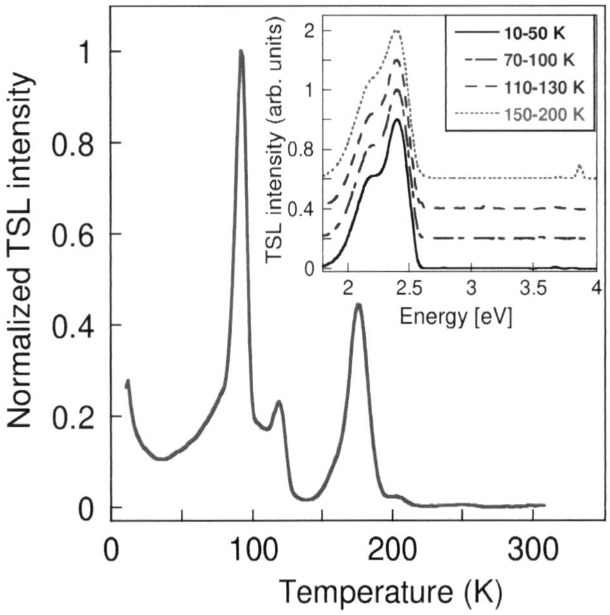

Fig. 1: TSL glow curve of YAG:Ce0.2% after x-irradiation at 10 K. In the inset are reported the TSL spectra observed in the temperature regions marked in the legend.

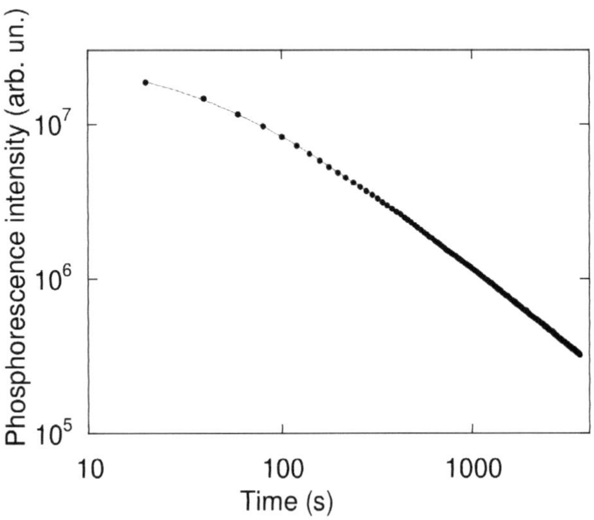

Fig.2: Phosphorescence decay of YAG:Ce after x-irradiation at 30 K (full circles). The red continuous line is a fit to the data according to the formula: $y=a(x+b)^c$ with $a=1x10^9$, b=44, c=-1.

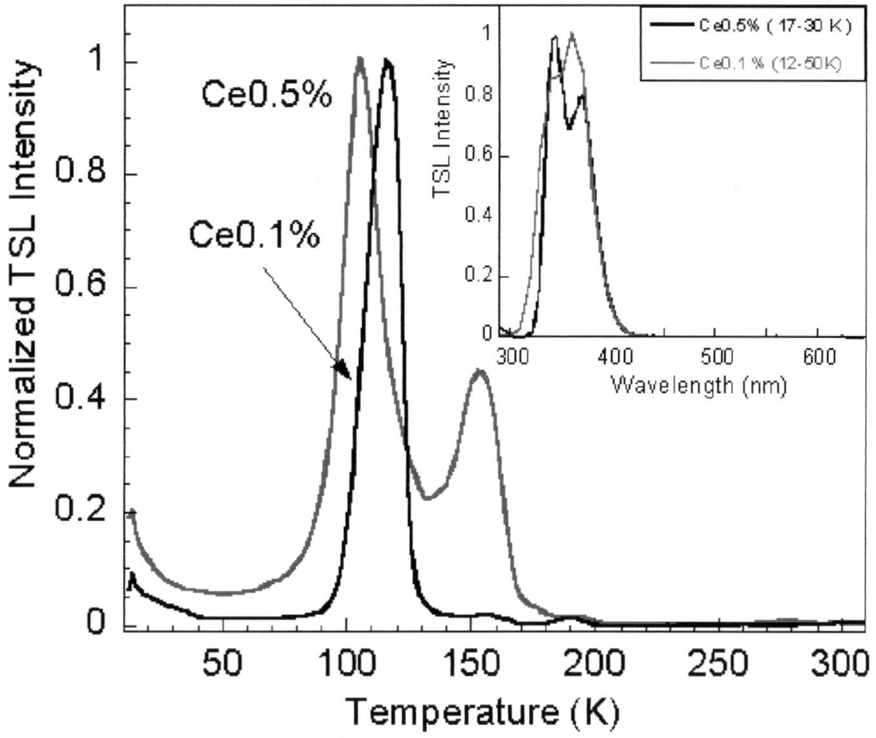

Fig. 3: TSL glow curves of YAP:Ce after x-irradiation at 10 K. The inset shows the TSL spectra of both samples in the temperature interval of athermal tunnelling.

3.2. Ce-doped aluminum perovskites

In Fig. 3 the TSL glow curves and their emission spectra are shown for YAP:Ce samples with different content of Cerium. Similarly to aluminum garnets, a temperature region governed by athermal tunnelling can be noticed below approimately 80 K, already evidenced also by phosphorescence decays [8]. It is worth to note that the relative intensity of the tunnelling process increases with Ce concentration: such an effect was observed also in LuAG:Ce [7]. In Fig. 4 the spectrally unresolved scintillation decay of YAP:Ce0.5mol% sample is displayed. We attempted to fit this decay by the same formula as in the case of LuAG:Ce, but the exponent in the power law component is far off (p= 3.9) the range predicted in [3] where p is expected to be approximately between 0.95 and 1.5. In contrast to Ce-doped garnets in YAP:Ce there can be a competition for the hole capture between Ce ions and lattice perturbed oxygen sites [9]. As a result, a mix of different recombination processes can easily occur in the scintillation decay curve and obscure the presence of tunneling recombination. Moreover, the very strong dependence of the TSL glow curve pattern on the rare earth ion dopant [10] requires further investigation to understand the scintillation mechanism in YAP host in greater detail.

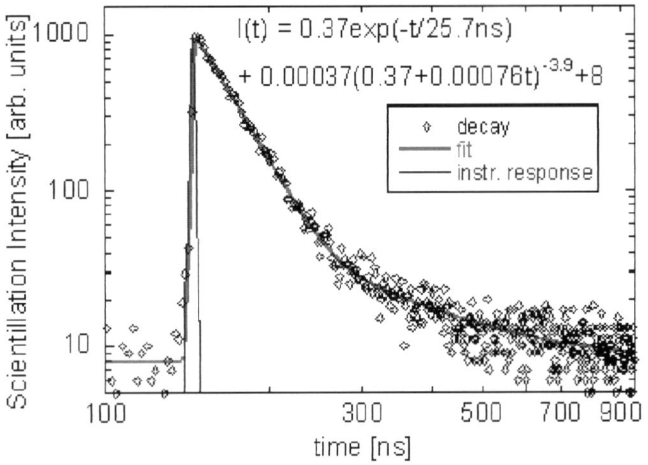

Fig. 4: RT spectrally unresolved scintillation decay of Cz-grown YAP:Ce0.5% . The solid line is the convolution of the instrumental response and the I(t) function reported in the figure.

3.3. Ce-doped orthosilicates

In the case of LSO:Ce, it was soon recognized that it presents a fairly strong afterglow at RT [11] persisting for several minutes up to hours. Fundamental studies, aimed at the comprehension of the microscopic physical mechanism governing afterglow were thus carried out in order to find possible technological solutions to suppress it. The activation energy of the process was found to be approximately 1 eV [11]; it is in accordance with the calculated trap depth of a TSL peak at 375 K (using a heating rate of 6 K/s) so that afterglow appears to be due to RT carrier detrapping from the trap responsible for this peak followed by radiative recombination at Ce^{3+} luminescent centers. Actually the 375 K peak is the first one of a series of as many as 6 peaks observed in the glow curve above RT, whose spectral emission coincides with Ce^{3+} 5d-4f transition [11] (Fig. 5). Annealing experiments in reducing or oxidizing atmosphere led to the suggestion that traps could be related to oxygen vacancies [12] and this hypothesis was further confirmed by finding an F^+ center by EPR whose temperature stability is correlated with the TSL glow curve above RT [13].

The TSL glow curve pattern of the samples from different producers and of substantially different intensity does show a striking similarity (Fig. 5). The thermal depth of electron traps was investigated by TSL measurements using the partial cleaning procedure followed by the analysis of glow curves with the initial rise method [14]. LYSO:Ce samples were first irradiated at RT, then heated up to a fixed temperature (T_{stop}), cooled to RT again, and further heated up to 400 °C to record the glow curve. In each case, the trap depth could be evaluated by the fit of the initial portion of the glow curve with the relation:

$$I(T) = n_o s \exp\left(\frac{-E}{kT}\right) \qquad (1)$$

where n_o is the initial number of trapped electrons after irradiation, s is the frequency factor, E is the trap depth , k is Boltzmann constant, and T is the absolute temperature. Fig. 6 reports the trap depth values obtained for partial cleanings performed on LYSO:Ce: in the 40-320 °C extended temperature range; a constant value of E = 0.97 eV ± 0.08 eV was obtained. This value is therefore representative of the thermal depths of the whole series of peaks at 78 °C, 135 °C, 181 °C, and 236 °C. A constant trap depth was revealed also in the TSL of (Lu-Y)AP:Ce. It was explained by a thermally assisted tunnelling process of electrons from oxygen vacancies to Ce^{3+} ions where holes are localized, at different lattice distances [15]. A similar explanation could also apply in the present case, [16].

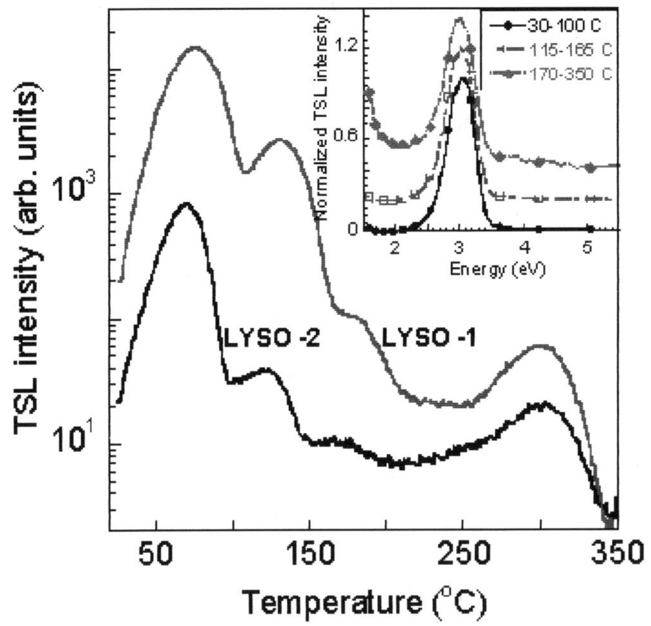

Fig. 5: TSL glow curves of two LYSO:Ce0.5% samples from different production laboratories in China (G. Ren, LYSO-1) and USA (B. Chai, LYSO-2), respectively, after irradiation at RT. In the inset the TSL spectra of sample no. 1 in different temperature regions are shown. The signal below 2 eV in the 170-350 °C interval is due to black body radiation.

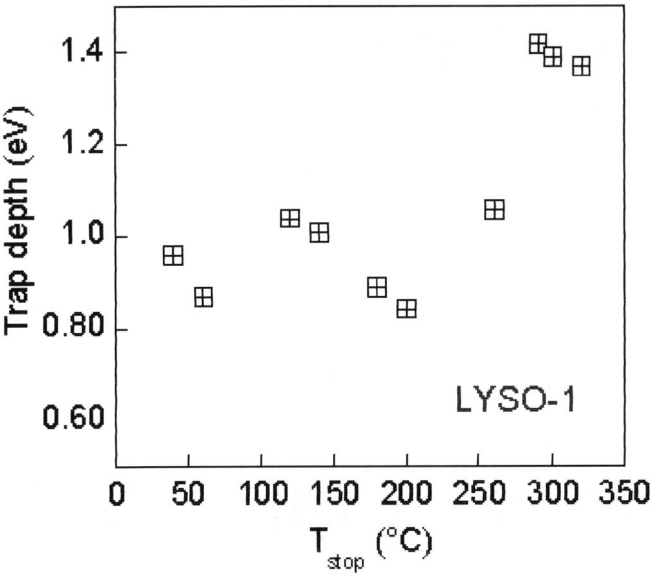

Fig. 6: Trap depths of the TSL glow peaks of LYSO:Ce0.5% no. 1 evaluated by the initial rise method.

3.4. $PbWO_4$

$PbWO_4$ single crystal became of interest for scintillation purposes in the early nineties due to its intended applications in high energy physics detectors [17]. Soon it was found that there are intense slow decay components both in the photoluminescence and scintillation decays especially in the extrinsic "green emission component" centered at about 500 nm [18]. In the green emission two different sub-components were resolved [19]. The so called G(I) component was observed mainly in the Mo-doped crystals, while the G(II) is based on the existence of isolated oxygen vacancies (i.e. WO_3 groups), persists up to higher temperatures with respect to G(I) and is the dominant source of the mentioned slow decay components at RT [20]. The correlated increase of both the G(II) emission and the overall scintillation efficiency was observed in the crystals grown from non-stoichiometric PbO-poor melt [21]. The radioluminescence spectra measured on samples taken from both the beginning and the end of the $PbWO_4$:Ba (2000 ppm) crystal are displayed in Fig. 7. Ba^{2+} cation has no role in the optical processes due to its stable charge state and no absorption transitions available within the $PbWO_4$ band gap. However, replacing PbO in the melt and having a considerably small segregation coefficient, BaO can induce a gradually increasing PbO-deficiency along the crystal growth axis. At the beginning of the crystal the spectrum shows the dominant blue emission component which originates from self-trapped excitons and is typical for stoichiometric crystals. However, at its end the intense green G(II) emission is observed due to increasing PbO-deficiency in the melt; so, the existence of WO_3 groups is enabled by the arising crystal nonstochiometry. The PL decay measured on the sample from the crystal end shows at 500 nm a very intense slow decay component up to tens of microseconds which has been observed in many $PbWO_4$ crystals [18, 22]. The decay of PWO:Ba at 500 nm is successfully analyzed by the combination of exponential and power law functions: the slow decay part follows a $\sim t^{-0.99}$ course (Fig.8), clearly revealing a tunnelling-driven radiative recombination. At variance, at 400 nm the dominant decay component shows an exponential course with a decay time of a few ns while slower decay components are very weak.

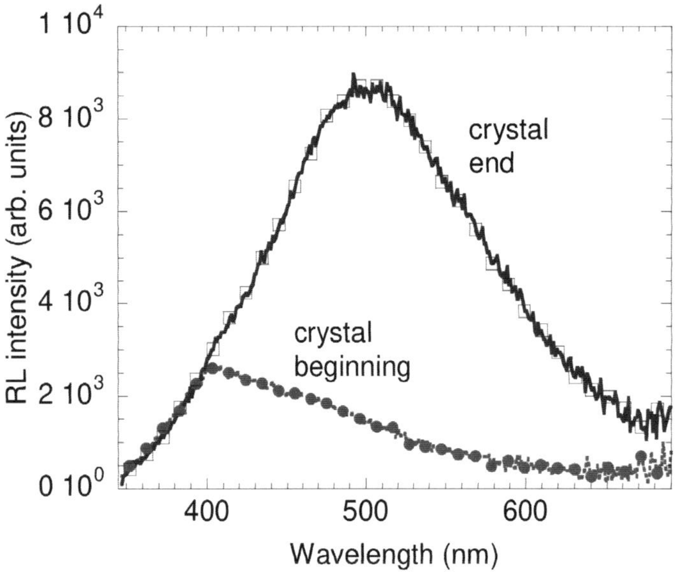

Fig.7: Radioluminescence spectra of Ba-doped (2000 ppm of BaO in melt) PWO at RT. Samples close to the crystal beginning and end are indicated.

Tunelling-driven radiative recombination in the green part of the spectrum suggests a spatially correlated localization of electron and hole pairs after the excitation. Excitation at 308 nm results in the creation of self-trapped excitons which are thermally disintegrated at RT so that various localized excitons appear [23]. Due to this process, excitons localized around oxygen vacancies or just electrons localized inside oxygen vacancies can arise. Such trapped electrons are localized at least for some milliseconds or even much more [2] and can serve as an anchoring element in the radiative recombination process around WO_3 group giving rise to the G(II) emission. The remaining question is how the nearby localization of the hole counterpart occurs in order to enable the tunnelling process: hole pairs can be in principle localized around lead vacancies [24]. Moreover, in partially La^{3+}-compensated $PbWO_4$ single O^- holes have been recently evidenced by EPR [2]. The requirement of the vicinity between electrons and holes for tunnelling-driven radiative recombination process can be fulfilled in spatially correlated lead and oxygen vacancy pairs which were indicated by EPR measurements [25].

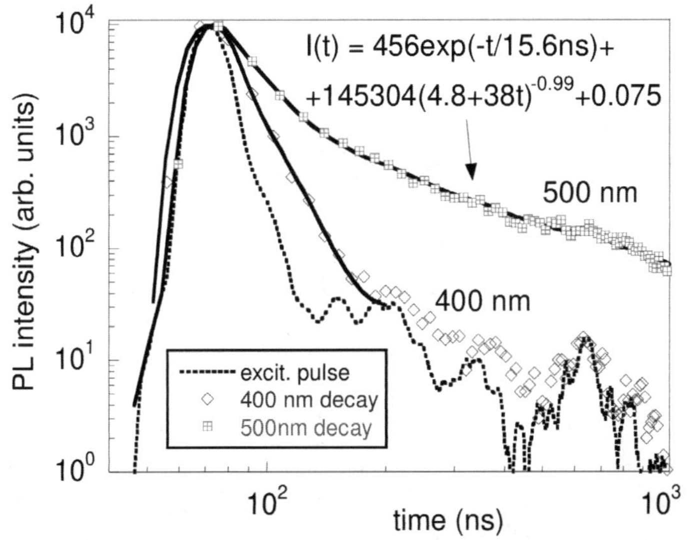

Fig. 8 PL decays in PWO:Ba (close to crystal end) at RT measured around the maximum of the blue (400 nm) and green (500 nm) emission bands. The former component decay is approximated by two exponential components (4.3ns (60%) and 17.6 ns (40%) decay times), while the latter is successfully fit by the sum of one exponential and one power law function *I(t)* reported in the figure. The numerical reconstruction of the decays are reported as black solid lines.

4. Conclusions

We reported several examples of tunnelling-driven radiative recombination processes in complex oxide scintillators. Apparently, this luminescence mechanism is considerably important in explaining the observed emission characteristics. However, our present measurements cannot really support the occurrence of important aggregation processes between traps and recombination sites, due to the difficulty to quantify the contribution of tunnelling recombinations with respect to the contribution of other delocalized processes to the luminescence/scintillation output.

Shallow electron traps adjacent to Ce^{3+} ions in YAG:Ce and LuAG:Ce enable a tunnelling process towards holes localised at Ce^{3+} ion. In YAP:Ce, the tunnelling process might be analogous, but for its detailed understanding other studies are necessary due to the more complicated charge carrier capture mechanism. The tunnelling process is evidenced by a $\sim t^{-1}$ form of the phosphorescence decay at low temperatures and might be responsible for slower scintillation decay components in the sub-μs time scale at RT.

In TSL, deep oxygen vacancy-based electron traps adjacent to Ce^{3+} ions in (Lu-Y)AP:Ce and (Lu-Y)SO:Ce give rise to thermally assisted tunnelling-driven radiative recombinations; this can explain the constant trap depth of several TSL peaks above RT, taking into account the occurrence of discrete distances between electron and hole traps defined by the crystal structure. Interestingly, no such a process was found in Ce-doped LuAG [26] possibly due to its more complex crystal structure obscuring the Ce-V_O tunnelling path.

In the G(II) green photoluminescence of $PbWO_4$ the tunnelling process in radiative recombination is evidenced from $\sim t^{-1}$ decay course at long times at RT. The anchoring role of electrons stored at RT temporarily around oxygen vacancies is probably the crucial element for its existence.

Acknowledgement
The financial support of Czech GA AV IAA100100810, MSMT KONTAKT ME953, ME08034, and AV CR-CNR joint projects is gratefully acknowledged. Thanks are due to B. Chai for providing us a sample of LYSO:Ce.

Literature

[1] Nikl M 2006 *Meas. Sci. Technol.* **17** R37
[2] Nikl M, Laguta V V and Vedda A, 2008 *Phys. Stat. sol. (b)* **245** 1701
[3] Huntley D J, 2006 *J. Phys. Cond. Matter* **18**, 1359
[4] Chang I F and Thioulouse P, 1982 *J. Appl. Phys.* **53**, 5873
[5] Zych E, Brecher C and Glodo J, 2000 *J. Phys. Cond.Matter* **12**, 1957
[6] Mihóková E, Nikl M, Mareš J A, Beitlerová A, Vedda A, Nejezchleb K, Blažek K, D'Ambrosio C, 2007 J. Lumin. **126**, 77
[7] Nikl M, Vedda A, Fasoli M, Fontana I, Laguta V V, Mihokova E, Pejchal J, Rosa J, Nejezchleb K, 2007 *Phys. Rev. B* **76**, 195121
[8] Fasoli M, Fontana I, Moretti F, Vedda A, Nikl M, Mihokova E, Zorenko Yu, Gorbenko V I, 2008 IEEE Trans. Nucl. Sci. **55**, 1114
[9] Nikl M, Laguta V V, Vedda A, 2007 *Phys.stat.sol. (a)* **204**, 683
[10] Vedda A et al, presented at ICDIM2008, August 25-29, 2008, Aracaju, Brasil.
[11] Dorenbos P, van Eijk C W E, Bos A J J, Melcher C L, 1994 *J. Phys.Cond. Matter* **6**, 4167
[12] Visser R, Melcher C L, Schweizer J S, Suzuki H, Tombrello T A 1994 *IEEE Trans. Nucl. Sci.* **41**, 689
[13] Cooke D W, Blair M W, Smith J F, Bennett B L, Jacobsohn L G, McKigney E A and Muenchausen R E, 2008 *IEEE Trans. Nucl. Sci.* **55,** 1118
[14] Mc Keever S W S, *Thermoluminescence of Solids* (Cambridge University Press, Cambridge, 1985).
[15] Vedda A, Martini M, Meinardi F, Mares J A, Mihokova E, Chval J, Dusek M, Nikl M, 2000 *Phys. Rev. B* **61**, 8081
[16] Vedda A et al, to be published
[17] Kobayashi M, Ishii M, Usuki Y and Yahagi H, 1993 *Nucl. Instrum. Methods Phys. Res. A* **333**, 429
[18] Nikl M, Nitsch K, Polak K, Mihokova E, Dafinei I, Auffray E, Lecoq P, Reiche P and Uecker E 1996 *phys.stat.sol. (b)* **195**, 311
[19] Babin V, Bohacek P, Krasnikov A, Nikl M, Stolovits A and Zazubovich S, 2007 *J. Lumin.* **124,** 113
[20] Laguta V V, Nikl M and Zazubovich S, 2008 *IEEE Trans. Nucl. Sci.* **55**, 1275
[21] Mares J A, Beitlerova A, Bohacek P, Nikl M, Solovieva N, D'Ambrosio C, 2005 *Phys. Stat. Sol. (c)* **2,** 73

[22] Pazzi G P, Fabeni P, Nikl M, Bohacek P, Mihokova E, Vedda A, Martini M, Kobayashi M and Usuki Y 2003 *J. Lumin.* **102–103**, 791
[23] Krasnikov A, Nikl M and Zazubovich S, 2006 *Phys. Stat.sol. (b)* **243** 1727
[24] Annenkov A A, Korzhik M V and Lecoq P, 2002 *Nucl. Instr.Meth. Phys. Research A* **490**, 30
[25] Laguta V V, Martini M, Vedda A, Rosetta E, Nikl M, Mihokova E, Rosa J and Usuki Y 2003 *Phys. Rev. B* **67**, 205102
[26] Vedda A, Di Martino D, Martini M, Laguta V V, Nikl M, Mihokova E, Rosa J, Nejezchleb K and Blazek K 2003 *phys. stat. sol. (a)* **195**, R1

EPR of radiation defects in lithium-oxyfluoride glass ceramics

A Fedotovs, U Rogulis, A Sarakovskis, L Dimitrocenko

Institute of Solid State Physics, University of Latvia, Kengaraga st. 8, LV-1063, Riga, Latvia

E-mail: andris-f@navigator.lv

Abstract. We studied oxyfluoride composites based on lithium silicate glasses with yttrium fluorides and rare-earth dopants. The electron paramagnetic resonance (EPR) has been used to obtain information about radiation induced defects in these materials. Spectra have been measured before and after X-ray irradiation at room temperature and at liquid nitrogen temperature. Fluoride crystallites within samples were created by means of thermal treatment at specific temperatures. EPR spectra of radiation induced defects in oxyfluoride glass ceramics, in which crystallites have not been yet created, show no explicit hfs interaction of fluorine nuclei. However, in glass ceramics, which already contains fluoride crystallites, the hfs characteristic to fluorine nuclei appears in the EPR spectra. EPR hyperfine structure could be explained within a model of an F-type centre in YF_3 crystalline phase.

1. Introduction

Composite materials, consisting of a glass matrix with introduced micro/nano-crystallites (called also glass-ceramics), can possess properties either of the glass matrix or the crystallites, as well as new properties, that extend the scope of their applications. Fluoride nano-crystals have been obtained recently in fluorozirconate glasses, a novel application of fluorozirconate glass-ceramic has been proposed as an x-ray storage phosphor [1]. Less investigated are fluoride crystallites in oxyfluorides, new interesting properties of the so-called up-conversion have been reported in [2,3]. Most of the studies concern the up-conversion process in oxyfluoride glass-ceramics doped with rare earth element activators. After a specific treatment, rare earth ions incorporate into the fluoride crystallites. The intensity level of the luminescence is higher than in the glass material. However, electron paramagnetic resonance studies are rare in this field.

2. Experimental

In the present work, we studied oxyfluoride composites based on lithium silicate glasses with fluorides, obtained at the Institute of Solid State Physics of University of Latvia. Oxyfluoride glass has been prepared from the following ingredients mixture containing 50 mol% SiO_2, 25 mol% Li_2CO_3, 20 mol% YF_3, 3 mol% ErF_3 and 2 mol% YbF_3. Fluoride crystallites within samples have been created by means of thermal treatment at approximately 555°C temperature for 20 minutes. To obtain information about radiation induced defects in these materials we used a standard X-band electron paramagnetic resonance (EPR) spectrometer. Spectra were measured before and after X-ray irradiation at room temperature (RT) and at liquid nitrogen temperature (LNT).

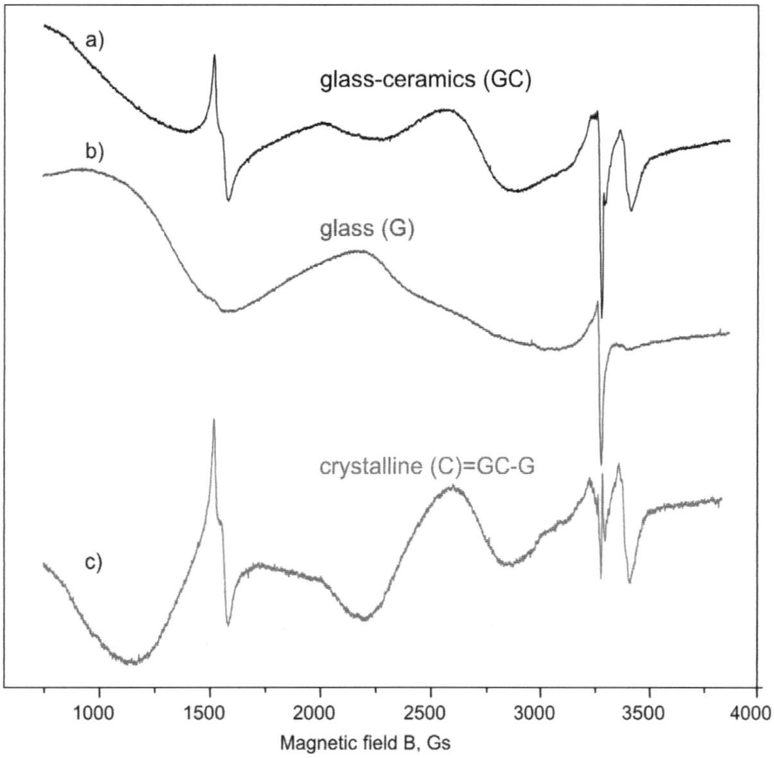

Figure 1. Experimental X-band EPR spectra of radiation defect in a) oxyfluoride glass-ceramics (GC), b) oxyfluoride glass (G) and c) in crystalline phase of glass-ceramics (C) measured at 77K after X-irradiation at RT.

3. Results

EPR spectra of radiation induced defects were obtained in both glass and glass-ceramics materials. No EPR signal could be detected prior to irradiation. After the irradiation various absorption lines appeared in the EPR spectra (see figure 1). Comparing EPR spectra of the glass and the glass-ceramics, considerable qualitative additional features could be observed in the glass-ceramics: 1) intense line at 1500 Gs, could probably be attributed to a Fe^{3+} impurity in oxide glasses, 2) two broad bands between 1500 Gs and 3000 Gs 3) two broad lines at 3250 Gs, one with a g-factor g = 1.975 obscuring partially resolved hyperfine structure (hfs) of the spectrum (see details in figure 2). A sharp line at 3200 Gs is common in both material states and is attributed to some typical irradiation defect in silicate glasses. Radiation induced defects in oxyfluoride glass ceramics, in which crystallites have not been yet created, show no explicit hfs interaction of fluorine nuclei. Hyperfine lines could be resolved only at 77 K and after warming to RT the hfs spectrum disappears.

4. Discussion

Our analysis of the obtained EPR spectra is based on our previous knowledge about radiation induced defects in fluoride single crystals. EPR spectra of such defects often showed well-pronounced hyperfine structure (hfs) caused by interaction with neighboring flourine nuclei (LiF [4], $LiBaF_3$ [5] and others).

To separate EPR spectra of the glass and the glass-ceramics spectra the glass spectrum was subtracted from the spectra of the glass ceramics (see figure 1c) so distinction between both is seen in more details.

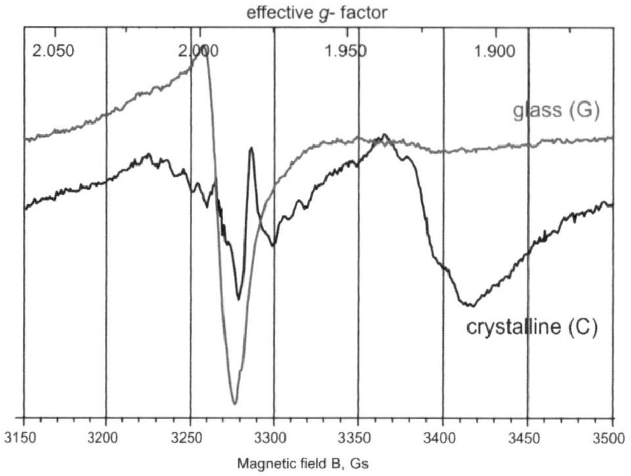

Figure 2. Part of the experimental EPR spectra of radiation defects in oxyfluoride glass (G) and glass-ceramics, whereby only the crystalline part (C) is shown.

There are several possibilities for creation of crystalline phases. Most probable phases are as following: oxides – $LiYSiO_4$, Y_2SiO_5, Li_2SiO_3 and fluorides – YF_3, LiF and $LiYF_4$. It is well known that hfs interaction with fluorine nuclei often leads to a well-pronounced hfs structure in the EPR spectra. EPR of intrinsic defects induced with irradiation in fluoride single crystals are well studied [4,5,6]. g-factor <2 indicates that our discussed defect is rather an electron centre. In our case, an F-type centre as a most probable model could be considered. Similar centres have been reported in LiF and $LiBaF_3$ [4,5].

Spin Hamiltonians (1) and (2) could describe the interaction of a paramagnetic electron with its nearest surrounding ions with nonzero nuclear spins in the case of interaction with fluorine and lithium ions:

$$\hat{H} = \mu S\hat{g}B + \sum_{i=1}^{N_F} SA_i^F I_i^F + \sum_{j=1}^{N_{Li}} SA_j^{Li} I_j^{Li} \tag{1}$$

$$\hat{H} = \mu S\hat{g}B + \sum_{i=1}^{N_F} SA_i^F I_i^F + \sum_{j=1}^{N_Y} SA_j^Y I_j^Y \tag{2}$$

Nuclear spins of possible fluoride crystalline components are $I^Y=1/2$, $I^F=1/2$, $I^{Li}=3/2$ (only Li^7 isotope is taken into account). A number of specific nuclei is denoted accordingly with N_F, N_Y and N_{Li}. For evaluation of a proper number of nearest neighbors a crystalline structure of possible crystalline phases should be considered. Hyperfine structure in this case should be caused by a certain number of equidistant fluorine nuclei N_F, equidistant yttrium nuclei N_Y and/or equidistant lithium nuclei N_{Li}. Since the hfs of the spectrum is partially obscured by the two broad EPR lines, we cannot precisely resolve the number of lines. Thus we base our simulated spectra on a qualitative coincidence with the common width of the spectrum and relative intensities of the hfs lines. The most probable model of the crystalline phase would be that of YF_3. It allows us to consider a hfs interaction of the electron with $N_F=10$ and $N_Y=4$ nuclei. Resulting simulated spectrum shows a rather good coincidence with the experimental spectrum (see figure 3). Other possibilities, which included an hfs interaction with Li nuclei, do not seem so reasonable. In any case, by adding of two lithium nuclei to the model, the number of fluorine nuclei should be decreased by at least three nuclei, which is not reasonable in any considerable model of the crystalline phase.

The hfs of the F-centre in LiF includes hf interaction with six Li^7 and two F^{19} nuclei [4], but F-type center in $LiBaF_3$ possess hf interaction with two Li^7 and eight F^{19} nuclei [5]. Thus the Li hf interaction has been excluded from our present analysis. The XRD data on our samples confirm possibility of both YF_3 and $LiYF_4$ phases in the material. However, it is still not clear, which crystallographic structure obtained crystallites could possess.

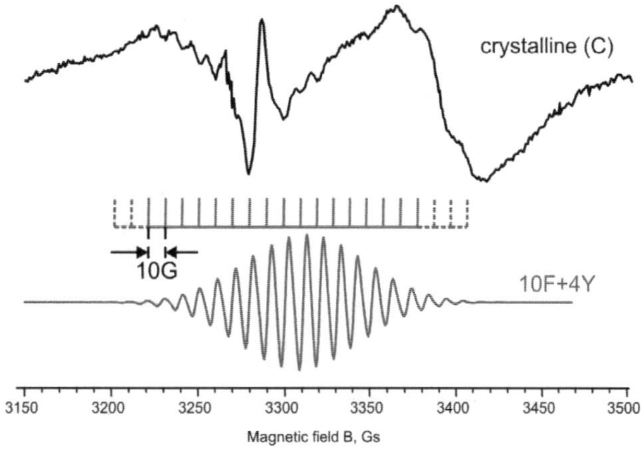

Figure 3. Experimental EPR spectra of radiation defect in oxyfluoride glass and glass-ceramics compared with a simulated EPR spectrum taking into account hfs with 10 fluorine and 4 yttrium nuclei.

5. Conclusions

Radiation defects have been studied in X-irradiated glass-ceramics crystalline phases, where clearly distinguishable hfs interaction in the EPR spectrum could be observed. Hyperfine structure of the spectrum could be explained within a model of an intrinsic radiation defect - F-type centre in the YF_3 crystalline phase. Following isotropic parameters have been estimated: g =1.975 , A_F = 56 MHz, A_Y = 23 MHz.

References

[1] Edgar A, Williams G V M, Sagar P K D, Secu M, Schweizer S, Spaeth J M, Hu J, Newmann P J and MacFarlane D R 2003 *J. Non-Cryst. Solids* **326-327** 489

[2] Dejenka M J 1998 *MRS Bulletin* **23** 57

[3] Chen D, Wang Y, Yu Y, Ma E, Bao F, Hu Z and Cheng Y 2006 *Materials Chem. and Phys.* **95** 264

[4] Kaplan R and Bray P J 1963 *Phys. Rev* B **129** 1919

[5] Fedotovs A, Elsts E, Rogulis U, Gulans A, Tale I, Nikl M, Ichinose N and Shimamura K 2007 *Phys. Status Solidi* (c) **4/3** 1284

[6] Fedotovs A, Rogulis U and Dimitrocenko L 2006 *Latv. J. of Phys. and Techn. Sciences* **6** 60

Growth and properties of epitaxial Ce-doped YAG and LuAG films for scintillators

M Kučera[1], K Nitsch[2], M Nikl[2], S Daniš[1], M Hanuš[1]

[1] Charles University, Faculty of Math. & Physics, Ke Karlovu 5, 12116 Prague 2, Czech Republic
[2] Institute of Physics v.v.i., Academy of Sciences CR, Cukrovarnická 10, 16200 Prague 6, Czech Republic

E-mail: kucera@karlov.mff.cuni.cz

Abstract. We have studied the growth and properties of the Ce^{3+} - doped epitaxial garnet films. The single crystalline films were prepared by liquid phase epitaxy. Various fluxes were used with the goal to reduce the flux-related impurities, which could have a negative influence on the optical and scintillation properties. Crystallographic properties, film morphology and emission properties were studied in detail.

1. Introduction

Thin epitaxial films of Ce^{3+}-doped $Y_3Al_5O_{12}$ (yttrium aluminum garnet, YAG) and $Lu_3Al_5O_{12}$ (lutetium aluminum garnet, LuAG) are promising candidates for 2D scintillation imaging screens with submicrometer resolution, where they can replace polished thin plates of single crystals [1]. Here we report on films grown by the liquid phase epitaxy. The advantage of epitaxial films compared to Czochralski (CZ) grown single crystals is much lower growth temperature (T_g ~ 1000 °C) and consequently lower structural disorder and lower number of intrinsic crystal defects [2, 3]. Furthermore, higher content of Ce^{3+} ions can be incorporated into the garnet lattice [4]. This is of fundamental importance in order to obtain reasonable light yield from films of micrometer thickness. On the other hand, various impurities can enter the lattice from the flux and other factors (such as substrate orientation, film-substrate lattice match, morphology of film-substrate interface, flux composition, purity of starting chemicals, possible second phases) can have a negative impact on the emission and scintillation properties of films. The films prepared from the standard $PbO\text{-}B_2O_3$ flux have very good structural and surface properties, however, they contain divalent and tetravalent Pb and Pt ions [5]. As a result, a variety of point defects are created, which result in electron traps or appearance of the slow component in the luminescence decay of Ce^{3+} centers [6]. Moreover, shallow electron traps due to the antisite Lu_{Al} defects (which are typical for single crystals grown from the melt at high temperatures) are the most probable origin of the submicrosecond slow component in the scintillation decay [7].

Recently, we have shown that the Ce-doped garnet films prepared from the lead-free BaO-based flux exhibit excellent optical and emission properties and much higher Ce concentration can be achieved in garnet films compared to Czochralski grown crystals [4]. However, the BaO flux suffers from high viscosity, which has negative effect on the film surface morphology. The aim of this work is study of crystallographic, surface, optical and emission properties of epitaxial garnet films prepared from various fluxes. An important question is the morphology of both the film surface and film/substrate interface, which may be critical especially in micrometer thick films.

2. Growth of epitaxial garnet films

The single crystalline epitaxial garnet films were grown by the isothermal dipping liquid phase epitaxy (LPE) onto single crystalline YAG and LuAG substrates of (111) and (110) crystallographic orientation. Starting raw materials of 5N purity were used. Several series of Ce-doped YAG and LuAG films were grown from various fluxes, namely, PbO-B_2O_3, BaO-B_2O_3-BaF_2, and MoO_3-Li_2MoO_4 fluxes were used. We refer these fluxes here as *PbO flux*, *BaO flux*, and *MoO_3 flux*, respectively. The growth temperature was in the range from 930 to 1080 °C and the growth rates were from 0.04 to 1.8 μm/min in dependence on used flux and melt composition. The epitaxial films prepared from the PbO and BaO fluxes were grown from supercooled melt at undercooling 3 – 60 K, the films from the MoO_3 flux were grown at a temperature gradient. Details of LPE growth from PbO and BaO fluxes were published elsewhere [4]. The thickness of films was in the range from 2 to 24 μm. Ce concentration in films was from 0.04 to 0.8 at. % depending on the growth conditions and used melt. All the films show an intense green luminescence.

The epitaxial films prepared from the PbO flux exhibited excellent crystallographic properties, see the next sections, and high quality smooth surface. Films with superior luminescent properties were obtained from the BaO flux. The photoluminescence decay curves do not indicate any energy loss or quenching of luminescence. Besides, due to higher segregation coefficient of Ce ions, higher Ce content can be achieved in films grown from the BaO flux compared to the Czochralski single crystals. However, high viscosity and high surface tension of this flux resulted in more complicated film growth and worse surface morphology.

In order to eliminate the charged impurities (typical for the PbO melt) and to improve the melt viscosity (drawback of the BaO melt) we used, as a novelty, the MoO_3 based flux. This flux has essentially zero volatility, low reactivity with the Pt crucible, low viscosity and provided films with smooth surface. The surface morphology was comparable with films grown from the PbO flux, however, the growth rate are rather small, < 0.1 μm/min, and concentration of Ce ions in films was smaller compared to films grown from other two fluxes.

In this work we will study in more detail also the garnet films grown onto various substrates, which differ in lattice constant. Namely, Ce:LuAG films were grown either on LuAG substrates – we shall refer this growth as *homoepitaxial* (the lattice constants of the substrate and film are virtually the same, $\Delta a < 0.02$ %). The Ce:LuAG films grown on YAG substrates we refer as *heteroepitaxial* (the lattice mismatch is rather high, $\Delta a = 0.8$ %).

3. Properties of epitaxial films

We have studied the crystallographic properties of the LPE films using the X-ray diffraction (XRD), the surface morphology and surface profiles using the optical microscopy and the interferometric surface profiler ZYGO. Optical properties were studied using the optical absorption spectroscopy, luminescence emission and excitation spectroscopy, and Ce^{3+} photoluminescence decay was studied, too. Scintillation characteristics and photoelectron yield of the films are published elsewhere [8].

3.1. X-ray diffraction

The crystallographic structure of the LPE films was studied by means of XRD on PANalytical high-resolution X-ray diffractometer X'Pert Pro MRD (Cu $K_{\alpha 1}$ peak). The X-ray beam was conditioned using the x-ray mirror followed by four bounce Bartels monochromator (4xGe220). Diffracted beam was registered using a three-bounce channel-cut analyser (Ge220) and proportional point detector.

We shall first discuss homoepitaxially grown films – Ce:YAG films grown on YAG substrates and Ce:LuAG film on LuAG substrates. The diffraction profile of a symmetric 444 diffraction of a Ce:YAG film (with ~0.13 at.% of Ce) grown on YAG substrate from the PbO flux is shown in figure

1. Lower intensity of substrate diffraction is due to radiation absorption in 11 μm thick film. From the position of the 444 diffraction peak in this sample, 2θ = 52.7583(1) deg, the lattice constant in the direction normal to the film surface was determined as $a_\perp$ = 12.0114(1) Å. The measured lattice constant of the YAG substrate was a_S = 12.0093(1) Å. The increased lattice parameter of films corresponds to doping of large Ce^{3+} ions (ionic diameter 1.143 Å), which are substituted for Y^{3+} ions (1.016 Å) in dodecahedral sites.

On the other hand, the XRD of Ce:LuAG films shows, that the measured lattice parameter of films is even lower than that of LuAG substrates. This can be explained due to increased number of antisite Lu_{Al} defects in LuAG substrates, which is several times higher than that in YAG substrates. Namely, due to the small ionic radius of Lu^{3+} ions (0.975 Å), some part of these ions occupies octahedral sites instead of regular Al^{3+} ions, and these defects increase the lattice parameter of LuAG substrates compared to the stoichiometric crystal. In epitaxial films, however, the concentration of these antisite defects is presumed to be much smaller [2, 6, 7] due to much lower growth temperature of films compared to that of the Czochralski grown crystals.

Observed halfwidths of 444 diffraction peak in epitaxial film, 12 ang. sec, are comparable to those of substrate, 10 ang. sec, giving evidence of high crystallographic quality of the epitaxial films.

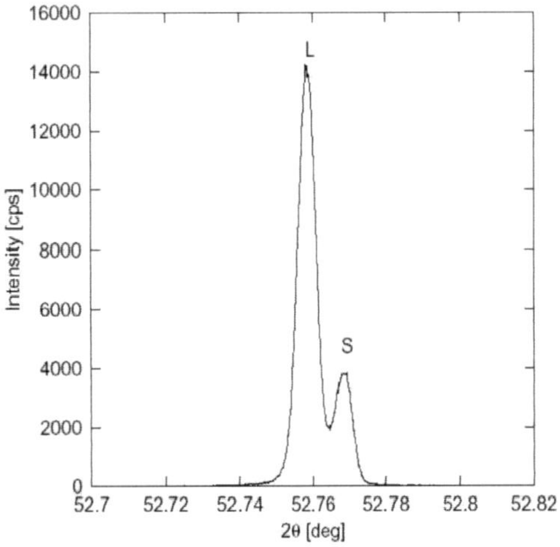

Figure 1. Diffraction profile of 444 diffraction (Cu-$K_{\alpha 1}$) of 10.7 μm thick Ce:YAG film grown on YAG substrate from the PbO-flux, L and S are respective diffractions from layer and substrate.

In order to explore the relaxation state of studied layers, asymmetrical reciprocal maps (RSM) were measured. The RSM around the asymmetric diffraction 664 suggests, figure 2, that homoepitaxially grown samples, i.e. Ce:YAG/YAG or Ce:LuAG/LuAG, showed pseudomorphic film on substrate (the intensities of scattered X-rays in the reciprocal space exhibit the same Q_x coordinate in the RSM), i.e. the film lattice parameter in the plane was the same as that of substrate, $a_\parallel = a_{YAG}$. Since $a_\perp$ differs < 2×10^{-4}, any misfit dislocations are not expected in the film/substrate interface and a macroscopic elastic strain is present in the film.

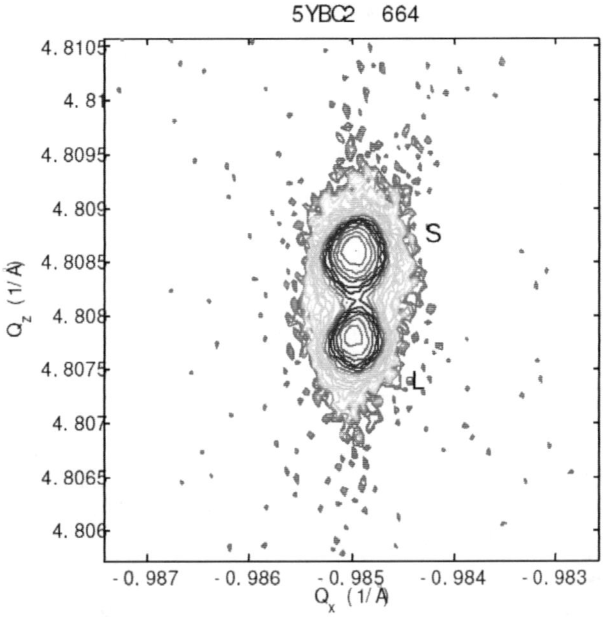

$a_{\perp}$ = 12.010(1) Å
$a_{\parallel}$ = 12.009(2) Å
a_{YAG}= 12.009(1) Å

Figure 2. Asymmetrical reciprocal map of 664 diffraction of Ce:YAG layer grown on (111) oriented YAG substrate from the BaO flux, thickness t = 6.9 μm, S = substrate, L = layer. A pseudomorphic layer was grown on substrate.

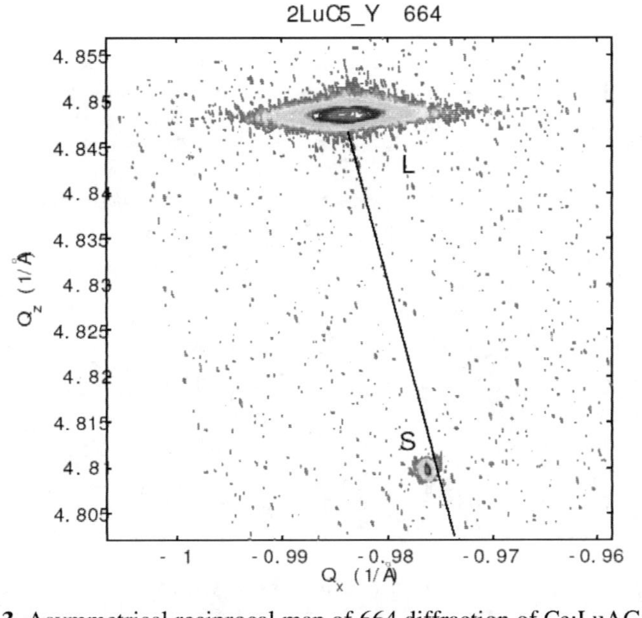

$a_{\perp}$ = 11.918(1) Å
$a_{\parallel}$ = 11.918(2) Å
a_{YAG}= 12.009(1) Å

Figure 3. Asymmetrical reciprocal map of 664 diffraction of Ce:LuAG layer grown on (111) oriented YAG substrate from the PbO flux, t = 13.3 μm, S = substrate, L = layer. A relaxed layer was grown on substrate.

Quite different situation was observed for heteroepitaxial growth of Ce:LuAG film grown onto YAG substrate where the lattice mismatch is too large, $\Delta a = 0.8$ %, for the pseudomorphic growth. In this case a relaxation of a film was observed and the lattice parameters of film and substrate differ. The RSM around the asymmetric diffraction 664 for Ce:LuAG/YAG system is shown in figure 3. The lattice parameter of Ce:LuAG film is virtually the same as that of LuAG crystal, a = 11.916 Å. Therefore the misfit dislocations that reveal misfit strain are expected in the film/substrate interface. The RSM of film diffraction is broad compared to the substrate and it shows that macroscopic defects are present in films. Nevertheless, the XRD proved that the films are of single garnet phase and the films exhibit a distinct 111 orientation.

3.2. Film surface and film-substrate interface properties

The film surface properties were studied using the optical microscopy and optical interferometry. The surface roughness of the films was measured by means of a noncontact 3D surface profiler Zygo, model New View 5022. It uses scanning white light interferometry to acquire ultrahigh z-resolution images. Typical surface profile of Ce:LuAG epitaxial film grown onto high quality LuAG substrate from the PbO flux is shown in figure 4a. Rather smooth surface was observed, see also figure 5a, and rms surface roughness is only a few nanometers.

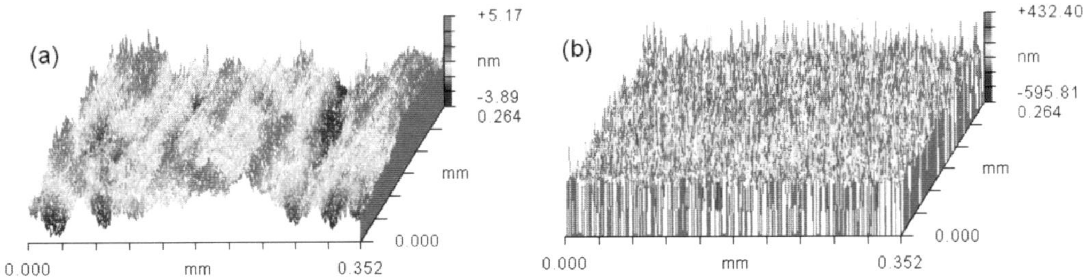

Figure 4. The surface profile measured using the ZYGO interferometer. (a) Ce:LuAG film on LuAG substrate (homoepitaxial growth), smooth surface with rms surface roughness of ~ 2 nm is observed; (b) Ce:LuAG film on YAG substrate (heteroepitaxial growth, lattice misfit $\Delta a = 0.09$ Å), rms surface roughness ~ 200 nm.

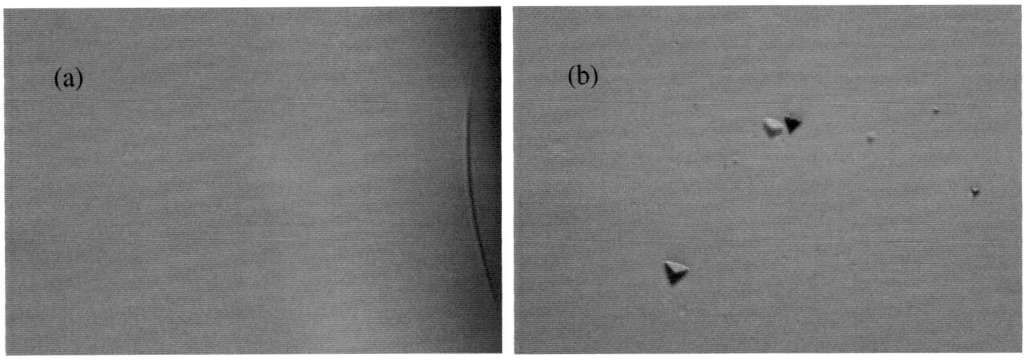

Figure 5. Optical microscopy of the surface of Ce:YAG films grown on (111) oriented YAG substrate, magnification 300×. (a) Defect-free surface of a film grown from the PbO flux (for comparison, the sample edge is shown on the right side of the picture). (b) Triangular pits in a film grown from the MoO_3 flux. Such pits are typical for 111 orientation and come from the substrate defects or defects induced during polishing of the substrate.

However, surface properties of epitaxial films are strongly dependent on crystallographic quality and final polishing procedure of used substrates. Any substrate defect, such as dislocation or inclusion, is replicated as a defect into the LPE film. Figure 5b shows triangular pits observed typically in films grown on 111 oriented substrates. These defects of μm dimensions come from the substrate defects, either intrinsic or induced during the polishing. In films grown from the PbO flux on high quality and well polished substrates such defects can be practically eliminated. However, increased number of these pits was observed in films grown from the BaO flux presumably due to high viscosity of this flux.

At heteroepitaxial growth (Ce:LuAG film on YAG substrate, where the lattice mismatch is 0.8 %), the rms surface roughness can be as high as several hundreds of nm, figure 4b, and surface of such films observed by naked eye is matte and strongly scatters light. In order to compare optical properties of these heteroepitaxial films with homoepitaxial ones, the uppermost surface 3 μm of films were polished out. After this repolishing a glossy surface was obtained. The transmission spectra of Ce:LuAG film grown on LuAG and YAG substrates from the PbO flux are presented in the next section, cf. figure 7.

An important issue is the morphology of the film surface and film/substrate interface. Large lattice mismatch between the Ce:LuAG film and YAG substrate at heteroepitaxial growth, $\Delta a = -0.09$ Å, is the main problem in obtaining films of high perfection. Of real importance, especially for thin films, is formation of a transient layer at the film-substrate interface. In the transient layer the chemical composition changes and a crucial issue is its contamination by impurity ions from the flux and presence of other defects. For low lattice mismatch < 0.02 Å, this transient layer is usually sharp and for homoepitaxial films its thickness is of the order of nanometers. For higher lattice mismatch the morphology of the transient layer is more complicated. Optical microscopy of edge of samples cleaved normal to the film-substrate interface is shown in figure 6. Sharp film-substrate interface is observed for homoepitaxial growth of the Ce:LuAG/LuAG system, where lattice constants of the film and substrate merge. Due to almost the same chemical composition, the optical contrast between film and substrate is very weak and the film-substrate interface is hardly observable. On the other hand at heteroepitaxial growth of Ce:LuAG/YAG system, where the lattice mismatch is rather high, the film-substrate interface is profound and oriented columnar like structure of the layer seems is observed. The morphology of the transient layer may be critical especially in micrometer thick films and detailed study of its structure is called for.

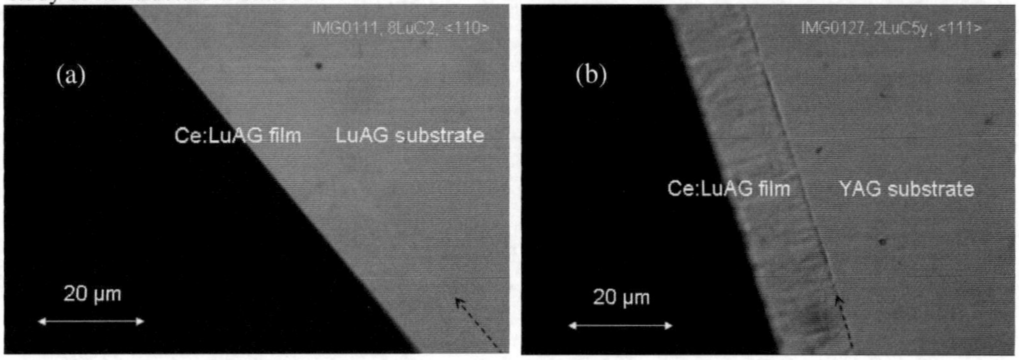

Figure 6. Edge of fractured samples observed in optical microscope; the arrows indicate the film/substrate interface. (a) Homoepitaxially grown Ce:LuAG film on LuAG substrate, lattice constants of the film and substrate are almost the same, $\Delta a < 0.02$ %, and sharp film-substrate interface is observed. (b) Heteroepitaxially grown Ce:LuAG film on YAG substrate, the lattice mismatch is rather high, $\Delta a = 0.8$ %, the film-substrate interface is profound. Both films were grown from the PbO flux.

3.3. Optical absorption and photoluminescence

Optical transmission measurements were performed in the spectral range from the ultraviolet to the near infrared, 200 – 1100 nm. Comparison of absorption spectra of Ce:YAG films grown from the PbO and BaO fluxes was published elsewhere [4]. We only mention here, that the Pb^{2+} ions, always present in film grown from the PbO flux, give rise to the strong absorption below 280 nm and optical transparency of films grown from the BaO flux is better. In figure 7 the optical transmission spectra of homo- and heteroepitaxial Ce^{3+}:LuAG films grown either on LuAG or YAG substrates are compared. The Ce^{3+} absorption bands are clearly resolved at 445 and 345 nm, but in the ultraviolet spectral range the Ce spectral bands are overlapped by Pb^{2+} absorption. The absorption peak situated near 262 nm originates from Pb^{2+} impurity ions [9, 6]. Owing to high undercooling (40-60 K) of the melt during growing of these two samples the relative concentration of lead ions is rather high, hundreds of ppm. The first results indicate that the content of Ce^{3+} ions and also the lead ions is higher in heteroepitaxial films compared to homoepitaxial ones grown under the same conditions. Surprisingly very close relation of both the spectra in figure 7 is observed. We note that the transmission spectra were not normalized and the same intensity of 445 nm spectral band is accidental.

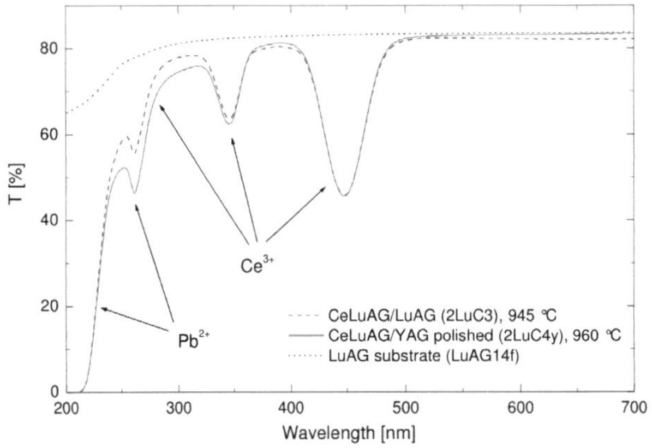

Figure 7. Optical transmission spectra of epitaxial Ce:LuAG films grown on LuAG substrate (dashed line, sample 2LuC3, t = 8.9 μm) and on YAG substrate (solid line, sample 2LuC4y, t = 14 μm) grown from the PbO flux. Absorption peaks from Ce^{3+} and impurity Pb^{2+} ions are indicated. Transmission spectrum of Czochralski grown LuAG substrate is shown for comparison, dotted line. The surface of Ce:LuAG film grown on YAG was polished before measurements.

The photoluminescence (PL) excitation spectra of Ce^{3+} emission, λ_{em} = 530 nm, of Ce:YAG samples grown from the MoO_3 flux are displayed in figure 8a. The structure in the spectra comes from the Ce^{3+} ions and it shows a close correspondence with the absorption curves in figure 7. In spite of strong absorption in the UV range, the Ce^{3+} contributions in the PL excitation spectra at 220 – 270 nm are distinguished. The PL emission spectra at λ_{exc} = 340 nm are shown in figure 8b. Below 470 nm the PL emission has a negligible intensity.

3.4. Decay kinetics of photoluminescence

In order to obtain more detailed information on the role of impurities in the PL emission of epitaxial garnet films, we measured the PL decay kinetics. Decay curves were measured by single photon counting method, the samples were excited at 340 nm and the PL Ce^{3+} emission was collected at 530 nm.

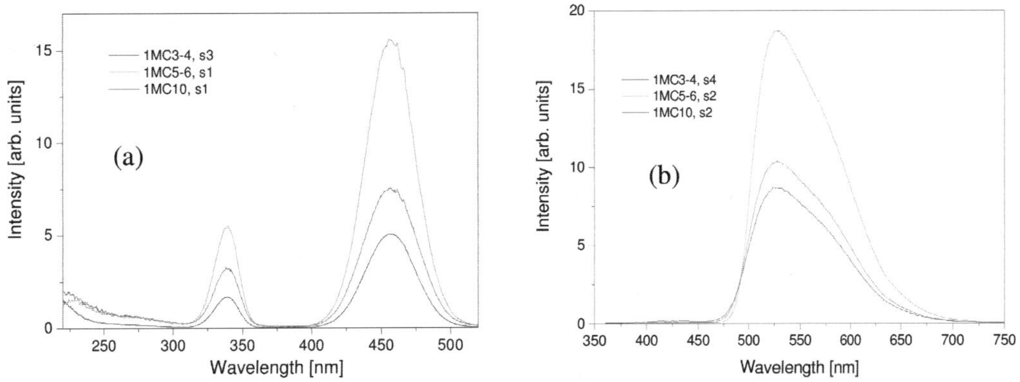

Figure 8. Excitation spectra (left) measured at λ_{em} = 530 nm and emission spectra (right) for λ_{exc} = 345 nm of Ce:YAG epitaxial films grown from the MoO_3 flux with various concentration of Ce^{3+} ions.

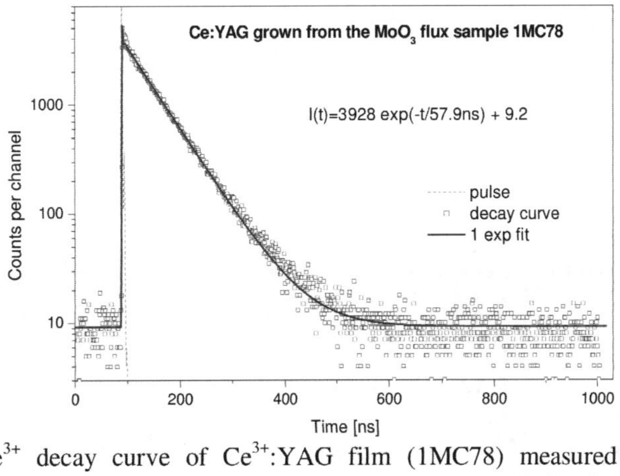

Figure 9. The Ce^{3+} decay curve of Ce^{3+}:YAG film (1MC78) measured at 530 nm at room temperature, excitation at 340 nm. The solid line is convolution of the fitting curve I(t) with the instrumental response.

The PL decay curve for a typical Ce^{3+} doped YAG films grown from the MoO_3 flux is shown in figure 9. The curve exhibit 1-exponential decay with τ ~ 58 ns, which is in close relation with values obtained for high purity Ce-doped single crystals.

The PL decay from Ce:YAG and Ce:LuAG films grown from PbO and BaO fluxes were reported elsewhere [4]. We summarize only that detailed inspection of Ce decay curves of films grown from the PbO flux shows on some nonlinearity and a two-exponential fit provides better correspondence. This indicates a nonradiative transfer of part of energy between Ce^{3+} and impurity ions or defect states. The decay curve of Ce:YAG films grown from the lead-free BaO flux exhibit linear dependence over three orders of magnitude indicating 1-exponential decay and the decay time τ ~ 60 ns.

Therefore we can conclude that in films grown from a flux, which does not contain lead, the PL decay curves do not indicate any loss of energy in contrast to the films grown from the PbO flux.

4. Conclusions

The garnet films were grown by the liquid phase epitaxy from three different fluxes: $PbO\text{-}B_2O_3$, $BaO\text{-}B_2O_3\text{-}BaF_2$, and $MoO_3\text{-}Li_2MoO_4$. The films grown from the PbO flux exhibit relatively high structural perfection and high quality surface. However, divalent Pb^{2+} ions in concentrations 0.01 – 0.1 at. % (depending on the growth conditions and undercooling) are always incorporated into the crystal lattice They cannot be easily charge compensated in YAG or LuAG and, as a consequence, lower emission efficiency is observed in such films. Quite the opposite, the films prepared from lead-free fluxes based either on BaO or MoO_3 do not contain any charged impurities, which influence the emission properties. However, the surface morphology of films grown from BaO is inferior due to high viscosity of the melt. The MoO_3 flux has low reactivity with the Pt crucible, low viscosity, and the films do not contain any solvent-related impurities, which would quench the luminescence. Very good emission properties of these films and one-exponential decay of 530 nm emission of Ce^{3+} with 58 ns decay time were observed.

Important advantages of epitaxial films compared to Czochralski grown single crystals are, first, lower growth temperatures and, as a consequence, lower number of intrinsic crystal defects [2, 3] and, second, higher content of Ce^{3+} ions (by a factor of 2-5), that can be incorporated into the garnet lattice, cf. ref. 4. This is of fundamental importance in order to obtain reasonable light yield from films of micrometer thickness. The films grown from the BaO or MoO_3 fluxes exhibit much higher purity and PL decay curves do not indicate any loss of energy in contrast to the PbO-grown films.

Acknowledgments

This work was supported by the Czech science foundation, grant KAN 300100802, and from the Ministry of education CR, research plan MSM21620834.

References

[1] Martin T, Koch A 2006 *J. Synchrotron Rad.* **13** 180

[2] Zorenko Yu, Gorbenko V, Konstankevych I, Voloshinovskii A, Stryganyuk G, Mikhailin V, Kolobanov V, Spassky D 2005 *J. Lumin.* **114** 85

[3] Nikl M, Mihokova E, Pejchal J, Vedda A, Zorenko Y, Nejezchleb K 2005 *phys. stat. sol. (b)*, **242** R119

[4] Kučera M, Nitsch K, Kubová M, Solovieva N, Nikl M, Mareš J A 2008 *IEEE Trans.Nucl.Sci.* **55** 1201

[5] Robertson J M, van Tol M W 1984 *Thin Solid Films* **114** 221

[6] Babin V, Gorbenko V, Makhov A, Mares J A, Nikl M, Zazubovich S, Zorenko Yu 2007 *J. Lumin.* **127** 384

[7] Nikl M, Vedda A, Fasoli M et al. 2007 *Phys. Rev. B* **76** 195121

[8] Prusa P, Nikl M, Mares J A, Kucera M, Nitsch K, Beitlerova A 2009 *phys. stat. sol. (a)* **206** 1494

[9] Scott G B, Page J L 1977 *J. Appl. Phys.* **48** 1342

Journal of Physics: Conference Series **249** (2010) 012021 doi:10.1088/1742-6596/249/1/012021

Stability and Influence of Dose Rate in the Signal Response of the Alanine Gel Dosimeter

Cléber Feijó Silva, Letícia Lucente Campos

Centro de Metrologia das Radiações, Instituto de Pesquisas Energéticas e Nucleares, São Paulo, Brazil

E-mail: cfeijos@yahoo.com.br, lcrodri@ipen.br

Abstract. Nowadays, the three dimensional (3D) dosimetry has been studied by many researchers for application in radiotherapy and radiosurgery treatments, because this kind of dosimeter enable to evaluate the 3D dose distribution. The alanine gel dosimeter was developed at IPEN to be applied to 3D dose evaluations using magnetic resonance imaging. This dosimeter is a gel material based on the transformation of the ferrous ions in ferric ions, which presents significant improvement on previous alanine systems developed by Costa (1994). The DL-Alanine ($C_3H_7NO_2$) is an amino acid tissue equivalent that improves the production of ferric ions in the solution. The ferric ions concentration can be measured by spectrophometry technique. This work aims to analyse the stability and influence of the dose rate in the optical signal response of alanine gel dosimeter, since these two properties are very important for characterization and standardization of any dosimeter.

1. Introduction

In current days, an useful tool for the verification of radiation treatments in water-equivalent tissues is the gel dosimetry, because the determination of absorbed dose is very important in the delivered dose quality control to the interest volume. The first publication in Gel Dosimetry area was in 1984 by Gore et al [1], when the Fricke solution was incorporated into a gel matrix and this system was combined with magnetic resonance imaging to make possible three-dimensional radiation dosimetry. Therewith it was born the modern gel dosimetry [2].

Gel dosimeters have been studied using different compositions of the dosimetric solution and gel materials such as organic gels or polymer gels [3,4]. The High Dose Laboratory of IPEN developed a alanine gel dosimeter based on the alanine dosimetric solution proposed by Costa (1994) [5] using spectrophotometry and electronic paramagnetic resonance – EPR evaluation techniques and improved by Mizuno (2007) [6] with the addition of gelatin at the dosimetric solution and using spectrophotometry as evaluation technique aiming to obtain a gel dosimeter enable to evaluate 3D dose distribution using MRI technique. In this dosimeter the production of ferric ions radiation induced in the solution is improved with addition of the alanine and, consequently, the dose response signal is higher than Frick gel dosimeter dose response.

This work aims to analyze the stability and influence of dose rate in the signal response of the alanine dosimeter developed by Mizuno (2007) [5] irradiated with ^{60}Co gamma radiation, because these dosimetric properties are of crucial importance for characterizing and standardizing a dosimetric system [7]. The concentrations of ferrous (Fe^{2+}) and ferric (Fe^{3+}) ions radiation induced in the dosimetric system were measured by means of spectrophometry technique.

2. Materials and Methods

2.1 Alanine gel solution

The method described by Mizuno (2007) using 300 Bloom gelatin was followed to prepare the dosimetric solution. The table 1 shows the chemical composition of the studied solution.

After preparation the solution was conditioned in plastic cuvettes (1 x 1 x 4.5 10^{-6} m^3) and optical path of 10^{-2} m.

Table 1: Chemical composition of Alanine gel solution.

Compound	C (mol/L)
Ferrous Ammonium Sulfate	0.001
Xylenol	0.0002
Sulfuric Acid	0.2375
DL-Alanine	0.6735
Tri-distilled water	5.55
Gelatin (300 Bloom)	10 % of the tri- distilled water volume

2.2 Gamma Irradiation

Different batches of gel solutions were prepared and maintained at low temperature during 12 h to solidification. Before irradiation the samples were maintained during 1 h at room temperature. The gamma irradiations were performed in air and electronic equilibrium conditions with doses between 0.5 and 50 Gy.

2.3 Spectrophotometric evaluation

According to Gore (1984), the absorbed dose is proportional to the variation on the optical density (OD) given by equation 1:

$$D = \frac{N_A.e}{\rho.l.G(Fe^{3+})} \cdot \frac{OD(D) - OD(0)}{\varepsilon_m} \tag{1}$$

where D is the absorbed dose, $G(Fe^{3+})$ is the chemical yield of Fe^{3+} (expressed in ions produced per 100 eV), ρ is the density in kg/L, N_A is Avogadro's number, e is the number of Joules per electron volt, l is the optical path length (width of the cuvette holding the solution), OD(D) and OD(0) are the optical densities at 585 nm of the irradiated and non-irradiated dosimeter, respectively, and ε_m is the molar extinction coefficient for Fe^{3+} [8].

The optical density was measured using a Shimadzu UV-2101 PC spectrophotometer using the following set up parameters, see the table 2:

Table 2: Spectrophotometer set up parameters.

Parameters	
Wavelength range (nm)	400 ~ 700
Light source	Tungsten and Deuterium
Slit width (nm)	2
Absorbance (%)	-9.999 ~ +9.999
Transmittance (%)	-999.9 ~ +999.9
Scan speed (nm/min)	1600 (fast and 2nm interval)
Precision (nm)	0.1

To study the stability the measures were always performed in regular time intervals after irradiation during 25 days, which the alanine gel response presents saturation.

To study the dose rate dependence each batch was composed of 24 cuvettes filled with alanine gel solution, separated in 6 groups; each group was irradiated with same dose (5 Gy) and different dose rates between 21,8 and 250 Gy/h, except one that was not irradiated, considered as reference (background).

The measures were always performed 1 h after irradiation. Each presented value is the average of 5 measures.

3. Results and Discussion

3.1 Dose Rate Response

According to the obtained results, the alanine gel solution presents a dose rate dependence better than 1,8 % (1σ), while the Fricke Gel developed at IPEN presents 10 % (1σ) described by Cavinato [9], indicating that the signal response can be considered independent of the dose rate.

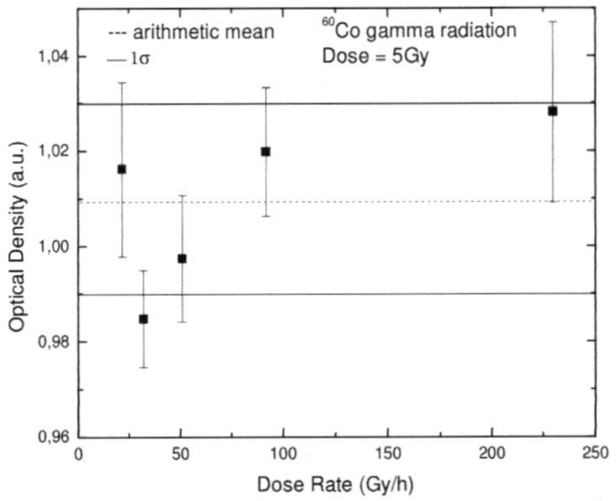

Figure 1. Dose rate response of alanine gel solution to ^{60}Co gamma radiation

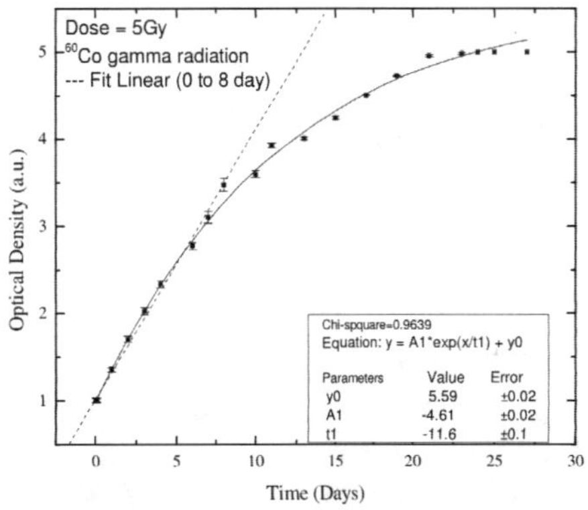

Figure 2. Signal response of alanine gel solution as a function of storage time after irradiation

3.2 Stability Test

The optical response is stable between approximately 1 to 2 h after irradiation. After this time, it presents an increasing linear behavior until 8th day, a sublinear behavior between 8th and 25th days and saturation in 25th day, see the figure 2. It was considered as day zero the measurement performed two hours after irradiation.

4. Conclusion

It was observed that the optical response is stable between approximately 1 to 2 h after irradiation, so the measures should be performed in this period to obtain reproducible results. After this time, it is necessary to correct the obtained response, while the signal response can be considered dose rate independent. The obtained results indicate that the alanine gel dosimeter presents good performance and can be useful as dosimeter in the radiotherapy area using MRI technique for 3D dose distribution evaluation.

Acknowledgements
The authors are thankful to Comissão Nacional de Energia Nuclear (CNEN), Fundação de Amparo à Pesquisa do Estado de São Paulo (FAPESP), Conselho Nacional de Desenvolvimento Científico e Tecnológico (CNPq) and Coordenação de Aperfeiçoamento de Pessoal de Nível Superior (CAPES) by the financial support, and the Radiation Technology Centre of IPEN for assistance during samples irradiation.

References

[1] Gore J C, Yang Y S and Schulz R I 1984 *Phys Med and Biol* **29** 1189-97
[2] Baldock C 2006 *J. Phys* **56** 14-22
[3] Uusi–Simola J, Heikkinen S S, Kotiluoto P, Serén T, Seppälä T, Auterinen I and Savolainen S 2007 *J. Appl Phys* **8** 114-123
[4] Guillerminet C, Gschwind R, Makovicka L, Spevacek V, Soukoup M and Novotny J 2005 *Rad Meas* 39 39-42
[5] Costa Z M 1994 *Desenvolvimento de Sistemas de DL-Alanina para dosimetria da Radiação Gama de Elétrons* Thesis (Master degree) – Instituto de Pesquisas Energéticas e Nucleares – IPEN/CNEN-SP, São Paulo.
[6] Mizuno E Y 2007 *Desenvolvimento e caracterização de um gel de alanina para aplicação na medida da distribuição da dose de radiação usando a técnica de espectrofotometria* Thesis (Master' degree) – Instituto de Pesquisas Energéticas e Nucleares – IPEN/CNEN-SP, São Paulo.
[7] Attix F H 1986 *Introduction to radiological physics and radiation dosimetry* (New York: John Wiley)
[8] Schreiner L J 2004 *J. Phys* **3** 9-21

[9] Cavinato, C.C. 2007 Study of the stability, reproducibility and dose rate dependence of the fricke gel dosimeter developed at IPEN *Proc 2007 Int Nuclear Atlantic Conf (Santos, BR, September 30 to October 5 2007)* vol 1 pp 1-6.

Radiation effects on TL and EPR of sodalite and application to dosimetry

N F Cano[1,2], S Watanabe[1], A R Blak[1], J M Yauri[2]

[1]Institute of Physics, University of São Paulo, Rua do Matão, Travessa R, 187, CEP 05508-900, São Paulo, SP, Brazil

[2]Professional School of Physics, National University of San Agustín, Av. Independencia S/N, Arequipa, Peru

Email: nilocano@dfn.if.usp.br ; nfcano@gmail.com

Abstract. Sodalite, $Na_8Al_6Si_6O_{24}Cl_2$, a natural silicate mineral, has been investigated concerning its thermoluminescence (TL) and electron paramagnetic resonance (EPR) properties. Samples, both natural and heat treated at 500°C for 30min, present an EPR signal at g=2.01132. The irradiation induces two signals at g=2.0008 and a group of 11 lines due to an O^- ion in an intermediate position with respect to two adjacent Al nuclei, that are superimposed on the signal at g=2.0008. An annealing at 500 °C/30min produces a broad signal around g~2.0, probably due to Fe^{3+} magnetic dipole interactions. The samples were annealed at 500 °C/30min and then irradiated with gamma-ray doses varying from 1 Gy to 20 kGy. All the samples have shown TL peaks at 110, 230 and 270 °C. Similar results were obtained for natural samples. The accumulated dose (AD) was determined to be AD~2.5 kGy and the annual dose of (0.595±0.001) mGy/year was obtained from the concentration of U, Th, and K_2O in the sample. Therefore, the age obtained for natural sodalite was $(4.2\pm0.2)10^6$ years.

1. Introduction

A natural sodalite sample from Minas Gerais, Brazil, has been investigated concerning its electronic paramagnetic resonance (EPR) and thermoluminescence (TL) properties and also the effects of the ionizing radiation and annealing at high temperatures.

Sodalite, of chemical formula $Na_8Al_6Si_6O_{24}Cl_2$, is a mineral of the aluminum silicate family (tectosilicates). This mineral has been investigated by many authors due to its color and interesting photochromic and cathodochromic properties [1-3].

The crystal structure of sodalite consists of an ordered framework of AlO_4 and SiO_4 tetrahedra, in such a way that cubo-octahedral cages are formed, where each atom of Al or Si is located at the center of a tetrahedrical cage of oxygens. In each cage there is a Cl ion tetrahedrally coordinated by Na ions, forming a body-centered cubic structure with lattice parameter between 0.0887 and 0.892 nm [4].

Hodgson et al. [5] studied an artificial sodalite with impurities of S using the EPR technique. Its photochromic property was activated by heating the sample in a hydrogen atmosphere and the hyperfine interaction was observed with 13 lines in the spectrum. The lines are separated by 30.6 G and centered at g=2.002 when the sample is exposed to UV irradiation. Under these conditions the sample acquires a magenta color and an absorption band at 530 nm is then observed. The incidence of visible light produces a loss of the color and the disappearance of the 13 lines of the EPR spectrum and of the optical absorption bands. Hodgson and collaborators [5] proposed that the color center is an electron trapped in a Cl^- vacancy interacting with the nuclear spins (I=3/2) of the four Na ions.

Van den Brom et al. [6] investigated the behavior of electric dipoles and proposed a model where the dipole is associated with the alkali metal ion (in general Na^+) that provides the necessary charge compensation resulting from the substitution of Al^{3+} for Si^{4+} in the SiO_4 tetrahedron. The arising Al-Na complex or $[AlO_4/M^+](M^+=Na^+)$ center is affected by the X-ray irradiation which liberates an alkali ion leaving an aluminum center $[AlO_4/h]$, where h is a hole.

Hassib et al. [7], investigated natural sodalites and verified that samples heat treated at 900 °C in air and X-ray irradiated, show a pink color that is bleached when illuminated with visible light. After some cycles of coloration and bleaching processes the sample presents the characteristic blue color of the natural samples and a growth of the isotropic EPR line around g=2.0112 and ΔH=10 G is observed. The collapse of the F centers at room temperature into colloidal aggregates of Na generates the blue color of the sodalite. The isotropic EPR line (g=2.0112) linked to the blue color disappears with the disappearance of the color. Another EPR signal occurring at g=1.9978 was also observed by Annersten and Hassib [8] on an irradiated blue sodalite single crystal. The spectrum was anisotropic with maximum resolution of a hyperfine structure along the orientation [1 1 0] parallel to the magnetic field. The 11 observed hyperfine lines support the assignment of the O^- ion to a position intermediate with respect to two Al nuclei (I=5/2), one in a lattice site and one replacing a Si^{4+} of the crystal structure breaking the usual alternation of aluminum and silicon that occurs in the perfect lattice.

Hassib [9] analyzed the allowed (Δm=0) and prohibited (Δm=±1) hyperfine Mn^{2+} transitions in the sodalite. The six hyperfine lines of the central transition (m=|+1/2> → |-1/2>) were observed. The obtained g-value and hyperfine structure constant were g=2.033 and |A|=83±1 G respectively. EPR spectral investigations of natural sodalite containing Fe^{3+} and Fe^{2+} ions in tetrahedral coordination were performed by Ravikumar et al. [10]. The signal near g=2 was attributed to Fe^{3+} in Al^{3+} sites in the sodalite crystal lattice.

No paper was found in the literature concerning thermoluminescent studies on natural sodalite crystal.

2. Experimental procedure

The natural sample of blue sodalite investigated in our work is from Bahia, Brazil.

A portion of the sample was crushed and sieved and the grains between 0.180 and 0.080 mm in diameter were selected for TL and EPR measurements and the grains smaller than 0.080 mm were used for fluorescence and X-ray diffraction analysis.

The X-ray diffraction measurements have been carried out at the Laboratory of X-Ray Diffraction of the Institute of Physics, University of São Paulo, while the X-Ray Fluorescence analysis at the Institute of Geosciences at the University of São Paulo.

The irradiations were carried out at the center of technology of radiations of the Institute of Nuclear an Energy Research using a ^{60}Co γ-source with a dose rate of 0.37 kGy/h and a Gammacell with a dose rate of 5.50 kGy/h at room temperature and under conditions of electronic equilibrium.

The TL measurements were carried out in a Daybreak TL reader, Model 1100, equipped with a bialkali photomultiplier EMI 9235QA for light detection with filters Corning 7-59 and Schott BG-39 used in front of the PMT. The used heating rate was 4°C/s in nitrogen atmosphere.

EPR measurements were performed in a Bruker EMX spectrometer with a rectangular cavity (ST ER4102) using a microwave frequency of 9.75 GHz (X–band), microwave power of 20 mW and a modulation field of 100 kHz.

3. Results and discussion

Table 1 shows the composition in weight % of the oxide components of the sodalite and the concentration of some impurities in ppm.

Table 1. Composition of natural sodalite using X-ray fluorescence analysis.

Compound	SiO_2	Al_2O_3	Na_2O	CaO	Fe_2O_3	K_2O	MnO	MgO	Ti_2O	P_2O_5
(w.%)	38.93	32.27	25.68	0.21	0.08	0.07	0.035	0.02	0.014	0.01
Elements	U	Th								
(ppm)	0.985	0.54								

The EPR spectra of natural and additionally irradiated with several γ-doses samples are shown in Fig. 1. In Fig. 2 the samples were submitted to a thermal treatment at 500 °C for 30 minutes prior to the irradiation. An intense signal at g=2.01132 (3468 G) was observed in both cases. The irradiation induces two signals at g=2.0008 and a group of 11 lines completely superimposed on the signal at g=2.0008. These lines are associated with an O^- ion in an intermediate position with respect to two adjacent Al nuclei as discussed in the introduction. The behavior of the EPR signal at g=2.01132 for five different γ-doses is presented in Fig. 2(a). In Fig. 2(b) a sublinearly growth starting from 1 kGy of dose is observed and for doses larger than approximately 5 kGy the EPR intensity becomes linear with the dose.

For high irradiation doses, above 10 kGy, 11 hyperfine lines are now observed, due to the fact that high irradiation is able to form a large concentration of O^- centers. On the other hand, in most of the silicate minerals the line at g=2.01132 is due to E'_1 centers, which are oxygen vacancies that have captured electrons.

The fact that we did not observe the 11 hyperfine lines below 5 kGy of irradiation probably means that, in the present case only small concentrations of O^- centers are created.

Fig. 3 presents the changes in the EPR spectra of samples thermally treated for a period of 30 min at temperatures from 200 to 900 °C respectively before being γ-irradiated with a dose of 2 kGy. The signal at g=2.01132 decreases with temperature up to 600 °C, but for T>600 °C the intensity of the signal increases quickly. On the other hand as the annealing temperature increases, the signal around g=2 becomes broad. Furthermore for the 900 °C anneal, the intensity of the signal at g~2 grows in an intense way. This may indicate the oxidation of iron from Fe^{2+} to Fe^{3+} after heat treatment.

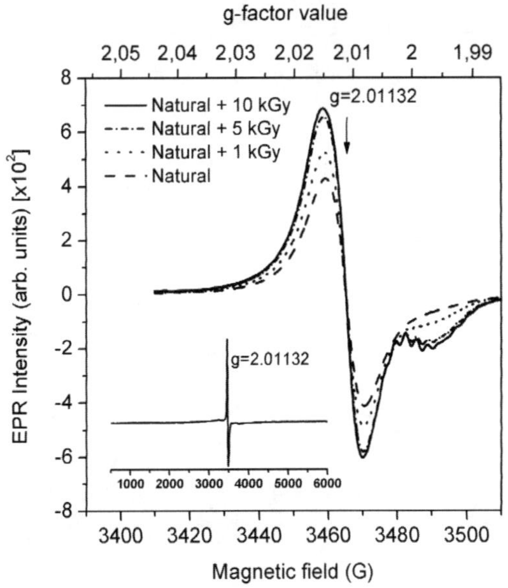

Figure 1. Behavior of the lines at g=2.01132 and g=2.0008 in the EPR spectra of natural and additionally irradiated samples.

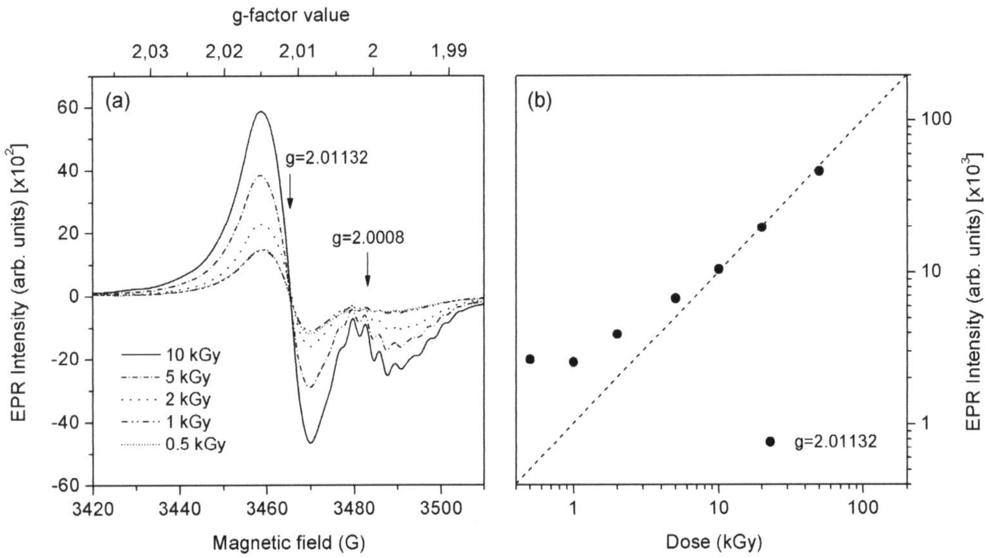

Figure 2. (a) EPR spectra of samples annealed at 500 °C/30min and then irradiated with five gamma doses. (b) EPR intensity vs. dose.

Isochronal annealing results for natural samples and isochronal annealing results for natural samples receiving a prior heat treatment at 500 °C for 30 min being subsequently irradiated with a γ-ray dose of 2 kGy are shown in Fig. 4. The respective thermal stability of the EPR signal at g=2.01132 observed in both types of samples was investigated during these isochronal annealing at temperatures varying from 50 to 390 °C for 15 min periods. The EPR signal intensity at g=2.01132 decrease substantially above 290 °C, for both types of samples indicating that for the samples irradiated at the laboratory the paramagnetic centers are stable at room temperature. Fig. 4 also shows the decay during annealing of the g=2.0008 EPR signal for the sample, in this case the signal decays substantially above 70 °C.

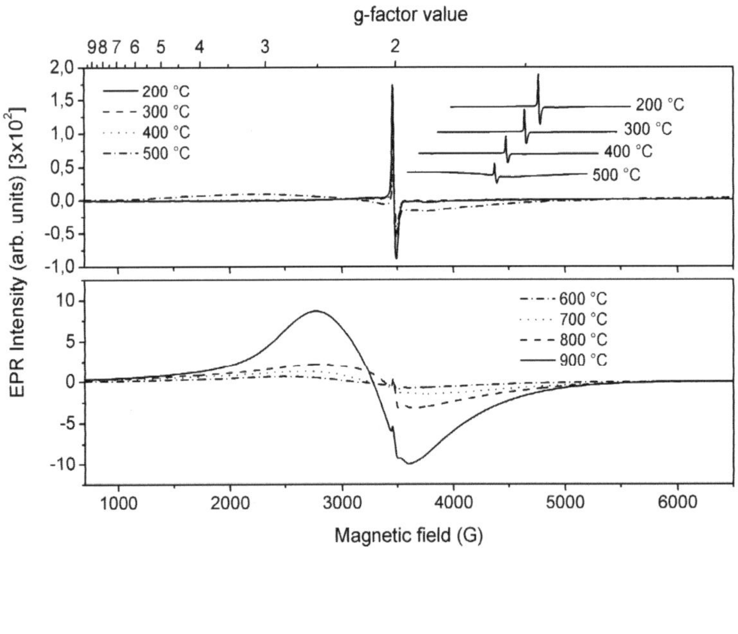

Figure 3. EPR spectra of natural sodalite annealed from 200 up to 900 °C before the γ-irradiation.

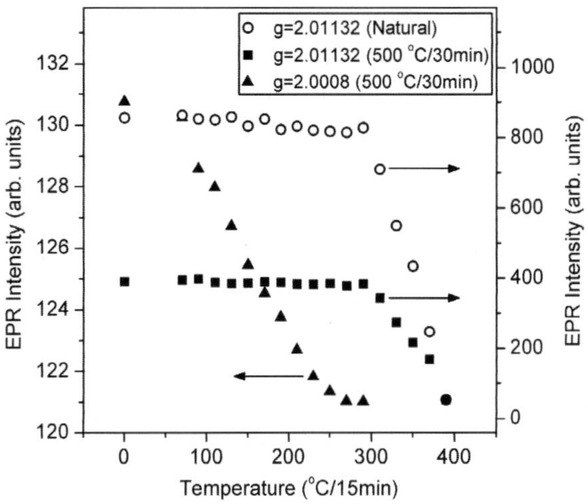

Figure 4. Behavior of the lines at g=2.01132 and g=2.0008 with isochronal annealing for the indicated samples indicated.

Natural sodalite samples were each subjected to thermal treatment at temperatures varying from 300 to 700 °C/30min (see Fig. 5) respectively before being γ-irradiated with a dose of 2 kGy. The inset in Fig. 5 having an enlarged ordinate scale shows the TL glow curves for as-received natural sample as well as those of the samples pre-heated to 300 and 400 °C respectively. The main figure shows the results for all samples and together with the inset illustrates the great increase in the TL intensity with heat treatment. The TL peaks at 365 and 445 °C observed in the as-received natural samples almost disappear with the rapid growth of the peaks at 110, 230 and 270 °C with heat treatment and irradiation.

In Fig. 6(a) the glow curves of as-received natural samples irradiated to additional γ-ray doses varying from 1 Gy to 20 kGy are shown. Fig. 6(b) present the sublinear growth of the 110 °C peak from 100 Gy to 1 kGy. Beyond 1 kGy the peak saturates. The 365 and 445 °C peaks grow much more slowly than the 110 °C peak. However the 365 °C peak can be used for radiation dosimetry and geological dating as will be discussed below. A very weak TL peak around 200 °C becomes noticeable for high doses.

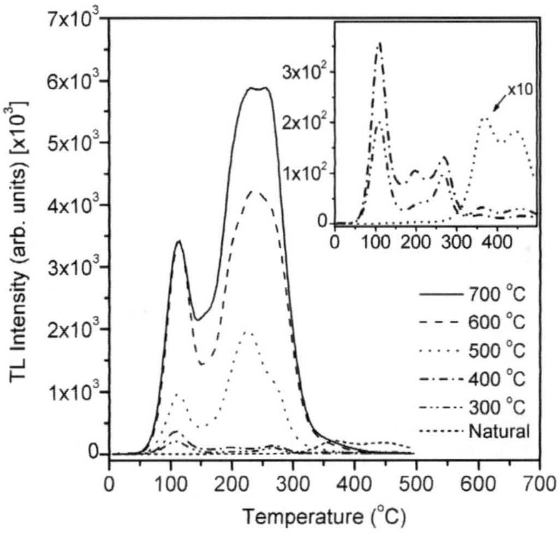

Figure 5. TL glow curves of natural sodalite samples subjected to preheating to the temperature indicated and then gamma irradiated with a dose of 2 kGy. The glow curve of as-received sample is shown. The inset provides results in more detail for the as-received sample and those for the lower temperature preheating.

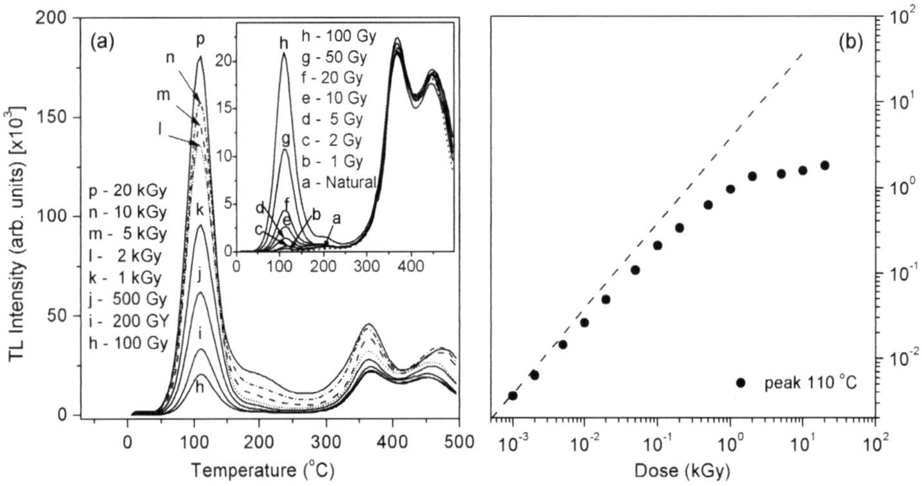

Figure 6. (a) Glow curves of natural sodalite samples plus additional γ-doses up to 20 kGy. (b) TL response curve of 110 °C peak as function of dose.

Fig. 7(a) shows the glow curves of samples previously annealed at 500 °C/30min and irradiated to several doses. Fig. 7(b) presents the TL response curve as a function of dose. It is interesting to notice that the 500 °C annealing brought as a consequence the increase of the 110 °C TL peak by a factor of more than 4 times and furthermore the weak peaks observed in the glow curves shown in Fig. 6(a), now appear very strong, presenting a peak intensity growth of a factor close to 600. It is known that in silicate crystals pre-irradiation annealing at high temperatures (500 - 1000 °C) sensitizes the TL peaks [11, 12]. However in the present case, the peaks around 230-280 °C that are almost invisible in Fig. 6(a), become greatly enhanced as a result of the 500 °C pre-annealing. This fact has not been yet explained. Fig. 7(b) shows that both 110 and 230 °C peaks grow linearly with dose up to about 1 kGy, saturating subsequently.

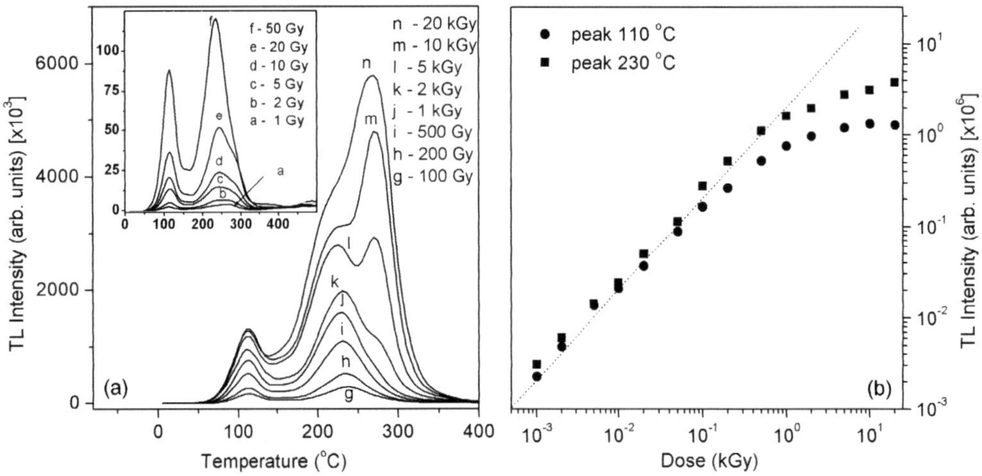

Figure 7 (a) TL glow curves of the natural samples pre-annealed at 500 °C for 30 min and then irradiated at different γ-ray doses. (b) The intensities of the TL peaks at 110 and 230 °C as a function of γ-ray dose.

Fig. 8 shows the isochronal decay of the peaks at 230 (from Fig. 7a) and 365 °C (from Fig. 6a). Both peaks are stable at room temperature allowing us to use the natural sodalite to calculate geological ages as well as the 500 °C/30min thermal treated sample for dosimetry.

Fig. 9(a) shows the obtained activation energies (E) using the E-Tstop method [13, 14]. With these results and applying the equations proposed by Kitis et al. [15] we obtain the deconvolution of the TL glow curve of natural sodalite thermally pre-heated at 500 °C/30min and then γ-irradiated with a dose of 2 kGy. Fig. 9(b) shows the five TL peaks separated using first- and second- order kinetics in the region between 50 and 350 °C. The positions, frequency factors (*s*), activation energies (E) and lifetimes (τ) are presented in Table 2.

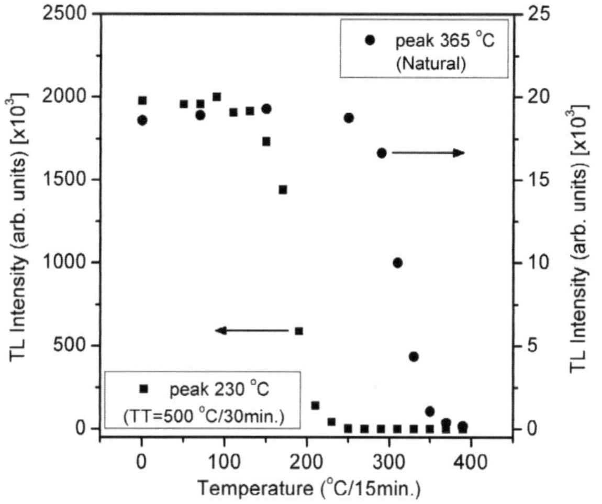

Figure 8. Effect of the isochronal annealing on the TL intensity of the sodalite samples.

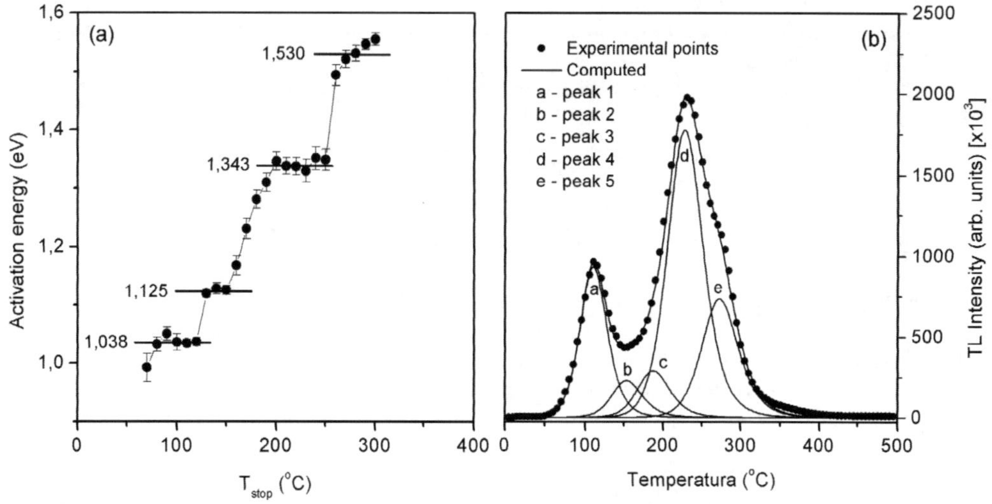

Figure 9. (a) Activation energy vs. T_{stop} obtained by the E-T_{stop} method. (b) TL glow curve from natural sodalite pre-annealed at 500 °C/30min and subsequently receiving a γ-ray dose of 2 kGy. A good fit between the experimental glow curve (circles) and the simulated glow curve (full line) can be achieved by assuming the presence of five peaks.

Table 2. Energy (E), frequency factor (*s*) and lifetime (τ) of TL traps of sodalite.

TL peak (°C)	E (eV)	$s(s^{-1})$	τ(years) (at 15 °C)
110	1.038±0.007	$(1.478±0.011)10^{3}$	$(3.091+0.007)10^{-3}$
153	1.125±0.005	$(5.805±0.029)10^{12}$	(0.262±0.002)
*187	1.250	$1.558x10^{13}$	17.352
227	1.343±0.008	$(8.509±0.053)10^{12}$	$(1.162±0.010)10^{3}$
272	1.530±0.024	$(3.337±0.052)10^{13}$	$(5.525±0.123)10^{5}$

* In the E-T_{stop} method the error could not be evaluated.

4. Conclusions

The EPR spectra show the g=2.01132 and g=2.0008 lines and 11 hyperfine lines due to an O^- ion in an intermediate position with respect to two Al nuclei (I=5/2), superimposed on the g=2.0008 signal.

The TL glow curve of natural samples shows high temperature peaks at 365 and 445 °C. Samples heat treated at 500 °C/30min and irradiated with different gamma doses present peaks at 110, 230, 270 and very weak TL peaks at 365 and 445 °C. The glow curve deconvolution shows that in fact, in the region varying from 50 to 350 °C, five overlapping TL peaks are observed.

The TL peak at 365 °C and the EPR signal at g=2.01132 probably are due to a same defect center, and they yield ages of ~$(4.2±0.2)10^{6}$ year for natural sodalite.

Acknowledgments

The authors wish to thank Ms. E. Somessari and Mr. C. Gaia, Instituto de Pesquisas Energeticas e Nucleares (IPEN), Brazil, for kindly carrying out irradiation of the samples. This work was carried out with financial support of the "Fundação de Amparo à Pesquisa do Estado de São Paulo - FAPESP".

References

[1] Lee O I 1936 *American Mineralogist* **2** 764
[2] Medved D B 1954 *American Mineralogist* **39** 615
[3] Kirk R D *American Mineralogist* **40** 1955
[4] Deer E A, Howie R A, Zussman J 1996 *An introduction to the rock forming minerals*, Addison Wesley, London
[5] Hodgson W G, Brinen J S and Williams E F 1967 *J. Chem Phys.* **47** 3719
[6] Van den Brom W E, Kerssen J and Volger J 1974 *Physica* **77** 1
[7] Hassib A, Beckman O and Annersten H 1977 *J. Phys. D: Appl. Phys.* **10** 771
[8] Annersten H, Hassib A 1979 *Can. Mineral.* **17** 39
[9] Hassib A 1980 *Phys. Chem. Of Mineral* **6** 31
[10] Ravikumar R V S S N, Chandrasekhar A V, Yamauchi J, Reddy Y P, Rao P S 2005 *Radiat. Eff. Def. Sol.* **160** 109
[11] Cano N F, Yauri J M, Watanabe S, Mittano J C R and Blak A R 2008 *J. Lumin.* **128** 1185
[12] Yauri J M, Cano N F and Watanabe S 2008 *J. Lumin* **128** 1731
[13] McKeever S W S 1980 *Physica Status Solidi* **62** 331
[14] McKeever S W S 1985 *Thermoluminesce of solids*, London, Cambridge
[15] Kitis G, Gomez-Ros J M and Tuyn W N 1998 *J. Phys. D: Appl. Phys.* **31** 2636

Journal of Physics: Conference Series **249** (2010) 012023 doi:10.1088/1742-6596/249/1/012023

Thermoluminescence and Optical Absorption of Lizardite crystals

René Rojas Rocca[1]*, Sonia Hatsue Tatumi[2,1] and Shigueo Watanabe[1]

[1]Physics Institute, University of São Paulo, S. Paulo, Brazil.

[2] Dept. of General Education, Faculty of Tecnology of Sao Paulo, Praça Cel.Fernando Prestes, 30, 01124-060, São Paulo, SP, Brazil.

renerr@usp.br

Abstract. Pellets of lizardite crystals (Mg_3 $[Si,Fe]_2O_5$ $(OH)_4]$) and teflon powder have been investigated to identify the centers related to thermoluminescence (TL) emission. All the Tl peaks are well fitted by using a second order kinetic model and the related trap parameters and activation energy are evaluated. Moreover, the measured optical absorption bands, in the visible and UV ranges, and in the infrared region, are attributed to charge transfer involving Mn and Fe ions and to vibration modes of Mg-O-H groups, respectively

1. INTRODUCTION

Lizardite Antigorite and chyisotile are silicate minerals belonging to a group called serpentine. Their chemical formula is $Mg_3[Si_2O_5](OH)_4$, however they differ from each other by crystal structure. Lizardite crystallizes into trigonal, antigorite into monoclinic and chrysotile into orthorhombic system.

Most of papers on these crystals are of mineralogical nature nothing on their thermoluminescence was found in the literature. Hunt et al. [1] investigated reflectance spectra in visible and near infrared region of 87 silicates minerals, including two samples one called serpentine 318B and the other asbestos 323B – Arizona which presented bands at 0.65, 0.45 and 0.8 μm due to ferric ion and also bands at 1.4 and 2.34 μm related to hydroxyl ion. The lizardite is the most abundant and the least is crysotile, but this one has been most important source of asbestos.

An antigorite $[Mg_{3-x}(Si_2O_5)(OH)_{4-2x}]$ sample from India was investigated by Reddy et al, [2], using Optical absorption technique. They found bands related to Fe^{3+} ion at 15428, 17852, 19602, 22773, 24869 and 26448 cm^{-1} and the others at 9122, 10635, 11679, 15380 and 22929 cm^{-1} due to Fe^{2+}. Using EPR they verified the presence of Fe^{3+}, with g-values ranging from 5.4 to 3.1, and a sextet lines of the Mn^{2+} ion, which substitute Mg^{2+} in the crystalline lattice, with g-values near 2.0.

Götze et al, [3], investigated clay minerals using cathodoluminescence (CL), EPR, X-ray diffraction, scanning electron microscopy and trace element analysis; the selected mineral groups are the serpertine-kaolin, talc-pyrophyllite, smectite and illite. They concluded that the minerals kaolinite, dickite, nacrite, halloysite and pyrophyllite are characterized by a blue CL emission (400 nm), and EPR results indicated that this blue emission can be related to radiation induced defects centres, as electron holes trapped on apical oxygen atoms (Si-O centre) or located at the Al-O-Al group. They also verified that serpentine does not show visible CL.

In this work, we investigate the thermoluminescence and optical absorption properties of Lizardite mineral specimens from Brazil, collected in the state of Goiàs.

2. MATERIALS AND METHODOLOGY

Our specimens exhibit the characteristic green coloration of the commercial Lizardite crystal, which is named "Chinese Jade" because of its Jade-like coloration. The triclinic Lizarite structrure (Mg_3 [$(Si,Fe)_2O_5$]$(OH)_4$) of our samples was confirmed by X-rays diffraction measurements.

It is known in the literature that the lizardite changes to forsterite, when it is heated up to 350 °C [4].

By using ICP technique the major elements concentrations and impurities were identified in the sample. Their list in weight unit is: SiO2 = 35.76; Al_2O_3 = 0.09; Fe_2O_3 = 0.49; MgO = 45.24; CaO < 0.01; Na_2O < 0.01; K_2O = 0.93; P_2O_5 = 0.034; MnO = 0.028. The silicon in the SiO_4 tetrahedron can be replaced by aluminum and the magnesium by aluminum, ferric and ferrous ions in different environments.

For thermoluminescence (TL) investigations samples were prepared as follow. Specimen parts were first grind with a mortar and a pestle, then sieved to separate grains between 75 and 150 μm. They were subsequently pressed with teflon powder producing pellets of 5 mm diameter and 1 mm thickness. Finally, the samples were irradiated with γ-rays in a ^{60}Co source; doses from 20 to 500 Gy were applied with a rate of 253.7 Gy/h, high doses until 172 kGy were applied with a rate of 2,82 kGy/h. TL measurements were performed in an oxygen-free nitrogen atmosphere using a *Daybreak Nuclear and Medical Systems Inc, Model 1100-series*, with a photomultiplier *EMI 9235QA* coupled to optical filters: Corning 7-59 and Schott BG–39, resulting in a transmittance at 335-470 nm and heating rate was 10 °C/s.

Optical absorption (OA) measurements, in the range of 300-2700 nm, were carried out with a Carry 500 spectrometer on 2 mm thick polished samples.

The initial results using only Lizardite grains were not reproducible, On the other hand, reproducible results could be obtained using teflon pellets along with Lizardite grains. In this configuration, the grains are embedded on the upper surface of the teflon pellet.

3. RESULTS AND DISCUSSION

Two prominent TL peaks at 153 and 250 °C were observed in the TL glow curve of irradiated pellets, as shown in figure 1.

The TL glow curves can be fitted by using second order kinetics model [5, 6], the equation for TL intensity (*I*) for each peak is:

$$I(T) = sn_o \exp(-E/kT)\left[(s/\beta)\int_{T_o}^{T} \exp(-E/kT')dT' + 1\right]^{-2} \quad (1)$$

where s is the frequency factor, n_0 is the initial concentration of trapped electron, E is the activation energy, k is the Boltzmann constant, T is the absolute temperature, β is the linear heating rate.

An excellent theoretical fit using second order kinects model was obtained for all the peaks, however the theoretical deconvolution showed two additional peaks at 204 and 311 °C, as is shown in figure 1. The presence of 4 peaks is also confirmed by *Tm-Tstop* method [7].

Electron traps parameters s and E were evaluated through peak shape, initial rise and Glow Curve Deconvolution methods [8, 9]. All the results are summarized in table 1.

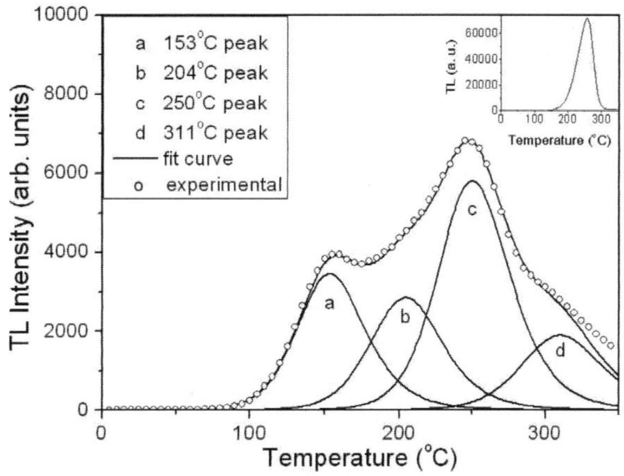

Figure 1. Theoretical and experimental TL glow curves of Lizardite-teflon pellets and the inset shows the TL from only telfon pellets.

Table 1. Calculated activation energies (E) and frequency factors (s) of the TL peaks found in Lizardite-Teflon pellets. A= Glow Curve Deconvolution, B= Peak Shape, C= Initial Rise.

	A		B		C	
Tm (°C)	*E* (eV)	*s* (10^{12} s^{-1})	*E* (eV)	*s* (10^{12} s^{-1})	*E* (eV)	*s* (10^{12} s^{-1})
153	0.95	0.11	1.00	0.09	0.95	0.14
204	1.12	0.35				
250	1.35	5.24	1.39	18.79	1.53	22.76
311	1.49	3.28				

Figure 2 shows the intensity of TL peaks 153, 204 and 311 °C peaks increasing up to 2 kGy almost linearly and after this dose the intensity falls off due to the onset of saturation; however the 250 °C one still increased up to a dose of 172 kGy. Actually, measurements on pellets only prepared with Teflon demonstrated that the 250 °C peak is not related to the Lizarite but to the Teflon itself (as seen on figure 1) , and must be therefore considered as an artifact in the mixed pellets.

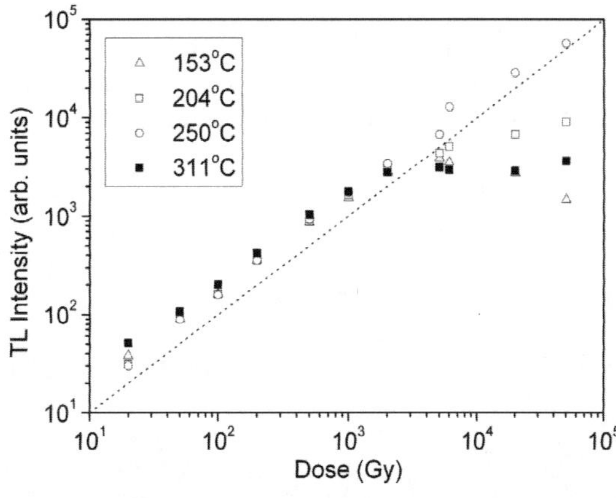

Figure 2. TL growth curves of 153, 204, 250 and 311 °C peaks irradiated with γ-rays.

The individual mean life of the peaks were deduced by using formula 2 and the E and s values listen I table 1 for T= 27 °C.

$$\tau = \frac{\exp\left(\frac{-E}{kT}\right)}{s} \qquad (2)$$

We found the following values of τ : 1 day, 200 days and 93 kyears for 153, 204 and 311 °C peaks respectively for lizardite sample.

The OA spectra of a polished Lizardite crystal sample recorded from 300 to 2700 nm is given in figure 3. The increase of the absorbance for wavelengths lower than 1000 nm can be attributed to charge transfer bands of 3d transition ions [10], in particular of Fe^{3+}, and Mn^{2+}, in agreement with our ICP results showing the presence of these ions. Moreover, several bands are observed in the near and infrared intervals, due to primary and secondary combinations of modes of Mg-OH groups, while the fine structure observed at 1400 nm can be assigned to the first overtone of the O-H stretching mode [11, 12]. It is worth noting that irradiation with γ-rays does not affect the OA bands intensity for the used doses (20-750 Gy).

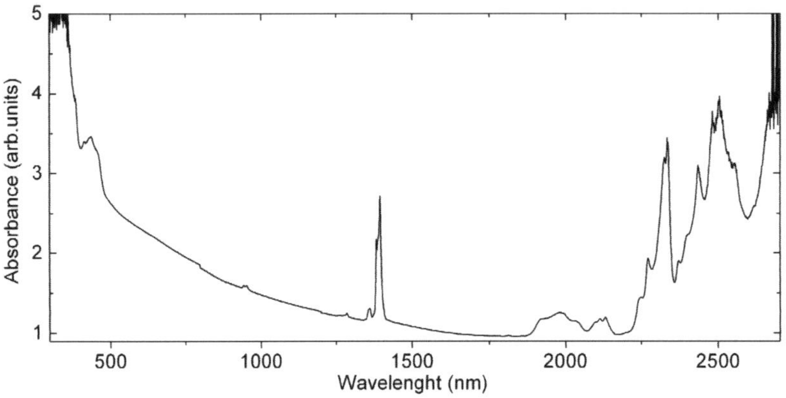

Figure 3. Optical absorption spectra of lizardite showing several peaks.

4. CONCLUSIONS

Our optical absorption measurements show that the green coloration of the studied Lizardite mineral specimen is due to the high absorbance in the UV and blue region. The OA bands intensity does not change with γ-irradiation and cannot be therefore related to the TL emission bands measured in the visible range.

Moreover, the observed linear relationship on more than three decades between the TL bands and the ionizing irradiation dose could be applied both for dating and dosimetry. Thanks to its very high estimated thermal stability (93 k years), the high-temperature TL peak (at 311°C) is particularly suitable for high dose dosimetry applications (up to 2 KGy).

Acknowledgments

Authors wish to thank Conselho Nacional de Desenvolvimento Científico e Tecnológico (CNPq) and Fundação de Amparo à Pesquisa do Estado de São Paulo (FAPESP) for providing financial support.

Journal of Physics: Conference Series **249** (2010) 012023 doi:10.1088/1742-6596/249/1/012023

References

[1] Hunt G R, Salisbury J W and Lenhoff C J 1973 *Modern Geology* **4** 85.

[2] Reddy S N, Ravikumar R V S S N, Reddy B J, Reddy Y P and Rao P S 2001 *Neus Jahrbuch fur Mineralogy-Monatshefte* **6** 261.

[3] Götze J, Plötze M, Götte Th, Neuser R D and Richter D K 2002 *Miner. Petrol.* **76** 195.

[4] Deer W A, Howie R A, Zussman J 1966 *An Introduction to the Rock-Forming Minerals* (Longman London).

[5] Garlick G F J and Gibson A F 1948 *Proc. Roy. Soc London* **A60** 574.

[6] Chen R, Kristianpoller N and Davidson Z 1981 *Journal of Luminescence* **23** 293.

[7] McKeever S W S 1985 *Thermoluminescence of solids* (Great Britain, Cambridge U. Press.).

[8] Chen R, 1969 *J. Appl. Phys.* **40** 570.

[9] Kitis G, Gomez-Ros J M and Tuyn J W N 1998 *Journal of Physics D: Aplied Physics* **31** 2639.

[10] Manning P G 1970 *Canad. Mineral.* **10** (4) 677.

[11] King T V V and Clark R N 1989 *Journal of Geophysics Research* **24** **(B10)** 13997.

[12] Clark, R.N., T.V.V. King, M. Klejwa, G. Swayze, and N. Vergo 1990 *J. GeophysRes.* **95**, 12653-12680.

Effects of monochromatic UV radiation on the thermoluminescence of annealed KBr:Cu^+ crystals.

R. Pérez-Salas[1], T. Piters[1], R. Aceves[1], Héctor Riveros R[2] and R. Rodríguez-Mijangos[1]

[1] Dpto. Investigación en Física, Universidad de Sonora. 83190 Hermosillo, Son, Méx.

[2] Instituto de Física, Universidad Nacional Autónoma de México. México DF

rperez@cajeme.cifus.uson.mx

Abstract. Effects of monochromatic UV radiation on the thermoluminescence glow curve of annealed KBr:Cu crystals have been studied at low temperature. The crystals were irradiated at 20K and heated up to 300K at a rate of $0.1Ks^{-1}$, approximately. For all the irradiation wavelengths used, the glow curves consist of two peaks around 170 and 270K. Their intensities have been observed to depend on the wavelength of irradiation. Both emissions resulted to be more intense when the irradiation wavelengths was near the absorption peak of the Cu^+ ion or near the band to band absorption of CuBr. Their intensities increase with increasing annealing time. The results are explained in terms of the formation of CuBr nanoparticles during the annealing treatment. The thermally stimulated recombination luminescence appears to be associated to the $3d^9 4s$-$3d^{10}$ transitions of the Cu^+ ion. So two possible mechanisms of the e-h pair formation could be used to explain the results: ionization of Cu^+, $Cu^+ \rightarrow Cu^{2+} + e$, or ionization of halogen ions that are within the CuBr as $Br^- \rightarrow Br^0 + e$. In the first case, the recombination luminescence is obtained from the trapping of electrons by the Cu^{2+} ions, while in the second case, the *e-h* recombination energy would be transferred to the Cu^+ ions.

1. Introduction

When Cu^+-ion is introduced into an alkali halide crystal, the formation of CuX (X=Cl, Br, I) nanocrystals [1] are performed. The growing rate of the nanocrystals formation depends on several physical conditions such as the annealing temperature, Cu^+-ion concentration, and the stress applied on the crystal. For instance, it has been reported that 120°C is the optimal temperature for the aggregation process of CuCl in NaCl. The electronic properties of CuX nanocrystals resembled that of CuX bulk material, but additionally show quantum effects [2], such as the shift of the exciton absorption to higher energies due to decreasing size [3]. However, as in other crystals, the formation of electron-hole pairs in CuX nanocrystals should be expected during its exciton absorption or band to band electronic transitions. It is believed that the spectral holes appearing in the optical absorption of CuBr embedded in glasses are to the photoionization of CuBr nanocrystals when they are exposed to laser light [4, 5]. Photostimulated luminescence of UV-laser irradiated nanocrystals confirms the presence of the laser light induced electron-hole pairs in CuCl nanocrystals [6]. In spite of that, few studies of trapped defects in this type of systems have been performed. In this work, monochromatic UV radiation effects on annealed KBr:Cu crystals have been studied at low temperature by thermoluminescence (TL) in order to gain insight into the UV radiation effects in the CuX nanocrystals.

2. Experimental

The crystals were grown by the Czochralski method in Argon atmosphere. CuBr was added to the melt in 0.2% concentration. Before experiments the crystals were quenched from 500 °C to room temperature by contact with an aluminium block. The sample was fixed to the cold finger of a closed cycle helium cryostat H2-APD Cryogenics Inc. The crystals were irradiated at 20K, and heated up to 300K at a rate of $0.1 Ks^{-1}$, approximately, with a homemade temperature controller. For irradiation, a 150W Xenon lamp has been used. During the TL running, the emission spectra were recorded at a rate of $10 nm.s^{-1}$. The emission spectra were taken with a Fluoromax of Spex Industries. The formation of aggregates has been induced with annealing at 100ºC, a temperature near of the optimum temperature given to the NaCl:CuCl [2].

3. Results

Figure 1 shows a typical excitation spectrum obtained with emission at 413 nm of an annealed sample. This spectrum shows two broad bands, one about 265 nm and the other in the range from 320 to 400 nm with a relative maximum about 377nm. The 265nm band is the characteristic absorption associated to the $3d^{10} - 3d^{9}4s$ electronic transition of the Cu^{+} ions. The structured broad band observed in the region from 320 to 400 nm appears only when the Cu^{+} ions are found in aggregates of CuBr. In their right edge the Z_3 and $Z_{1,2}$ exciton absorptions are found, while the other bands in this region correspond to the electronic transitions from the valence band to the conduction band of the nanocrystals. Excitations carried out with light of this region produce free exciton emissions and emissions of less energy associated with different types of precipitates.

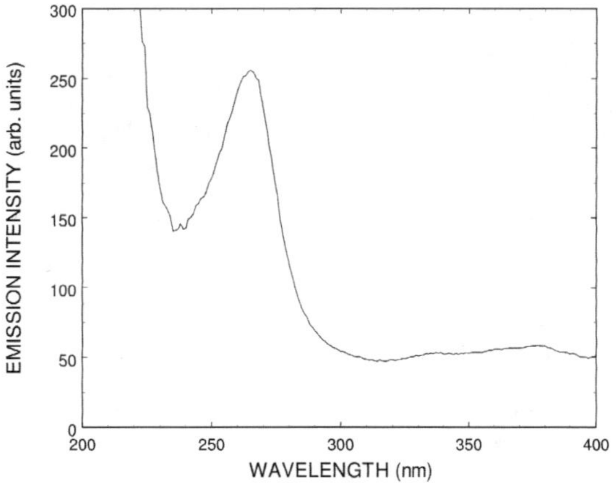

Figure 1. Excitation spectrum with fixed emission at 413nm of an annealed sample.

The temperature dependence of the spectral measurements during the thermoluminescence process of an irradiated sample is displayed in Fig. 2. During the heating, the emission spectra were scanned at a rate of $10 nm.s^{-1}$ while the heating rate was $0.1\ K.s^{-1}$, approximately. Then during a complete spectrum scanning, the temperature increases 2 K, approximately. All of the emission spectra show a broad band centred at 388 nm which were obtained using a monochromator slit of 7.00 mm. These spectra are very similar to the emission band of the photoluminescence (PL) spectrum obtained under excitation of 260 nm and similar conditions of the monochromator slit.

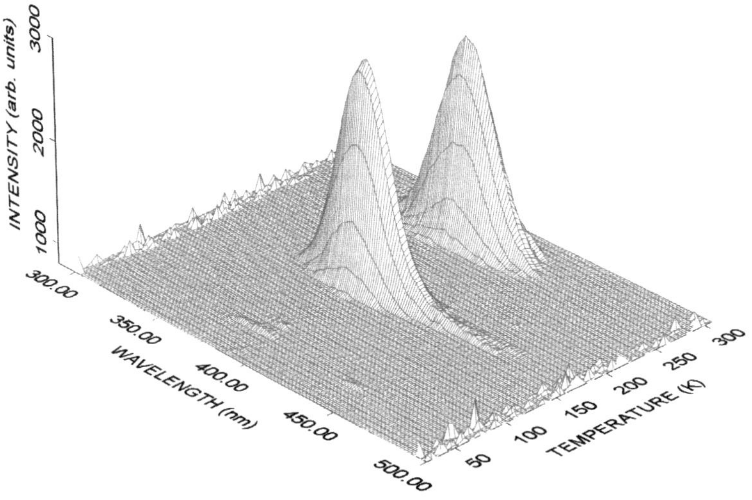

Figure 2. Emission spectra of the TL glow peaks at 170 and 270K

Particularly, the emission spectra corresponding to temperatures of 170 and 270K are compared in Fig. 3. The high similarity between the spectral position and width of both spectra indicates that they have a common emission centre. The resemblance of these bands with the emission obtained from the same crystal when is excited around 260 nm led us to assume that these emissions originate in the $3d^9 4s$-$3d^{10}$ electronic transitions of the Cu^+ ion, which suggest that Cu^+ acts as an active center during the thermoluminescence.

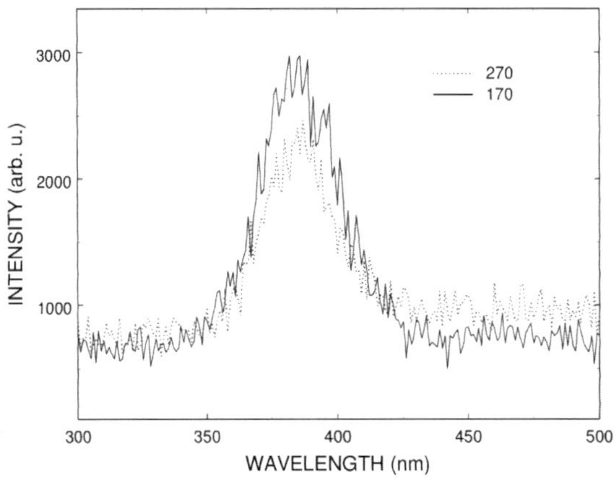

Figure 3. Excitation spectrum with fixed emission at 413nm of an annealed sample.

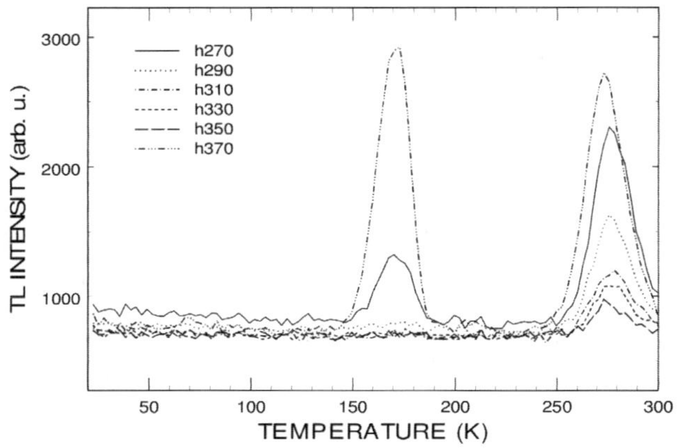

Figure 4. TL glow curves of KBr:Cu obtained after 300 s UV irradiations with different wavelengths between 270 nm and 370 nm at 20K.

The thermal dependence of the emitted light intensity was obtained with the integrated intensity of the emission spectra between 350 and 420nm. A typical behaviour is shown in Fig. 4. TL is only observed in annealed samples in which nanocrystals have been formed As can be seen from Fig. 4, each glow curve present two peaks about 170 and 270 K, but their relative intensities depend on the wavelength of irradiation. This dependence is shown in the Fig. 5 for irradiation wavelengths in the 250-370 nm spectral range applied during 300 s. The higher intensities resulted for 270 and 370 nm irradiations, but with inverted relative intensities. These results suggest that both emissions are more intense when the irradiation wavelengths are near the Cu^{+}-ions absorption peak or near the band-to-band absorption of the nanocrystals CuBr.

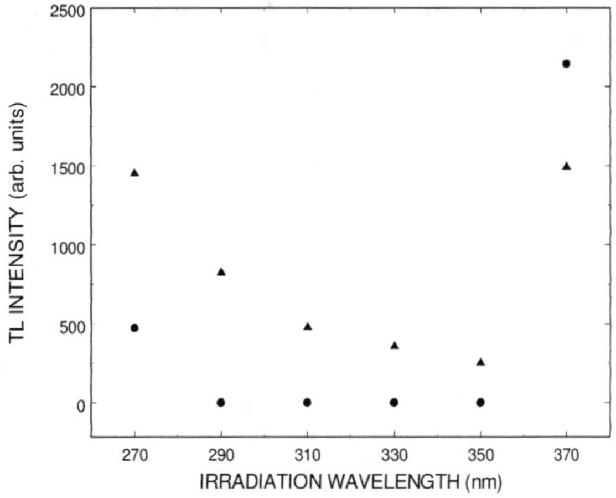

Figure 5. a) 170K (full circles) and b) 270K (full triangles) glow peak intensity for different irradiation wavelengths corresponding to the glow curves of Fig. 4.

4. Discussion

The detection of two glow peaks suggest that the aggregation of Cu^{+} produce two types of traps for electrons or holes induced by irradiation, which is independent of the irradiation wavelength. The TL emission in 388 nm (Fig.3) as well as the emission obtained with the excitation at the 260 nm band of the optical excitation spectrum indicate that the thermostimulated process induce the $3d^{9}4s \rightarrow 3d^{10}$ electronic transitions of the Cu^{+}-ion. This gives evidence that the recombination centres are near or contain Cu^{+} ions. Our observation that TL is only observed in annealed samples in which nanocrystals have been formed (Fig 4.) suggests that ionisable centres in KBr:Cu are related with CuBr nanocrystals. A possible explanation for the TL in CuBr nanocrystals is that the 370 nm UV ionizes the Cu^{+} ions to form stable Cu^{2+} and free electrons that could be trapped at halogen vacancies V_A or Cu^{+} ions to form Cu^{0} as proposed by Masumoto and Ogasawara [6] who consider that the nanocrystals are surrounded by Cu^{+} monomers and dimmers. According to this mechanism the electrons would be released from their traps (Cu^{0} or V_A centres) and recombine with the Cu^{2+} centres to produce the characteristic Cu^{+} emission. Alternatively one could assume that the Br^{-} ions in the CuBr nanocrystal are ionized since the ionization energies of Br^{-} and Cu^{+} in CuBr are almost the same. In that case neutral Br^{0} atoms could be trapped at interstitial sites or at the crystal interface while the electrons could be trapped at the V_A centres left behind by the Br^{0} atoms. According to this mechanism the trapped Br^{0} atoms would be released during the TL and recombine with the electron at the V_A centres transferring the recombination energy to a neighbour Cu^{+} ion which causes the characteristic Cu^{+} luminescence.

In summary, KBr:Cu is ionized by (370nm) UV photons at the CuBr nanocrystals and the TL indicates that there are two types of traps for the mobile defect which could be a free electron or a halogen atom depending on the mechanism assumed. Which of the above mentioned mechanisms is dominant is still a subject under investigation.

5. Acknowledgements.

This work was partially supported by CONACyT under project U 40497- F and PIFI-2007 under project P/CA51 2006-26-18.

References

[1] A. Pérez-Rodríguez, M. Flores-Acosta, R. Rodríguez-Mijangos, and R. Pérez-Salas. Rev. Mex. Fis. **52** (2) 151–4 (2006).
[2] L.E. Brus. J. Chem. Phys. **80**, 4403–7 (1984).
[3] M. Haselhoff and H.J. Weber. Phys. Rev. B **58**, 5052-61 (1998).
[4] J. Valenta, J. Moniatte, P. Gilliot, B. Honerlage, J. B. Grun, and R. Levy. Phys. Rev. B **57**, 1774-83 (1998).
[5] Yasuaki Masumoto, Tadashi Kawazoe, and Tetsuya Yamamoto. Phys. Rev. B **52**, 4688-91 (1995).
[6] Yasuaki Masumoto and Seitaro Ogasawara. Jpn. J. Appl. Phys. **38**, L623-5 (1999).

Preparation and characterization of thermoluminescent aluminium oxide doped with Tb^{3+} and Tb^{3+}-Mg^{2+}

V S M Barros[1], W M Azevedo[2], H J Khoury[1], P Linhares Filho[2,3]

[1] Grupo de Dosimetria e Instrumentação, DEN, UFPE, Recife-PE, 50740-540, Brazil, e-mail: vdbarros@terra.com.br
[2] Laboratório de Química do Estado Sólido, CCEN, UFPE, Recife-PE, 50540-740, Brazil
[3] Curso de Ciência dos Materiais, CCEN, UFPE, Recife-PE, 50540-740, Brazil

Abstract. This paper presents the preparation method and the thermoluminescence analysis of aluminium oxide doped with Tb^{3+} and Tb^{3+}-Mg^{2+} obtained by Combustion Synthesis (CS). An aqueous solution containing stoichiometric amounts of aluminium, terbium, magnesium nitrates and urea were mixed and introduced in a muffle furnace pre-heated to 500℃. After combustion, the samples were thermally treated at 1300°C and irradiated with a Co-60 gamma radiation source. The TL glow curves of the annealed Al_2O_3:Tb and Al_2O_3:Tb,Mg samples presented a well defined TL peak at approximately 200 °C, whereas the samples without heat-treatment presented a large number of TL peaks in the range from 150 to 500°C. These peaks were attributed to amorphous and phase impurities (γ-Al_2O_3 mixed with the α-phase) still present in the sample. Dose response analysis showed a linear response in the dose range from 0.5 to 5 Gy. These results strongly suggest that CS is a suitable technique to prepare doped aluminium oxide for TL dosimetric applications.

1. Introduction

Anion deficient aluminium oxide (α-Al_2O_3:C) is an attractive material for thermoluminescent (TL) dosimetry of ionizing radiation due to its high sensitivity when compared to LiF:Mg,Ti (40-60 times higher), low effective atomic number (10.2), low fading and a relatively simple glow curve consisting of an isolated TL peak at ~190°C [1,2,3,4]. The high TL sensitivity of this material is attributed to a high concentration of F^+ centers (oxygen vacancies with one captured electron), induced by C^{2+} heterovalent impurities which substitute the Al^{3+} ion in the Al_2O_3 crystal structure. The commercial fabrication processes of Al_2O_3:C dosimeters utilize Czolchralsky, Vernuil or Stepanov crystal growth techniques [5], where furnaces at high temperature (2050ºC) and highly reducing atmospheres are required. For this reason few research groups are able to prepare this material. On the other hand, the Combustion Synthesis (CS) process has become an attractive alternative route to synthesize a wide range of advanced ceramics, composites, alloys, intermetallics and nanomaterials [6,7,8,9]. The characteristic feature of the CS method is the fact that the material is synthesized by use of the heat released from the highly exothermic reduction-oxidation chemical reaction, where the temperature can, in some cases, reach 2000 – 4000ºK [9]. Recently, in our laboratory, the CS method has been successfully used to prepare Al_2O_3 doped with Eu, Tb and Si for thermoluminescent (TL) dosimetry

[1] Vinícius S M Barros - vdbarros@terra.com.br.

applications [10]. This paper presents the preparation, the rare earth concentration analysis, the results of thermoluminescence properties of aluminium oxide doped with Tb^{3+} and Tb^{3+}-Mg^{2+} and the dose response of the samples prepared by Combustion Synthesis.

2. Materials and methods

The aluminium oxide samples doped with Tb^{3+} and Tb^{3+}-Mg^{2+} were prepared by mixing stoichiometric amounts of aluminium nitrate, urea and terbium nitrate. The excess water from the reactants was evaporated in a glass beaker on a hot plate until a gelatinous pre-combustion mixture was obtained. The solution was then transferred to a muffle furnace pre-heated to 500°C to start the reaction. After a few minutes, the reactants ignited spontaneously resulting in a fluffy white powder. The resulting powders were pelleted and annealed at 1300°C for 10 h. In order to study the effect of the Tb^{3+} concentration on the TL sensitivity, the Tb:Al relative concentrations were varied from 0.17 to 1.0 mol%. For the Tb^{3+}-Mg^{2+} samples, the concentrations of Tb:Al and Mg:Al were fixed at 0.17 and 0.5 mol%, respectively.
X-ray diffraction (XRD) measurements were made with a RIGAKU model D/MAX 2200 diffractometer with a Cu K_α (λ=0.1540 nm) radiation and surface morphology was studied with a JEOL scanning electron microscope (SEM), model JSM-6460. To evaluate the TL response, the pellets were irradiated with a Co-60 gamma radiation source in air, under electronic equilibrium, with absorbed doses in the range of 0.5 to 5 Gy. The TL measurements were carried out using a Harshaw-Bicron TL reader, model 3500, equipped with a Hamamatsu R6094 photomultiplier tube, which has a maximum efficiency at 420 nm. The glow curves were acquired in the temperature range of 50 to 350 ºC with a heating rate of 2°C/s.

3. Experimental results and discussion

Figs. 1 and 2 show the XRD diffractogram for the Tb^{3+} doped aluminium oxide and Mg-Tb^{3+} doped aluminium oxide samples respectively, prepared by combustion method, before and after heat treatment at 1300°C for 10 h. The patterns indicate that the "as sintered" sample shows a broad peak characteristic of amorphous phase and diffraction peaks associated with the disordered crystalline γ phase of alumina and the stabilized crystalline α phase of alumina which can be attributed to an incomplete combustion reaction and to remaining material from the decomposition of the reagents. On the other hand the XRD of the samples after the heat treatment shows a considerable decrease of the amorphous and γ phases, indicating a γ to α phase transformation. This transformation usually occurs in the temperature range of 1100-1200°C [11].

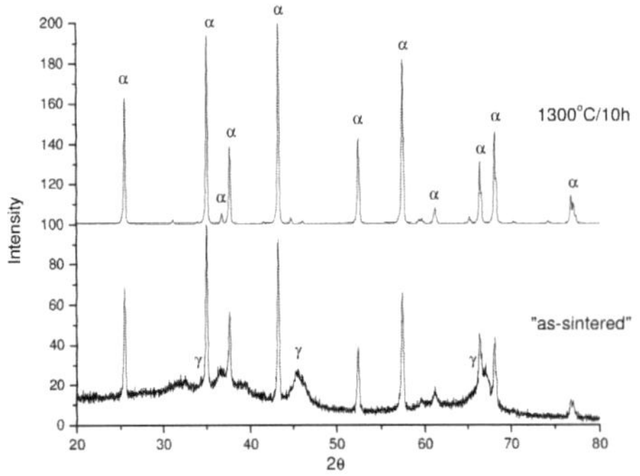

Figure 1. X-ray diffraction pattern of Al_2O_3:Mg,Tb immediately after combustion ("as-sintered") and after heat treatment.

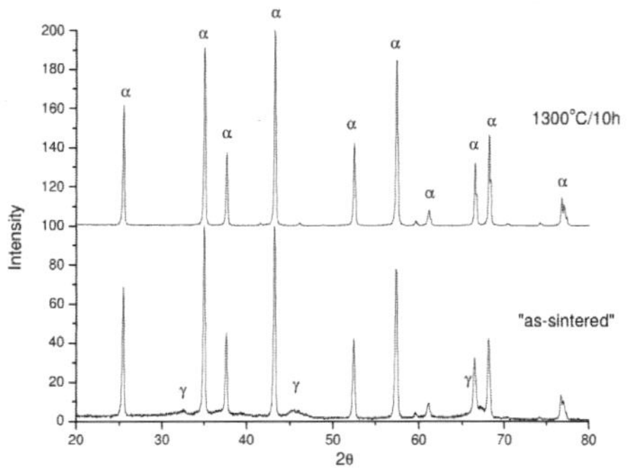

Figure 2. X-ray diffraction pattern of Al_2O_3:Tb immediately after combustion ("as-sintered") and after heat treatment.

The scanning electron micrographs of Al_2O_3:Tb,Mg obtained by CS and annealed at 1300℃ is shown in Fig. 3. This result shows a smooth-surface microstructure with grain sizes in the range 0.5–3 μm. This behaviour could be attributed to the strong heat involved in the combustion reaction, when urea is the combustion fuel. This fuel forms an intermediate stable phase capable of keeping the heat, increasing the particle size [12, 13].
Fig. 4 shows the TL glow curve for the Al_2O_3:Tb sample without heat treatment, and treated at 1000℃ and 1300℃. It can be seen that only the samples sintered at 1300℃ presented an isolated TL peak at around 200°C, which can be used for radiation dosimetry. The "as-sintered" samples and the samples heat-treated at 1000℃ exhibited a large number of peaks between 150 and 500°C that can be attributed to the presence of the amorphous and γ-phase Al_2O_3.

Figure 3. SEM micrograph of the Mg-Tb doped (0.17mol%) samples after thermal sintering at 1300°C.

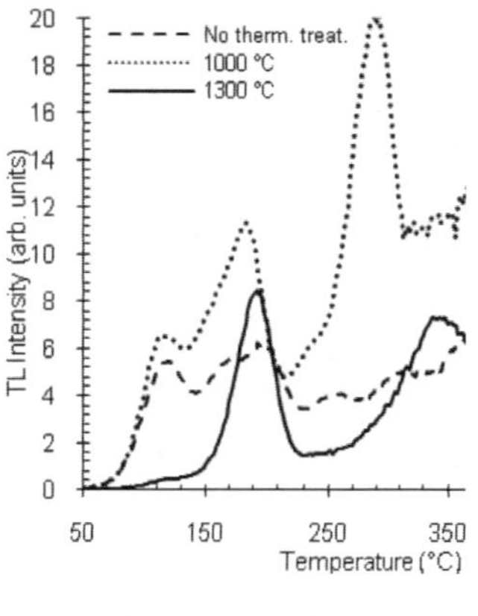

Figure 4. TL glow curves of three Tb doped (0.17mol%) samples: without thermal treatment, sintered at 1000°C and at 1300°C.

The TL glow curves for Al_2O_3 sample doped with Tb and Tb-Mg analyzed as a function of the dose are shown in Fig. 5 and 6, respectively. It can be seen that in Tb-Mg there is an enhancement of the TL peaks at 110 and 310°C. At the current stage of the development, the TL peak at 200℃ can be used in both samples to measure Co-60 gamma doses as low as 50 mGy. The dose response calculated from the areas under the 200℃ TL peak versus dose, shown in Fig. 7, indicate that the Tb and Tb-Mg dose response curves have approximately the same slope indicating a similar sensitivity to ionizing radiation.

The effect of the Tb^{3+} ions on the TL sensitivity was analyzed as a function of the concentration of Tb samples irradiated at 5 Gy. The TL intensity as a function of the Tb^{3+} concentration shown in Fig. 8 clearly indicate that the highest sensitivity was found for the sample with 0.17mol% Tb. A similar result was found for Al_2O_3 doped with Eu^{+3} ions [10] where the highest TL sensitivity was obtained with 0.5mol% Eu^{+3} concentration. The reason for this TL behaviour is still unknown; however, it may be associated with the fact that Tb^{3+} or Eu^{3+} ions prevent the formation of the α phase as the concentration of the dopant ion increases [14].

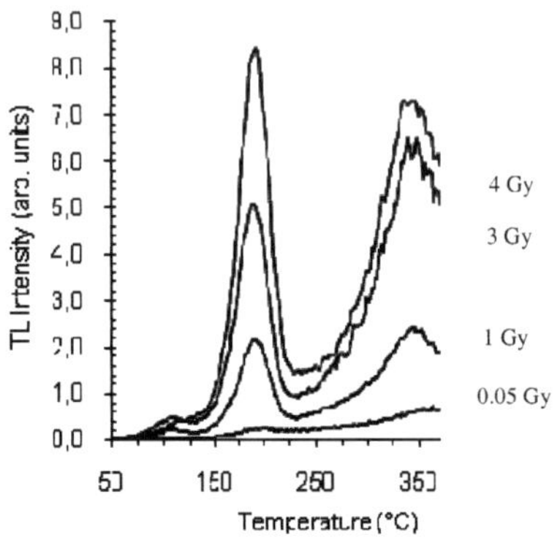

Figure 5. TL glow curves of Al_2O_3:Tb (0.17 mol%) irradiated at different Co-60 doses.

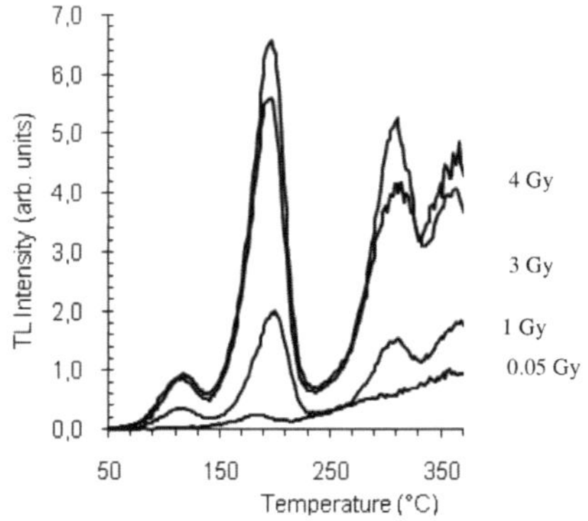

Figure 6. TL glow curves of Al_2O_3:Tb,Mg irradiated at different Co-60 doses.

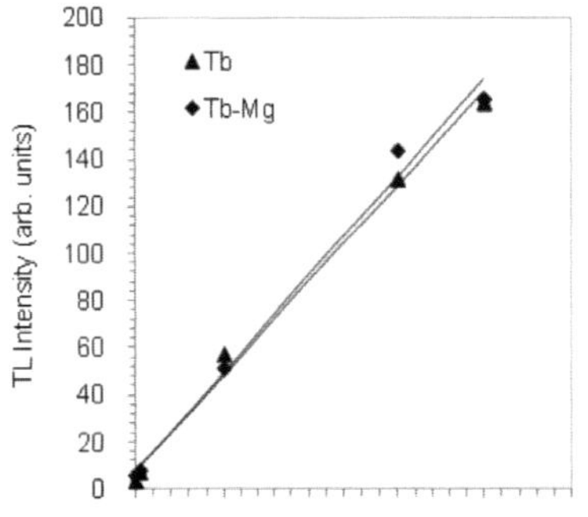

Figure 7. TL dose response curves of Tb and Tb-Mg doped Al_2O_3.

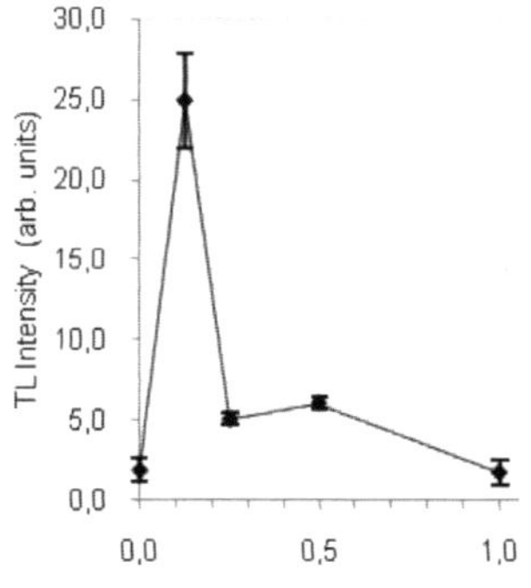

Figure 8. TL intensity versus Tb concentration of samples irradiated at 5 Gy.

4. Conclusions

From the results above we can conclude that CS is a suitable method to prepare Al_2O_3 and Al_2O_3 doped with rare-earth ions for dosimetric applications. This technique is particularly attractive due to its low cost, short reaction times, high purity products and straightforwardness to prepare samples with different dopant concentrations. The results showed that 0.17mol% rare earth optimizes the TL sensitivity and that a thermal treatment at 1300°C is necessary to obtain a well defined TL peak at 200℃ for dosimetric applications. The TL response for both samples presented a linear response with dose up to 5Gy, which strongly indicates that this material is suitable for dosimetric application. More work is under way to optimize the sensitivity of the material to lower dose measurements.

5. Acknowledgments
The authors thank the financial support of the Laboratório de Metrologia das Radiações Ionizantes (LMRI/DEN-UFPE) and the Brazilian agencies Conselho Nacional de Desenvolvimento Científico e Tecnológico (CNPq), Coordenação de Aperfeiçoamento de Pessoal de Nível Superior (CAPES) and to Renami network and Inct-Inami.

6. References

[1] Akselrod M S, Kortov V S, Kravetsky D J, Gotlib V I 1990 *Rad. Prot. Dosim.* **32** 15
[2] Akselrod M S, Gorelova E A *Nuc. Tra. Rad. Meas.* 1993 **21** 143
[3] Mckeever S W S, Akselrod M S, Colyott L E, Agersnap-Larsen N, Polf J C, Whitley V 1999 *Rad. Prot. Dosim.* **84** 163
[4] Larsen N A 1999 *Dosimetry based on thermally and optically stimulated luminescence* (RISØ National Laboratory: *PhD Thesis n °RISØ-R-1090*)
[5] Bøtter-Jensen L, Mckeever S W S, Wintle A G 2003 *Optically Stimulated Luminescence Dosimetry* (Amsterdam: Ed. Elsevier)
[6] Merzhanov A G *Russ. Chem. Bulletin* 1997 **46** 1
[7] Minami T *Resonance* 2000 **2** 50
[8] Varma A 2001 *Sci Am* **283** 44
[9] Patil K C, Aruna S T, Mimani T 2002 *Curr. Opin. Solid State Mater. Sci.* **6** 507
[10] Barros V S M, Azevedo W M, Khoury H J, Linhares P 2008 *Rad. Meas.* **43** 345
[11] Monteiro M A F, Brito H F, Felinto MC, Brito G E S, Teotonio E E S, Vichi F M, Stefani R. 2008 *Microp Mesop Mat* **108** 237
[12] Aruna S T, Rajam K S. 2004 *Mater. Res. Bull.* **39** 157
[13] Sarikaya Y, Ada K, Alemdaroglu T, Bozdogan I. 2002 *J. Eur. Ceram. Soc.* **22** 1905
[14] Hirata G, Perea N, Tejeda M, Gonzalez-Ortega J A, Mckittrick 2005 J *Optical Materials* **27** 1311

Journal of Physics: Conference Series **249** (2010) 012026 doi:10.1088/1742-6596/249/1/012026

Study of the effects of heat treatments in the paramagnetic centers of alkaline feldspars and their implications in the luminescence properties

S H Tatumi[1,*], J F S Bitencourt[2], D M da Silva[2], J S Lyra[3], A Kinoshita[4] and H S L Sullasi[5]

[1] Faculdade de Tecnologia de São Paulo, Depto. de Ensino Geral, pça. Cel. Fernando Prestes, 30, 01124-060, São Paulo, SP, Brazil.
[2] Escola Politécnica da Universidade de São Paulo, Depto. de Engenharia de Sistemas Eletrônicos, av. Luciano Gualberto, 380, 05508-970, São Paulo, SP, Brazil
[3] Escola Politécnica da Universidade de São Paulo, Depto. de Engenharia Metalúrgica e de Materiais, av. Mello Moraes, 2463, 05508-900, São Paulo, SP, Brazil.
[4] Universidade do Sagrado Coração, r. Irmã Arminda, 10-50, 17011-160, Ribeirão Preto, SP, Brazil.
[5] Universidade Federal de Pernambuco, 50740-540, Recife, PE, Brazil.

Abstract. The present work shows a detailed study of the effects of preheating in ESR signals of four alkaline feldspar samples. ESR results showed Fe^{3+}, Al-O^{-}Al, Ti^{3+} and $[Pb\text{-}Pb]^{3+}$ centers, When the samples were gamma-irradiated, the concentration of the Fe^{3+} centers deceased in all the samples, but the Al-O –Al center concentration increased after about 50 Gy while that of the Ti^{3+} and $[Pb\text{-}Pb]^{3+}$ centres became almost constant. The effects of preheating were observed in samples previously irradiated with doses of 0.2 and 3 kGy after heating at 100, 200, 300 and 400 °C, respectively for 10 minutes. In the orthoclase sample FELW, the Fe^{3+} signal remained almost constant up to 300 °C and increased at 400 °C, while that of the Al center decreased slowly between 100 and 300 °C, but increased at 400 °C. In the microcline sample FELG an increase of the Al coupled with $[TiO4]^{-}$ centres was observed up to 300 °C, while a decrease occurred for higher temperatures. Fe^{3+} center concentration increased up to 200 °C and became nearly constant for higher temperatures. According to our results, even preheating at temperatures as low as 200 °C can change the concentration of Fe, Al and Ti centers in the sample; this fact can modify the luminescence response of the samples with the dose, because all these centers are associated with one or more emission spectra bands.

1. Introduction

It is known that heating occasionally can change the luminescence sensitivity of crystals. A large number of luminescence studies carried out to analyze the origin of the sensitivity change, but few works correlating luminescence with ESR (Electron Spin Resonance) experiments have been done.

* To whom any correspondence should be addressed. tatumi@fatecsp.br

Nowadays, the SAR protocol [1] is largely applied in IRSL (Infrared Stimulated Luminescence) of feldspar for determining accrued radiation doses in unheated and heated samples, both for dating and accident dosimetry purposes. This protocol can correlate the luminescence sensitivity of the sample in regeneration method.

In the regeneration method using a single aliquot, the sample was submitted to several irradiation, optical stimulation and preheating cycles, that can promote changes in luminescence sensitivity. Another example of luminescence sensitivity change occurs in the optical dating of materials previously submitted to heating, such as pottery, bricks and tile.

The main object of this work is to present a detailed study of the effects of preheating in ESR signals of alkaline feldspar samples and its implications with ESR and optical dating results.

2. Experimental Procedure

Natural specimens of microcline and orthoclase, identified by X-rays diffraction, were investigated; Table I shows some characteristics of each sample. The samples named FELW, FELG and FELP were previously analyzed with TL, ESR, X-ray fluorescence and Neutron Activation Analysis; the results were published by Bitencourt et al., 2006 [2].

Table I. Some crystallographic and structure information, colour and origin of the sample.

Sample	Structure	Compositions	Colour
FELW	Orthoclase	$Or_{86.99}Ab_{12.97}An_{0.04}$	White
FELG	Microcline	$Or_{87.63}Ab_{12.33}An_{0.04}$	Gray
FELP	Microcline	$Or_{81.36}Ab_{15.45}An_{3.19}$	Pink
FELAM	Microcline	$Or_{86.99}Ab_{12.97}An_{0.04}$	Green

The fourth sample named FELAM is an amazonite collected from the southeast region of Brazil (Minas Gerais State). X-ray fluorescence analysis were performed and trace elements of La < 28 ppm, Ce < 35 ppm, TiO_2 < 0.007 mol%, Pb = 378 ppm, Fe_2O_3 = 0.1 mol%, MnO = 0.008 mol% and Al_2O_3 = 18.35 mol% impurities were found. Similar results were obtained for the three previously cited samples, except for the Pb content, which was not detected in FELW, FELG and FELP samples.

For ESR and emission spectra experiments, the specimens were carefully ground with a mortar and pestle, and the fraction between 88 and 180 μm was selected. The selected grains were treated with 10% HCl for 10 minutes, in order to reduce the grind effects. After that, they were washed extensively with distilled water and acetone. All the gamma-rays irradiations were performed at RT with ^{60}Co source (0.3906 $kGy.h^{-1}$).

ESR spectra were taken at 77K, below saturation, by means of the homodyne X-band Varian E-4 spectrometer with a magnetic field modulation of 100 kHz, a microwave power of 20 mW and a TE011 mode cavity. The modulation amplitude used was 2.5 G peak-to-peak and the scan speed was 2.0 and 100 G in 60 seconds. The g-values were determined with a DDPH standard signal.

3. Results and Discussion

Fig. 1 shows the ESR spectrum for the natural FELAM sample and the spectra resulting for gamma irradiation of the sample with various doses. The spectra show the presence of Fe^{3+}, Al-O –Al, $[Pb-Pb]^{3+}$ and $[TiO4]^{-}$ centers in all samples. Table II shows the g-values found for the FELAM sample and their related centers. The values g = 2.946 and g = 1.802 can be related to $[Pb-Pb]^{3+}$ centres [3,4], g =2.012 to Al centers and g = 2.946 and g=3.819 to Fe^{3+} centers.

The effects of preheating are observed in samples previously irradiated with doses of 0.2 and 3 kGy and after preheating at 100, 200, 300 and 400 °C, respectively, for 10 minutes (Figure 2). In the orthoclase sample FELW, the Fe^{3+} signal became almost constant up to 300 °C and increased at 400

°C, while the Al centers decrease slowly between 100 and 300 °C and increase for the 400 °C preheating

In the microcline sample FELG an increase in the concentration of the Al coupled with $[TiO4]^-$ centres was observed up to 300 °C, while a decrease occurred for higher temperatures (Figure 3). Fe^{3+} center concentration increased up to 200 °C and became nearly constant for higher temperatures

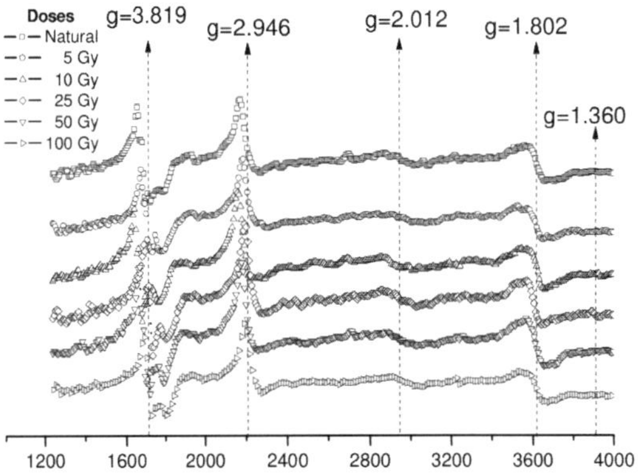

Figure 1. ESR spectra of microcline sample FELAM, irradiated with gamma-rays.

Table II. G-values found for FELAM sample and their respective probable relates center.

Sample	g-values	Probable related center
FELAM	3.819	Fe^{3+}
	2.946	
	2.012	Al-O$^-$-Al
	1.802	$[Pb\text{-}Pb]^{3+}$
	1.360	

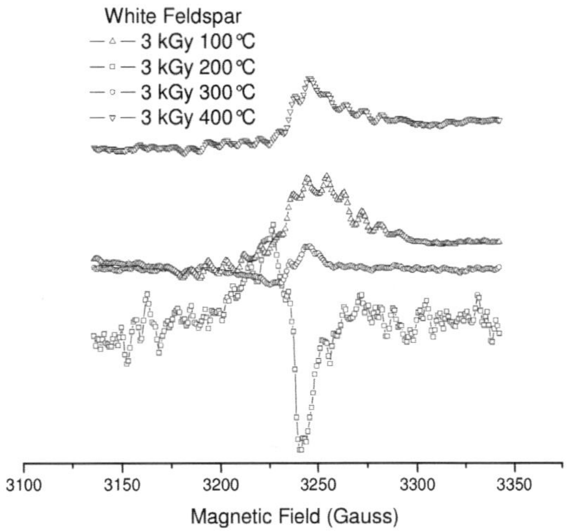

Figure 2. Effect of preheating in ESR spectra of the Al-O^--Al center of FELW sample, irradiated with 3 kGy of gamma-rays.

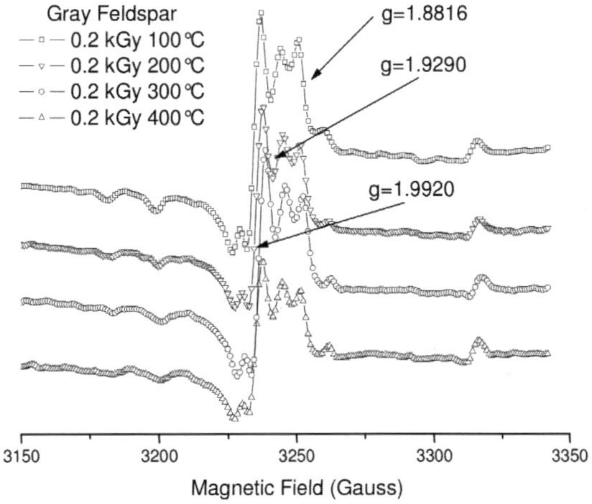

Figure 3. Effect of preheating in ESR spectra of the Al/Ti^{3+} centers of the FELG sample, irradiated with 200 Gy of gamma-rays.

4. Conclusions

According to our results, the concentrations of Fe^{3+}, Al-O^-Al and $[TiO_4]^-$ centers can change, even with the use of low temperature preheating. It is know that these impurities are important luminescence centers [5,6], which are used for sediments optical dating [7,8].

All the centers can be associated with emission bands obtained with infrared stimulation. Thus precautions are advised with heat treatments of dating materials, taking into account the results presented.

5. Acknowledgements

We would like to thank FAPESP and CNPq (Brazilian agencies) for financial support and Ph.D. Oswaldo Baffa for allowing the use ESR equipment.

References

[1] Jacobs Z, Wintle A G, Duller G A T 2006 Evaluation of SAR procedures for De determination using single aliquots of quartz from two archaeological sites in South Africa Radiation Measurements vol. 41 issue 5 pp.520-533.

[2] Bitencourt J F S, Silva D M, Silva P H, Kinoshita A, Munita C A and Tatumi S H 2006 Luminescence and ESR properties of Brazilian feldspars Radiation Measurements vol. 41 issues 7-8 pp. 948-953.

[3] Petrov I, Mineeva R M, Bershov L V and Agel A 1993 EPR of $[Pb\text{-}Pb]^{3+}$ mixed valence pairs in amazonite-type microcline American Mineralogist vol. 78 pp. 500-510.

[4] Mittani J C R, Cano N F and Watanabe S 2005 Use of $[Pb\text{-}Pb]^{3+}$ center of the amazonite for dating Applied Radiation and Isotopes vol. 62 pp. 251-254.

[5] Poolton N R J, Mauz B, Lang A, Jain M and Malins A E R 2006 Optical excitation processes in the near band-edge region of $KAlSi_3O_8$ and $NaAlSi_3O_8$ feldspar Radiation Measurements vol. 41 pp. 542-548.

[6] Krbetschek M R, Götze J, Dietrich A and Trautmann T 1997 Spectral information from mineral relevant for luminescence dating Radiation Measurements vol. 27 pp. 695-748.

[7] Argyilan E P, Forman S L, Johnston J W and Wilcox D A 2005 Optically stimulated luminescence dating of late Holocene raised strandplain sequences adjacent to lake Michigan and Superior, Upper Peninsula, Michigan, USA Quaternary Research vol. 63 pp. 122-135.

[8] Hashimoto T, Yawata T and Takano M 2005 Comparison of naturally accumulated radiation-doses between RTL, BTL, OSL and IRSL using white minerals from burnt archaeological materials and usefulness of RTL-dating from quartz extracts Geochemical Journal vol. 39 pp. 201-212.

Effect of particle size in the TL response of natural quartz sensitized with high gamma dose

A B Carvalho Jr[1]; P L Guzzo[2]; H L Sullasi[3] and H J Khoury[3]
[1] Graduate Program of Mechanical Engineering, Federal University of Pernambuco, 50740-530, Recife, PE, Brazil
[2] Department of Mining Engineering, Federal University of Pernambuco, 50740-530, Recife, PE, Brazil
[3] Department of Nuclear Energy, Federal University of Pernambuco, 50740-540, Recife, PE, Brazil

E-mail: alvarobcjr@yahoo.com.br

Abstract. The aim of this study is to investigate the effect of particle size in the thermoluminescence (TL) response of natural quartz sensitized with high gamma dose. For this, fragments of a single crystal taken from the Solonópole district (Brazil) were crushed and classified into ten size fractions ranging from 38 µm to 5 mm. Aliquots of each size fraction were sensitized with 25 kGy of gamma dose of ^{60}Co and heat-treated in a muffle furnace at 400°C. The non-sensitized samples were exposed to test doses between 50 Gy and 5 kGy and the sensitized samples were exposed to a unique test dose equal to 50 mGy. For non-sensitized samples, the TL peak near 325 °C increases with the particle size decreasing. However, in the case of sensitized samples, the TL output near 280 °C increases with the increasing of particle size up to mean grain size equal to 308 µm. Above 308 µm, an abrupt reduction in the TL intensity was noticed. These effects are discussed in relation to the specific surface area and the different interaction of high gamma doses with fine and coarse particles of quartz.

1. Introduction

Many natural crystals exhibit thermoluminescent (TL) properties that are suitable for TL dating amongst which quartz is one of the most important ones because it is currently found in fragments of ancient ceramics and geological sediments [1,2]. Recently, natural quartz has been investigated for TL dosimetry applications [3]. Even though there is a large number of studies dealing with TL properties of quartz and silica, it is believed that natural quartz has not been used as a dosimeter for ionizing radiation. The reason for this is that its glow curve is very dependent on the geological conditions in which quartz was grown [1,3]. In addition, the point defects acting as electron traps and recombination centers in crystalline quartz are not yet fully characterized.

An important effect that can be observed in quartz is the increase in the sensitivity of a given glow peak following the absorption of high radiation doses. This sensitization process, also know as pre-dose effect, is an important feature in radiation dosimetry where the enhancement in sensitivity is highly desirable. Until now, systematic studies on the sensitization in quartz have been restricted to the first glow peak around 110 °C by ionising radiation and heat-treatment. This sensitization process was explained using a phenomenological model, which involves an increase in the probability of radioactive recombination at the luminescence sites [4,5]. On the other hand, a recent study reported the sensitization of a peak between 250 and 300 °C induced by high gamma irradiation [6]. This study showed that the sensitization of the peak near 280 °C allows the detection of doses in range of few mGy. Thus, the sensitization of quartz glow peaks appearing above 110 °C can be useful for applications in TL dosimetry due to the no effect of thermal fading.

There are several materials produced for TL dosimetry using powdering procedures and it is well known that TL intensities are affected by the particle size [7-9]. Studies have been performed with the objective to understand the behavior of TL materials of different grain sizes [8,10,11]. In the case of silica powders, it was reported that the intensity of the TL peak at 180 °C increases with decreasing grain size when silica is irradiated with gamma doses of 23 Gy [11]. In addition, efforts have been made to produce TL dosimeters using quartz grains prepared from sensitized natural quartz [12]. Therefore, it seems important to better understand the behavior of TL glow curves as a function of quartz particles with different grain sizes. Thus, the aim of this study is to investigate the effect of particle size in the TL response of non-sensitized and sensitized quartz particles.

2. Experimental procedure

For this study, a quartz single crystal extracted from the Solonópole district located in the Ceará State of Brazil was used. Crystal plates cut with a diamond saw were manually crushed using an agate mortar and pestle. After crushing, the particles were classified into fine (300x425 μm, 150x315 μm, 150x300 μm, 75x150 μm, 38x75 μm and < 38 μm) and coarse fractions (2.8x4.8 mm, 2.0x2.8 mm, 1.4x2.0 mm and 0.85x1.7 mm) by using standard Tyler sieves. The particle size distributions of the fine fractions were measured with a laser particle size analyzer, model Malvern Mastersizer 2000 with the Hydro 2000MU sample dispersion accessory. For these fractions, the mean particle size (D_m) corresponds to the aperture associated with 50% of passing material. For the coarse fractions, D_m corresponds to the average of the lengths measured along three directions with a digital microscope.

Aliquots of fine and coarse grains were sensitized with 25 kGy of ^{60}Co in a gamma cell irradiator with a dose rate of 10 kGyh^{-1}. Three heat treatments were successively performed in order to guarantee the release of charge carriers from the trap levels. The heat-treatments were performed in a muffle furnace as follows: continuous heating up to 400 °C, annealing for one hour at 400 °C, cooling, annealing for two hours at 100 °C and cooling. This thermal cycle was adopted as the standard procedure throughout this work. This heat-treatment was performed only once for non-sensitized samples.

For the study of TL response as a function of particle size, batches containing three aliquots of non-sensitized and sensitized samples were prepared. The mass of each aliquot (~ 5 mg) was determined with an analytical balance accurate to 0.0001 g. The non-sensitized samples were exposed to test doses of 0.05, 0.5, 2 and 5 kGy of ^{60}Co called here test doses to distinguish them from the sensitization doses. The sensitized samples were exposed to test doses of 50 mGy. The TL glow curves were record from 50 to 400 °C using a Harshaw 3500 reader with a heating rate equal to 2 °Cs^{-1}. The TL signal was normalized regarding weight and dose. The TL intensities were obtained integrating the TL glow curves between 175 and 390 °C.

To analyze the severity of the crushing procedure on the crystalline structure of quartz particles, aliquots of three fine fractions (75x150 μm, 38x75 μm and < 38 μm) were examined by X-ray diffraction (XRD). Standard XRD patterns were obtained with a θ-2θ diffractometer between 10° and 60° using Cu-Kα radiation and the full width at half maximum (FWHM) of the diffraction peaks related to $(10\bar{1}0)$, $(10\bar{1}1)$ and $(11\bar{2}2)$ were measured. Afterwards, scanning electron microscopy (SEM) was used to observe the shape and the fracture pattern of crushed particles.

3. Results

The TL curves of the non-sensitized samples with particle sizes ranging from 38 to 425 μm show glow peaks near 90, 215 and 325 °C. An intense TL peak was observed at 90 °C which corresponds to the well documented 110 °C peak of quartz. In the present study, the peak at 90 °C was not considered due to its unstable behavior at room temperature. Figure 1 shows the characteristic TL curves for samples with different particle size irradiated with 50 Gy. These results show that the TL intensity of the peak near 325 °C increases with decreasing particle size. The TL glow curves of samples irradiated at 500 Gy are shown in Figure 2. Besides the 110 and 325 °C glow peaks, a new peak is observed near 215 °C. In this case, an opposite effect is observed with respect to particle size, i.e., the TL intensity at 215 °C decreases with decreasing particle size.

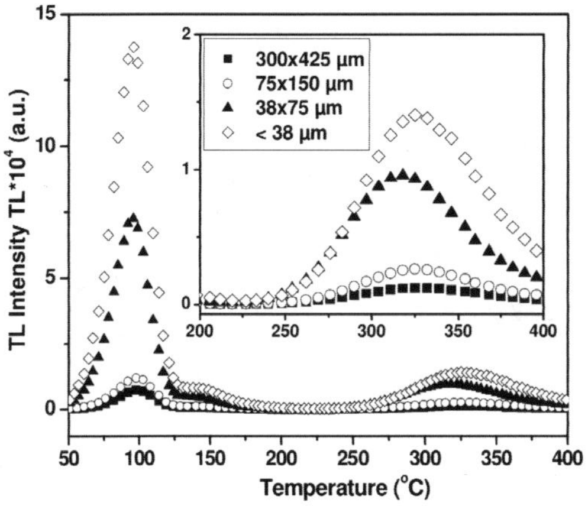

Figure 1. TL glow curves of quartz particles with different sizes exposed to 50 Gy.

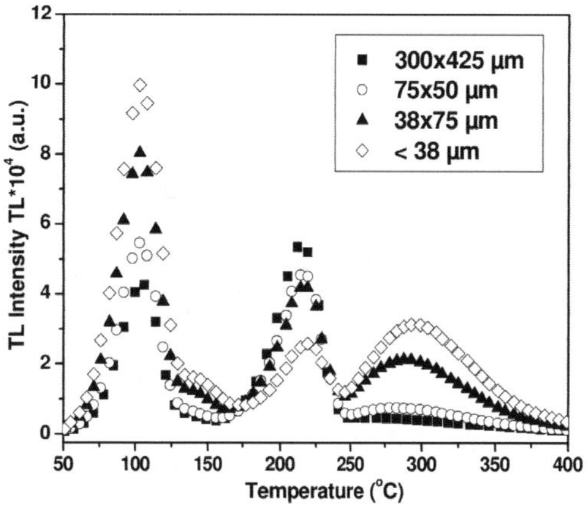

Figure 2. TL glow curves of quartz particles with different sizes exposed to 500 Gy.

Figure 3 summarizes the behavior of the TL intensity integrated from 175 to 390 °C as a function of the mean particle size. In this figure, the TL intensity was normalized in relation to the intensity measured for the size fraction < 38 μm. According to Figure 3, the samples irradiated with 50 and 500 Gy show a similar behavior in relation to the mean particle size (D_m), i.e., the increase in TL intensity is observed for samples with D_m < 150 μm. In case of samples irradiated at 2 and 5 kGy, the TL intensity is not clearly correlated with D_m due to the increase of the peak near 215 °C. As shown in Figure 2, this peak increased with increasing particle size. Thus, the increase of the 325 °C peak with decreasing particle size was completely surpassed by the increase of the 215 °C peak.

In order to better observe the effect of particle size in the TL response, the sensitized samples were divided in two groups. Figure 4 shows the TL glow curves for fine fractions and Figure 5 shows the TL glow curves for coarse fractions. In Figure 4, an increase in the TL intensity of the peak near 300 °C is observed for larger particle sizes. This behavior is similar to that observed for the peak near 215 °C occurring in non-sensitized samples (Figure 2). In addition, it is observed that the TL peak near 300 °C shifts to 280 °C for particles classified as coarse fractions. In principle, it would be expected that this shift should be noticed in the opposite sense because it is believed that the temperature set is reached in a smaller amount of time when fine particles are distributed over the heating planchet of the TL reader. This statement is based on the assumption that the heating flow rate through a mass of insulating material is higher for fine grains. In this way, the onset of the luminescence should occur at lower temperature for fine grains. Therefore, the temperature shift shown in figures 4 and 5 cannot be explained by the effect of the particle size on the heat flow during the TL readings. Certainly, the temperature shift of the second peak is connected with the sensitization of a broad TL peak occurring between 200 and 300 °C when quartz is irradiated with high gamma doses.

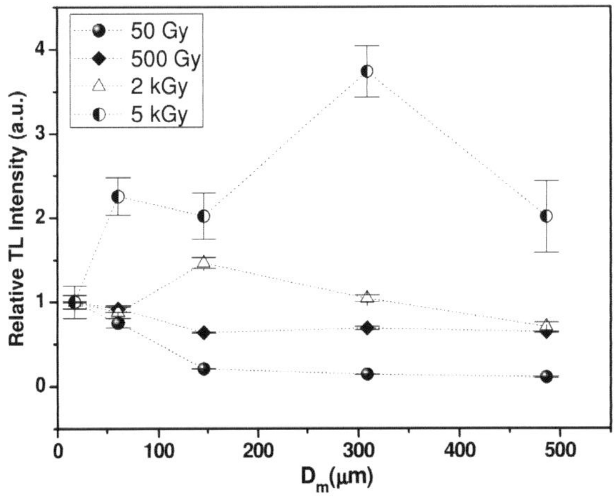

Figure 3. Relationship between TL intensity integrated from 175 to 390 °C and the mean particle sizes (D_m) for non-sensitized quartz.

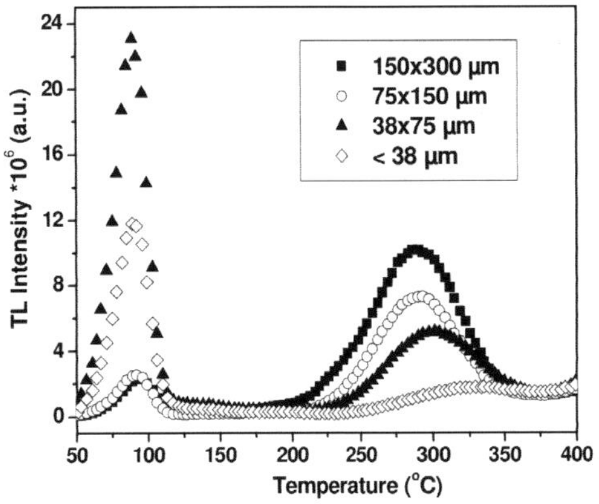

Figure 4. TL glow curves of fine particles of sensitized quartz (test dose: 50 mGy).

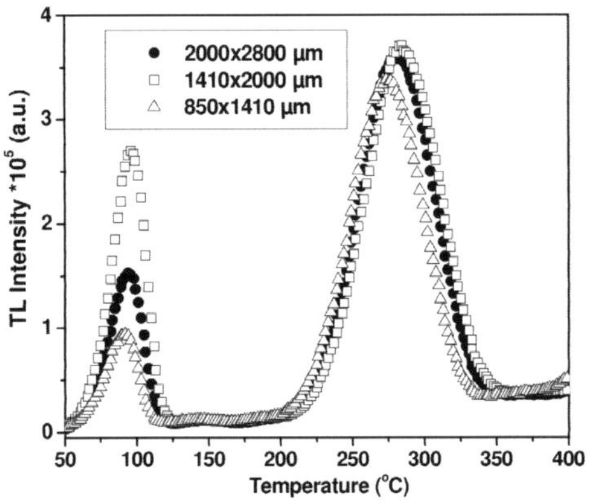

Figure 5. TL glow curve of coarse particles of sensitized quartz (test dose: 50 mGy).

Figure 6 shows the TL signals integrated from 175 to 390 °C as a function of the mean particle size for sensitized quartz. It can be seen that the TL intensity increases as the particle size increases from 17 to 308 μm. For particles larger than 308 μm, an abrupt decrease in the TL intensity is noticed and no important change in TL intensity is observed for particle sizes larger than 2 mm. In this figure, the sample with the largest D_m is a solid disc with diameter of 6 mm and 1 mm thick. This disc was prepared directly from the crystal plate and it was not submitted to the crushing procedure. These measurements suggest a reliable similarity between the TL responses of coarse particle sizes and the TL response of a single crystal prepared from the same sensitized quartz.

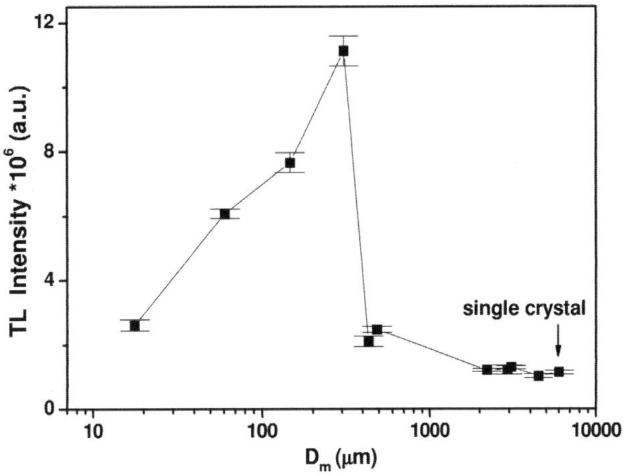

Figure 6. Relationship between the TL intensity integrated from 175 to 390 °C and the mean particle size of sensitized quartz (test dose: 50 mGy).

4. Discussion

Previous studies showed that the manual crushing of quartz particles with mortar and pestle can induce mechanical damage or amorphization into the crystal structure when fine quartz grains are submitted to high pressures [13,14]. However, any change in the width of the $(10\bar{1}0)$, $(10\bar{1}1)$ and $(11\bar{2}2)$ diffracting peaks was observed with the FWHM analysis carried out in 75x150, 38x75 and < 38 µm particle fractions. Based on this analysis, the assumption that the decreasing TL intensity for D_m < 308 µm shown in Figure 6 would be associated with amorphization or mechanical damage can be ruled out. Similarly, the SEM analysis did not reveal any substantial difference in the morphology of the quartz particles. For instance, figures 7 and 8 show typical images obtained with particles sizes of 308 µm and 432 µm mean grain sizes. Besides the size range and the presence of very fine grains deposited on the surface of 308 µm particles, no difference is observed in fracture patterns that could be associated with the abrupt change of TL intensity shown in Figure 6. It is observed that brittle microcracking was the principal mechanism of size reduction related to the crushing procedure performed here.

Figure 7. SEM micrograph of quartz particles with D_m = 308 µm.

Figure 8. SEM micrograph of quartz particles with D_m = 432 µm.

The effect of the particle size in the TL response of non sensitized quartz can be explained by the increase in the specific surface area for fine quartz grains. Figure 9 shows the relationship between the TL intensity and the specific surface area of several particle fractions. For test doses 50 and 500 Gy, it is observed that the TL intensity increases with increasing specific surface area. A similar behavior was previously reported for particles of potassium iodate (KI) and amorphous silica exposed to similar doses [10,11]. As stated before, the TL output increases with decreasing particle size because a much higher surface area of material is exposed during the TL reading.
The effect of the specific surface area on the TL intensity was not observed when test doses such as 2 and 5 kGy were administrated to quartz grains due to the onset of peak near 215 °C. As shown in Figure 2, this peak increases with increasing particle size.

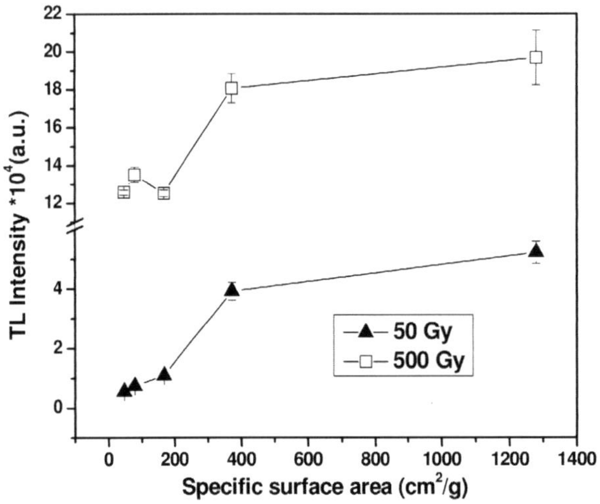

Figure 9. Relationship between TL intensity integrated from from 175 to 390 °C and the specific surface area of non sensitized quartz samples with different particles sizes.

For sensitized samples, no additional information beyond that shown in Figure 6 was observed when the integrated TL signal was plotted against the specific surface area. A small increase in TL intensity is observed when the TL responses of single crystal and fragments of few millimeters are compared with the response of particles with D_m = 400 μm. This effect can also be associated with the increase of the specific surface area. In the sequence, the abrupt increase in TL response for D_m = 308 μm and the subsequent decrease noticed for finer grains cannot be merely explained by the effect of specific surface area. As a first approximation, these results can be explained by the effect of particle size during the irradiation of quartz with high gamma doses. For some reason that is not clear, the sensitization with 25 kGy is much more effective to create paramagnetic centers acting as electron traps and recombination centers in those particles with to D_m = 308 μm. It is believed that finer particles do not absorb the radiation in the same way as particles with grain size larger than 308 μm. Thus, the 280 °C TL peak noticed for quartz particles with D_m = 308 μm could be explained by the concomitant effects of the absorption of high gamma doses and the specific surface area of the particles within this size range.

5. Summary

For non-sensitized quartz of fine grain sizes, the increase in TL intensity integrated from 175 to 390 °C is associated with the increase of the peak near 325 °C and can be explained by the increase of the specific surface area. In this study, this effect was clearly observed at test doses ranging from 50 to 500 Gy. For test doses in the range of a few kGy, the effect of the specific surface area was not observed due to the onset of an additional peak near 215 °C . After sensitization, the effect of the specific surface area was noticed only when the TL intensity of fragments of a few millimetres is compared to that particles of mean grain size close to 300 μm. The subsequent decrease in TL intensity observed for finer particles was tentatively explained by their lower effectiveness in absorbing high gamma doses during the sensitization process. Solonópole quartz with grain size near to 300 μm has suitable properties for future use in TL dosimetry.

Acknowledgments

The authors are thankful to Profa. Márcia Carvalho A. Fantini (Physics Institute, USP-São Paulo) for the XRD analyses. One the authors (ABCJr.) is grateful for the financial support provided by CAPES.

References

[1] McKeever S W S 1984 Thermoluminescence in quartz and silica Radiat. Protec. Dosim. **8** 81
[2] Wintle A G 1997 Luminescence dating: laboratory procedures and protocols Radiat. Meas. **27** 769
[3] Guzzo P L, Khoury H J, Souza C P, Souza A M, Souza Jr, Schawrrtz M O and Azevedo W M 2006 Defect analysis in natural quartz from brazilian sites for ionising radiation dosimetry Radiat. Protec. Dosim. **119** 168
[4] Zimmerman, J 1971 The radiation-induced increase of the 110°C thermoluminescence sensitivity of fired quartz J. Phys. C **4** 3265
[5] Yang X H and Mckeever S W S 1990 The pre-dose effect in crystalline quartz J. Phys. D. Appl. Phys. **23** 237
[6] Khoury H J, Guzzo P L, Brito S B and Hazin C A 2007 Effect of high gamma doses on the sensitisation of natural used for thermoluminescence dosimetry Radiat. Effects & Defects in Solids **162** 101
[7] Mahesh K, Weng P S and Furetta C 1989 Thermoluminescence in solids and its applications Nuclear Technology Publishing 306
[8] Driscoll C M H and McKinlay A F 1981 Particle size effects in thermoluminescent lithium fluoride Phys. Med. Biol. **26** 321
[9] Carlson G A and Lorence L 1990 Particle size effect in CaF_2:Mn/Teflon TLD response at photons energies from 5-1250 keV IEEE Trans. on Nuclear Sci. **37** 1560
[10] Dhoble S J, Sahare P D and Moharil S V 1991 Thermoluminescence and colour centres in KI: particle size effect J. Phys. **3** 1189
[11] Ranjbar A H, Durrani S A and Randle K 1999 Electron spin resonance and thermoluminescence in powder form of clear fused quartz: effect of grinding Radiat. Meas. **30** 73
[12] Carvalho Jr A B, Guzzo P L, Khoury H J 2007 Obtenção de discos policristalinos de quartzo natural para dosimetria das radiações ionizantes XII Encontro Nacional de Tratamento de Minérios e Metalurgia Extrativa 767
[13] Hazen R M, Finger L W, Hemley R J, Hemley R J and Mao H K 1989 High-presure crystal chemistry and amorphization of α-quartz Solid State Communications **72** 507
[14] Kingma K J, Meade C, Hemley R J, Mao H and Veblen D R 1993 Microstructural observations of α-quarz amorphization Science **259** 666

Luminescence properties of feldspars from the Northeast region of Brazil

S T Santana[1]; H J Khoury[1]; H L Sullasi[1] and P L Guzzo[2]
[1] Department of Nuclear Energy, Federal University of Pernambuco, 50740-540, Recife, PE, Brazil
[2] Department of Mining Engineering, Federal University of Pernambuco, 50740-530, Recife, PE, Brazil

E-mail: sergiotsantana@hotmail.com

Abstract. The aim of this study is to investigate the thermoluminescence (TL) and infrared-stimulated luminescence (IRSL) properties of three types of feldspars from the districts of Solonópole and Parelhas located in the Northeast region of Brazil in order to propose a preheat procedure to minimize the anomalous fading effect in these materials. The feldspar samples were characterized by X-ray diffraction (XRD) and X-ray fluorescence spectrometry (XRF). The XRD analysis showed that the feldspar from Solonópole and one of the samples from the Parelhas district were classified as microcline (K-feldspar). The other sample from Parelhas was classified as albite (Na-feldspar) mixed with low concentration of muscovite and quartz. Studies of the fading effects in TL and IRSL signals for 30 days and preheating effects from 50 to 250°C for 10 minutes were carried out with these samples. The results show that preheating at 125°C is sufficient to avoid the 30 days fading. The TL and IRSL dose responses for ^{60}Co gamma radiation were studied in the range of 1 to 400 Gy. The results showed a linear response in the range of 1 to 100 Gy for the two types of Parelhas feldspars and from 50 to 400 Gy for the Solonópole feldspar. The low sensitivity showed by the feldspar from Solonópole is discussed in relation to its mineralogical composition. It is concluded that the TL and IRSL sensitivities are not associated with K- and Na- types of feldspar. The results of this paper will contribute to a standardized moderate annealing treatment for retrospective dosimetry and dating.

1. Introduction

The luminescence emission properties of quartz and feldspars have been widely used in dosimetry. Natural quartz and feldspar are generally chosen as dosimeters for radiation accidents since they are substances abundantly found on the earth´s crust and are used in the fabrication process of building materials such as roofs, bricks, tiles and many ceramic objects [1, 2]. In the last decade, most retrospective luminescence dosimetry research has focused on developing methods based on quartz. It was observed that the quartz dose response curve generally saturates at 200 Gy, limiting its use for dating studies to approximately 100-150 ka [3].

Feldspars are another type of natural luminescence dosimeters. They have many advantages compared to quartz, that include; a higher intrinsic luminescence sensitivity resulting in a lower detection limit; a luminescence signal which has been shown to saturate at a higher dose than that of quartz, and, unlike quartz, they are sensitive to IR stimulation [4, 5]. Unfortunately, the feldspar luminescence signal decreases with time faster than expected from a consideration of trap depth and ambient temperature. This feldspar anomalous fading is weakly dependent on temperature [6, 7].

The aim of this paper is the evaluation of TL and IRSL responses of three types of Brazilian feldspars and the effect of preheat temperature in TL and IRSL curves, in order to propose a preheat procedure to minimize the anomalous fading effect in these materials. The results of this paper will contribute to a standardized moderate temperature treatment for retrospective dosimetry and dating.

2. Materials and methods

The feldspars used in this study are from Brazil; they were taken from two deposits located in the Parelhas district (Rio Grande do Norte State) and from one deposit located in Solonópole district (Ceará State). The rocks were crushed with a laboratory grinder and carefully pulverized using a mortar and pestle. The grains were sieved and the particle size classified. Sizes under 75 μm were used for mineralogical characterization by X-ray diffraction (XRD) and X-ray fluorescence spectrometry (XRF). The mineralogical phases were determined by XRD with a Bruker-AXS D5005 instrument using Co Kα radiation. The XRD patterns were obtained scanning from 5° to 80° (2θ) in steps of 0.02° with a measurement time of 1 s per step. Each pattern was analyzed using a Bruker DiffracPlus software comparing the sample with a standard in the PDF02 database (ICDD, 1996) [8]. The semi-quantitative analysis of the oxides in each sample was carried out with a Rigaku-RIX 3000 XRF spectrometer, equipped with a Rh tube.

Feldspar fractions between 75 and 150 μm were selected for TL and IRSL studies. These samples were annealed at 500 °C for 1 hour in a programmable temperature-controlled muffle-furnace at atmospheric pressure. The samples were separated in aliquots of 150 mg, and placed in opaque bags, to be protected from the ambient light. The feldspar aliquots were irradiated with a ^{60}Co Gammacell.

For TL and IRSL fading studies, 1000 mg of each type of feldspar were irradiated with 100 Gy and were stored in the dark, at room temperature. The TL and IRSL signals of the 150 mg irradiated aliquots were read after different periods, ranging from 1 hour to 30 days. In order to evaluate the preheat temperature effect on TL and IRSL signals, a sample of each feldspar was irradiated with 100Gy. The aliquots of 150 mg were thermally treated for 10 minutes at different temperatures in the range of 50 °C to 250°C. After each heat treatment, the luminescence signals were read under the same conditions described above. To evaluate the dose response of the TL and IRSL curves, the aliquots were given a preheat treatment at 125°C for 10 minutes following each irradiation. The dose response curve for feldspars from Parelhas and Solonópole were, respectively, analysed at 1 to 100 Gy and at 50 to 400 Gy.

The TL signal was read using the Harshaw 3500 reader, with a heating rate of $4°C.s^{-1}$ in a temperature range from 50 to 400°C. The IRSL measurements were performed at room temperature with a continuous wave stimulation (CW) instrument generated by infrared LEDs (λ=850 nm). The luminescence detection was made with a bialcaline Hamamatsu photomultiplier tube (PMT) using a BG-39 filter 2 mm thick [9]. The IRSL signal was defined as the integrated counts from the beginning of stimulation until reaching 50% of the maximum.

3. Results and discussion

The XRD patterns are shown in Fig.1. Using the PDF02 database it was possible to obtain the phase indexation of the feldspar samples. The results show that the feldspars from Solonópole (MS) and one of the feldspars from Parelhas (MP) present a microcline phase, and were classified as microcline ($KAlSi_3O_8$). The other sample from Parelhas (AP) was classified as albite ($(Na,Ca)Al(Si,Al)_3O_8$) mixed with low concentration of muscovite ($KAl_2[Si_3AlO_{10}](OH,F)_2$) and α-quartz ($SiO_2$).

Table 1 shows the XRF analysis. The MP and MS samples showed high concentration of K_2O confirming the potassium characteristic of these feldspars. The feldspar from Solonópole shows higher Fe_2O_3 concentration compared with the other samples. The high concentration of SiO_2 showed in the

AP sample is due to the presence of quartz as well as of muscovite, observed in the XRD analysis. The presence of muscovite in the AP sample can also explain the high percentage of K_2O compared with Na_2O in this sample.

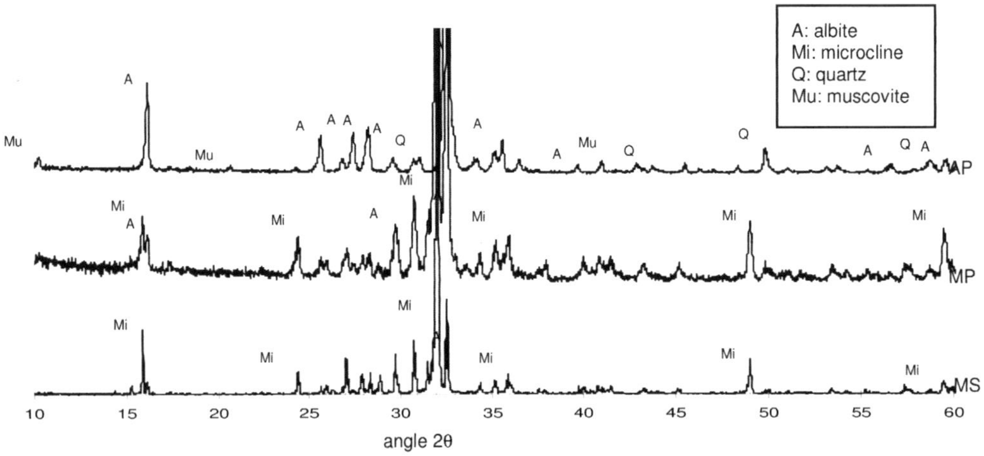

Fig. 1 - XRD patterns of feldspars from Parelhas (AP and MP) and Solonópole (MS) deposits (Co-Kα radiation).

Table 1 - Chemical composition of major constituents found in feldspar samples obtained by XRF

Sample	Oxides (weight %)										
	SiO_2	Al_2O_3	Fe_2O_3	MgO	TiO_2	CaO	Na_2O	P_2O_5	K_2O	RbO_2	Cr_2O_3
AP	72.2	16.7	0.5	-	-	1.9	1.0	2.7	2.0	0.02	-
MP	54.5	12.2	0.3	-	-	0.7	0.2	1.3	29.3	1.4	0.3
MS	64.7	18.3	1.5	0.2	0.02	0.07	1.4	0.04	13.9	-	-

- not determined

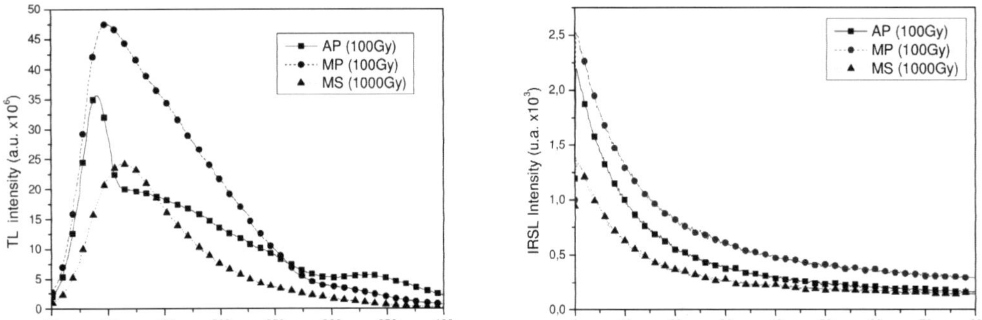

(a) (b)

Fig. 2 – (a) TL and (b) IRSL curves of feldspars AP, MP and MS irradiated with ^{60}Co.

Fig. 2 (a) shows typical TL curves obtained for the three feldspars studied. The microcline samples (MP and MS) present a large TL peak with a maximum at about 100°C and a slow decay composed of several overlap peaks between 100 and 400 °C, in agreement with the literature [10]. The albite sample (AP), presents a broad peak and two additional peaks observed at 90 °C and 340 °C. These peaks are probably due to the presence of quartz in this sample, as shown in the XRD pattern. Fig. 2(b) shows the IRSL decay curves for the feldspars, where it is possible to observe that the three samples present a typical IRSL signal decay.

Figure 3a shows the decay in the TL glow curve of the feldspar MP due to the fading at room temperature and in the dark, during the period of one month. The integrated TL signal from 50 °C to 400 °C after 30 days, was 50% lower than the signal on the first day for the MP sample, and 37% for the AP and 56% for the MS samples, respectively. These results are similar with those reported in the literature [6].

Fig. 3 (b) shows the MP feldspar IRSL signal decay as a function of the time elapsed after irradiation, where the fading effect is evident after 10 seconds. A similar behaviour was observed for the AP and MS samples. Integrating the IRSL signal counts from the beginning of the stimulation until the signal reached 50% of the maximum, the unwanted fading effect is minimized. In fact, after 30 days storage at room temperature, the integral IRSL signals did not change for any of the feldspar samples studied.

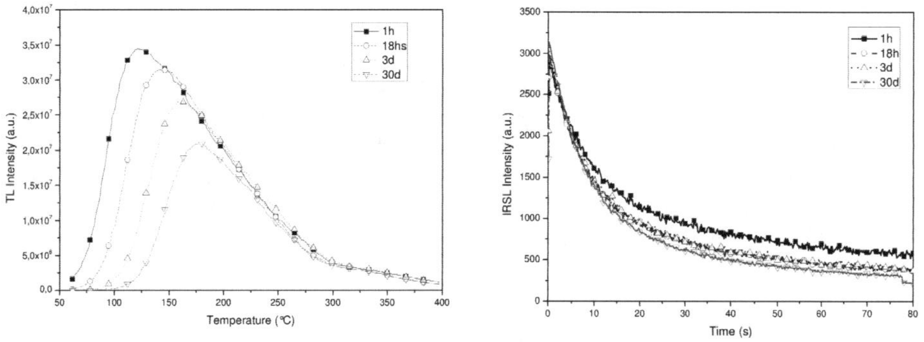

(a) (b)

Fig. 3 –Fading effect at various storage times in (a)TL and (b) IRSL curves of MP feldspar.

Fig. 4 (a) shows the behaviour of the TL glow curve for the feldspar MP sample after different preheat treatments, ranging from 50 °C to 250 °C for 10 minutes, after irradiation. The effect of preheating in the other feldspars was normalized and summarized in Table 2. Preheating temperatures higher than 150 °C diminish significantly the TL signal of all the samples, and, hence, may only be used for applications with a long exposure time, such as in archaeological dating [11]. The effect of preheating in the IRSL signals of the MP feldspar is shown in Fig. 4 (b). It is observed that until reaching a preheating temperature of 150 °C, there is no significant alteration of the IRSL signal during the first 10 seconds of the luminescence decay. Preheat temperatures from 200 to 250°C affect significantly all the IRSL signals.

Table 2 – The effect of preheat in the feldspar AP, MP and MS TL signal

Sample	Preheat treatment for 10 minutes (% of the TL signal)					
	50°C	100°C	125°C	150°C	200°C	250°C
AP	100	53	44	38	17	10
MP	100	35	26	21	6	2
MS	100	32	22	15	5	1

Since the IRSL signal was defined as the integrated counts from the beginning of the stimulation until reaching 50% of the maximum, it is possible to conclude that fading (30 days after the irradiation) doesn't influence significantly the integrated IRSL signal. However, for TL measurements, a preheat treatment at 125°C for 10 minutes was chosen as part of the protocol for these feldspar samples to simulate natural fading.

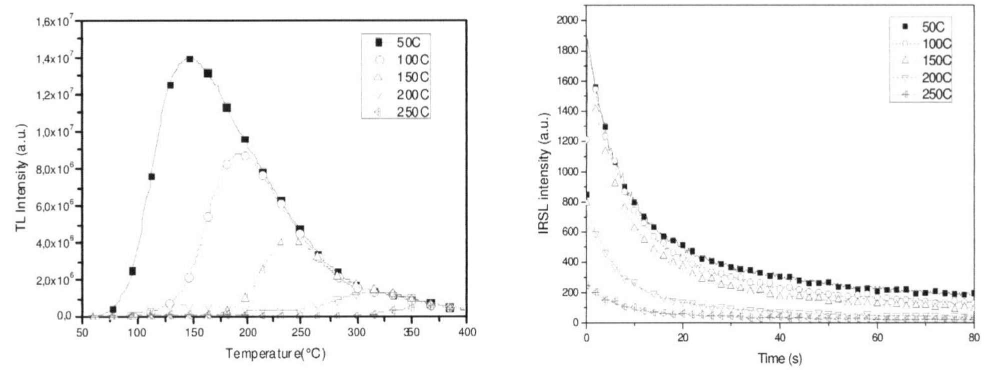

(a) (b)

Fig. 4 – Preheat effect at various temperatures for 10 minutes in (a)TL and (b) IRSL curves of MP feldspar.

Fig. 5 shows the TL (5a) and the IRSL (5b) dose response curves. Results show a linear dose response, indicating that these feldspars can be used as natural dosimeters regardless of the type or origin. However, the AP and MP feldspars have higher sensitivity than the MS feldspar.

For doses above 25 Gy, a sub-linear behaviour of the IRSL curves for the Parelhas feldspar (MP and AP) samples was observed. This effect was eliminated reducing the PMT high voltage from 700 to 600 V. Thus, the sub-linear behaviour observed in Fig. 5 (b) was attributed to a decrease in the efficiency of the PMT rather than an effect of sample saturation. The PMT high voltage adjustment opens the possibility to use the equipment in a large range of doses and permits the investigation of samples with different sensitivities. In the MS feldspar case, which shows a weaker luminescence than the other samples, it is possible to measure the IRSL signal at doses under 50 Gy just by increasing the PMT high voltage.

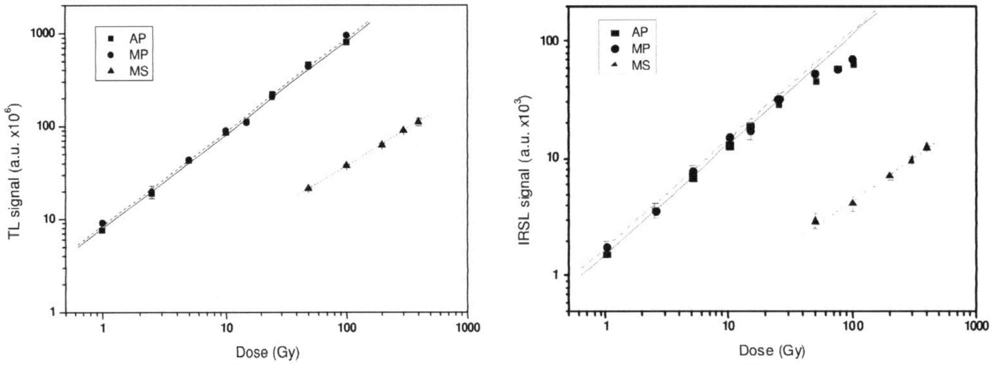

(a) (b)

Fig. 5 – (a) TL and (b) IRSL doses curve response for feldspars AP, MP and MS irradiated with ^{60}Co and preheated at 125°C for 10min following each irradiation.

4. Conclusions

The three feldspars showed many overlapping TL peaks between 50 and 400 °C. In addition, two peaks were observed in the AP samples due to the presence of quartz. The preheat study showed that treating the samples at 125 °C for 10 minutes eliminates part of TL unstable signals and thus minimize the fading effect for a 30 day storage time. This treatment of 125 °C for 10 minutes may be used to standardize both TL and IRSL techniques, despite the fact that a restriction on the integration of the IRSL curve would be enough.

The results showed linear TL and IRSL responses of feldspars from Parelhas (AP and MP) at dose ranges from 1 to 100 Gy and from 50 to 400 Gy for the feldspar from Solonópole (MS). We believe that the higher concentration of iron oxide measured in the MS sample can explain its lower sensitivity because, with increasing total Fe_2O_3 concentration, the substitution of Fe^{3+} at tetrahedral positions increases, whereas the concentration of $Al\text{-}O^{-}\text{-}Al$ centres decreases [12]. Many papers [13, 14, 15, 16, 17] reported that the $Al\text{-}O^{-}\text{-}Al$ centres were correlated with the UV-blue emission band and the Fe^{3+} in the feldspar is associated with the red-IR emission. Due to the high concentration of Fe_2O_3, the TL emission will have a high contribution of the red-IR band, located in the lower PMT efficiency region.

The standardized moderate annealing treatment proposed in this study will facilitate TL and IRSL techniques using different feldspar types of different origin, thus contributing to retrospective dosimetry and archaeological dating.

5. Acknowledgments

The authors are grateful to Dr. Reiner Neuman (CETEM, RJ) for the XRD and XRF analyses and to the CNPq and the LMRI-DEN/UFPE for their financial suport.

References

[1] Wintle A G and Murray A S 2006 A review of quartz optically stimulated luminescence characteristics and their relevance in single-aliquot regeneration dating protocols *Radiat. Meas.* **41** 369

[2] Botter-Jensen L and Murray A S 1999 Developments in optically stimulated luminescence techniques for dating and retrospective dosimetry *Radiat. Prot. Dosim.* **84** 307

[3] Göksu H. Y. and Bailiff I K 2006 Luminescence dosimetry using building materials and personal objects *Radiat. Prot. Dosim.* **119** 413

[4] Clarke M L and Rendell H M 1997 Infra-red stimulated luminescence spectra of alkali feldspars *Radiat. Prot. Dosim.* **27** 221

[5] Spooner N A, Aitken M J, Smith B W, Franks M and Mcelroy C 1990 Archaeological dating by infrared stimulated luminescence using a diode array *Radiat. Prot. Dosim.* **34** 83

[6] Visocekas R 2000 Monitoring anomalous fading of TL of feldspars by using far-red emission as a gauge *Radiat. Meas.* **32** 499

[7] Wallinga J, Bos A J J, Dorenbos P, Murray A S and Schokker J 2007 A test case for anomalous fading correction in IRSL dating *Quat. Geochron.* **2** 216

[8] Luz A B and Coelho J M 2005 Feldspato in: Rochas & Minerais Industriais CETEM/MCT, Rio de Janeiro 413

[9] Santana S T 2006 *Desenvolvimento de um equipamento para dosimetria por luminescência oticamente estimulada (LOE)* (Recife: UFPE)

[10] Blair M W, Yukihara E G and McKeever S W S 2005 Experiences with single-aliquot OSL procedures using coarse-grain feldspars *Radiat. Meas.* **39** 361

[11] Wintle A G 1997 Luminescence dating: Laboratory procedures and protocols *Radiat. Meas.* **27** 769

[12] Krbetschek M R, Götze J, Dietrich A and Trautmann T 1997 Spectral information from minerals relevant for luminescence dating *Radiat. Meas.* **27** 695

[13] Petrov I 1994 Lattice-satabilized CH_3, C_2H_5, NO_2, and O^{1-} radicals in feldspars with different Al-Si order *Ram. Min.* **78** 221

[14] Rae D A and Chambers A D 1988 Metasomatism in the North Qôroq centre, south Greenland: cathodoluminescence and mineral chemistry of alkali feldspars *Trans. Roy. Soc. Edinburgh* **79** 1

[15] Finch A A, Hole D E and Townsend P D 2003 Orientation dependence of luminescence in plagioclase *Phys. Chem. Min.* **30** 373

[16] White W B, Masako M, Linneham D G, Furukawa T and Chandrasekhar B K 1986 Absorption and luminescence of Fe^{3+} in single crystal orthoclase. *American Mineralogist* **71** 1415

[17] Kirsch Y and Townsend P D 1988 Speculations on the blue and red bands in the TL emission spectrum of albite and microcline *Nucl. Tracks* **14** 43

Journal of Physics: Conference Series **249** (2010) 012029 doi:10.1088/1742-6596/249/1/012029

Total mass attenuation coefficient evaluation of ten materials commonly used to simulate human tissue

C C Ferreira[1*], R E Ximenes [1], C A B Garcia[2], J W Vieira[3,4], A F Maia [1]
[1]Federal University of Sergipe, Physics Department,São Cristovão, Sergipe, Brazil
[2]Federal University of Sergipe, Chemistry Department,São Cristovão, Sergipe, Brazil
[3]Centro Federal de Educação Tecnológica de Pernambuco (CEFET-PE), Recife, Brasil.
[4]Escola Politécnica de Pernambuco, Universidade de Pernambuco (EPP / UPE), Recife, Brasil.

E-mail: ccosta@fisica.ufs.br

Abstract. To study the doses received by patient submitted to ionizing radiation, several materials are used to simulate the human tissue and organs. The total mass attenuation coefficient is a reasonable way for evaluating the usage in dosimetry of these materials. The total mass attenuation coefficient is determined by photon energy and constituent elements of the material. Currently, the human phantoms are composed by a unique material that presents characteristics similar to the mean proprieties of the different tissues within the region. Therefore, the phantoms are usually homogeneous and filled with a material similar to soft tissue. We studied ten materials used as soft tissue-simulating. These materials were named: bolus, nylon®, orange articulation wax, red articulation wax, PMMA, modelling clay, bee wax, paraffin 1, paraffin 2 and pitch. The objective of this study was to verify the best material to simulate the human cerebral tissue. We determined the elementary composition, mass density and, therefore, calculated the total mass attenuation coefficient of each material. The results were compared to the values established by the International Commission on Radiation Units and Measurements – ICRU, report n° 44, and by the International Commission on Radiation Protection – ICRP, report n° 89, to determine the best material for this energy interval. These results indicate that new head phantoms can be constructed with nylon®.

1. Introduction

To study the doses received by patient submitted to ionizing radiation, several materials are used to simulate the human tissue and organs for long time [1]. These materials are named tissue equivalent materials. Tissue equivalent materials have been often used in diagnostic and therapeutic radiology [2-4]. In the energy interval 10 keV to 150 keV (diagnostic radiology), electrons are ejected through X-ray photons interactions with the matter. The presence of this interstitial defect increases proportionality with the total mass attenuation coefficient. The total mass attenuation coefficient can be calculated through the photon energy and constituent elements of the material [2,5]. The total

[*] To whom any correspondence should be addressed

mass attenuation coefficient is a reasonable way for evaluating the usage in dosimetric applications of these materials.

The phantoms are usually homogeneous and filled with a material similar to soft tissue. In diagnostic radiology, acrylic (PMMA) is the most equivalent tissue material used. Nowadays, two acrylic phantoms are used in the CT dosimetry. They are used for simulating the patient's head and the patient's abdomen. To perform dosimetric procedures in CT is extremely important due the high doses delivered in these examinations [6].

The ICRU n° 44 has compiled the results of various studies about the utilization of tissue equivalent materials in radiodiagnostic and radiotherapy applications. The report brings the total mass attenuation coefficients, and other quantities, calculated in the X-ray energy range of 1 keV to 50 MeV. Especially in radiodiagnostic applications (10 to 150 keV), materials should have attenuation properties substantially similar from those of body tissues, since photoelectric effect is much observed [3].

Jones *et al* have developed tissue-equivalent materials for simulating the soft, lung and bone tissue of newborn, children and adult patients in CT examinations [2]. These materials were compared with the reference materials found in ICRU n° 44.

We studied ten materials often used as soft tissue-simulating. These materials were named: bolus, nylon®, orange articulation wax, red articulation wax, PMMA, modelling clay, bee wax, paraffin 1, paraffin 2 and pitch. The objective of this study was to verify which the best material to simulate the human cerebral tissue, in terms of total mass attenuation coefficient and mass density in the energy range of 10 to 150 keV. The results were compared to the values established by the International Commission on Radiation Units and Measurements – ICRU, report n° 44 [3], and International Commission on Radiation Protection – ICRP, report n° 89 [4].

2. Materials and Methods

The materials analized here were bolus, nylon®, orange articulation wax, red articulation wax, PMMA, modelling clay, bee wax, paraffin 1, paraffin 2 and pitch. All are commercially available. Before calculating the total mass attenuation coefficient, it was necessary obtain the elemental compositions (mass fraction of each element). For obtaining the mass fraction of the hydrogen (H), carbon (C), nitrogen (N) and sulphur (S) contained in each sample, it was used the Flash EA 1112 Series CHNS analyzer. The mass fraction remaining in each sample was considered as oxygen (O). These materials are basically composed of these elements (H, C, O, N, S) [2-4].

The total mass attenuation coefficient can be obtained through the total cross section for photons by the rule given in Hubbell (2006):

$$\frac{\mu}{\rho} = \frac{\sigma_{tot}}{m_u A} \qquad (1)$$

where μ/ρ is the total mass attenuation coefficient, σ_{tot} is the total cross section for photons, m_u is the atomic mass unit and A is the relative atomic mass of the material.

To calculate the total mass attenuation coefficient, the software WinXCom was utilized [7-9]. The mass fraction of each element was insered as input parameter and WinXCom has calculated the total mass attenuation coefficient through the available tables in Hubbel and Seltzer [10]. The mass density of each sample was calculated using the Archimedean principle.

3. Results and discussion

3.1. Mass density

The table 1 shows the results of the mass density. It also is showed the percentage difference of each material in relation to ICRU's brain and ICRP's brain, respectively. The mass density obtained for each material indicated a good concordance with the mass density of the ICRU's brain and ICRP's brain. The major absolute value of the percentage difference was 23.6% for modelling clay in relation to ICRP's brain. The better values for mass density were encountered for the bee wax.

Table 1. Mass density of the tissue material equivalent and percentage differences in relation to ICRU's brain and ICRP's brain.

Tissue equivalent materials	Mass density (g/cm^3)	Percentage difference (%) (substitute/ICRU's brain)	Percentage difference (%) (substitute/ICRP's brain)
Bee wax	0.964	-7.3	-6.4
Red articulation wax	0.911	-12.4	-11.6
Paraffin 1	0.959	-7.8	-6.9
Bolus	1.112	6.9	8.0
Nylon®	1.160	11.5	12.6
Orange articulation wax	0.931	-10.5	-9.6
Modelling clay	1.273	22.4	23.6
PMMA	1.178	13.3	14.4
Pitch	1.148	10.4	11.5
Paraffin 2	0.918	-11.7	-10.9
ICRU's brain[a]	1.040	c	c
ICRP's brain[b]	1.030	c	c

[a] ICRU reference [3].
[b] ICRP reference [4].
[c] not applicable.

3.2. Total mass attenuation coefficient

The table 2 shows the elemental compositions obtained for each material. The results for each material were used as input data on the WinXcom software. Thus, the total mass attenuation coefficient for each material was calculated.

The total mass attenuation coefficient for each material was divided by the total mass attenuation coefficient of the human brain (ICRU and ICRP) in the energy range of 10 to 150 keV. The figure 1 and figure 2 show the ratios of total mass attenuation coefficient for the modelling clay and the nylon®, respectively. The table 3 shows the mean and the standard deviation of the ratios between each material and the human brain (ICRU and ICRP). In this way, the material to simulate the human cerebral tissue will be indicated by the mean that presents the value closer to 1. Thus, modelling clay and nylon® present good agreement with human brain in terms of total mass attenuation coefficient. Nylon® presents also good agreement in terms of mass density. For this reason, it can be concluded that, among the materials evaluated in this work, nylon® is the most indicated for simulating the human cerebral tissue. Although the mass density suggests the bee wax as substitute material of the human brain, it did not present good agreement in the total mass attenuation coefficient, which is the more indicated quantity for dosimetric comparisons [2-5].

Table 2. Elemental composition for tissue equivalent materials.

Tissue equivalent materials	C (%)	N (%)	S (%)	H (%)	O (%)
Bee wax	75.25	8.42	0.19	1.87	14.27
Red articulation wax	80.17	11.23	0.09	0.36	8.14
Paraffin 1	81.73	0.74	0.10	0.61	16.81
Bolus	82.22	0.78	0.09	0.50	16.41
Nylon®	59.49	4.34	3.96	0.63	31.58
Orange articulation wax	82.00	7.37	0.08	2.73	7.82
Modelling clay	19.76	0.86	3.55	0.00	75.83
PMMA	94.96	4.71	0.10	0.24	0.00
Pitch	42.18	0.42	0.46	0.19	56.76
Paraffin 2	79.61	9.63	0.14	0.68	9.94

Table 3. Mean of the ratios of total mass attenuation coefficients of the tissue equivalent materials in relation to ICRU's brain and ICRP's brain, respectively.

Tissue equivalent materials	Mean of the Ratio (substitute/ICRU's brain)	Mean of the Ratio (substitute/ICRP's brain)
Bee wax	0.782 ± 0.138	0.773 ± 0.148
Red articulation wax	0.760 ± 0.147	0.752 ± 0.156
Paraffin 1	0.772 ± 0.137	0.764 ± 0.147
Bolus	0.771 ± 0.138	0.762 ± 0.148
Nylon®	0.995 ± 0.040	0.941 ± 0.037
Orange articulation wax	0.756 ± 0.163	0.756 ± 0.173
Modelling clay	1.046 ± 0.136	1.029 ± 0.118
PMMA	0.732 ± 0.173	0.725 ± 0.182
Pitch	0.885 ± 0.022	0.873 ± 0.042
Paraffin 2	0.766 ± 0.144	0.758 ± 0.154

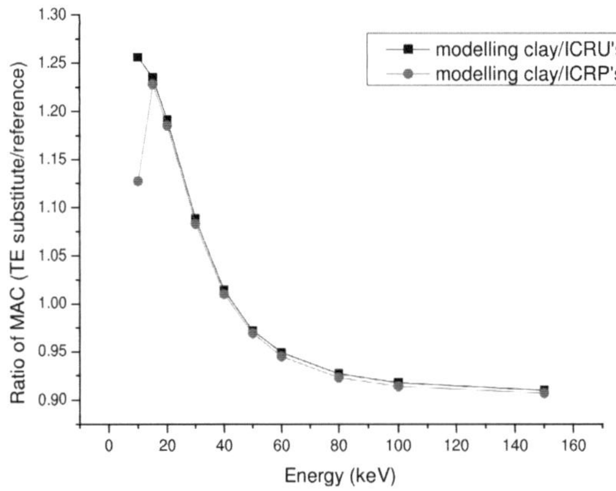

Figure 1. Ratio of total mass attenuation coefficients (MAC) for the modelling clay to their corresponding reference values (ICRU's brain and ICRP's brain).

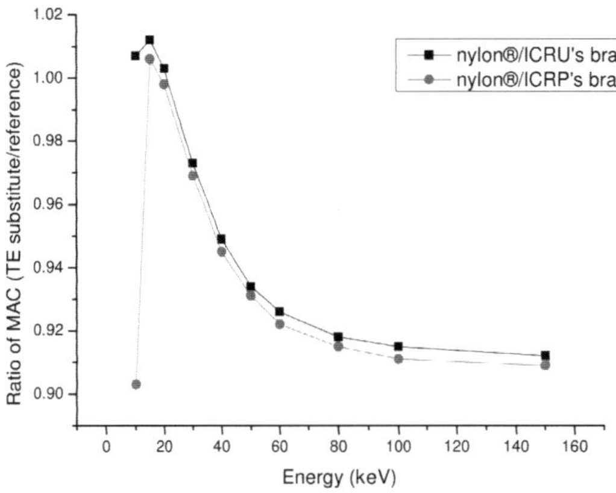

Figure 2. Ratio of total mass attenuation coefficients (MAC) for the nylon® to their corresponding reference values (ICRU's brain and ICRP's brain).

The figures show that the values of μ/ρ differ of their corresponding values mainly in low energies (10 up to 40 keV). This is due to the difference of elemental composition that influences the photoelectric effect which is the major responsible by the total mass attenuation coefficient in this energy range.

4. Conclusions

A more reasonable way for evaluating tissue equivalent materials is the total mass attenuation coefficient. The total mass attenuation coefficients results indicated that the modelling clay and nylon® could be used as tissue equivalent of human brain. But, only nylon® has found good agreement with the human brain (ICRU and ICRP) in terms of total mass attenuation coefficient and mass density. Therefore, the utilization of this material for constructing human phantoms is a good and low-cost alternative. For example, a nylon® CT head phantom has been constructed and its cost has been about fifteen times less than acrylic CT head phantom actually used in dosimetry.

References

[1] Kienböch R 1906 On the quantimetric method *Arch Roentgen Ray* **11** 17

[2] Jones A K, Hintenlang D E and Bolch W E 2003 Tissue-equivalent materials for construction of tomographic dosimetry phamtons in pediatric radiology *Med. Phys.* **30** 2072-2081

[3] ICRU 1989 Tissue substitutes in radiation dosimetry and measurement *ICRU Report* 44 International Commission on Radiation Units and Measurements Bethesda, MD, USA

[4] ICRP 2003 Basic Anatomical and Physiological Data for Use in Radiological Protection: Reference Values *ICRP Publication 89*

[5] Hubbell J H 2006 Review and history of photon cross section calculations *Phys. Med. Biol.* **51** 245-262

[6] Rehani M M, Bongartz G, Golding S J, Gordon L, Kalender W, Murakami T, Shrimpton P, Albrecht R and Wei K 2001 *Annals of the ICRP Report 87* **30** 4

[7] Gerward L, Guilbert N, Jensen K B and Levring H 2004 WinXCom – a program for calculatin X-ray attenuation coefficients *Radiat. Phys. Chem.* **71** 653-654

[8] Gerward L, Guilbert N, Jensen K B and Levring H 2001 X-ray absorption in matter. Reengineering XCOM *Radiat. Phys. Chem.* **60** 23-24

[9] Berger M J and Hubbell J H 1987/99 XCOM: Photon Cross Sections Database, Web Version 1.2, available at http://physics.nist.gov/xcom. National Institute of Standards and Technology, Gaithersburg, MD 20899, USA. Originally published as NBSIR 87-3597 "XCOM: Photon Cross Sections on a Personal Computer"

[10] Hubbel J H, Seltzer S M 1995 Tables of x-ray mass attenuation coefficients and mass energy-absorption coefficients 1 keV to 20 MeV for elements Z = 1 to 92 and 48 additional substances of dosimetric interest, *National Institute of Standards and Technology.*

Dosimetric Characterization of LiF Doped with Unusual Dopants

M A P Chagas, L J Parente and D N Souza

Universidade Federal de Sergipe, Departamento de Física (UFS / DFI)
Av. Marechal Rondon, s/n, Rosa Elze, 49100-000 São Cristóvão , SE Brasil

E-mail: dnsouza@fisica.ufs.br

Abstract. LiF with different dopants has been one of the most investigated materials to use as thermoluminescent dosimeter. Even so, in a previous work we observed that pellets produced with pure LiF have applicability to ionizing radiation dosimetry, showing dosimetric peaks in temperatures high enough to overcome most of fading problems. In this work we studied the thermoluminescent properties of pellets prepared with LiF and unusual dopants. As dopants were tested rare earth metals like europium, terbium, dysprosium and neodymium. The more intense TL response was showed by samples doped with Nd. Although they are bluish due to presence of the Nd, the pellets produced with LiF:Nd exhibited a TL emission two times more intense than the one in pure LiF.

1. Introduction

Of the many types of thermoluminescent detectors (TLD) available, LiF:Mg,Ti (TLD-100) remains one of the most widely used dosimeters in routine personal and environmental monitoring and space dosimetry[1].

TLD is a versatile tool for assessing the dose of ionizing radiation. The variety of ceramic materials and their different physical characteristics allow the determination of quantities of radiation at levels of μGy dose of the kGy, enabling the timing of materials [2-4]. Among the potential termoluminescente dosimeters, materials based on LiF are used in personal dosimetry and deserve attention because of low energy dependence and their equivalence of human tissue [5]. LiF-100H TLD, for example, has a density of 2.6 g/cm^3 with effective atomic number about 8.2**,** close to the 7.4 value of human tissue**.** Lithium fluoride is beyond not soluble in water.

The intrinsic efficiency is defined as the ratio of light energy emitted during heating to energy absorbed during gamma irradiation [6]. LiF presents an intrinsic efficiency of thermoluminescent of approximately 0.04%. This characteristic make it widely used in medical applications. In the LiF:Mg,Ti the excess of positive charge in the grid of lithium fluoride crystal, which implies in the formation of new traps for electrons, is a consequence of the substitution of one lithium ion to other of magnesium. Of all the peaks of issue of LiF: Mg, Ti, about ten, only the so-called peaks 4 and 5 are used in dosimetry, those that occur between the temperatures of 160 and

190°C [11]. The lithium fluoride presents a whole series of dosimetric characteristics that define their applications in specific areas of dosimetry.

In this context, we are searching about the possibilities to produce a new dosimeter based on LiF using unusual dopants. In a previous study our time presented results about thermoluminescence of pellets produced with this material, undoped [2] in this work we will show the firsts obtained with Dy, Eu, Nd and Tb as doping of LiF for use in thermoluminescent (TL) dosimetry. Initially, a study of the possibilities of preparation of LiF:Nd was carried by following others routes of preparation without sulphuric acid. After that, comparisons between the responses to beta radiation and x-ray were made using thermoluminescent techniques.

2. Methodology and Materials

The samples were prepared in the Laboratory of Materials Preparation and Characterization Department of Physics of Federal University of Sergipe. It was used LiF with chemical purity of 99% doped with 2% of EuO_3, NdO_3 and TbO_3, all them by Aldrich with purity of 99.9%. The composites were mixed with 10 ml of distilled water and taken to a magnetic agitator where they remained for 30 minutes. After that, drying took place in oven at a temperature of 100°C for 24 h. After the drying, the powder was homogenized again in mortar with the polyvinyl alcohol addition. The pellets were submitted to a uniaxial presage of 100 kgf/cm2 producing pellets of size 6mm diameter and 1mm thickness after sintering at temperatures of 650, 700 or 750°C for 1h, with a rate of heating of approximately 10°C per minute, and a free rate cooling down to room temperature.

The composites were irradiated with a beta source ($^{90}Sr+^{90}Y$). After each exposition, the pellets were submitted to thermal treatment at a temperature of 300ºC for 30 min. The measures of thermoluminescent were carried out up to 6 h after irradiation. TL measurements were made with a Harshaw 3500 thermoluminescent reader using a heating rate of 10°C/s.

The analysis of the homogeneity of chosen samples let witch the uncertainties in this study were always below 5%.

3. Results

In order to compare the influence of the sintering temperature it produced same pure LIF samples and of the dopant in the TL response it was initially prepared a set of pure LiF samples. Figure 1 shows the glow curves the pure LiF samples produced at 650, 700 and 750°C. TL emission of pure LiF pellets irradiated to 22.6 Gy with beta radiation exhibits two very well defined peaks around 170 and 250°C. The first TL peak was assumed as a dosimetric peak. The sintering at the 750°C has resulted in a more intense TL response to the absorbed dose.

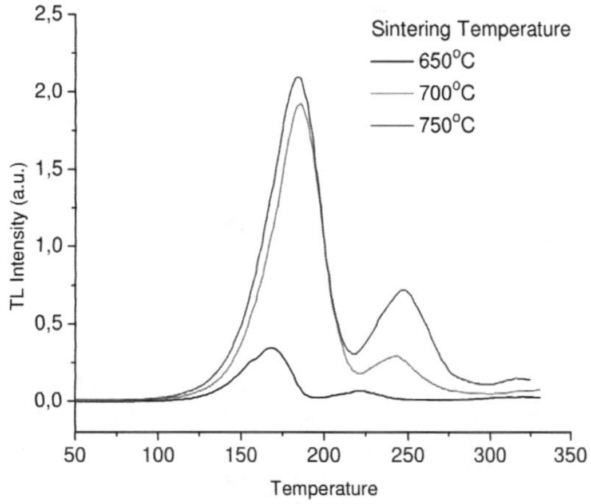

Figure 1. Typical TL emissions of LiF composites sintered at 650, 700 and 750°C beta irradiated (22.6 Gy, $^{90}Sr+^{90}Y$).

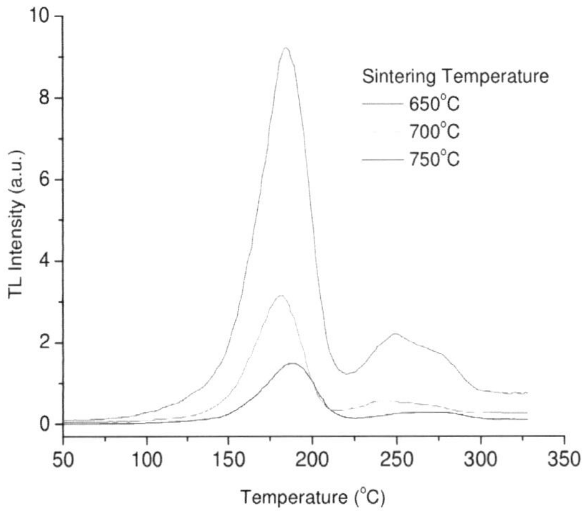

Figure 2. Typical TL emissions of LiF:Nd composites sintered at 650, 700 and 750°C beta irradiated (22.6 Gy, $^{90}Sr+^{90}Y$).

In figure 2, the TL emission curves of the LiF:Nd are compared regarding of sintering temperature influence in emission curves of the dosimeters The main difference between the TL emissions of the three composites irradiated with beta rays is in the TL intensities. The topaz composites sintered at 750°C displays a high sensitivity thermoluminescence as compared with the others. The shape of the emission curve is similar to the pure LiF. Meanwhile, they are bluish due to presence of the Nd and for the same absorbed dose, the TL intensity of the sample sintered in this temperature is four times higher than that one showed by pure LiF in the same produced in the same conditions. It is possible to observe too that the TL intensity of the peak of this emission is six times more intense than in pellet sintered at 650°C and three times when compares with pellet made at 700°C.

Figure 3 shows the dose response of LiF:Nd. This calibration curves presented responses proportional to absorbed dose between 0.5 Gy and 16.0 Gy for the beta radiation; the curves are useful in the whole tested dose range. No dose saturation was observed.

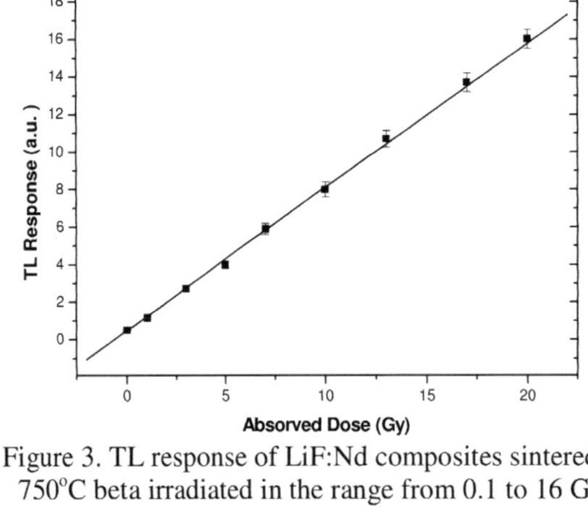

Figure 3. TL response of LiF:Nd composites sintered at 750°C beta irradiated in the range from 0.1 to 16 Gy.

The pellets doped with Eu sintered at 750°C exhibit a TL peak at 150°C witch has a low intensity if compared at the TL emission of LiF:Nd. In the same way the LiF:Dy not displays a performance of a good dosimetric material, although has shown a TL peak more intense and at higher temperature that the LiF:Dy (figure 4). The LIF:Tb (Figure 5) presents the lowest emission TL. Furthermore the glow curve presents many TL peaks probably due the colour of the pellets which was dark yellow.

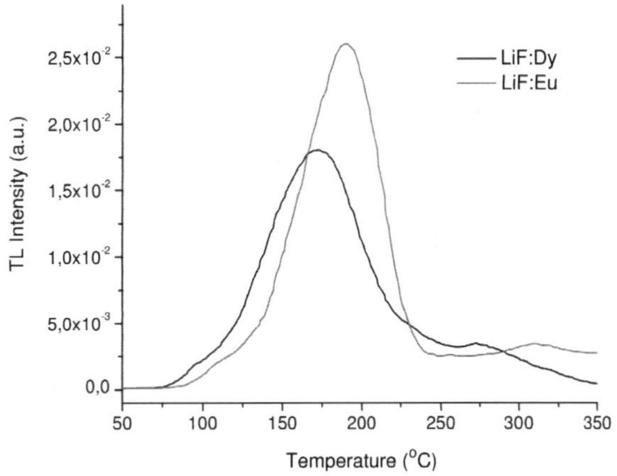

Figure 4. Typical TL emissions of LiF:Dy and LiF:Eu composites sintered at 750°C beta irradiated (11 Gy, $^{90}Sr+^{90}Y$).

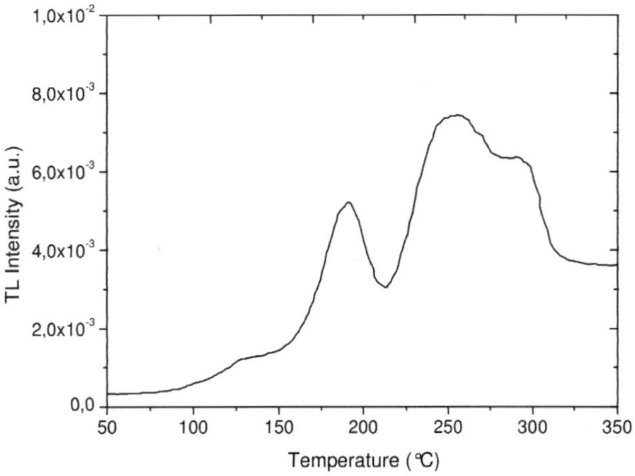

Figure 5. Typical TL emissions of LiF:Tb composite sintered at 750°C beta irradiated (11 Gy, $^{90}Sr+^{90}Y$).

4. Conclusion

In this study it could be observed that the LiF doped with neodymium is a promising material for dosimetric purposes because presents a good response to dose and the temperature of occurrence of its higher peak is adequate o use in TL dosimetry. The other dopants, Dy, Eu and Tb, did not showed characteristics that could be useful to dosimetry in the dose range used in this work.

Due the previously mentioned properties associated with their low cost, the described detectors can be used in teaching laboratories as a complement of didactic practices involving the use of other kinds of radiation detectors, for example. Further techniques for preparation of the described composites, as well as other radiation sources for testing the pellets, can be within the scope of future works.

Acknowledgments
The authors are thankful to Fundação de Amparo à Pesquisa e à Inovação Tecnológica do Estado de Sergipe (FAPITEC/SE) and Coordenação de Aperfeiçoamento de Pessoal de Nível Superior (CAPES), Brazil, for partial financial support.

References

[1] Baldacchini, G, Davidson, AT, Kalinov, V S, Kozakiewicz, A G, Marolo,T, Montecchi, M, Montereali, R M, Nichelatti, E, Voitovich, A P,2003. *Opt. Mater.* **24** 129
[2] Ribeiro, D R S, Souza, D N, Maia, A F, Baldochi, S L and Caldas, L V E 2007. *Radit. Measur.* **23** 667
[3] Baldacchini, G, Davidson, A T, Kalinov, V S, Kozakiewicz A G, Montereali R M, Nichelatti E and Voitovich A P, 2007 *J. Lumin.* **122** 373
[4] Duggan, B M, Johnson, B L, 2000. *Med. Phys.* **28** 601.
[5] Ertl A, Zehetmayer M, Schoggl A, Kindl P and Hartl R 1997 *Phys. Med. Biol.* **42**, 2137.
[6] Eakins J S, Bartlett D T, Hager L G, Molinos-Solsona C and Tanner R J 2007 *Radiat. Prot. Dosim.* **128** 21

Thermoluminescent dose reconstruction using quartz extracted from unfired buildings

Simara S. Campos[1], Geângela M. Almeida[1], Leandro X. Cardoso[1], Leonardo L. de Lima[1], Carlos A.[2], Susana O. de Souza[1]

[1]Physics Department, Federal University of Sergipe

[2]Chemistry Department, Federal University of Sergipe

simasantos@hotmail.com

Abstract. The use of thermoluminescence (TL) is very useful for application in retrospective dosimetry, that is, for inference of the dose of radiation undergone for some material. The calculation of the lifetime of TL peaks can help to clarify which TL intensity is due to the dose of environmental radiation and which is proceeding from the accident with radiation. In this work, the kinetic order (b), activation energy (Et) and the frequency factor (s) were obtained for quartz samples that had been extracted from unfired buldings. It was observed the occurrence of peaks in the 150 °C – 250 °C range, that corresponds to a range commonly used in retrospective dosimetry, due to its shorter lifetime when compared with TL geological signal. The analysis of the peaks was done through several well known methods, namely the peak shape (PS), curve fitting (CF) and heating rate (HR) methods. To confirm the structure of the sample, additional X-ray diffraction was made.

Keywords: Themoluminescence, quartz, kinetic parameters, retrospective dosimetry

1. Introduction

The retrospective dosimetry for luminescence using ceramics, or materials that had undergone burning during manufacture, as is the case of the brick, roofing tile, amongst others, is one of the kinds to find the cumulative dose due to an accident involving ionizing radiation. Those materials contain quartz (SiO2), a thermoluminescent sensitive one that belongs to the family of silicates [1].

Thermoluminescence has been applied for dosimetry of ionizing radiation since 1940's, when the number of workers exposed to this type of radiation was increasing and efforts had been made in developing new types of dosimeters [2]. However, materials not annealed, as mortar and concrete, are commonly found in industrial places and, particularly, in nuclear installations. These materials also contain quartz, that is normally less thermoluminescent sensitive than the quartz found in annealed materials. Therefore, now the interest is come back toward the use of construction materials, to extend the application scope, being one of the new challenges of retrospective dosimetry. Thus, the main goal of retrospective dosimetry is the development of a trustworthy methodology for the reconstruction of the dose received by construction materials in a radioactive accident [3].

The kinetic parameters of the thermoluminescent peaks are important to characterize the material, as they allow getting the lifetime of those peaks. These parameters are: the kinetic order (b), the activation energy (E_t) and the frequency factor (s). There are three primary requirements for the usage of these natural dosimeters:

- Absence of an accumulated saturated natural dose from the geological period;
- The ability to hold back a register of the exposition to the accident, with worthless fade-out, in the interval between exposition and measurement;
- The ability to determine the cumulative dose of natural sources of radiation.

Quartz is one of materials that match all those requirements.

This work has for purpose to present a methodology for attainment of the extracted quartz grains from mortar and techniques to evaluate the sample, as well as kinetic parameters, in the temperature range of 150 – 250 °C, commonly used in retrospective dosimetry. Through X-ray diffraction (XRD), it was possible to determine the constituents of the sample, and the kinetic parameters had been evaluated using the peak shape (PS), curve fitting (CF) and heating rate (HR) methods, that allow calculating the lifetime of the peaks.

2. Materials and methods

2.1. Sample preparation and TL measurement

The first stage in retrospective dosimetry requires the separation of the present quartz in the sample. In this work, common mortar of construction was chosen. So, the mortar was pulverized to allow a better penetration of chemical agents used in removal of undesirable residues. The powder was then treated in a concentrated HCl + HNO_3 solution during 2 hours, to dissolve the present organic material. To this followed a wash with distilled water and an acid treatment to dissolve calcite, often found in this kind of sample. A new wash in distilled water was made. Finally, the sample was dry in a greenhouse at room temperature for 24 hours.

Grains with sizes between 75 – 200 μm were chosen and then shared in two parts, one for measurement of natural TL emission, increased of an artificial dose of 300 Gy of beta radiation, and another for measurement of TL emission after an annealing of 400 °C for 1 hour, plus additional dose of 300 Gy of beta radiation. The samples had been irradiated with a 90Sr/90Y, with dose rate of 0,376 Gy/min during 13,3 hours.

Thermoluminescent signal of samples was taken immediately after irradiation in a homemade TL reader. The reader is composed by a platinum plate, whose function is store the sample for the heating. This plate is enclosed in a drawer to prevent noise from external light. The plate is connected to the exit of a transformer, connected in the potency module. That potency module is linked to a signal converter. With the aid of software developed at the Physics Department, the current is controlled with a constant heating rate during the TL measurement. The current is controlled and monitored through a Chromel-Alumel thermocouple, that sends the tension sign to the computer through signal converter, the material emits light, which is captured by a EMI 9789-B photomultiplier. This photomultiplier is fed by a source of high variable tension (0 - 3000 KV).

2.1.1. *Kinetic order.* By the PS method it was found that the peaks obey first order kinetics. This is in concordance with descriptions found in literature [2].

2.1.2. *Peak shape methods.* The method based on the shape of the peak utilizes just two or three points from the glow-curve. Usually, these are the maximum of the peak Tm and either, the low and high temperature half-heights at T1 and T2 (see figure1).

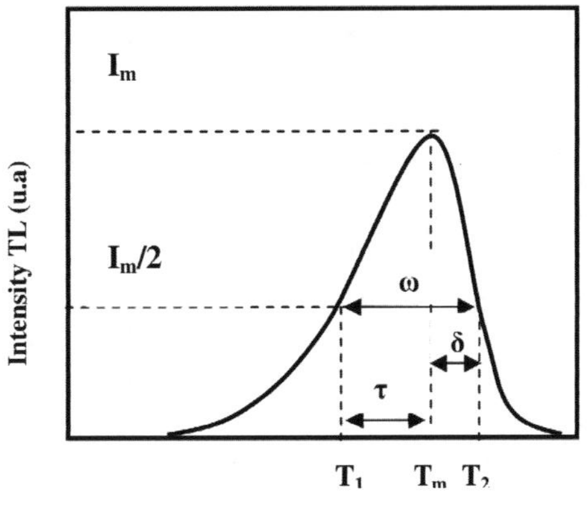

Figure 1 A schematic thermoluminescence peak showing the maximum temperature T_m , the low and high temperatures of half intensity T_1 and T_2 and the half widths $\omega= T_2 - T_1$, $\delta=T_2-T_m$ and $\tau= T_m-T_1$.

However, because the shape of the peak is strongly affected by kinetics order then the method is dependent on the order of kinetics. Starting from a great number of formulas for the cases of first and of second order several of these methods were appearing.

It was Chen (1969) who analyzed those methods and obtained a complete expression [5]. Chen's equation will be quoted here. Chen (1997) gives three equations each for first- and second-order peaks, relating the trap depth to either the full width of the peak at its half-height $(\omega=T_2-T_1)$, its low-temperature half-width $(\tau=T_m-T_1)$, or its high-temperature half-width $(\delta=T_2-T_m)$.

A general formula for *E* was given namely:

$$E = c_y (kT_m^2 / y) - b_y (2kT_m) \qquad (1)$$

where y is ω, τ or δ. The constant c_y and b_y for the three methods, for general order kinetics, are given in table 1. The geometrical factor $\mu_g=\delta/\omega$ will give a value of 0.42 for 1st order kinetics and 0.52 for 2nd order kinetics[3].

Table 1 – cγ and bγ values for first order kinetics

	ω	τ	δ
c_γ	2,52	1,51	0,976
b_γ	1,0	1,58	0

2.1.3. Curve fitting method. The most popular model of thermoluminescence is based on the two levels model, on trap-one recombination centre, which allows writing equations describing a single TL peak, i.e., the first- and the second-order kinetics equations [5]. In the first order it is assumed that only recombination is present, while which in the second it is assumed that recombination and retrapping are both present. Since in this work only 1st order kinetics will be analyzed, only the first-order kinetics equation will be given:

$$I(T) = n_0 s \exp\left(-E / kT\right) \times \exp\left[-(s / \beta)\int_{T_0}^{T} \exp(-E / kT) dT\right] \qquad (2)$$

where *I(T)* is the glow-peak intensity, *E* (eV) the activation energy, *s* (s^{-1}) the frequency factor, n_0 (cm^{-3}) the initial concentration of the trapped charges, *k* ($eV\ K^{-1}$) the Boltzmann constant, T (K) the absolute temperature, T_0 *(K)* the initial temperature and β (K s^{-1}) is the constant heating rate.

For numerical calculations of TL intensity it is necessary to evaluated the integral

$$F(T,E)=\int_{T_0}^{T}\exp\left(-\frac{E}{kT}\right)dT' \quad (3)$$

This integral can be solved by successive integrations by parts, giving then

$$F(T,E)=T\exp\left(-\frac{E}{kT}\right)\sum_{n=1}^{\infty}\left(\frac{kT}{E}\right)^{n}(-1)^{n}n! \quad (4)$$

However, in the most practical cases only the following simple expression is considered:

$$F(T,E)=\frac{kT}{E}\exp\left(-\frac{E}{kT}\right)\left(1-\frac{2kT}{E}\right) \quad (5)$$

The computed curve is then compared with the actual experimental curve and a root-mean-square (RMS) deviation between the two is calculated. The procedure continues by sequentially changing the *E*, *s* and *b* values until a minimum value of the RMS deviation is reached [5].

2.1.4. *Heating rate method.* This method considers the temperature of maximum of one peak. Varying the heating rate, the peak position shifts to higher values. Applying several heating rates, one can obtain the values of E and s by plotting

$$\ln\left(\frac{T_m^{\ 2}}{\beta}\right)x\left(\frac{1}{T_m}\right) \quad (6)$$

and calculating the values of the angular and linear coefficients of the straight line given by (6).

In the others cases, the frequency factor can be calculated through

$$s=\frac{\beta E_\gamma}{kT_m^2}\exp\left(\frac{E_\gamma}{kT_m}\right) \quad (7)$$

and the lifetime is given by

$$\tau=s^{-1}\exp(E/kT) \quad (8)$$

3. Results and discussion

Figure 2 shows the TL glow curve of quartz extracted from cement, with thermal annealing of 400 °C/1h, and without annealing, both irradiated with a dose of 300 Gy and held in a constant heating rate of 5 K/s.

3.1. Peak Shape

Figure 3 shows the TL glow curves of the peaks with temperature of maximum between 150-250 °C. For attainment of these curves, it was necessary to make a thermal cleaning [5], with intention of isolate the TL peaks for analysis, since quartz present several overlapping peaks. After the thermal cleaning, the geometrical factor was calculated, resulting in a value of 0.42 for both curves, indicating 1st order kinetics. The values of activation energy, frequency factor and lifetime of those peaks are shown in table 2.

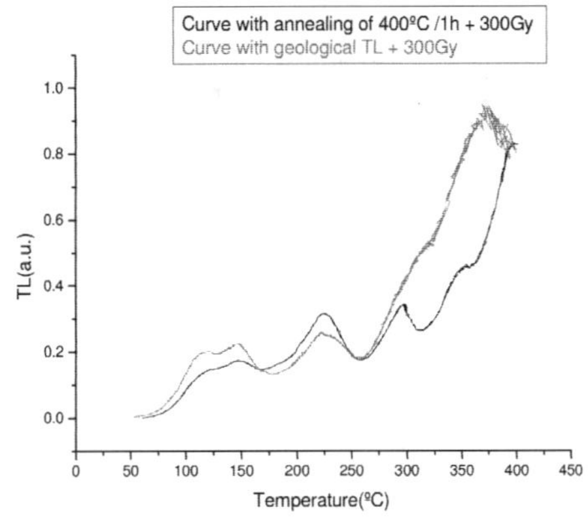

Figure 2 – TL glow curve of quartz, irradiated with a dose of 300 Gy.

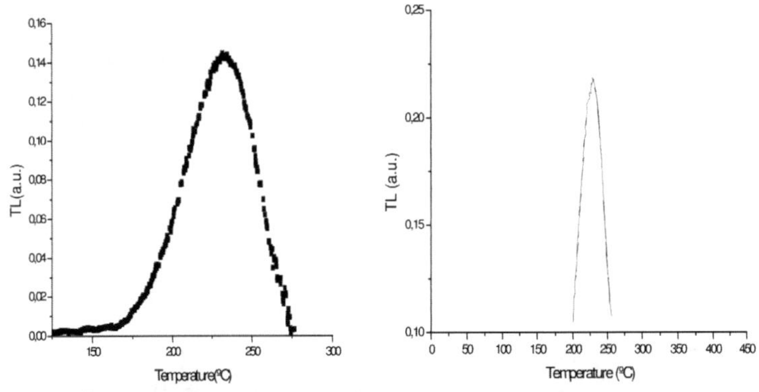

Figure 3 – Quartz TL glow curve with and without thermal annealing, respectively.

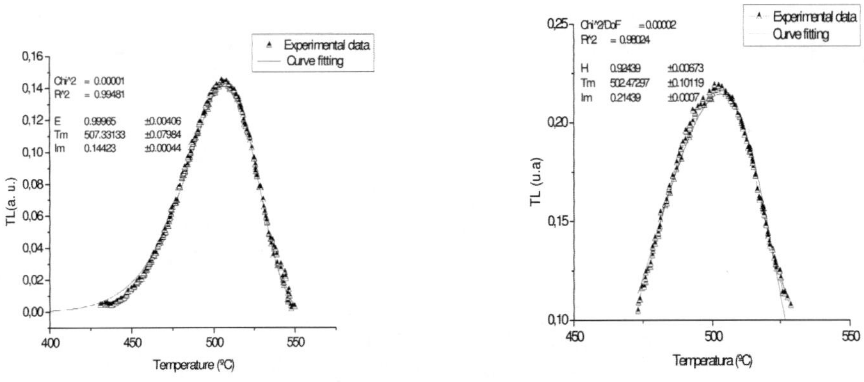

Figure 4 – Adjusted TL glow curve for the peaks between 150 and 250 ºC

3.2. Curve fitting

Figure 4 shows the experimental curve and the adjusted curve for 1st order kinetics. The correct application of this method demands that the peak is completely isolated of the adjacent peaks, as well as in the peak shape method. The frequency factor was gotten indirectly through the equation (7). The values are also shown in table 2.

3.3. Heating rate method

The biggest advantage of this method is that it's needed only the data of maximum intensity (I_m, T_m), with no necessity of a thermal cleaning process as in the previous methods. To the application of this method, heating rates of 2ºC/ s^{-1}, 4ºC/ s^{-1}, 5ºC/ s^{-1}, 8ºC/ s^{-1}, 10ºC/ s^{-1} were utilized.

Figure 5 shows the resulting straight lines for the peaks with and without annealing, respectively

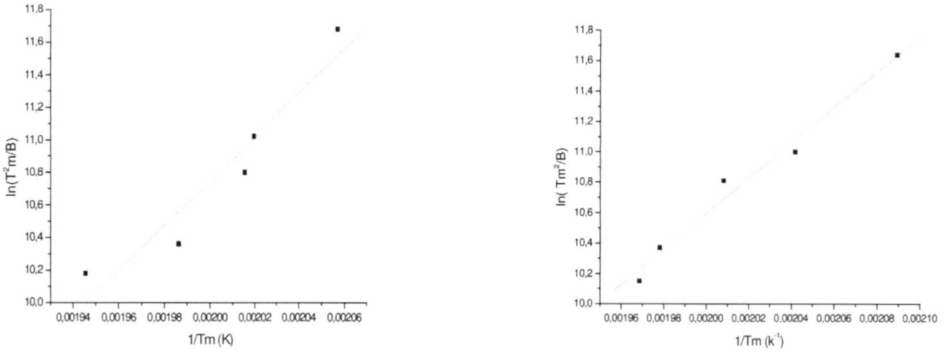

Figure 5 – Heating rate method for the quartz samples with and without annealing, respectively.

4. X- ray diffraction

The verification of the crystalline phases of the sample was made through x-ray diffraction measurements. The Rietveld method [8] was used to make the refinement with the powder diffraction pattern, and software (DBWS9807) was utilized as described by Young et al. [9]. The full width at half maximum of all peaks, as well as asymmetric coefficients, scale factor, lattice parameters and polynomial parameters of the background line were refined.

The refinements of data obtained through analysis of the XRD pattern are shown in figure 6. One of the samples utilizes the mechanism described in [3] and [12], as well as [10]. In another sample, we have utilized the method described here.

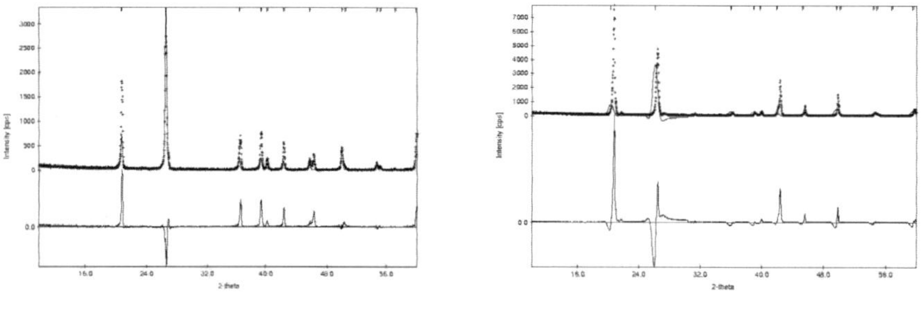

Figure 6 – XRD experimental curve with theoretical refinement for the samples

Quartz belongs to a trigonal crystalline system, $P3_221$ spatial group. The data from refinement indicates the following crystallographic parameters: a = b= 4.9130(5) Å, c = 5.4050(2) Å and $\alpha = \beta$ = 90°, γ = 120°. The values of expected error (R) and obtained error (R-WP) had been for the sample with regal-water, 4.05% e 13.45%, respectively, and for the conventional method, 38.05% e 46.45%, respectively. The R-WP/R ratio was 3.31, a very good value due to the crystalline degree of quartz. The obtained error indicates that the calculated diffractogram for the sample with regal-water presents 88.55% of similarity with the experimental diffractogram.

The tables 2, 3-a and 3-b summarizes the results obtained in this work. As can be seen, there exist a good agreement between the results through the applied methods. Comparing our results with those found in literature, a good agreement is also found.

Table 2- Kinetic parameters of peaks between 150°C e 250°C of quartz extracted from mortar without annealing.

Peak temperature (°C)	Activation energy (eV)	Frequency factor (s^{-1})	Method	Lifetime at 15°C (years)
234±1	0,96±0,02	(4,03±1,73)x 10^8	PS	(2,8±0,6)
234±1	0,99	(1,52)x10^9	CF	4,3
-	1,17	(2,1)x10^{11}	HR	47,1

Table 3-a - Kinetic parameters of peaks between 150°C e 250°C of quartz extracted from mortar with annealing (400 °C/1h)

Peak temperature (°C)	Activation energy (eV)	Frequency factor (s^{-1})	Method	Lifetime at 15°C (years)
232±1	0,910±0,002	(2,53±0,15)x10^8	PS	(1,08±0,04)
229±1	0,92	(3,64)x10^8	CF	1,09
-	1,00	(3,88)x10^9	HR	3,10

Table 3-b - Kinetic parameters of 225°C peak of not annealed quartz.

Peak temperature (°C)	Activation energy (eV)	Frequency factor (s^{-1})	Method	Lifetime at 15°C (years)
225±1	1,16±0,06	(1,2±1,6)x10^{11}	PS	(50±40)
225±1	1,16±0,06	(1,17±0,20)x10^{11}	CF	(50±10)

5. Conclusions

The methodology developed in the sample preparation of the extracted mortar of construction materials was efficient due to the oxidation power of the regal-water. The methods applied in the attainment of kinetic parameters have shown good results. The geological TL present in quartz in the range of temperature commonly used for retrospective dosimetry was very small, due to its very short lifetime. Through refinement of the data of XRD by Rietveld method, we could verify just one crystallographic phase and to calculate the grain size of this material. The peak found between 150 and 250 °C has demonstrated sufficient stability for its use in dosimetry of radioactive accidents. The usage of construction material not burnt, as mortar and concrete, can be used in a trustworthy way in retrospective dosimetry.

6. Acknowledgements

The authors acknowledge to the Fundação de Apoio a Pesquisa e a Inovação Tecnológica de Sergipe (FAPITEC-SE), Coordenação de Aperfeiçoamento de Pessoal de Nível Superior (CAPES), Conselho Nacional de Desenvolvimento Científico e Tecnológico (CNPq) for the financial support and to the Instituto de Pesquisas Energéticas e Nucleares (IPEN-SP) for the sample irradiation.

References

[1] BAILIFF, I.K. ***Retrospective Dosimetry With Ceramics***. Rad. Meas. 27 (1998), 923 – 941

[2] BULL, Nucl. R. K. Tracks Radiat. Meas. 11 (1986) 105.

[3] GOKSU, H.Y., Bailiff, I.K. V.B. Mikhailik, V.B. (2003) ***New approaches to retrospective dosimetry using cementitious building materials*** Rad. Meas. 37 (2003), 323 – 327

[4] MCKEEVER, Stephen W.S. ***Thermoluminescence of solids***. 1st ed. Cambridge: Cambridge Uni. Press, 1997.

[5] CHEN, Reuven & MCKEVEER, Stephen W.S. ***Theory of thermoluminescence and related phenomena.*** 1st ed. Singapore:World Scientific,1997.

[6] OLIVEIRA, Luiz C. de, BONFIM, Ana Paula S. SOUSA, Susana O.de: ***Estudo dos Parâmetros Cinéticos do Pico Termoluminescente em 225°C do Quartzo Natural.*** Scientia Plena. 2005.

[7] PAGONIS, Vasilis, KITIS, George, FURETTA, Claudio. ***Numerical and Practical Exercises in Thermoluminescence.*** 2006, XXII, Hardcover.

[8] WILL, G. Powder Diffraction: The Rietveld Method and the Two-Stage Method. 1 ed. Stürtz, Würzburg: Springer, 2005. 224p.

[9] YOUNG, R. A. DBWS-9411 - an upgrade of the DBWS programs for Rietveld Refinement with PC and mainframe computers. J. Appl. Cryst. 28 (1995) 366.

[10] AITKEN, Martin Jim. ***Thermoluminescence Dating***. 1st ed. London: Academic Press, 1985.

[11] VERONESE. I. ***The trap parametres of electrons in intermediate energy levels in quartz.*** Rad. Meas. 38 (2004), 743 – 746.

[12] GOKSU. H.Y. ***Luminescence Dosimetry Using Building Materals and Personal Objects.*** Radiation Protection Dosimetry. 119 (2006), 413-420.

Effect on the insulation material of a MOSFET device submitted to a standard diagnostic radiation beam

C M S de Magalhães[1,2], L A P dos Santos[2], D do N Souza[1], A F Maia[1]

[1]Departamento de Física, Universidade Federal de Sergipe 49100-000 Aracaju SE, Brazil

[2]Laboratório de Instrumentação Nuclear, CNEN/CRCN-Recife 50740-540 Recife PE, Brazil

E-mail: cinthiamsm@yahoo.com.br

Abstract. MOSFET electronic devices have been used for dosimetry in radiology and radiotherapy. Several communications show that due to the radiation exposure defects appear on the semiconductor crystal lattice. Actually, the structure of a MOSFET consists of three materials: a semiconductor, a metal and an insulator between them. The MOSFET is a quadripolar device with a common terminal: gate-source is the input; drain-source is the output. The gate controls the electrical current passing through semiconductor medium by the field effect because the silicon oxide acts as insulating material. The proposal of this work is to show some radiation effects on the insulator of a MOSFET device. A 6430 Keithley sub-femtoamp SourceMeter was used to verify how the insulating material layer in the structure of the device varies with the radiation exposure. We have used the IEC 61267 standard radiation X-ray beams generated from a Pantak industrial unit in the radiation energy range of computed tomography. This range was chosen because we are using the MOSFET device as radiation detector for dosimetry in computed tomography. The results showed that the behaviour of the electrical current of the device is different in the insulator and semiconductor structures.

1. Introduction

The use of metal oxide semiconductor field effect transistor (MOSFET) as dosimeters has been increased. Their small physical size, good spatial resolution, simple use, dose integration properties and the availability of real time data make them attractive detectors for radiation dosimetry. The main disadvantage, however, is their limited useful life due to the damage from the accumulated dose [1]. The principle in operation of MOSFET dosimetry is based on the generation of electron-hole pairs in the semiconductor region due to the interaction of the ionizing radiation. The changing in the semiconductor material occurs each time it is irradiated and its electrical characteristics can demonstrate how its response is. The behaviour of the device response when it is submitted to a radiation beam allows the health physical scientists make analysis about the entrance dose in a patient undergoing a radiodiagnostic examination and radiotherapy, for while [2-4]. The purpose of this study was to investigate the radiation effects on the insulator of a MOSFET device applying radiation beam normally used in the computed tomography.

2. Materials and methods

There are several MOSFET type devices commercially available. One of most used is the N channel MOSFET. Such a device has a *p*-type silicon semiconductor substrate separated from a metal gate by an insulating oxide layer. Two of its terminals called the source (S) and the drain (D) corresponds to two negatively doped (n type) silicon parts. The third terminal is the gate (G) (Fig. 1a). Five identical BSS138N MOSFET (Fig. 1b) devices were used aiming to make analysis how this detector type responds to radiation beam chosen, and one of them was not irradiated for monitoring the devices. BSS138N device has a Surface Mount Device (SMD) encapsulation [5].

An HF320 Pantak X-ray unit was used to generate filtered X-ray beam (Fig. 2a): the RQR9 standard radiation quality, chosen based on the IEC 61627, which is employed in computed tomography application. For all measurements the devices were positioned with the black epoxy layer opposite to the incident beam for direct fixation at the centre of the respective applicators (closed).

There are two ways to test MOSFET devices in radiation field: active and passive mode [6]. The last one was used in this work. The procedure of make measurements followed the steps: 1) reading of the gate electrical current (I_{GS}) using a 6430 Keithley sub-femtoamp SourceMeter and reading of the drain electrical current (I_{DS}) using a 2400 Keithley SourceMeter; 2) irradiation of the device with 25 gray accumulated dose; 3) reading after irradiation of I_{GS} and I_{DS} (Fig. 2b). This procedure was repeated systematically until 100 Gy dose. Each reading is an average of five data points. A 47Ω resistor was used to bias the drain terminal at 10V and avoid to breakdown them. The gate-source voltage V_{GS} was chosen to be 10V.

A RC6 ionization chamber coupled to a 6517 Keithley was used to measure the dose at detector position.

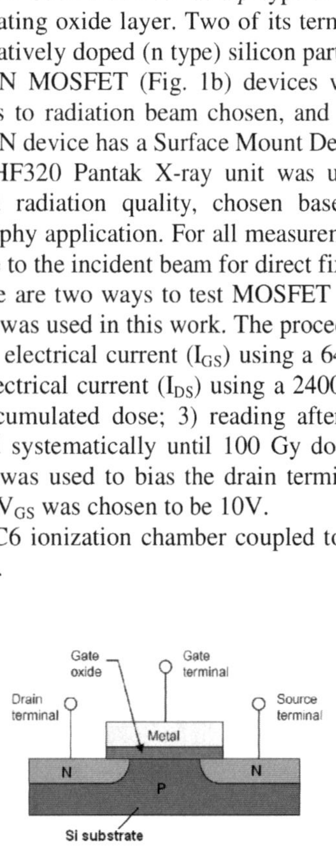

Figure 1.a) Schematic diagram of a MOSFET.

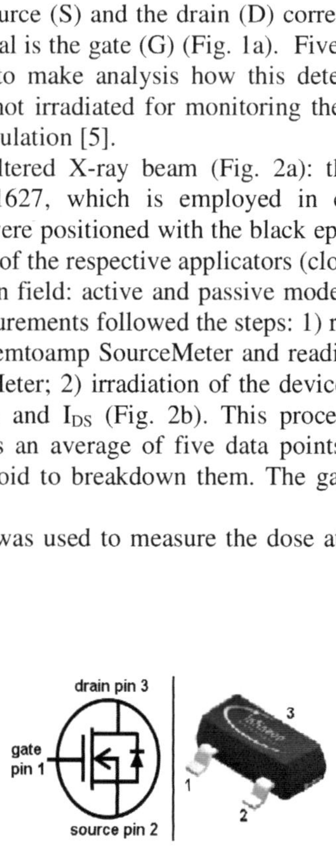

Figure 1.b) BSS138N MOSFET device [6].

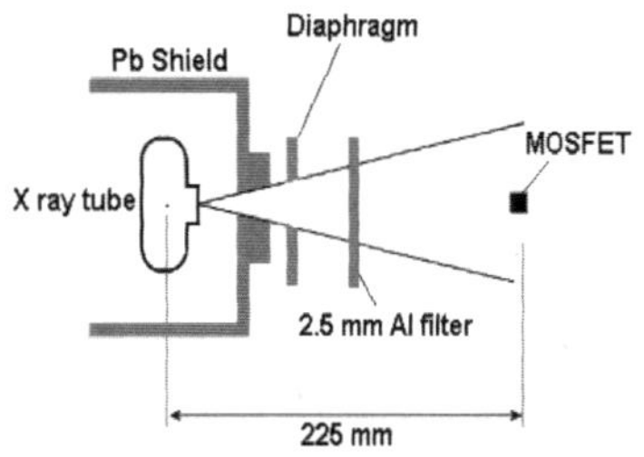

Figure 2. a) Experimental arrangement (not to scale).

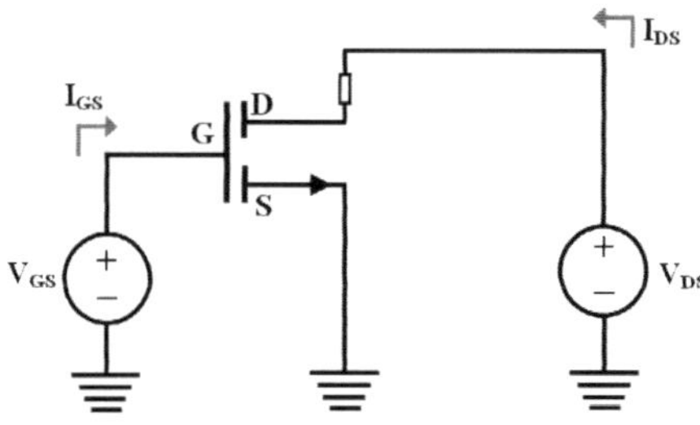

Figure 2. b) Schematic circuit of readings.

3. Results and discussion

Observing Table 1, which shows I_{GS} and I_{DS} data obtained before the irradiation and after 100 Gy accumulated dose, one may be noted that I_{GS} practically did not change remaining at picoampere scale. However, the semiconductor current (I_{DS}) varied significantly, from picoampere up to microampere scale. This suggests that the silicon dioxide did not have its structure altered due to ionizing radiation from RQR9 beam quality. In fact the oxide is considered as a region with high density of defects whose charge state can be altered by irradiation. The damage to the SiO_2 structure caused by the interaction of radiation manifests as a build up of positive charge due to the semipermanent trapped holes when radiation-generated electrons are diffused out of the insulator [7].

Table 1. Values of I_{GS} and I_{DS} current before the irradiation and after 100 Gy dose.

	I_{GS} (pA) (*Dose = 0 Gy*)	I_{GS} (pA) (*Dose =100 Gy*)	I_{DS} (pA) (*Dose = 0 Gy*)	I_{DS} (μA) (*Dose = 100 Gy*)
MFT01	0.7 ± 0.6	1.0 ± 0.5	19.4 ± 3.2	5.632 ± 0.006
MFT02	0.5 ± 0.1	4.3 ± 0.1	7.3 ± 4.7	2.080 ± 0.005
MFT03	1.7 ± 0.3	5.4 ± 1.2	14.3 ± 1.5	9.473 ± 0.014

As a result, Figures 3 and 4 present the glow curve after the cumulative dose irradiation with 25 Gy intervals. The graphics are on the same scale to show more clearly the behaviour of current with cumulative dose. The radiation damage in the insulator structure of the device did not appear in form of changing in its bulk electrical characteristic (Fig. 3). However, according to Figure 4 one observes that the glow varied with the dose and that this feature can be used to relate the device response with the absorbed radiation. The total cumulative dose chosen represents more than one thousands head computed tomography examinations.

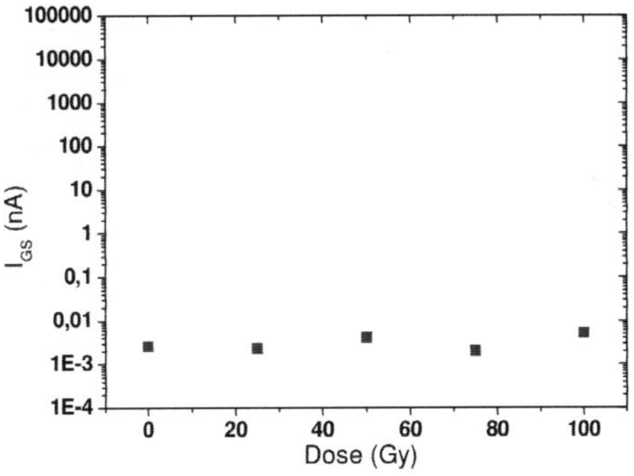

Figure 3. The I_{GS} current variation as a function of the dose.

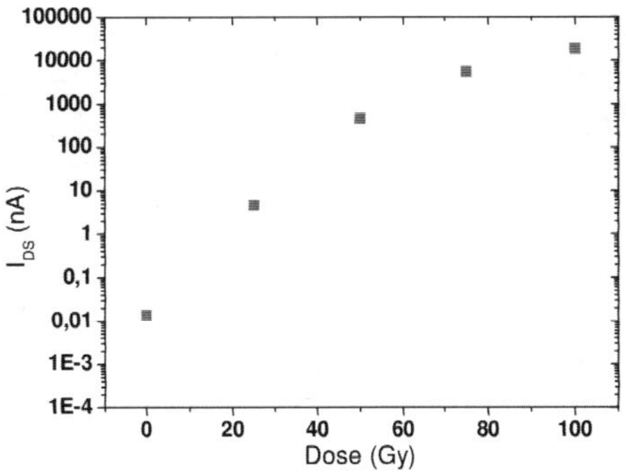

Figure 4. The I_{DS} current variation as a function of the dose.

4. Conclusions

On the one hand the variation of the gate current (I_{GS}) as a function of the cumulative dose applied to the device was not significant. For the computed tomography radiation field used to verify the electrical response of passive mode MOSFET as X-ray detector, one can conclude that the mechanism of oxide radiation damage did not alter at all the insulator structure. On the other hand, the crystal semiconductor structure of the device (drain-source and substrate) showed a variation of several orders of magnitude larger than that of insulator (10^6). This result is expressed in terms of the semiconductor conductivity after the device to have been submitted to a 100 Gy cumulative dose.

5. Acknowledgments
The authors wish to thank FACEPE and CNPq agencies for financial support.

References
[1] Braranicharan G, Manigandan D, Devan K, et al. 2005 *Medical Dosimetry* **30** 213
[2] Chuang C F, Verhey L J, Xia P 2002 *Medical Physics* **29** 1109
[3] Peet D J, Pryor M D 1999 *The British Journal of Radiology* **72** 562
[4] Bower M W, Hintenlang DE 1998 *Health Physics* **75** 197
[5] Infineon 2007, Small-Signal-Transistor: BSS138N
[6] Rosenfeld A B 2007 *Radiation Measurements* **41** S134
[7] Lutz G 1999 *Semiconductor Radiation Detectors* (Berlim: Springer)

An insight into the local $O_h \rightarrow T_d$ instability in $BaF_2:Mn^{2+}$

P Garcia-Fernandez[1], JA Aramburu[1], MT Barriuso[2] and M Moreno[1,*]
[1]Departamento de Ciencias de la Tierra y Física de la Materia Condensada, Universidad de Cantabria, Avda. de los Castros s/n. 39005 Santander, Spain
[2] Departamento de Física Moderna, Universidad de Cantabria, Avda. de los Castros s/n. 39005 Santander, Spain

E-mail: morenom@unican.es

Abstract. While most complexes in fluorite-type lattices (CaF_2, SrF_2, CdF_2) containing Mn^{2+} impurities remain cubic at all temperatures, electron paramagnetic resonance (EPR) measurements have shown that the one in BaF_2 is tetrahedral below 50K. This surprising behaviour is intrinsic to the centre and not associated to any close defect or pure lattice phase-transition. Through the use of density functional (DFT) calculations we show that the distortion is linked to the unexpected low force constant along the tetrahedral distortion mode with a_{2u} symmetry in these complexes and the large metal-ligand distance existing in $BaF_2:Mn^{2+}$. Ultimately, these facts reflect that, for substitutional impurities in fluorite-type crystals, the ligand-ligand interaction is dominant over the metal-ligand one.

1.Introduction

Magnetic resonance data [1,2] on $BaF_2:Mn^{2+}$ reveal the existence of a surprising phenomenon not observed either in $MF_2:Mn^{2+}$ (M: Ca, Sr, Cd) or in Mn^{2+} doped cubic fluoroperovskites. Indeed electron paramagnetic resonance (EPR) spectra of $BaF_2:Mn^{2+}$ detected at temperatures T>50 K are consistent with a cubic local symmetry around an *isolated* Mn^{2+} impurity as the one found in $MF_2:Mn^{2+}$ (M: Ca, Sr, Cd) at and below room temperature [3]. However this situation is not longer true for $BaF_2:Mn^{2+}$ because below 50 K the superhyperfine (shf) interaction shows the existence of a local tetrahedral symmetry involving two different Mn^{2+}-F^- bonds [1,2]. This fact just means that at T = 0K the cube involving eight F^- ions in the corners is distorted giving rise to four short Mn^{2+}-F^- bonds with a length equal to R_s and four long bonds with a length equal to R_l (figure 1). The isotropic superhyperfine constants, $A_s(s)$ and $A_s(l)$ associated with short and long bonds respectively have been measured to be equal to $A_s(s)$=29.1 MHz and $A_s(l)$=10.5 MHz [1,2]. From these values it has been estimated R_l- R_s = 0.25 Å [4]. All these conclusions were supported by subsequent electron nuclear double resonance (ENDOR) spectra obtained on $BaF_2:Mn^{2+}$ from which no evidence of any nearby defect to an *isolated* Mn^{2+} impurity was encountered [2].

The $O_h \rightarrow T_d$ instability is puzzling first because there is not any structural phase transition in the host lattice below 1200 K at zero pressure to which the phenomenon could be attributed [5]. On the other hand such instability involves the *softening* of a local a_{2u} mode (figure 1) which has a *stretching* character.

* To whom any correspondence should be addressed.

The present work is aimed at clearing out the origin of the $O_h \rightarrow T_d$ instability observed in $BaF_2:Mn^{2+}$ by means of density functional theory (DFT) calculations carried out on this system as well as on $CaF_2:Mn^{2+}$ and $SrF_2:Mn^{2+}$. In Section 2 a short outline of the methods of calculation employed in this work is provided while the main results are shown in section 3. A short account of the main conclusions and some final remarks are given in the last section.

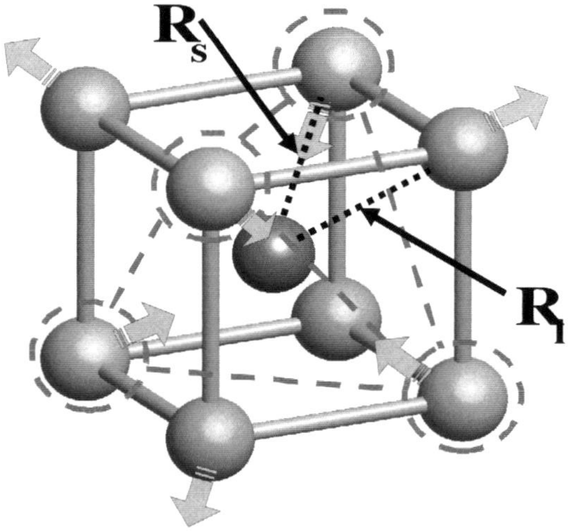

direction of displacement of the ligands along the a_{2u} mode and a red outline shows the tetrahedral distortion. The short and long metal ligand distances are denoted by R_s and R_l, respectively. The four ligands with red outline move towards the central Mn^{2+} ion along <111> type directions giving rise to short bonds.

2. Computational details

Due to the strong localization of the active electrons in the MnF_8^{6-} complex we have been able to simulate the system using the cluster approximation. The main results have been obtained employing a 21-ion cluster (MnF_8Ba_{12}). Further calculations have been carried out in a 57-ion cluster ($MnF_8Ba_{12}F_{24}Ba_6$). The validity of both clusters to study impurities in fluorite-type crystals has been already checked in previous works [6, 7]. The effect of the long-range potential due to the rest of the lattice over the clusters has been introduced by including a large number of point charges at the ions' positions. Finally, and due to the presence of heavy atoms like barium and transition metals like manganese in the clusters, we checked the effect of scalar relativistic corrections in our calculations which were negligible.

3. Results and Discussion

In a first step the value of the equilibrium Mn^{2+}-F^- distance, R, for $CaF_2:Mn^{2+}$ and $SrF_2:Mn^{2+}$ has been calculated. These values are reported in table 1 and compared to the M^{2+}-F^- distance (M:Ca,Sr) of the host lattice designated by R_0. It is worthwhile to remark that the value R = 2.29 Å derived for $CaF_2:Mn^{2+}$ essentially matches the figure R = 2.28 ± 0.01 Å measured by extended x-ray absorption fine structure (EXAFS) technique [14] and previous calculations by Pascual et al. [15]. Along this line the R values collected in Table 1 are not far from R = 2.27 Å and R = 2.31Å *estimated* from the analysis of the experimental isotropic superhyperfine (shf) constant, A_s, for $CaF_2:Mn^{2+}$ and $SrF_2:Mn^{2+}$ respectively [16]. Comparing the R_0 (the cation-anion distance in the pure lattice) and R values gathered in table 1 in $MF_2:Mn^{2+}$ (M = Ca, Sr) it can be noted that the $M^{2+}\rightarrow Mn^{2+}$ substitution gives rise to an inwards relaxation. Bearing in mind the ionic radii of Mn^{2+} and F^- as well as Pauling's rules that take into account the influence of the coordination number upon the equilibrium distance this fact is quite reasonable. Accordingly, when a Mn^{2+} ion is surrounded by *eight* F^- ions one would expect an equilibrium distance, R_8, close to 2.22 Å. The calculated R values for $MF_2:Mn^{2+}$ (M = Ca, Sr) in Table 1 thus imply a deviation of less than 5.5% with respect to R_8 = 2.22 Å.

For the sake of clarity the R value calculated for $BaF_2:Mn^{2+}$ *imposing* a hexahedral symmetry is also reported in table 1. The obtained value R = 2.47 Å thus implies a huge deviation (12%) with respect to R_8 = 2.22 Å. It is worth noting that this shift is certainly bigger than that found in the case of Mn^{2+} doped cubic fluoroperovskites entering an octahedral position where the difference between the obtained R values and R_6 = 2.13 Å never exceeds 4% [17].

In a second step, the frequencies $\nu(a_{1g})$ and $\nu(a_{2u})$, corresponding, respectively, to the symmetric a_{1g} local mode and to the a_{2u} mode(depicted in figure 1), have been derived. In the case of the $BaF_2:Mn^{2+}$ the $\nu(a_{1g})$ and $\nu(a_{2u})$ values reflect distortions around a *forced* hexahedral symmetry. From the results gathered in table 1 and figure 2 it can be noticed that:

i) The calculated $\nu(a_{1g})$ values are always found to be well above zero.

ii) Despite the a_{2u} mode has also a stretching character the calculated $\nu(a_{2u})$ value for $MF_2:Mn^{2+}$ (M = Ca, Sr) is found to be substantially smaller than the figure corresponding to $\nu(a_{1g})$.

iii)Similarly to what is observed for stretching local modes [18,19] under hydrostatic pressure a reduction on R favours an increase of both $\nu(a_{1g})$ and $\nu(a_{2u})$.

iv) In the case of $BaF_2:Mn^{2+}$, $\nu(a_{2u})$ is found to be imaginary, implying that the equilibrium geometry is not cubic but tetrahedral in agreement with experimental EPR and ENDOR data [1,2].

Table 1. Metal-ligand distance (R) and stretching frequencies, $\nu(a_{1g})$ and $\nu(a_{2u})$, of the MnF_8 complex as calculated using DFT in different lattices. The cation-anion distance in the pure lattice (R_0) is also shown for comparison.

	CaF_2	SrF_2	BaF_2
R_0 (Å)	2.37	2.51	2.69
R (Å)	2.29	2.34	2.47
$\nu(a_{1g})$ (cm^{-1})	357	326	267
$\nu(a_{2u})$ (cm^{-1})	168	132	59i

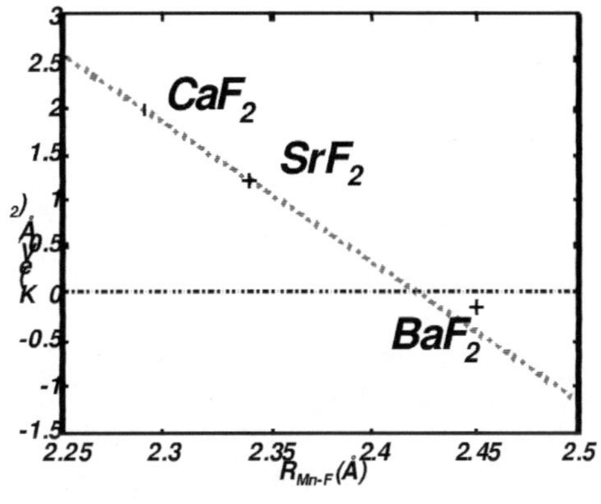

Figure 2. Variation of $K(a_{2u})$ with the metal-ligand distance.

Looking at table 1 it appears that the a_{2u} instability in $BaF_2:Mn^{2+}$ is somewhat related to the high R value embodied in table1. Let us write the two ν(j) frequencies (j: a_{1g}, a_{2u}) as

$$\upsilon(j)=\frac{1}{2\pi}\sqrt{\frac{K(j)}{M_L}} \quad (1)$$

where M_L denotes the ligand mass. The value of the calculated $K(a_{2u})$ force constant for $MF_2:Mn^{2+}$ (M = Ca, Sr) is portrayed in figure 2. It can be noticed that a simple linear interpolation of such values leads to a small but negative $K(a_{2u})$ value when R = 2.47 Å.

A central matter in the present analysis is thus to understand why $\nu(a_{2u})$ is significantly smaller than $\nu(a_{1g})$ for a given system despite both modes have a stretching character. Some insight on this key issue is gained writing both $K(a_{2u})$ and $K(a_{1g})$ approximately in terms of the two-body metal-ligand, K_{ML}, and ligand-ligand, K_{LL}, force constants which appear in a *phenomenological* description of vibrations considering only central forces[20]. The relation between $K(a_{2u})$ and $K(a_{1g})$ and K_{ML} and K_{LL} can easily be derived to be

$$K(a_{1g}) = K_{ML} + 2\,K_{LL} \quad ; \quad K(a_{2u}) = K_{ML} \quad (2)$$

It is worth noting that $K(a_{2u})$ *does not* depend on K_{LL}. This fact merely reflects that the distance between two closest ligands is unmodified through a small a_{2u} distortion. Consider for instance the ligands placed initially at (r,r,r) and (-r,r,r) which under the a_{2u} distortion their coordinates become equal to (r+ε,r+ε,r+ε) and (-r+ε,r-ε,r-ε) where ε<<r .The change of distance between the two ligands thus remains unchanged to first order of ε . Moreover if the $\nu(a_{1g})$ and $\nu(a_{2u})$ values obtained for $MF_2:Mn^{2+}$ (M = Ca, Sr) in table 1 are analysed by means of Eqs (1) and (2) it turns out that $K(a_{1g})$ is dominated by the ligand-ligand contribution, K_{LL}, which does not play any role in the a_{2u} mode. For instance for $SrF_2:Mn^{2+}$ the analysis of $\nu(a_{1g})$ and $\nu(a_{2u})$ values collected in table 1 lead to a ratio K_{LL}/K_{ML}= 2.6. Therefore in fluorite-type lattices the dominant contribution to the frequency of stretching modes is not K_{ML} but K_{LL} a fact which is thus contrary to what happens in cubic lattices with octahedral coordination [21].

A better insight on the instability in $BaF_2:Mn^{2+}$ can be gained plotting the ground state energy, E, as a function of the distortion coordinate $Q(a_{2u})$, whose absolute value is given by $\sqrt{2}\left(R_l - R_s\right)$.The result, displayed in figure 3, reveals the existence of a shallow double well with minima located at $Q(a_{2u})$ = ±0.41 Å. This value implies a figure R_L- R_S = 0.29 Å which is not far from the previous estimations derived from the experimental isotropic shf constant. As expected form the results displayed in figure 2 we have verified that the tiny barrier (of about 50 cm^{-1}) shown in figure 2 disappears if a small pressure that reduces the lattice parameter of BaF_2 by only 1.5% is applied.

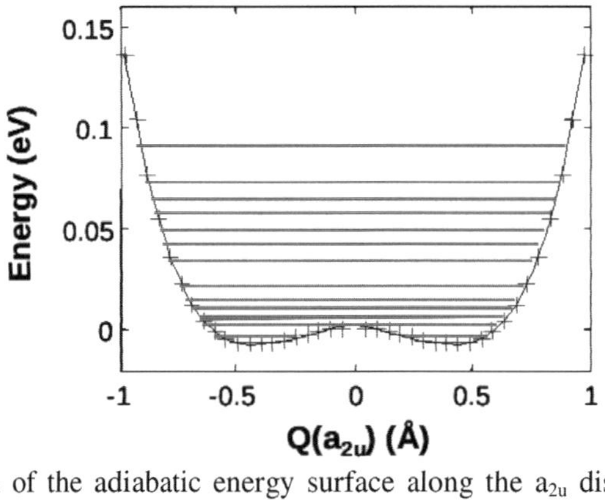

Figure 3. Cross section of the adiabatic energy surface along the a_{2u} distortion and the vibrational levels associated to it.

Despite the energy barrier obtained at zero pressure is only ~50 cm^{-1} it should be pointed out that the lowest vibrational state is found to be essentially localized in a given well. By contrast in all excited states the calculated wavefunction is delocalized. The relevance of this fact in connection with the $T_d \rightarrow O_h$ transition observed upon rising the temperature has recently been discussed [22].

4.Conclusions and Final Remarks

The present study has shown that the a_{2u} instability observed in BaF_2:Mn^{2+} is favoured by two factors. On one hand the Mn^{2+}-F^- distance for a MnF_8^{6-} complex in BaF_2 keeping the O_h symmetry is calculated to be unusually high. On the other hand $\nu(a_{2u})$ is not dependent on the ligand-ligand force constant, K_{LL}, which is clearly the dominant one for host lattices with fluorite structure.

It should be emphasized that the a_{2u} instability can not properly be understood considering Mn^{2+} and F^- as *rigid* ions. In fact, for this kind of phenomena arising from the so called pseudo Jahn-Teller effect it is crucial to take into account the *changes* in electronic density which take place on going from $Q(a_{2u}) = 0$ to a distorted geometry involving $Q(a_{2u})\neq 0$ [23,6,22].

The present results can also be of interest in the realm of structural phase transitions in pure insulating materials. Indeed it has been pointed out [24] that the occurrence of an on site instability favours the existence of a structural phase transition.

According to the results displayed in figure 3 the a_{2u} instability can also have an important influence for understanding the isotropic superhyperfine constant on the cubic phase. Indeed the distance between a given F^- ligand ion and the Mn^{2+} impurity would experience a fluctuation of at least ±0.15 Å with respect to its average value. This relevant fact could explain why A_s=21.2±1.5 MHz measured for BaF_2:Mn^{2+} at room temperature is identical to A_s=22.2±0.3 MHz for SrF_2:Mn^{2+} within experimental uncertainties [3]. Work on this relevant issue and on the influence of the vibronic admixture with excited states on the value of $\nu(a_{2u})$ for BaF_2:Mn^{2+}, CaF_2:Mn^{2+} and SrF_2:Mn^{2+} is currently under way.

Acknowledgements
The support by the Spanish Ministerio de Ciencia y Tecnología under project FIS2006-02261 is acknowledged.

References

[1] Badalyan AG, Baranov PG, Vikhnin VS, and Khramtson VA 1986 *JETP Lett.* **44** 110
[2] Soethe H, Vetrov VA, and Spaeth JM 1992 *J. Phys.: Condens. Matter* **4** 7927
[3] Richardson RJ, Lee S and Menne TJ, 1972 *Phys. Rev. B* **6** 1065
[4] Barriuso MT, Baranov PG and Moreno 1991 M, *Rad. Eff. Defects Solids* **119-121** 177
[5] Manasreh MO and Pederson DO 1998 *Phys. Rev. B* **38** 6270
[6] Garcia-Fernandez P, Aramburu JA, Barriuso MT, and Moreno M 2006 *Phys. Rev. B* **73** 184122
[7] Garcia-Fernández P, Aramburu JA, Barriuso MT, and Moreno M 2004 *Phys. Rev. B* **69** 174110
[8] Vosko SH, Wilk L and Nusair M 1980 *Can. J. Phys.* **58** 1200
[9] Becke AD 1988 *Phys. Rev. A* **38** 3098; Lee C, Yang W, and Parr RG 1988 *Phys. Rev. B* **37** 785
[10] te Velde, G, Bickelhaupt EF, Baerends EJ, Fonseca Guerra C, van Gisbergen SJA, Snijders JG and Ziegler T 2001 *J. Comp. Chem.* **22** 931
[11] Gaussian 98 (Revision A.9), Frisch MJ, Trucks GW, Schlegel HB, Scuseria GE, Robb MA, Cheeseman JR, Zakrzewski VG, Montgomery JA, Jr., Stratmann RE, Burant JC, Dapprich S, Millam JM, Daniels AD, Kudin KN, Strain MC, Farkas O, Tomasi J, Barone V, Cossi M, Cammi R, Mennucci B, Pomelli C, Adamo C, Clifford S, Ochterski J, Petersson GA, Ayala PY, Cui Q, Morokuma K, Malick DK, Rabuck AD, Raghavachari K, Foresman JB, Cioslowski J, Ortiz JV, Baboul AG, Stefanov BB, Liu G, Liashenko A, Piskorz P, Komaromi I, Gomperts R, Martin RL, Fox DJ, Keith T, Al-Laham MA, Peng CY, Nanayakkara A, Gonzalez C, Challacombe M, Gill PMW, Johnson BG, Chen W, Wong MW, Andres JL, Head-Gordon M, Replogle ES and Pople JA, Gaussian, Inc., Pittsburgh PA, 1998.

[12] Kaupp M, Schleyer PVR, Stoll H and Preuss H 1991 *J. Chem. Phys.* **94** 1360
[13] D.E. Woon and T.H. Dunning 1994 Jr. *J. Chem. Phys.* **100** 2975
[14] Barkyoumb JH and Mansour AN 1992 *Phys. Rev. B* **46** 8768
[15] Pascual JL and Seijo L 1995 *J. Chem. Phys.* **102**, 5368
[16] Barriuso MT and Moreno M 1984 *Chem. Phys. Lett.* **112** 165
[17] Barriuso MT, Aramburu JA, and Moreno M 1999 *J. Phys.: Cond. Matter* **11** L525
[18] Woods AM, Sinkovits RS, Charpie JC, Huang WL, Bartram RH and Rossi AR 1993 *J. Phys. Chem. Solids* **54** 543
[19] García-Lastra JM, Moreno M and Barriuso MT 2008 *J. Chem. Phys.* **128** 144708
[20] Ballhausen CJ 1962 *Ligand Field Theory* (New York: McGraw Hill) ch 8
[21] Barriuso MT, Moreno M, and Aramburu JA 2002 *Phys. Rev. B* **65**, 064441
[22] Garcia-Fernandez P, Aramburu JA, Barriuso MT and Moreno M 2008 *J. Chem. Phys.* **128** 124513
[23] I. Bersuker 2006 *The Jahn-Teller Effect* (Cambridge: Cambridge University Press).
[24] Bruce AD 1980 *Adv. Phys.* **29** 111

Journal of Physics: Conference Series **249** (2010) 012034 doi:10.1088/1742-6596/249/1/012034

Refractive index of glass and its dispersion for visible light

D Y Smith[1, 2] and W Karstens[3]

[1] Dept. of Physics, University of Vermont, 82 University Pl., Burlington, VT, USA
[2] Physics Div., Argonne National Laboratory, 9700 S. Cass Ave., Argonne, IL, USA
[3] Dept. of Physics, Saint Michael's College, Colchester, VT, USA

E-mail: dysmith@uvm.edu

Abstract. The classification of optical glass and empirical relations between the refractive index and its dispersion are discussed in terms of moments of the glass's IR and UV absorption spectra. The observed linear dependence of index on dispersion within glass families is shown to arise primarily from the approximately linear superposition of the electronic absorptions of glass former and glass modifiers. The binary classification into crown and flint glasses is also based primarily on electronic spectra: Crown glasses are "wide-gap" materials with excitation energies greater than ~12.4 eV, while flint glasses are their "narrow-gap" counterpart.

1. Introduction

The search for glasses to minimize chromatic aberrations in transmission optics has spurred extensive studies of the refractive index and its dispersion for optical glass for more than a century [1-6]. Traditionally, these quantities are summarized within the framework of empirical relations [1, 4, 7-8] that are suitable for lens design, but give little insight into processes determining the index. Here we employ a moment-analysis procedure [9, 10] based on the Kramers-Kronig relations to explore the physics underlying both the empirical relations and the common classification scheme used for optical glass. Specifically, the index and dispersion of various glass families trace out separate straight lines on plots of index *vs.* mean dispersion [4, 11]. We show that this is a consequence of the near linear additivity of the absorption spectra of glass formers and glass modifiers. Further, we show that glass classifications based on the Abbe number are directly related to the glass's optical excitation spectrum.

2. Background

The IR and UV absorptions of vitreous silica [12], a typical glass-forming wide-gap insulator, are shown in figure 1. The index of refraction, shown by the dashed curve, exhibits strong dispersion at energies approaching the absorptions at the low- and high-energy limits of the transparent region. However, for visible light, dispersion amounts to only a few percent. Since the optical response of a charge is inversely proportional to its mass, the total oscillator strength of the IR absorption, which involves ionic motion, is of the order of 10^{-4} times smaller than that of the electronic transitions in the UV. Moreover, ionic dispersion lies well below the low-energy cut off of visible light. Thus to a first approximation, the refractive index at visible wavelengths is determined by electronic transitions. However, a full discussion of dispersion requires that both IR and UV transitions be considered.

Optical glasses generally consist of a glass former such as SiO_2, B_2O_3, etc., and glass-modifying metallic oxides such as Na_2O, CaO, PbO, BaO, etc. In silicate glasses the modifiers originally served as fluxes, that is, as agents to lower the melting point of silica by interrupting the strong Si– O bonds,

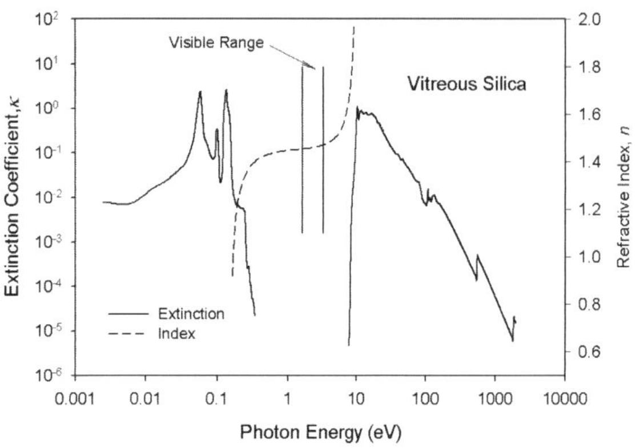

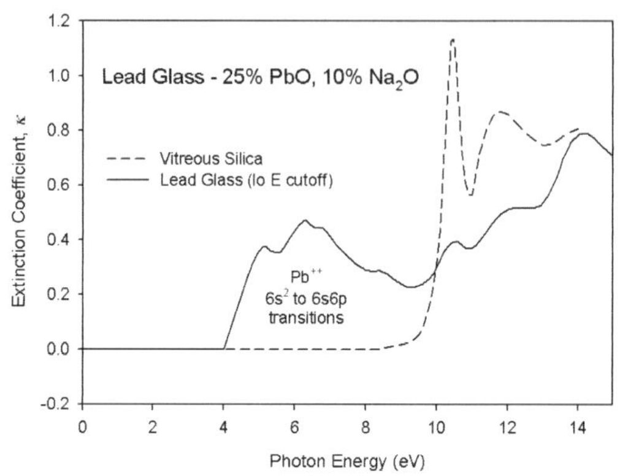

Figure 1. Absorption & dispersion in vitreous SiO_2. **Figure 2.** Absorption of lead glass [13].

but secondarily they alter other properties. In particular, the optical properties of glass-forming materials may be tailored to a limited extent by adding ions that introduce electronic absorptions in the UV. An example for PbO in silica [13] is shown in figure 2. Not only are new Pb^{2+} bands added, but the absorption of the silica host is altered by both dilution and perturbations caused by replacing Si–O bonds by Pb–O bonds. Changes in the net absorption in the IR are less pronounced (see [14]).

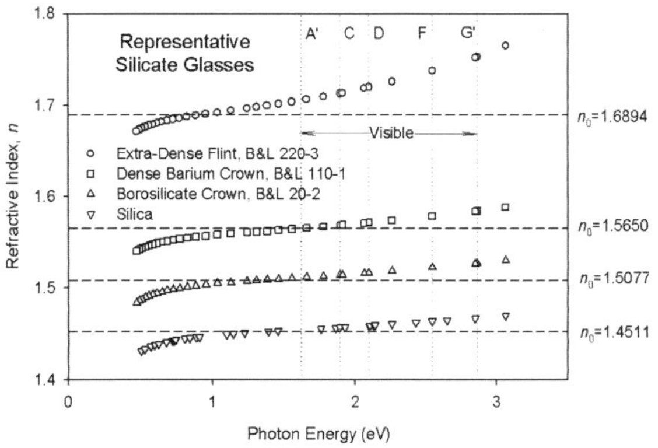

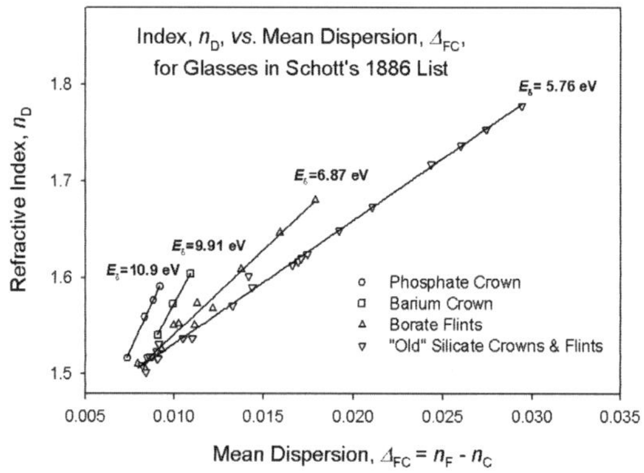

Figure 3. Refractive indices of silicate glasses. **Figure 4.** Index vs. dispersion in early glasses.

Concentrating on the transparent region, we give examples of the index at visible wavelengths for a range of commercial silicate glasses [12, 15] in figure 3. Here, pure vitreous SiO_2 has the lowest index and dispersion. As modifiers are added, the refractive index increases; the dispersion increases even more markedly, especially toward the blue. It is notable that the low-energy dispersion from ionic vibration is not as strongly dependent on doping.

3. Linear-response treatment of refraction

For insulators, the Kramers-Kronig dispersion expression for the refractive index, $n(E)$, may be divided into contributions from IR vibrational and UV electronic absorptions [9, 10],

$$n(E) - 1 = \frac{2}{\pi} P \int_{\mathrm{IR}} \frac{E'}{E'^2 - E^2} \kappa(E') dE' + \frac{2}{\pi} P \int_{\mathrm{UV}} \frac{E'}{E'^2 - E^2} \kappa(E') dE', \tag{1}$$

where $\kappa(E)$ is the extinction coefficient. For energies, E, in the transparent region between the IR and UV, the integrands may be expanded in power series, the first in the variable $(E'/E)^2$, the second in $(E/E')^2$. The index then takes the form of a Laurent series,

$$n(E) \approx 1 + \frac{2}{\pi}(\cdots - \mu_1 E^{-2} + \mu_{-1} + \mu_{-3}E^2 + \mu_{-5}E^4 \cdots), \tag{2}$$

where $\mu_1 = \int_{IR} E'\,\kappa(E')\,dE'$ is the 1st moment of the IR absorption, and $\mu_{-n} = \int_{UV} \frac{\kappa(E')}{E'^n}\,dE'$ is the inverse n-th moment of the UV electronic absorption. Similarly, the mean dispersion between the F and C lines is

$$\Delta_{FC} \equiv n_F - n_C \approx \frac{2}{\pi}(E_F^2 - E_C^2)\left[\frac{\mu_1}{E_F^2E_C^2} + \mu_{-3} + \mu_{-5}(E_F^2 + E_C^2)\right]. \tag{3}$$

Electronic absorption typically accounts for 97% or more of $n - 1$ at visible wavelengths, with the bulk of the contribution arising from the μ_{-1} term [10, 16]. Infrared contributions from the μ_1 term are commonly less than 0.3%. Though relatively unimportant for the index, the μ_{-3} term dominates the dispersion. The μ_1 and μ_{-5} terms are retained because of their contribution to partial dispersion. Higher-order terms can always be added for greater accuracy, but are not needed for the present semi-quantitative considerations. (Herzberger finds that the E^4 term must be retained for designing infrared lenses [17]).

4. First-order formulation: Index and mean dispersion

To a first approximation, the index, n_D, at the Fraunhofer D line (which is near the centre of the visible) and the mean dispersion, Δ_{FC}, between the C and F lines (which span the red and the blue) are

$$n_D - 1 \approx \frac{2}{\pi}\mu_{-1}, \tag{4}$$

and

$$\Delta_{FC} \approx \frac{2}{\pi}(E_F^2 - E_C^2)\mu_{-3}. \tag{5}$$

It should be noted that the index and dispersion are not entirely independent quantities since both depend on the shape of the UV absorption. However, this shape may be varied by changing the composition of the glass, which allows a limited degree of independence of n_D and Δ_{FC}.

Notwithstanding this, a remarkably strong correlation between index and dispersion holds for glasses within families of the same glass former. This is apparent in figure 4 that shows that index-dispersion pairs with the same glass former, such as SiO_2, B_2O_3, etc., fall on nearly straight lines. This linear dependence may be understood by considering a two-component glass. Designating the atomic fractions of the glass former and modifier by c_f and c_m, respectively, we require that (a) $c_f + c_m = 1$, and (b) assume the electronic absorption is additive so that the total electronic absorption becomes $\kappa(\omega) = c_f\kappa_f(\omega) + c_m\kappa_m(\omega)$. Additivity of host and dopant absorption is seen at low dopant concentration in colour centres [18], and appears to hold approximately at higher concentrations in glasses provided the dopant's effect on the electronic structure of the host glass former is taken into account. Then the index and dispersion of the composite glass becomes

$$n_D - 1 \approx \frac{2}{\pi}[c_f\,\mu_{-1,f} + c_m\,\mu_{-1,m}] \tag{6}$$

and

$$\Delta_{FC} \approx \frac{2}{\pi}(E_F^2 - E_C^2)[c_f\,\mu_{-3,f} + c_m\,\mu_{-3,m}], \tag{7}$$

where $\mu_{-i,f}$ and $\mu_{-i,m}$ are the inverse i^{th} moments of glass former and glass modifier, respectively.

Eliminating the concentrations between these equations and solving for the index as a function of the dispersion yields

$$n_D - n_{D,f} = \frac{1}{E_F^2 - E_C^2}\left(\frac{\mu_{-1,m} - \mu_{-1,f}}{\mu_{-3,m} - \mu_{-3,f}}\right)[\Delta_{FC} - \Delta_{FC,f}], \tag{8}$$

where we have abbreviated the index and dispersion of the glass former by

$$n_{\mathrm{D,f}} - 1 = \frac{2}{\pi}\mu_{-1,\mathrm{f}} \quad (9) \qquad \text{and} \qquad \Delta_{\mathrm{FC,f}} = \frac{2}{\pi}(\omega_{\mathrm{F}}^2 - \omega_{\mathrm{C}}^2)\mu_{-3,\mathrm{f}}\,. \quad (10)$$

Equation (8) shows that the index of the composite glass is linear in dispersion, starts at the index-dispersion point of the glass former, and has slope

$$\frac{1}{E_{\mathrm{F}}^2 - E_{\mathrm{C}}^2}\left(\frac{\mu_{-1,\mathrm{m}} - \mu_{-1,\mathrm{f}}}{\mu_{-3,\mathrm{m}} - \mu_{-3,\mathrm{f}}}\right) = \frac{< E_{\delta\kappa}^2 >}{E_{\mathrm{F}}^2 - E_{\mathrm{C}}^2}. \quad (11)$$

In writing the right-hand side of equation (11), we have interpreted the moments ratio as the change in average (square) excitation energy caused by the modifier,

$$< E_{\delta\kappa}^2 > = \frac{\mu_{-1,\mathrm{m}} - \mu_{-1,\mathrm{f}}}{\mu_{-3,\mathrm{m}} - \mu_{-3,\mathrm{f}}} = \frac{\int_{\mathrm{UV}} E'^{-1}\delta\kappa(E')\,dE'}{\int_{\mathrm{UV}} E'^{-3}\delta\kappa(E')\,dE'}. \quad (12)$$

This reasoning is based on the observation that the differences in spectral moments can be written as moments of the difference spectra, $\delta\kappa(\omega) = \kappa_{\mathrm{m}}(\omega) - \kappa_{\mathrm{f}}(\omega)$, that is, the absorption introduced by the glass-modifier and the accompanying decrease in glass-former absorption. An example of this difference is given in figure 5 for data derived [13] from reflectivity data currently available for a lead-silicate glass. In light of the limited reflectivity data available, details of this figure are provisional.

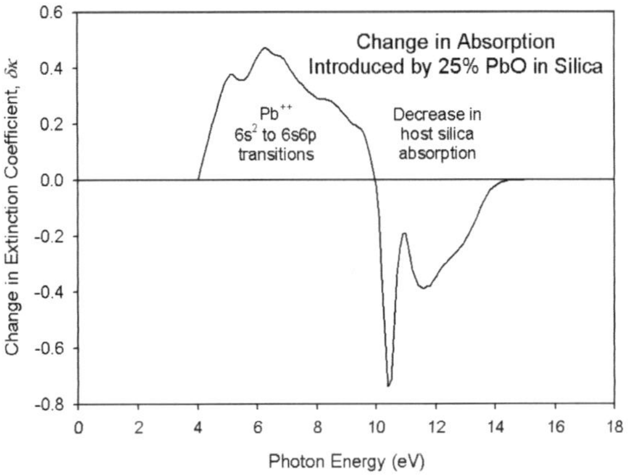

Figure 5. Change in absorption with Pb doping.

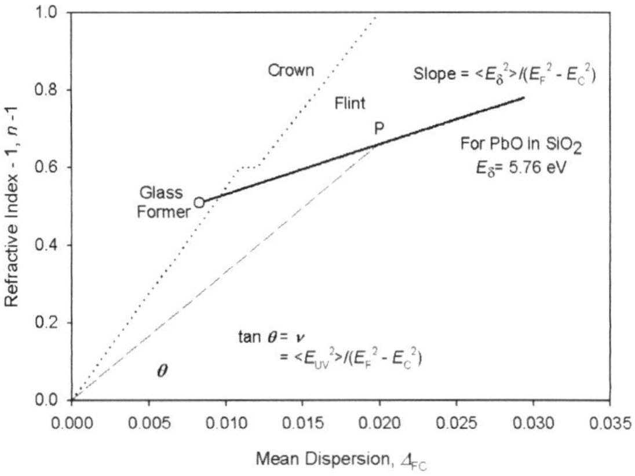

Figure 6. Schematic index *vs.* dispersion plot.

Thus, it seems reasonable to interpret the ratio of moments of the difference spectrum as the net change in excitation energy introduced by the glass former, a view that is in line with the intuitive picture of dopant atoms replacing host atoms. This formulation may be summarized in a schematic index-*vs.*-dispersion diagram, figure 6. For this example, measured values of index and dispersion for lead silicate glasses have been used. The modifier, PbO, introduces absorption bands in the region of transparency of the host, and lowers the average excitation energy. Starting from the pure glass former shown by the open circle, addition of PbO yields glasses with (n, Δ) pairs lying along the solid line, with greater distances from the glass-former origin correspond to greater modifier concentration. According to equation (11), the slope of the glass-family line is determined by the *net change* in excitation energy caused by the modifier, $< E_{\delta\kappa}^2 >^{1/2}$. A fit to experimental data for lead-silica glass gives $< E_{\delta\kappa}^2 >^{1/2}$ = 5.76 eV, which is consistent with the change in the extinction coefficient derived from reflectivity measurements shown in figure 5.

Journal of Physics: Conference Series **249** (2010) 012034 doi:10.1088/1742-6596/249/1/012034

5. Classification in terms of the Abbe number

Traditionally, glasses are classified by index and "reciprocal dispersion" or Abbe number, ν, a quantity that arises in lens-design. This corresponds to specifying the glass at point P in figure 6 by its index and the tangent of the angle θ.

$$\nu_{\mathrm{D}} = \frac{n_{\mathrm{D}} - 1}{n_{\mathrm{F}} - n_{\mathrm{C}}} = \tan\theta \approx \frac{1}{E_{\mathrm{F}}^2 - E_{\mathrm{C}}^2}\left(\frac{\mu_{-1}}{\mu_{-3}}\right) = \frac{< E_{\mathrm{UV}}^2 >}{E_{\mathrm{F}}^2 - E_{\mathrm{C}}^2}. \tag{13}$$

Here, in analogy with the treatment of the slopes of the glass-family lines, we have observed that the moments ratio on the right-hand side of equation (13), μ_{-1}/μ_{-3}, has units of energy squared, and that, by the mean-value theorem, it may be interpreted as the square of an average energy for electronic excitations of the *glass as a whole*, $< E_{\mathrm{UV}}^2 >$. Typical values of $< E_{\mathrm{UV}}^2 >^{1/2}$ range from 7.4 to 15.1 eV.

The historical division of optical glass into low-dispersion crown (soda-lime silicate) and high-dispersion flint (lead-alkali silicate) glasses has been formalized by defining crowns as having Abbe numbers of ~50 or more [4]. From equation (13) and the energies of the C and F Fraunhofer lines, this limit corresponds to $< E_{\mathrm{UV}}^2 >^{1/2} \approx 12.4$ eV. Thus, crown glasses are "wide-gap" materials with excitation energies greater than ~12.4 eV, while flint glasses are their "narrow-gap" counterpart.

6. Second-order formulation: Partial dispersion

The primary chromatic aberration of a single lens can be eliminated by using a crown-flint doublet with the appropriate index-dispersion combinations for the two lenses (for details see [1, 3]). The geometric-optics requirement is that two colours, here C and F, focus at the same point. However, higher-order aberrations remain since all other colours remain out of focus to some extent. The reason is that the slope of index *vs.* photon energy curve is not constant, but differs slightly from that associated with the mean dispersion, especially in the far-red and violet. Mitigation of the remaining aberration in second-order is achieved [1, 3] by requiring that three colours are confocal. In the thin-lens approximation this is achieved in a doublet if the relative partial dispersions of the two lenses,

$$P_{\mathrm{XY}} \equiv \frac{n_{\mathrm{X}} - n_{\mathrm{Y}}}{n_{\mathrm{F}} - n_{\mathrm{C}}} \approx \frac{E_{\mathrm{X}}^2 - E_{\mathrm{Y}}^2}{E_{\mathrm{F}}^2 - E_{\mathrm{C}}^2}\left\{1 - \frac{\mu_1}{\mu_{-3}}\left(\frac{1}{E_X{}^2 E_Y{}^2} - \frac{1}{E_C{}^2 E_F{}^2}\right) + \frac{\mu_{-5}}{\mu_{-3}}\left(E_X{}^2 + E_Y{}^2 - E_F{}^2 - E_C{}^2\right)\right\}, \tag{14}$$

are equal. The right-hand side of equation (14) has been obtained by substituting equation (2) for the index in the definition of P_{XY} and by using $1/(1 + a) \sim 1 - a$, $a << 1$, to simplify the small terms involving μ_1 and μ_{-1}. Had these small terms been neglected (as in our first-order treatment), the relative partial dispersion would be constants for all glasses, $P_{\mathrm{XY}} = (E_{\mathrm{X}}{}^2 - E_{\mathrm{Y}}{}^2)/(E_{\mathrm{F}}{}^2 - E_{\mathrm{C}}{}^2)$.

Equation (14) can be further transformed to reflect the relationship of the moments to the Abbe number by introducing ratios of moments in which the powers of energy within the various integrals sum to zero. For a distributed absorption, these factors are of the order of unity and vary slowly with compositions. Then,

$$P_{\mathrm{XY}} = \frac{E_{\mathrm{X}}{}^2 - E_{\mathrm{Y}}{}^2}{E_{\mathrm{F}}{}^2 - E_{\mathrm{C}}{}^2}(1 - \alpha_{ionic}\nu^2 + \beta_{electronic}\nu^{-1}), \tag{15}$$

where the coefficients α and β are nearly constant. The IR term is quadratic in ν; the electronic term is hyperbolic. At small Abbe numbers (large dispersion) where electronic effects dominate, the magnitude of the partial dispersion decreases rapidly with ν. In contrast, at large Abbe numbers (small dispersion) where electronic effects are small, the magnitude of the partial dispersion increases quadratically. For practical values of ν, the effective dependence is nearly linear as found by Abbe [1, 3, 5, 6, 19]. Fits of equation (15) to the observed P_{XY} are shown in figures 7 and 8. The curves for silicate and borate glasses are distinct since the ratio of IR and UV absorption strengths enters α_{ionic}.

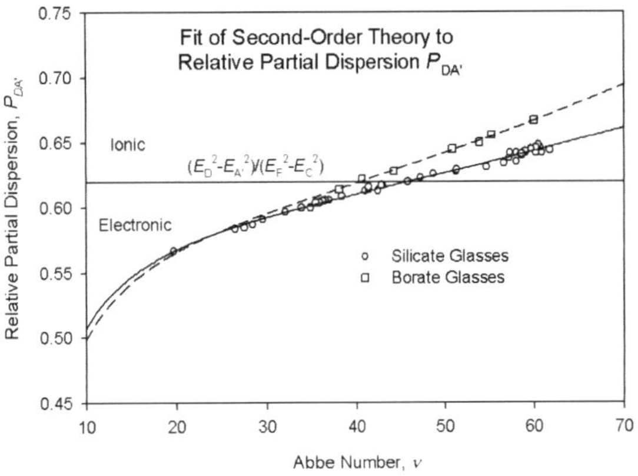

Figure 7. Relative partial dispersion in the far red.

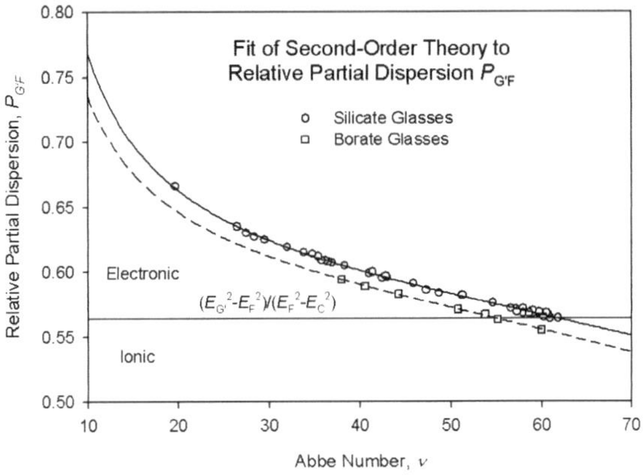

Figure 8. Relative partial dispersion in the violet.

7. Summary

A Laurent-series expansion of the refractive index over the region of transparency of optical materials provides a systematic means of treating dispersive phenomena in terms of moments of the material's absorption spectra. In first order the empirical linear relation between index and dispersion within glass families follows from the approximate additivity of the electronic absorptions of glass former and glass modifier. Partial dispersion requires a second-order treatment in which both vibrational absorption in the IR and higher-order electronic terms in the UV are considered.

Acknowledgments

This study was supported by the EMPOM Gift Fund of the University of Vermont, and the US Department of Energy, Office of Nuclear Physics, under contract DE-AC02-06CH11357.

References

[1] Hovestadt H 1902 *Jena Glass* (London: Macmillan)
[2] Morey G W 1954 *The Properties of Glass* (New York: Reinhold)
[3] Herzberger M 1958 *Modern Geometrical Optics* (New York: Interscience)
[4] Kingslake R 1989 *A History of the Photographic Lens* (San Diego: Academic)
[5] Vogel W 1991 Optical Properties of Oxide Glasses in *Optical Properties of Glass* ed D R Uhlmann and N J Kreidl (Westerville, OH: American Ceramics Society) pp. 1-36
[6] Efimov A M 1995 Optical Constants of Inorganic Glasses (Boca Raton, FL: CRC Press)
[7] Wright F E 1920 *J. Opt. Soc. Am.* **4** 148-59
[8] Nutting P G 1922 *J. Opt. Soc. Am.* **6** 109-11
[9] Smith D Y, Inokuti M and Karstens W 2001 *J. Phys.: Condens. Matter* **13** 3883-93
[10] Smith D Y, Inokuti M and Karstens W 2002 *Rad. Effects & Defects in Solids* **157** 823-28
[11] Herzberger M 1963 *J. Opt. Soc. Am.* **53** 661-71
[12] Philipp H R 1985 Silicon Dioxide (SiO_2) (Glass) in *Handbook of Optical Constants of Solids* vol 1 ed E Palik (New York: Academic) pp 749-63
[13] Smith D Y, Shiles E and Inokuti M 2005 *phys. stat. sol. (c)* **2** 310-13
[14] Smith D Y, Black C E, Homes C C and Shiles E 2007 *phys. stat. sol. (c)* **4** 838-42
[15] Kingslake R and Conrady H G 1937 *J. Opt. Soc. Am.* **27** 257-62
[16] Black C E, Karstens W and Smith D Y 2005 *Bull. Am. Phys. Soc.* **50** 1413
[17] Herzberger M 1962 *J. Opt. Soc. Am.* **52**, 420-27
[18] Schulman J and Compton W D 1962 *Color Centers in Solids* (New York: Macmillan)
[19] Izumitani T and Nakagawa K 1965 Comptes Rendus VIIe Congres International du Verre (Bruxelles, 28 Juin-3Juillet 1965) (Bruxelles: Institut National du Verre) paper 5.

First principles study of electronic and optical properties of the chromium doped BGO crystal

A F Lima and M V Lalic[1]

Universidade Federal de Sergipe, Departamento de Fisica, P.O.Box 353, 49100-000 Sao Cristovao – SE, Brasil

E-mail: mlalic@fisica.ufs.br

Abstract. Ab-initio calculations based on density functional theory have been employed to study electronic properties of Cr doped $Bi_4Ge_3O_{12}$ (BGO), as well as its optical absorption spectra in ultraviolet region. Two different situations have been analyzed: when the Cr is situated at the Ge position (Cr^{4+}), and when it resides at the Bi site (Cr^{3+}). For each case the band structure and absorption spectra were calculated, and the Cr contributions identified and interpreted. Comparison with the experimental data indicates that the Cr enters both the Ge and Bi sites during the doping.

1. Introduction

The bismuth ortho-germanate $Bi_4Ge_3O_{12}$ (BGO) has been the focus of numerous studies due to its excellent scintillation properties suitable for technological applications. As a pure crystal it is utilized as high-efficient scintillator in scientific research, as well as in non-linear optical devices and nuclear medicine [1,2]. When doped with Cr, the BGO exhibits enhanced photorefractive response [3] and shows potential to be used as a near infrared (NIR) solid state laser host [4]. Several experimental studies have been performed on this system so far, focusing its optical properties and the possible Cr accommodation within the lattice. It was not unambiguously determined which position the Cr impurity occupies in the BGO: the octahedral Bi^{3+} site (with trigonal symmetry) or tetrahedral Ge^{4+} site (tetrahedron symmetry). The EPR study of Bravo et al [4] indicates the Cr presence in both sites, with the approximately the same concentration of the Cr^{3+} and Cr^{4+} ions. Using the same technique, Chernei et al [5] concluded that the Cr should reside at the Ge position, and the same conclusion has been drawn from the optical absorption study of Mello et al. [6].

The objective of the present study is to calculate the optical characteristics of the Cr doped BGO in the ultra-violet region, as well as to discuss the possible Cr accommodation within its crystal lattice. We utilized the first-principles, density-functional theory based calculations in order to analyze the electronic and some optical properties of the BGO doped with substitutional Cr impurities situated either at the Ge^{4+} or at the Bi^{3+} site. By comparing these properties with the ones recently calculated for the pure BGO crystal [7] we were able to extract the effects caused by the Cr presence. The results of our study indicate that the Cr should be present at the both Bi and Ge sites in the BGO, thus supporting the conclusions of the experimental work of Bravo et al [4].

[1] Corresponding author

Journal of Physics: Conference Series **249** (2010) 012035 doi:10.1088/1742-6596/249/1/012035

2. Calculations

The BGO crystallizes in cubic symmetry, space group *I-43d*. Primitive unit cell contains two formula units (38 atoms) and does not have a center of inversion. Each Bi ion is surrounded by six oxygens arranged in a strongly distorted octahedron: three of them are situated nearer to the Bi than the other three. The Ge atoms are surrounded by four oxygens, all at the same distance, arranged in the vertices of a perfect tetrahedron.

In this work we have treated two systems: the BGO with the Cr substituting the Bi atom (BGO:Cr→Bi case) and the BGO with the Cr residing at the Ge position (BGO:Cr→Ge case). The doped systems were simulated using the primitive unit cell of the pure BGO from which the central Bi (Ge) atom has been removed and replaced by the Cr atom. The crystal is then generated by infinite repetition of this unit cell. Therefore, we have calculated doped systems with Cr concentration of the 1/38 (2.6%).

The spin-polarized, self-consistent band-structure calculations were performed by density-functional theory [8] based, full potential linear augmented plane wave (FP-LAPW) method [9] as embodied in WIEN2k computer code [10]. In this method, the electronic wave functions, charge density and crystal potential are expanded in spherical harmonics inside the non-overlapping spheres centered at each nuclear position (atomic spheres), and in plane waves in the rest of the space (interstitial region). The choice for the atomic sphere radii (in atomic units) was 2.3 for the Bi, 1.8 for the Ge, 1.45 for the O and 1.8 for the Cr atom. Inside atomic spheres the partial waves were expanded up to $l_{max}=6$, while the number of plane waves in the interstitial was limited by the cut-off at $K_{max}=7.0/R_{MT}$. The augmented plane waves were utilized as a basis set. The charge density was Fourier expanded up to $G_{max}=14$. For the k-space integration a mesh of 12 k-points in the irreducible part of the Brillouin zone was used. Exchange and correlation effects were treated by generalized-gradient approximation (GGA96) [11]. The Bi 5d, 6s, 6p, the O 2s, 2p, the Ge 3d, 4s, 4p and the Cr 3s, 3p, 3d and 4s electronic states were considered as valence ones, and treated within the scalar-relativistic approach, whereas the core states were relaxed in a fully relativistic manner.

The crystal structure of the doped systems was computationally optimized: all atomic positions around the impurity were allowed to relax by total energy minimization, while the optimized unit cell parameter was taken from the recently performed theoretical study of the pure BGO [7].

The optical response of the Cr doped BGO is determined by calculating its complex dielectric tensor ε. Imaginary part of this tensor is directly proportional to the optical absorption spectrum of the material. It can be computed from knowledge of the electronic band structure. In the limit of linear optics, neglecting electron polarization effects and within the frame of random phase approximation, the expression for the imaginary part of ε is the following [12]:

$$\mathrm{Im}\,\varepsilon_{\alpha\beta}(\omega)=\frac{4\pi^2e^2}{m^2\omega^2}\sum_{i,f}\int_{BZ}\frac{2dk}{(2\pi)^3}\left|\langle\varphi_{fk}|P_\beta|\varphi_{ik}\rangle\right|\left|\langle\varphi_{fk}|P_\alpha|\varphi_{ik}\rangle\right|\cdot\delta\left(E_f(k)-E_i(k)-\hbar\omega\right) \quad (1)$$

for a vertical transition from a filled initial state $|\varphi_{ik}\rangle$ of energy $E_i(k)$ to an empty final state $|\varphi_{fk}\rangle$ of energy $E_f(k)$ with the same wave vector k. ω is the frequency of the incident radiation, m the electron mass, P the momentum operator, and α and β stand for the projections x,y,z.

We computed the Im(ε) up to incident radiation energy of ħω = 40 eV, with a mesh of 55 (BGO:Cr→Ge case) and 68 k-points (BGO:Cr→Bi case) in the irreducible wedge of the first Brillouin zone. Owing to the cubic symmetry the dielectric tensor is diagonal, with $\varepsilon_{xx}=\varepsilon_{yy}=\varepsilon_{zz}=\varepsilon$. Thus, in this particular case, it is reduced to scalar function ε(ω).

Journal of Physics: Conference Series **249** (2010) 012035 doi:10.1088/1742-6596/249/1/012035

3. Results and the discussion

3.1. Electronic structure

In the figure 1 it is presented the total density of states (TDOS) for the BGO:Cr→Ge system.

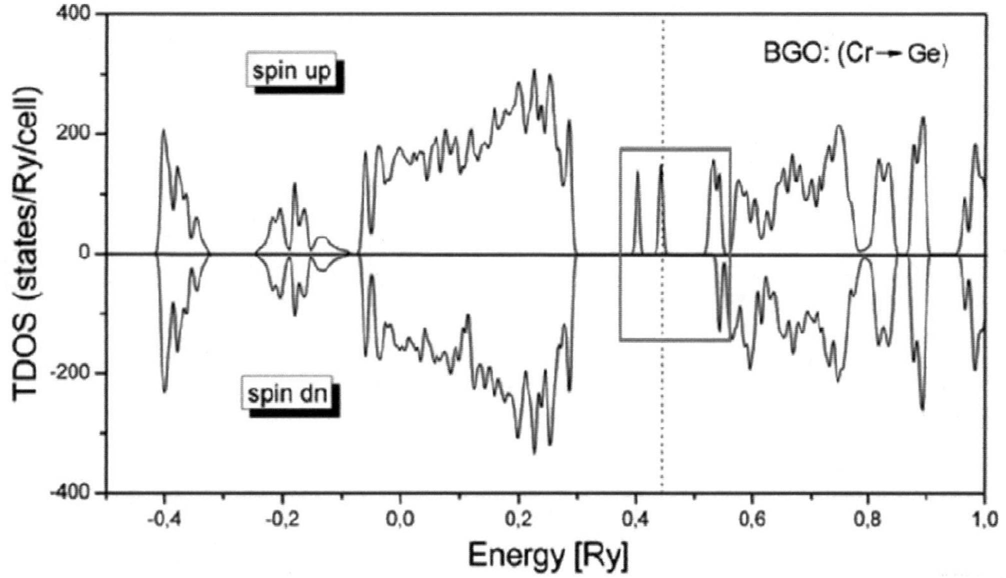

Figure 1: Total spin-polarized electronic density of states of the BGO doped with the Cr residing at the Ge position. The dashed line denotes Fermi level.

The basic difference between the TDOS of the doped BGO and the TDOS of the pure BGO (ref. [7]) consists of a presence of the bands situated within the fundamental gap of the pure system. These bands are emphasized by the square in the Fig. 1. When the Cr substitutes the Ge, it introduces two deep and one shallow band within the gap (the last is superposition of the spin up and down states). Deep bands are populated, and the shallow one is empty. The deep band of lower energy consists of the Cr d_z^2 states, where the z-axis is a symmetry axis of the tetrahedral surrounding of the Cr. The deep band of higher energy is superposition of the Cr d_{xy} and d_{x2-y2} states, while the shallow band is formed from the Cr d_{xz} and d_{yz} spin-up states and the d_z^2 spin-down states. All the Cr d-states within the gap are strongly hybridized with the p-states of nearest neighbor oxygen's, a fact which causes the broadening of bands. The majority states are the spin-up states, and the calculated magnetic moment of the Cr atom is $+1.58\mu_B$.

In the figure 2 it is presented the total density of states (TDOS) for the BGO:Cr→Bi system.

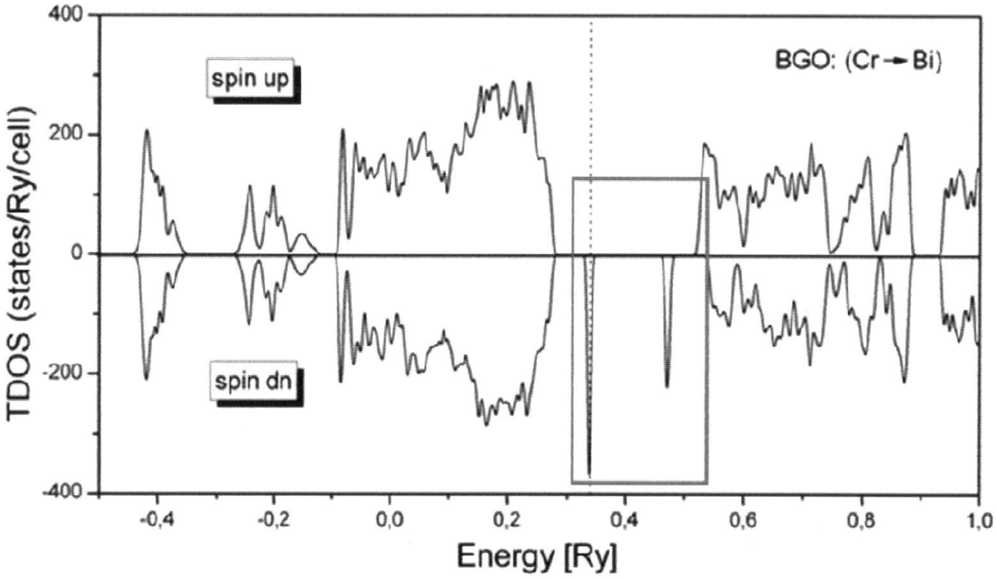

Figure 2: Total spin-polarized electronic density of states of the BGO doped with the Cr residing at the Bi position. The dashed line denotes Fermi level.

In the case when the Cr substitutes the Bi, it introduces two shallow bands within the gap and positions its states at the bottom of the conduction band (emphasized by the square in the Fig. 2). Both shallow bands are composed of the superposition of the Cr spin-down d_{xy} ,d_{x2-y2}, d_{xz} and d_{yz} states, strongly hybridized with the neighboring O's p-states. The spin-up states at the bottom of the conduction band have the dominant Cr d_z^2 character. Majority states are the spin-down states, and the calculated magnetic moment of the Cr atom is -2.44μ_B.

3.2. Optical absorption spectra

Figure 3 shows intensity of the optical absorption of the BGO:Cr→Ge system as a function of the wavelength of the incident light. In the region above the absorption threshold of the pure BGO there are three prominent peaks (structures) centred at the 355, 710 and 1080 nm. All the three are caused by the Cr presence. A more detailed analysis [13] shows that the 355 nm band originates from the electronic transitions from O-p (valence band top) to the Cr-d states (shallow band within a gap). The 710 nm band is dominated by the transitions from the Cr-d states (deep band of lower energy) to the Cr-d (shallow band) and the O-p states (conduction band bottom), while the 1080 nm band is due to Cr-d (deep band of higher energy) to Cr-d states (shallow band). It should be stressed that the d-d transitions are not forbidden owing to the fact that the BGO unit cell does not possess the inversion centre.

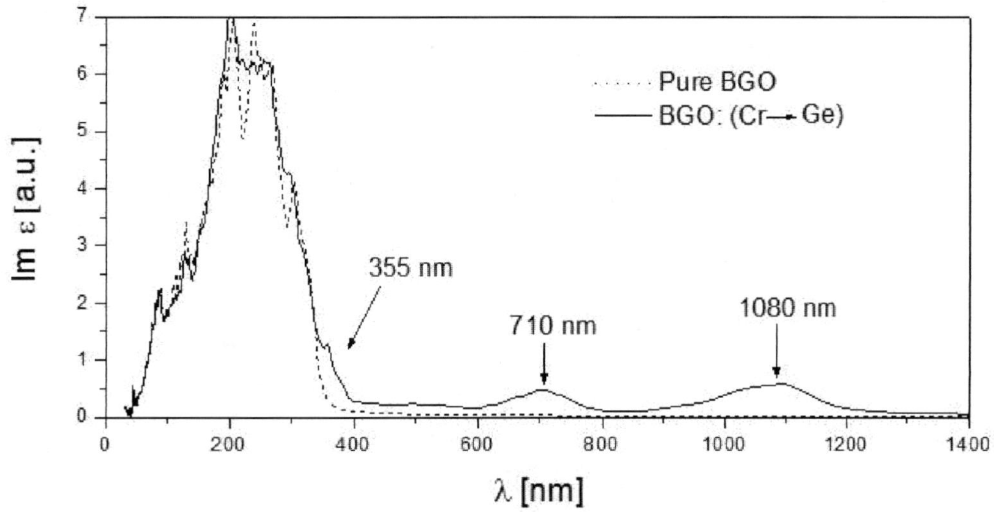

Figure 3: Optical absorption spectrum of the BGO:Cr→Ge system (in arbitrary units). The dashed curve represents the optical absorption of the pure BGO.

The absorption spectrum of the BGO doped with the Cr residing at the Bi site is shown in the figure 4. It is seen that the Cr presence induces three absorption bands above the absorption threshold of the pure BGO. The first two, centred at the 380 and 420 nm, originate from the electron transitions from the O-p states (valence band top) to the Cr-d states (empty shallow band within a gap). The 680 nm band is dominated by the transitions between the Cr-d states of different symmetry (from semi-populated shallow band to the empty shallow band).

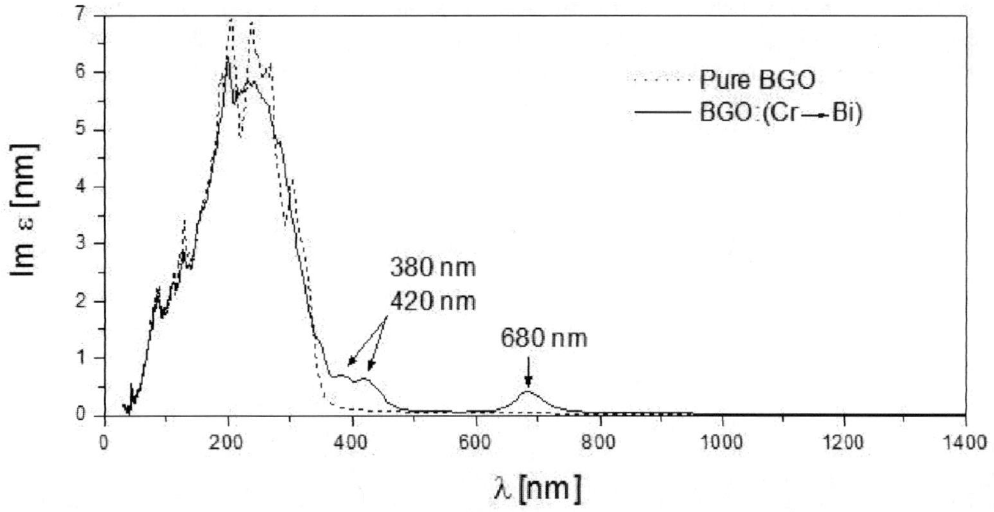

Figure 4: Optical absorption spectrum of the BGO:Cr→Bi system (in arbitrary units). The dashed curve represents the optical absorption of the pure BGO.

The experimentally recorded absorption spectrum of the Cr doped BGO at ambient temperature is characterized by the 6 peaks above the absorption threshold of the pure BGO [5]. Three of them lie in the range of 300-600 nm, relatively near the absorption threshold (which is about 300 nm), and the other three are positioned between the 600 and 900 nm. All the peaks are found to be very broad.

Tempting to directly compare the theoretical and experimental absorption spectra of the Cr doped BGO one should be cautious. The theoretical spectra refer to the ground state of the system, i.e. strictly speaking to the T=0K. Besides, the calculated band gap of the pure BGO is underestimated [7] due to the well known deficiency of DFT to correctly describe exchange and correlation effects. Thus, the calculated peak positions cannot be directly comparable with the experiment. On the other side, the electronic band structure is not expected to change very much from T=0 to T=300K if there is no phase transition in this temperature range. Since the absorption spectrum depends crucially on this band structure, it is also not expected to be much different at the ambient temperature. Therefore, one could trust that at ambient temperature the calculated absorption spectra would be more or less the same as presented in the figures 3 and 4, with eventual slight dislocation and rearrangement of the peaks. The superposition of the two Cr optical responses, from the Ge and the Bi site, would result in the 6 peaks above the absorption threshold. Keeping this in mind, we conclude that our calculations indicate that the experimental spectrum shown in the ref. [5] is probably composed of the optical response of the Cr residing at the both Ge and Bi positions, supporting thus the conclusion of the experimental work of Bravo et al [4].

4. Conclusions

We have analyzed theoretically the electronic and optical properties of the Cr doped BGO scintillator, taking into account the fact that Cr can substitute either Ge or Bi atom in the BGO lattice. As a computation tool it was employed the DFT based, full potential linear augmented plane wave method. The results show that the Cr modifies the electronic structure of the pure BGO introducing its states within the band gap: two deep and one shallow band if it resides at the Ge position and two shallow bands if it resides at the Bi position. In the first case the Cr magnetic moment is calculated to be $+1.58\mu_B$, while for the second $-2.44\mu_B$. Thus, in the external magnetic field the Cr moments are expected to align oppositely. The optical absorption spectra were calculated as functions of the incident light energy up to 40 eV. The absorption peaks caused by the Cr presence were identified and interpreted in terms of calculated electronic structure, for both possible positions of the Cr impurity. Comparison with the experimental data indicate that the Cr doped BGO should contain the Cr atoms in both Ge and Bi substitution sites.

Acknowledgements

The authors gratefully acknowledge the CNPq and FAPITEC-SE (Brazilian funding agencies) for financial support.

References

[1] Auffray E et al 2002 Nucl. *Instrum. Methods A* **486** 22
[2] Classe J M et al 2005 J. *Nuclear Med.* **46** 395
[3] Moya E, Contreras L and Zaldo C 1998 *J. Opt. Soc. Am. B* **5** 1737
[4] D. Bravo D and Lopez F J 1999 *Optical Materials* **13** 141
[5] Chernei N V et al 2005 J. *Struct. Chem.* **46** 431
[6] de Mello A C S et al 2007 *Phys. Stat. Sol. C* No. **3**, 980
[7] Lalic M V and Souza S O 2008 *Optical Materials* **30** 1189
[8] Hohenberg P and Kohn W 1964 *Phys. Rev.* **136** B864; Kohn W and Sham L J 1965 *Phys. Rev.* **140** A1133

[9] Andersen O K 1975 *Phys. Rev. B* **12** 3060; Singh D J 1994 *Plane Waves, Pseudopotentials and the LAPW method* (Dodrecht: Kluwer Academic)

[10] Blaha P, Schwarz K, Madsen G K H, Kvasnicka D and Luitz J 2001 ***WIEN2k****, An Augmented Plane Wave + Local Orbitals Program for Calculating Crystal Properties* (Karlheinz Schwarz, Techn. Universität Wien, Austria)

[11] Perdew J P, Burke S and Ernzerhof M 1996 *Phys. Rev. Letters* **77** 3865

[12] Lynch W D 1985 *Handbook of Optical Constants of Solids* (Palik E D, Ac. Press, New York)

[13] Lima A F and Lalic M V, to be published

Effects of transition metal impurities in alpha alumina: a theoretical study

J M Dantas, A F Lima and M V Lalic[1]

Universidade Federal de Sergipe, Departamento de Física, P.O.Box 353, 49100-000 São Cristovão – SE, Brazil

E-mail: mlalic@fisica.ufs.br

Abstract. Ab-initio calculations based on density functional theory have been employed to study the electronic and some optical properties of Y and Nb doped α-Al_2O_3 with corundum structure. The Nd presence elevates the static dielectric constant of the pure alumina, but changes significantly its band offset. Doping with Y preserves the band gap value and increases slightly the dielectric constant of the pure alumina, thus showing a potential to be used in semiconductor industry.

1. Introduction

In recent years the semiconductor industry has been searching for a material which could substitute the silicon dioxide (SiO_2) in order to construct a new type of dielectric gate in transistors [1,2]. The new material should meet two fundamental criteria: it needs to possess a wide band gap and a high dielectric constant at the same time. Various studies indicate the alumina (Al_2O_3) as one of the most promising candidates for such a substitution [3]: it has a wide band gap, similar band offset as the SiO_2 and high dielectric constant ($\kappa = 9$), which can be elevated even more by appropriate doping of the pure material.

In 2002 the theoretical study of Haverty et al [4] showed that a presence of certain transition metals (Y, Sc) in the alumina with orthorhombic structure (κ-Al_2O_3) maintains the gap value and increases the dielectric constant, while the presence of others (Zr, Nb) diminishes the gap. In 2003 these theoretical predictions have been confirmed by Lee et al [5], who studied experimentally the amorphous alumina doped with 6,25 % of Y, Sc, Zr and Nb.

The objective of the present work is to verify if the presence of the same impurities induces the same effects in the α-Al_2O_3 with corundum structure, which is a common material with a large scale of applicability. For this purpose we calculated the electronic and some optical properties of α-Al_2O_3 doped with the Y (as a representative of the first group of impurities) and with the Nb (representing the second group of impurities). The principal goal is to estimate if the doped materials satisfy the fundamental criteria of semiconductor industry or not.

---

[1] Corresponding author

2. Calculations

The α-Al_2O_3 crystallizes in a corundum structure, with the hexagonal Bravais lattice and the space group R-3C (167). The unit cell contains two molecular formulas, i.e. 10 atoms. The local coordination of the Al consists of 6 oxygen atoms arranged in the vertices of distorted octahedron [6].

The pure alumina has been modeled by infinite repetition of a supercell comprising the 8 original cells (80 atoms). This has been done in order to compare the calculated properties with the doped systems, and to be able to extract the impurity effects. The doped alumina crystals were simulated by substitution of one Al in the center of pure alumina supercell by Y and Nb respectively.

The self-consistent band-structure calculations were performed by density-functional theory [7] based, full potential linear augmented plane wave (FP-LAPW) method [8] as embodied in WIEN2k computer code [9]. In this method, the electronic wave functions, charge density and crystal potential are expanded in spherical harmonics inside the non-overlapping spheres centered at each nuclear position (atomic spheres), and in plane waves in the rest of the space (interstitial region). The choice for the atomic sphere radii (in atomic units) was 1.8 for the Al, Y and Nb, and 1.5 for O. Inside atomic spheres the partial waves were expanded up to $l_{max}=6$, while the number of plane waves in the interstitial was limited by the cut-off at $K_{max}=7.5/R_{MT}$. The augmented plane waves were utilized as a basis set. The charge density was Fourier expanded up to $G_{max}=14$. For the k-space integration a mesh of 6 k-points in the irreducible part of the Brillouin zone was used. Exchange and correlation effects were treated by generalized-gradient approximation (GGA96) [10]. The Al 2p, 3s and 3p, the O 2s and 2p, the Nb 4s, 4p, 4d and 5s, and the Y 4s, 4p, 4d and 5s electronic states were considered as valence ones, and treated within the scalar-relativistic approach, whereas the core states were relaxed in a fully relativistic manner.

The crystal structure of the pure and the doped systems was computationally optimized: all atomic positions around the impurities were allowed to relax until the equilibrium position were reached, while the unit cell parameters were optimized only for the pure crystal and kept unchanged for the doped systems.

The optical response of the studied systems is determined by calculating their complex dielectric tensors ε. Imaginary part of this tensor is directly proportional to the optical absorption spectrum of the materials. It can be computed from knowledge of the electronic band structure. In the limit of linear optics, neglecting electron polarization effects and within the frame of random phase approximation, the expression for the imaginary part of ε is the following [11]:

$$\Im m\,\varepsilon_{\alpha\beta}(\omega)=\frac{4\pi^2e^2}{m^2\omega^2}\sum_{i,f}\int_{BZ}\frac{2dk}{(2\pi)^3}\left|\langle\varphi_{fk}|P_\beta|\varphi_{ik}\rangle\right|\left|\langle\varphi_{fk}|P_\alpha|\varphi_{ik}\rangle\right|\cdot\delta(E_f(k)-E_i(k)-\hbar\omega) \quad (1)$$

for a vertical transition from a filled initial state $|\varphi_{ik}\rangle$ of energy $E_i(k)$ to an empty final state $|\varphi_{fk}\rangle$ of energy $E_f(k)$ with the same wave vector k. ω is the frequency of the incident radiation, m the electron mass, P the momentum operator, and α and β stand for the projections x,y,z.

We computed the Im(ε) up to incident radiation energy of ħω = 40 eV. The real part of dielectric tensor is then determined using Kramers-Kroning relation. Both real and imaginary parts of ε were calculated with a mesh of 38 k-points in the irreducible wedge of the first Brillouin zone, for all three studied systems. Owing to the hexagonal symmetry the dielectric tensor is diagonal, with two different values: ε_{zz} (along the c-axis of the crystal) and $\varepsilon_{xx}=\varepsilon_{yy}$ (in the plane perpendicular to the hexagonal c-axis).

3. Results and discussions

3.1. Electronic structure

The figure 1 shows the calculated total electronic density of states (TDOS) of the pure and doped alumina in the energy interval around the Fermi level. The electronic structure of the pure Al_2O_3 is characterized by 3 blocks of electronic states, two of them in the valence and one in the conduction region. A more detailed analysis [12] shows that the bands in the valence region are dominated by the oxygen's states: s- (lower energy block) and p- (higher energy block). The conduction band is composed of the hybridized s- and p- states of Al and O. The band gap value is calculated to be 6.0 eV, underestimated in comparison with the experimental value (8.7 eV [13]) due to approximate treatment of exchange and correlation effects (GGA) in the practical application of DFT.

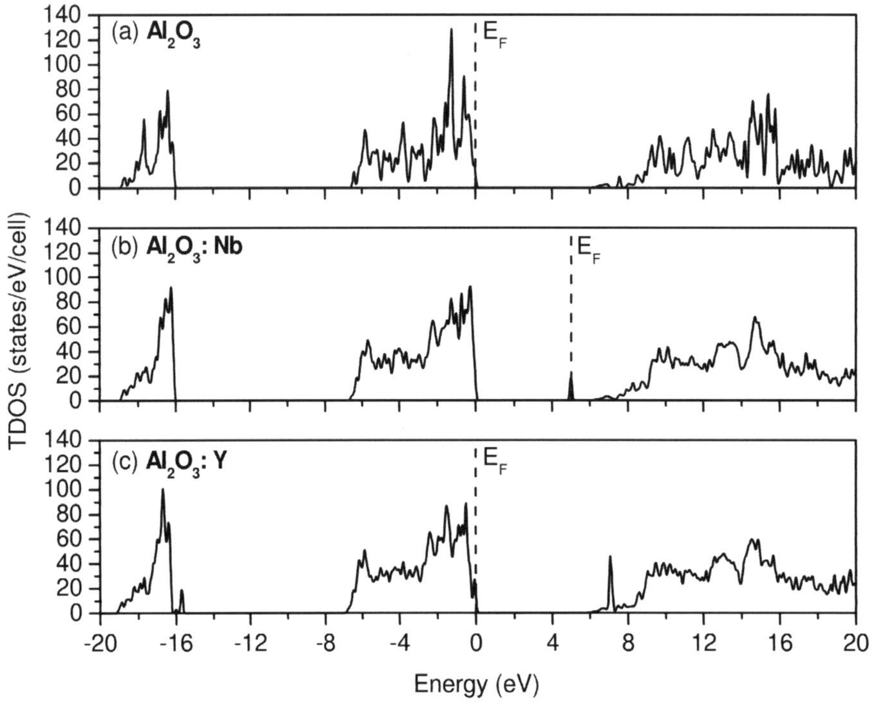

Figure 1: Total density of electronic states of (a) pure Al_2O_3, (b) Al_2O_3 doped with Nb and (c) Al_2O_3 doped with Y. Dashed lines denote Fermi levels.

Insertion of the Nb into the Al_2O_3 changes significantly its band structure (fig. 1b). The calculated band gap value is 4.8 eV, diminished in comparison with the gap value of the pure alumina. The Nb introduces one shallow band within the gap and one broad band at the bottom of the conduction band. These bands are composed of the Nd d-states, strongly hybridized with the nearest neighbour O's p-states [12]. The shallow band is situated approximately 1 eV below the bottom of the conduction band.

In the case when the Al_2O_3 is doped with Y, its electronic structure is much less changed (fig. 1c). The calculated gap value is practically the same as for the pure alumina (~6.0 eV), in accordance with the previously published theoretical value [14]. The Y introduces one sharp band situated at the very bottom of the conduction band. It is composed of Y d-states hybridized with the neighbouring O's p-states [12].

3.2. Dielectric constants

Figure 2 shows the calculated real part of dielectric tensor of the three studied alumina systems along the x- and z-directions.

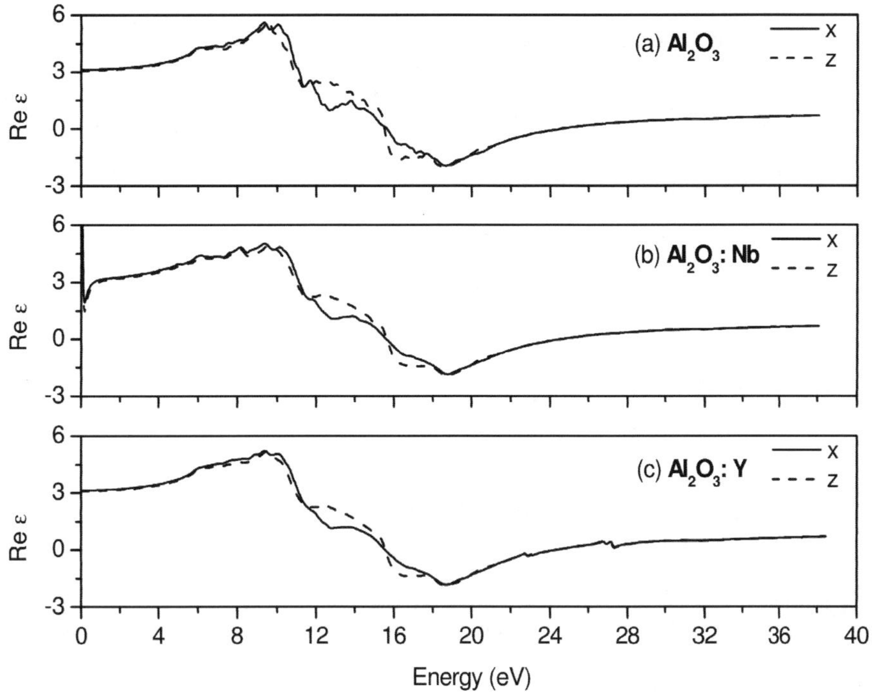

Figure 2: The real part of the dielectric tensor along the x- and z-axis of (a) pure Al_2O_3, (b) Al_2O_3 doped with Nb and (c) Al_2O_3 doped with Y.

The static dielectric constant ε(0) is calculated as a zero energy limit of the real part of dielectric tensor. For the pure Al_2O_3 the $\varepsilon_x(0)$ along the x-axis is calculated to be 3.12, in satisfactory agreement with the experimental value of 3.27 [6]. For the Y doped alumina $\varepsilon_x(0)$ is calculated to be 3.15, while for the Nb doped alumina 12.92. The similar conclusions were found for the static dielectric constant $\varepsilon_z(0)$ along the z-axis. For the pure alumina we calculated $\varepsilon_z(0)=3.08$, in good agreement with the already published value of 2.98 [6]. For the Y and Nb doped alumina $\varepsilon_z(0)$ is calculated to be 3.11 and 15.84 respectively.

The results show that the alumina exhibits a slight increase of its static dielectric constant when doped with Y, while this increase is drastic in the case of doping with the Nb. The latter originates from the intra-band electronic transitions between the populated and the empty part of the shallow band within the gap (Nb d-electrons are transferred to O-p states), and it is highly exaggerated by the approximate form of the formula (1).

4. Conclusions

We performed a theoretical study of the Y and Nb doped α-Al_2O_3 with corundum structure. The electronic and optical properties of the pure and doped systems were calculated on the basis of the first-principles, density functional theory based full potential augmented plane wave method. The principal goal was to verify if the doped materials satisfy some basic conditions for application in semiconductor industry, concerning their band gap values and the static dielectric constants.

The results of our study strongly indicate that the Nb presence elevates the dielectric constant value, but shrinks significantly the band gap of the pure alumina. Consequently, this system does not attend the semiconductor industry demands due to its low band gap value. On the other hand, the Y presence does not alter the band gap value and increases the dielectric constant of the alumina at the same time. This increase is rather small due to a low concentration of impurities in our computational model (1/80 or 1.25% versus 6.25 % used in experiment with κ-Al_2O_3 [5]). It, however, indicates that the α-Al_2O_3 doped with the Y has a potential to serve for a construction of new generation of dielectric gates in transistors.

Acknowledgements
The authors gratefully acknowledge the CNPq and FAPITEC-SE (Brazilian funding agencies) for financial support.

References

[1] Kingon A I, Maria J P and Streiffer S K 2000 *Nature* **406** 1032
[2] Gritsenko V A et al 2003 *Solid State Electron.* **47** 1651
[3] Robertson J 2000 *J. Vac. Sci. Technol. B* **18** 1785
[4] Haverty M et al 2002 *Appl. Phys. Lett.* **80** 2669
[5] Lee J J et al 2003 *Appl. Phys. Lett.* **83** 5226
[6] Shang-Di MO and Ching W Y 1998 *Phys. Rev. B* **57** 15219
[7] Hohenberg P and Kohn W 1964 *Phys. Rev.* **136** B864; Kohn W and Sham L J 1965 *Phys. Rev.* **140** A1133
[8] Andersen O K 1975 *Phys. Rev. B* **12** 3060; Singh D J 1994 *Plane Waves, Pseudopotentials and the LAPW method* (Dodrecht: Kluwer Academic)
[9] Blaha P, Schwarz K, Madsen G K H, Kvasnicka D and Luitz J 2001 ***WIEN2k****, An Augmented Plane Wave + Local Orbitals Program for Calculating Crystal Properties* (Karlheinz Schwarz, Techn. Universität Wien, Austria)
[10] Perdew J P, Burke S and Ernzerhof M 1996 *Phys. Rev. Letters* **77** 3865
[11] Lynch W D, *Handbook of Optical Constants of Solids* 1985 (Palik E D, Ac. Press, New York)
[12] Lalic M V, A. F. Lima A F and Dantas J M, to be published.
[13] Oetzel M and Heger G 1999 *J. of Appl. Crystallography* **32** 799
[14] Ching W Y et al 1997 *J. Am. Ceram. Soc.* **80** 3199

Computer simulation and spectroscopic study of inclusion complexes of cyclodextrins with luminescent porphyrins

George R S Andrade,Thiago dos S Rezende, Ledjane S Barreto, Luis E Almeida, Nivan B da Costa Jr and Iara de F Gimenez

Departamento de Química, CCET, Universidade Federal de Sergipe (UFS), Campus Universitário Prof. José Aloísio de Campos, Av. Marechal Rondon s/n, São Cristóvão – SE.

gimenez@ufs.br

Abstract. Here we report a computational study of the structure, thermodynamic and spectroscopic properties of 1:1 and 2:1 inclusion complexes of luminescent porphyrins in β-cyclodextrin. Semiempirical PM6 (Parametric Method 6) calculation allowed the optimization of the structure of the complexes, showing that the inclusion in the CD cavity changes significantly the porphyrin ring planarity for the 2:1 complexes. Thermodynamic calculations evidenced that the inclusion complex formation is slightly endothermic and that it is a non-spontaneous process in the absence of water molecules. Finally the calculated spectra were found to be in very good agreement to previously reported experimental ones.

1. Introduction

Organic and organo-inorganic molecular materials with functional properties are commonly useful as sensitive elements in sensors and smart devices [1]. However the application of functional molecules face some limitations related to the dependence of the desired properties on the solid-state structure, not to mention the limited stability and difficulties in construction of devices [2]. For instance, the application of functional molecules in solid devices it is not possible if those molecules aggregate prior to deposition with loss of property. Thus structural evaluation of potential functional elements is an important step in the study of advanced materials.

A class of compounds useful in sensor and other devices, having spectroscopic, catalytic, and biological interest are the porphyrins, macrocyclic species derived from the porphine ring [3]. Porphyrins are common in nature, playing a significant role in biological processes such as dioxygen transport and activation and electron- and energy transfer [4]. Their biological function depends on the nature of the axial ligands as well as on the presence of a metal ion coordinated to the porphyrin center. Furthermore, the supramolecular approach may be used to assemble different building blocks rendering sophisticated porphyrin supermolecules with important catalytic, photochemical and electrochemical properties. In this context we can mention the tetraruthenated H2(3-TRPyP) porphyrins with luminescent and electrocatalytic properties, which were previously described by Araki and co-workers [5]. Those systems have been obtained by attaching four [RuII(bipy)$_2$Cl] groups to the pyridyl substituents of meso-tetra(4-pyridyl)porphyrin and meso-tetra(3-pyridyl)porphyrin and its metallated derivatives [6]. These compounds display a rich electrochemistry and versatile catalytic and photochemical properties, associated with the ruthenium–bipyridine and metalloporphyrin complexes [7].

In general porphyrins exhibit aggregation behavior in solution, which can be prevented by inclusion complex formation with cyclodextrins [8]. Complex formation takes place by non-covalent interactions, which form the basis of a sustainable approach widely employed in materials preparation due to the low energy cost and waste minimization [9]. Cyclodextrins (CD's) are cyclic oligosaccharides able to form inclusion complexes by the accommodation of nonpolar molecules inside its hydrophobic cavity, in a process that modifies important properties of the guest species [10]. Water solubility, reactivity, stability, and optical properties are among the properties that can be controlled by inclusion in CD´s. One common target of this field is the modulation of spectroscopic properties by CD inclusion, which in this case is the basis of sensing process [11] A range of intermolecular interactions may be present in supramolecular structures, which can be evaluated currently by theoretical methods. Until the 1990's the use of those methods to the study of supramolecular systems has been limited due to the large size of the supermolecules as well as the fact that such systems are often studied in aqueous solution [12]. This has made computational studies very time consuming and force the use of assumptions which were not always realistic. Nowadays novel theoretical methods have been developed allowing the suitable application to large systems such as those studied here [13]. However, significant limitations remain on theoretical studies, especially those related to the explicit consideration of the presence of water molecules owing to the large size of the macromolecular systems [13].

Thus the objective of the present work was to simulate the structure of inclusion complexes of tetraruthenated H2(3-TRPyP) porphyrins with cyclodextrins, as well as the spectroscopic and thermodynamic properties, pointing out the possibility of formation of those species using the semiempirical PM6 computational method. The calculated absorption spectra were compared to experimental data reported previously [5-7]. Previous efforts have been described for device construction employing those porphyrins.

2. Calculation details

All calculation were carried out at the restricted Hartree-Fock level (RHF) by using PM6 [14] semiempirical method implemented in the MOPAC 2007 software. COSMO (Conduct-like Screening Model) [15] solvation model was used (the relative permittivity was set 78.4) in order to take into account the effects of water. β-CD initial geometry was taken from the ''Cambridge Structural Database " (CSD) with R-factor <5%, which indicates the high quality of the single crystal. The initial estimate of the porphyrin and of 1:1 and 2:1 (CD:porphyrin) inclusion complexes structures were generated from MM3 calculation in the Cache Worksystem 6.1. To estimate the charge transfer (induction) in the complexes, the individual net atomic charges obtained in calculation were summed up for estimation of the charge of the molecular fragments: β-CD and porphyrin. The electronic excited state calculations have been performed with the intermediate neglect of differential overlap/spectroscopic parametrization–configuration interaction singly replacements (INDO/S-CIS) method [16] implemented within the ZINDO program. The inclusion complex was held in the positions determined by each method. The line broadening of the calculated single transition energies and the intensities obtained from oscillator strengths were taken into account by fitting them to a Lorentzian line shape function with a half-maximum band width of 20–40 nm, which properly allows the comparison between the calculated and the observed absorption electronic spectra.

3. Results and Discussion

In figure 1 we show the molecular structure for β-CD modeled from crystallographic data and for H2(3-TRPyP). A previous study described the simulation of this structure using MM2 molecular mechanics method which is faster than more rigorous, time-consuming quantum mechanical methods. They employed ruthenium parameters described b y Norrby and co-workers [17] and had to test the method in simpler systems. Here we used PM6 quantum semiempirical method, which originally contains ruthenium parameters, finding a molecular structure for H2(3-TRPyP) that compared well to that described by Araki and co-workers [5]. In this structure, each $Ru(bipy)Cl^{+}$ groups present at opposite corners of the porphyrin ring are oriented above the ring plane while the other two are oriented below the plane.

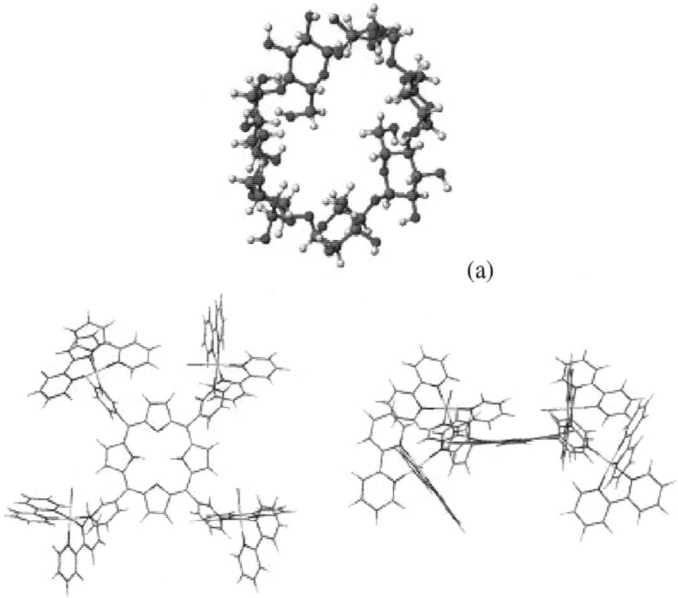

(a) (b)

Figure 1. Minimum-energy structures calculated with PM6 semiempirical method for β-cyclodextrin (a) and for H2(3-TRPyP) porphyrin (b), showing for the latter a front view and a side view from the ring.

Inclusion complexes with 1:1 and 2:1 CD:porphyrin stoichiometry, the latter more commonly observed for CD:porphyrin systems, were simulated and the minimum structures are shown in figure 2. Molecular modeling has shown that formation of both complexes is possible (thermodynamic study will be presented later) and that the porphyrin structure remains unchanged for the 1:1 complex, in contrast to the 2:1 complex for which the structure changes considerably. This can be more properly observed in the superposition of structures of the free and complexed porphyrin in both stoichiometries, figure 3. In the 2:1 complex, the planarity of the porphyrin ring is affected.

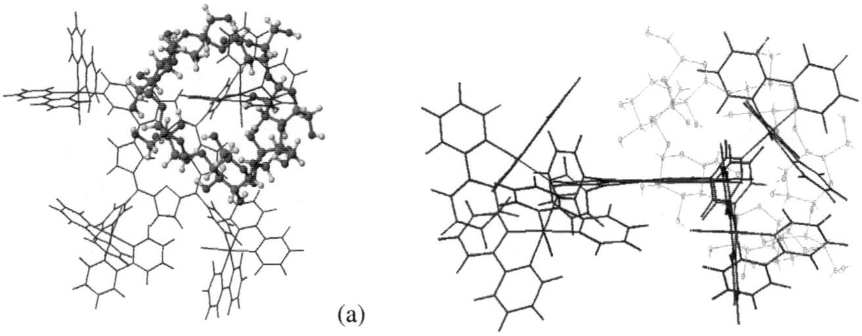

(a) (b)

Figure 2. Front view and view from the porphyrin ring plane

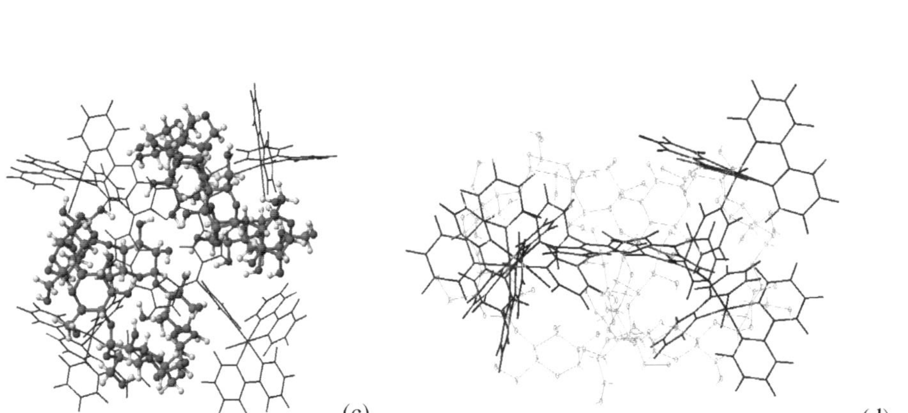

(c) (d)

of the structures of 1:1 (a,b) and 2:1 β-CD: porphyrin inclusion complexes.

In order to better evaluate the structural changes upon complexation the porphyrin structures were superimposed fixing the positions of the hydrogen atoms of the porphyrin ring. Superposition of the porphyrin structures in the 1:1 complex simulated in the vacuum and considering a continuum medium with the water dielectric constant (implicit solvent consideration) shows that the orientation of only one Ru(bpy)Cl+ group changed significantly, with slight changers for other two groups and only one of the remaining unchanged in vacuum. The consideration of the implicit water medium had no effect on the porphyrin conformation.

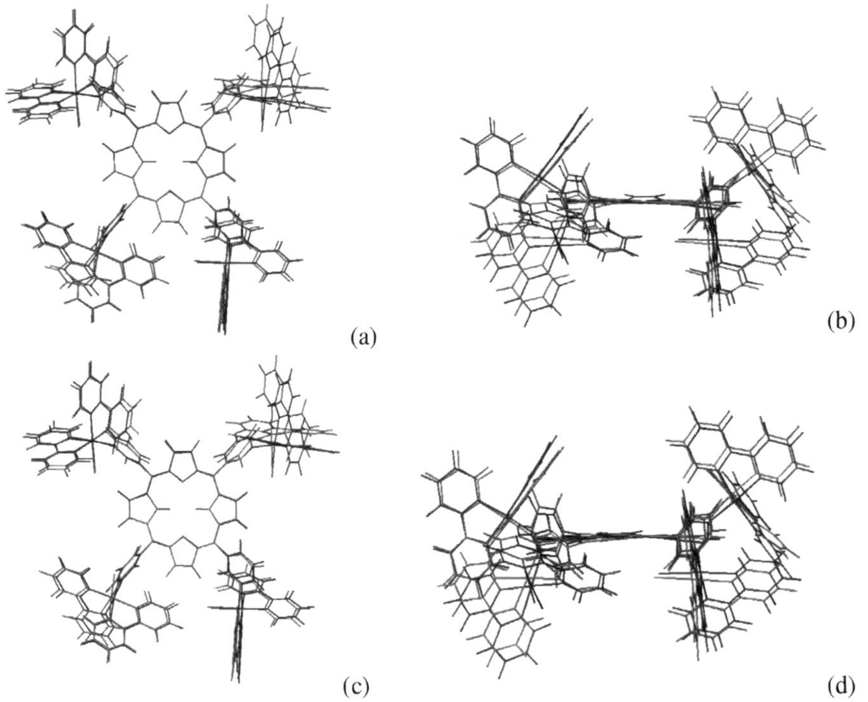

(a) (b) (c) (d)

Figure 3. Front view and view from the porphyrin ring plane of superimposed structures of free (blue) and complexed (pink) porphyrin in 1:1 stoichiometry in vacuum (a,b) and in an continuum aqueous medium (c,d).

When the porphyrin structures are superimposed for 2:1 complex, which is the dominating stoichiometry for porphyrin CD complexes, the orientation of all Ru(bpy)Cl^{+} changed significantly, in addition to a loss of planarity for the porphyrin ring. In this case the implicit solvent consideration was not possible for the complexes due to the high number of atoms. It is worth mentioning that this change is not reflected on the spectroscopic properties, which were preserved as we will see later.

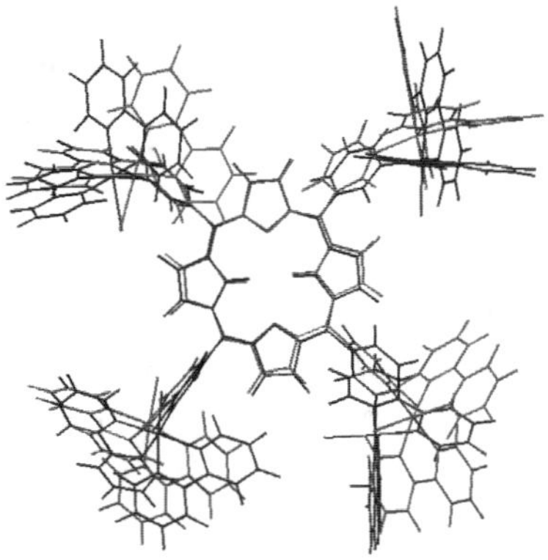

(a)

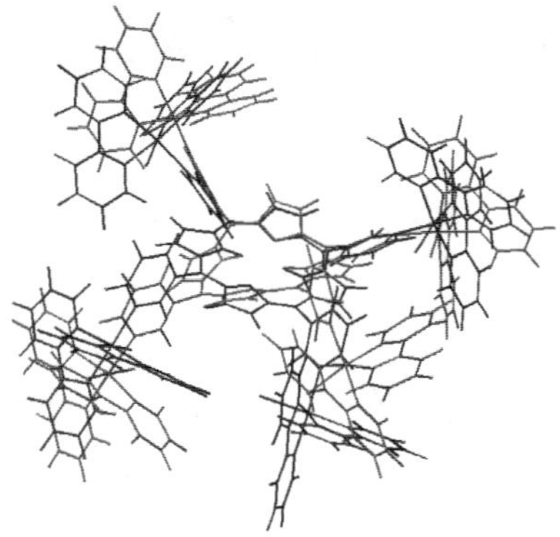

(b)

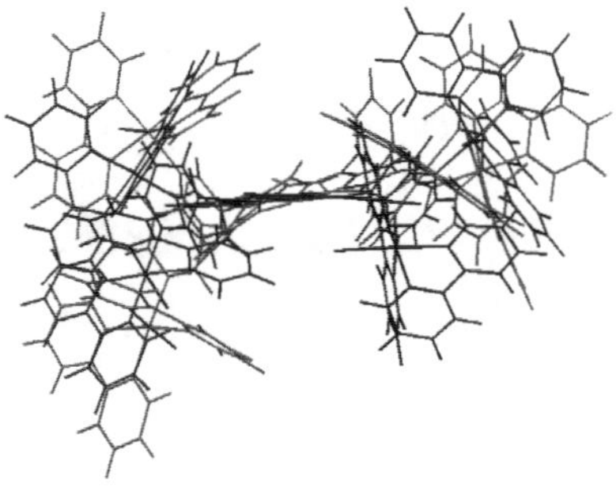

(c)

Figure 4. Front view (a), inclined view (b) and view from the porphyrin ring plane (c) of superimposed structures of free (blue) and complexed (pink) porphyrin in 2:1 stoichiometry in vacuum.

The calculation of thermodynamic functions was also carried out using PM6 method. Table I shows the values of the final charges of the supermolecules after complex formation showing that there is charge transfer interaction between guest and host in the inclusion complex, where the porphyrin transfers charge to CD. This fact contributes for the stability of the inclusion complex. Also, when the calculation is carried out in water, the 1:1 complex has an increase in the porphyrin-CD charge-transfer degree. These observations suggest that the presence of water contributes to the complex formation.

Table I: Final charge* for molecules in the inclusion complex

	H_2(3-TRPyP)	β-CD
In vacuum	+ 4,0259	- 0,0259
COSMO	+ 4,0331	- 0,0331

* PM6 – net atomic charge. The initial charge of the porphyrin is +4.

Thermodynamic functions such as entropy and free energy were calculated in the 273 K to 298 K temperature range for the 1:1 and 2:1 complexes, considering the chemical reactions in vacuum for the optimized structures:

$$\beta\text{-CD} + H_2\text{(3-TRPyP)} \rightarrow \beta\text{-CD} - H_2\text{(3-TRPyP)}$$
$$2\beta\text{-CD} + H_2\text{(3-TRPyP)} \rightarrow [\beta\text{-CD}]_2 - H_2\text{(3-TRPyP)}$$

Table 2 shows that both from ΔS and from ΔG, the inclusion complex formation is non-spontaneous processes in the vacuum, as negative ΔS and positive ΔG values were found at different temperatures for both stoichiometries. However both the charge transfer calculation and solubility increase (not shown in this work) evidence the formation of the inclusion complexes in solution. As positive ΔH values were found, the signal of ΔG, which would determine if the process is spontaneous, would depend on the signal and magnitude of ΔS in the equation ΔG = ΔH – TΔS. The positive ΔH values probably resulted from the positive charges of the porphyrins, since the presence of counterions was not considered in this calculation. Concerning the ΔS, in practice, the main driving force for the formation of inclusion complexes is the entropy increase caused by the release of water molecules originally present in the CD cavity, which were not considered explicitly in this work. The entropy change becomes less negative with increasing temperature, probably reflecting an increasing dynamic disorder of the molecular structure. Thus the calculation of thermodynamic functions pointing out nonspontaneous processes in vacuum reflects the absence of explicit water molecules in the calculation, whose consideration would demand a great deal of computational resource. Also the implicit solvent consideration doesn't yield better results and the explicit water consideration for these large systems is still too resource-consuming.

Table II: Thermodynamic parameters of formation of the inclusion complexes.

Parameter	1:1 complex		2:1 complex	
	273 K	298 K	273 K	298 K
$\Delta_f H$ / kcal mol^{-1}	2.25	2.37	11.60	12.24
ΔS / cal mol^{-1}	-69.84	-69.42	-118.86	-116.62
$\Delta_f G$ / kcal mol^{-1}	21.32	23.06	44.05	46.99

Finally the results of the calculation of absorption spectra for the 1:1 complex in the vacuum and implicit aqueous medium are shown in figure 5.

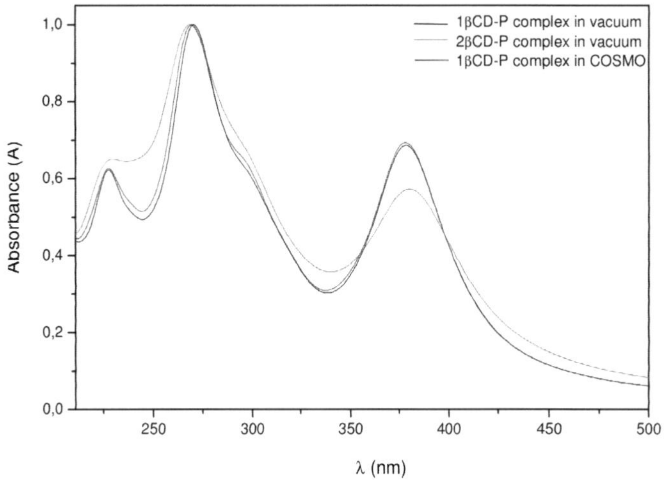

Figure 5. Calculated absorption spectra in vacuum and implicit aqueous medium (black and blue) for the 1:1 and 2:1 in vacuum (red) inclusion complexes.

It can be observed that the consideration of implicit aqueous medium had no effect at all on the spectrum of the 1:1 complex, as it is identical to the spectrum calculated in the vacuum. The band at 270 nm is assigned to a $\pi \rightarrow \pi^*$ transition from bpy ligands; the shoulder at 298 nm to a ligand field transition of RuII ions; the band at 380 nm is the Soret band characteristic of porphyrins. An interesting point is the absence in all spectra of a metal to ligand charge transfer band at 470 nm, observed experimentally by Araki and co-workers [5-7]. A comparison to the spectrum calculated for the 2:1 complex in vacuum shows that the intensity ratio of the $\pi \rightarrow \pi^*$ bpy to the Soret transitions changed slightly probably as a result of the structure distortion. Such changes are very subtle and it can be considered that the spectroscopic properties are practically preserved upon inclusion in CD cavity, evidencing that this process will not cause limitations to potential applications.

4. Conclusions

In conclusion it can be emphasized that the theoretical semiempirical study with PM6 method for the formation of inclusion complexes of the porphyrin H2(3-TRPyP) pointed out that the porphyrin structure is considerable changed in the 2:1 CD:porphyrin stoichiometry, in contrast to the 1:1 stoichiometry. Also the thermodynamic calculation carried out in the vacuum and considering an implicit aqueous medium indicated that the process is nonspontaneous, mainly due to the lack of explicit consideration of the water molecules responsible for the entropic contribution. This is in contrast to experimental evidences (not shown) of successful formation. Calculation of absorption spectra evidenced that for the 2:1 complex, the structural distortion of the porphyrin brought about by encapsulation causes modifications of the spectral profile.

5. Acknowledgements

Authors are grateful to Prof. K. Araki and H. E. Toma from IQ/USP-SP. GRSA and TSR acknowledge CNPq for fellowship.

References

[1] Balzani V, Credi A, Raymo F M and Stoddart J F 2000 Angew Chem Int Ed 39 3348
[2] Chia S, Cao J, Stoddart J F and Zink J I 2001 Angew Chem Int Ed 40 2447
[3] Santos S F, Santos M L, Almeida L E, Costa Jr N B, Gimenez I F, Araki K, Mayer I, Engelmann F M, Toma H E, Barreto L S 2007 J Colloid Interface Sci 305 264
[4] Deng W, Onji T, Yamaguchi H, Ikeda N and Harada A 2006 Chem Commun 4212
[5] Mayer I, Formiga A L B, Engelmann F M, Winnischofer, Oliveira P V, Tomazella D M, Eberlin M E, Toma H E and Araki K 2005 Inorg Chim Acta 385 2629
[6] Araki K and Toma H E 1993 Coord Chem Rev 3009
[7] Toma H E and Araki K 2000 Coord Chem Rev 196 307
[8] Leng X, Choi C-F, Luo H-B, Cheng Y-K and Ng D K P 2007 Org Lett 9 2497
[9] Araujo M V G, Vieira E K B, Lazaro G S, Conegero L S, Ferreira O P, Almeida L E, Barreto L S, Costa Jr N B and; Gimenez I F 2007 Bioorg Med Chem 15 5752
[10] Szejtli J 1998 Chem Rev 98 1743
[11] Leng X, Choi C-F, Lo P-C and Ng D K P 2007 Org Lett 9 231
[12] Lipkowitz K B 1998 Chem Rev 98 1829
[13] Liu L and Guo Q-X 2004 J Inclus Phenomena Macrocyclic Chem 50 95
[14] Stewart J J P 1989 J Comput Chem 10 209
[15] Klamt A and Schüürmann G 1993 J Chem Soc Perkin Trans 2 799
[16] Zerner M C 1990 ZINDO Manual, QTP, University of Florida, Gainesville
[17] Brandt P, Norrby T, Akermark B and Norrby P-O 1998 Inorg Chem 37 4120

Journal of Physics: Conference Series **249** (2010) 012038 doi:10.1088/1742-6596/249/1/012038

Monte Carlos studies of critical and dynamic phenomena in mixed bond Ising model

J. B. Santos-Filho[1], **N. O. Moreno**[1], **Douglas F. de Albuquerque**[2]
[1]Departamento de Física, Universidade Federal de Sergipe, 49100-000, São Cristovão, SE, Brasil
[2]Departamento de Matemática, Universidade Federal de Sergipe, 49100-000, São Cristovão, SE, Brasil

E-mail: nomoreno@fisica.ufs.br

Abstract. The phase transition of a random mixed-bond Ising ferromagnet on a cubic lattice model is studied both numerically and analytically. In this work, we use the Metropolis and Wolff algorithm with histogram technique and finite size scaling theory to simulate the dynamics of the system. We obtained the thermodynamic quantities such as magnetization, susceptibility, and specific heat. Our results were compared with those obtained using a new technique in effective field theory that employs similar probability distribution within the framework of two-site clusters.

1. Introduction

The study of the effects of disorder in magnetic systems has been an object of intense investigations during the last five decades.

The Monte Carlo technique is an useful tool which in many cases, gives better results regarding other methods from analytical approximations. The influence of quenched, random disorder on phase transitions is of great importance in a large variety of fields [1]. For pure systems exhibiting a continuous phase transition, Harris [2] derived the criterion that random disorder is a relevant perturbation when the exponent of the specific heat of the pure system is positive, $\alpha > 0$. In this case one expects that the system falls into a new universality class with critical exponents governed by a disordered fixed point. For $\alpha < 0$ disorder is irrelevant, and in the marginal case $\alpha = 0$ no prediction can be made.

Since for the three-dimensional (3D) Ising model it is well known that $\alpha > 0$, quenched, random disorder should be relevant for this model. In three dimensions (3D) most of the computer simulation studies have concentrated mainly on the site-diluted Ising model [3, 12].

In this work, we study the Ising model with mixed-bond by using of Monte Carlo simulation, applying the algorithm cluster of Wolff [5].

2. Model and Simulation Setup

We study the spin 1/2 ferromagnetic Ising mixed-bond model defined by the following Hamiltonian

$$\beta\mathcal{H} = \sum_{\langle ij \rangle} K_{ij}\sigma_i\sigma_j \qquad (\sigma_i = \pm 1), \tag{1}$$

where the sum extends over all pairs of neighboring sites on a cubic lattice of linear size L with periodic boundary conditions, $\beta = 1/k_BT$ and the exchange couplings K_{ij} are allowed to take two different values $K_{ij} = K \equiv J/k_BT$ and 0. The interactions are assumed to be independent random variables with the probability distribution

$$P(K_{ij}) = p\delta(K_{ij} - K) + (1-p)\delta(K_{ij} - \lambda K), \tag{2}$$

where p is the concentration of magnetic bonds in the system bonds such that $p = 1$ corresponds to the pure case and λ is the competition parameter with $|\lambda| \leq 1$.

The simulations were performed on a set of following lattice sizes L = 10, 16, 20, 26, 30, 36, 40 with periodic boundary conditions. The aim of the first set of simulations is to estimate the critical temperature of the model at different L. Due to the finite-size scaling theory [6], the finite system of linear size L will demonstrate an evidence of a critical behavior at a certain temperature $T_C(L)$ which differs from the critical temperature of the infinite system $T_C(\infty)$ [7].

$$T_C(L) = T_C(\infty) + \alpha L^{-1/v} + ..., \tag{3}$$

where the correction-to-scaling terms have been omitted.

The static thermodynamic quantities of interest include the average magnetization M and the magnetic susceptibility χ

$$M = \frac{1}{n}\sum_{i=1}^{n}\sigma_i, \tag{4}$$

$$\chi = \frac{1}{k_BT}[\langle M^2\rangle - \langle M\rangle^2]. \tag{5}$$

The phase diagram is obtained numerically from the maxima of a diverging quantity. Here we choose the magnetic susceptibility, since the stability of the disordered fixed point implies that the specific heat exponent is negative in the random system [8, 9]. Thus, the error in this quantity is larger than for the susceptibility. To get an accurate determination of the maxima of the susceptibility, we used the histogram reweighting technique with 2500 Monte Carlo sweeps (MCS) and between 2500 and 5000 samples of disorder. The number of Monte Carlo sweeps is justified by the increasing behavior of the energy autocorrelation time, τ_E, and we chose for

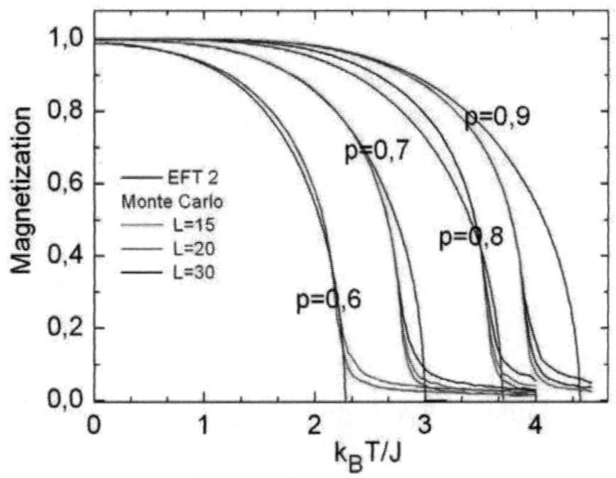

Figure 1. Magnetization vs. k_BT/J for the Ising model with L = 15, 20 and 30 in a cubic lattice with $\lambda = -0.1$.

each size at least 250 independent measurements of the physical quantities ($N_{MCS} > 250\ \tau_E$). The choice of N_{MCS} is justified by the increasing behavior of the energy autocorrelation time τ_E as a function of p and L. At the critical point of a second-order phase transition one expects a finite-size scaling (FSS) behavior $\tau_E \propto L^z$, where z is the dynamical critical exponent

3. Results

The figure 1 displays curves of magnetization versus temperature through computational simulation for $\lambda = -0.1$ and $L = 15$, 20 and 30. The critical temperature obtained when $p = 1$ was $T_C = 4.510$ it is close of the expected value $T_C = 4.51$.[10] We observe that the curves keep the same behavior, in spite of in the proximities of the critical point they move away each other. The critical point was estimated of the inflection of the curve. It can be notice that increasing the lattice size, i.e. the value of L, we get more precision to estimate the critical temperature. For comparison we have drawn, however, a simple effective-field approximation estimate of the temperature dependence of magnetization for several p-values, it gives very good agreement with the simulated transition line over the full temperature range.

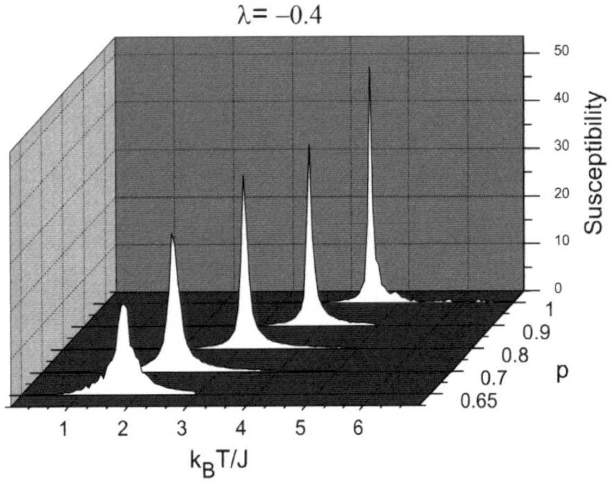

Figure 2. χ vs. k_BT/J for the Ising model with $\lambda = -0.4$ and L = 40 for several concentrations p.

The magnetic susceptibility as a function of temperature for different p values and for $\lambda = -0.4$ and L = 40 is shown in Fig. 2. The peaks are sharper for lower p values. In order to $p < 0.65$ and $\lambda = -0.4$ the susceptibility does not displays the peak associate with the magnetic transition due to competitions of the exchange interactions. When $\lambda \geq 0$ the system presents always ferromagnetic long range order. We used the histogram reweighting technique with 2500 Monte Carlo sweeps (MCS) and between 2500 and 5000 samples of disorder to get an accurate determination of the maxima of the susceptibility.

The phase diagram obtained from the location of the maxima of the susceptibility for the largest lattice size (L = 40) as a function of the concentration of magnetic bonds is shown in Fig. 3 for λ=0, 0.1, 0.2, 0.5, -0.2 and -0.4. Solid lines are the predictions of the effective field approximation. A very good agreement with the simulated transition line is obtained.

The reduced fourth-order Binder cummulants[11] supply an alternative method to estimated critical points that can be determined from the crossing point of the cumulants for different lattice sizes. It is calculated by using

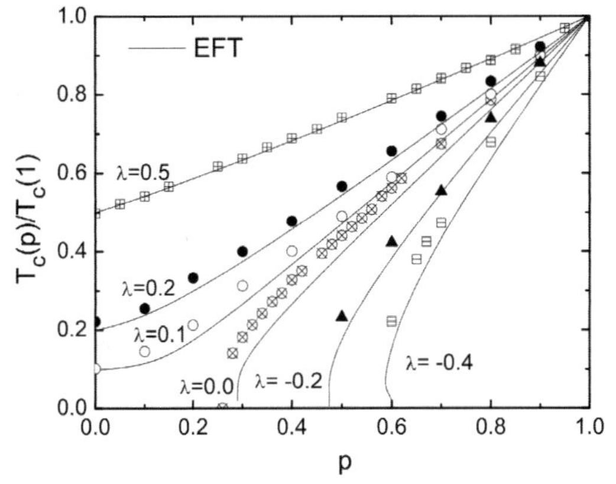

Figure 3. Phase diagram of the 3D mixed-bond Ising model compared with the effective field approximation.

$$U_4 = 1 - \frac{[\langle m^4 \rangle]}{3\left[\langle m^2 \rangle^2\right]},$$

where [...] denotes the average over disorder and $\langle ... \rangle$ refers at the thermal average. As an example, we show in Fig 4 the Temperature dependence of the reduced fourth-order Binder cummulant for $p = 1$ and for various lattice sizes. Critical temperature obtained from this figure is in agreement with those obtained of the maxima of the magnetic susceptibility.

The peak locations of the maxima susceptibility for each L are plotted versus $L^{-1/\nu}$, where the value of ν is determined of the linear fit of log-log plot $\frac{\partial U_L}{\partial T}$ vs L (not shown here). The critical temperature can be estimated from an infinite-size extrapolation in according with Eq.(3). We illustrate this procedure in Fig 5 for $\lambda = 0.5$ and $p = 0.4$, the linear fit gives $T_C = 3.1065(3)$.

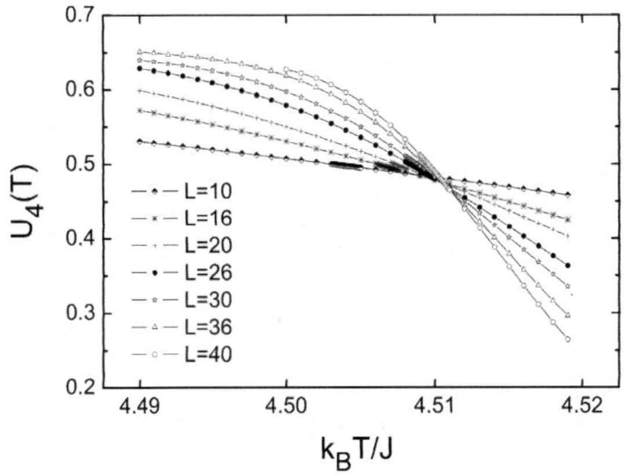

Figure 4. Fourth-order Binder cumulant U_4 vs. k_BT/J, for different lattice sizes as indicated in figure.

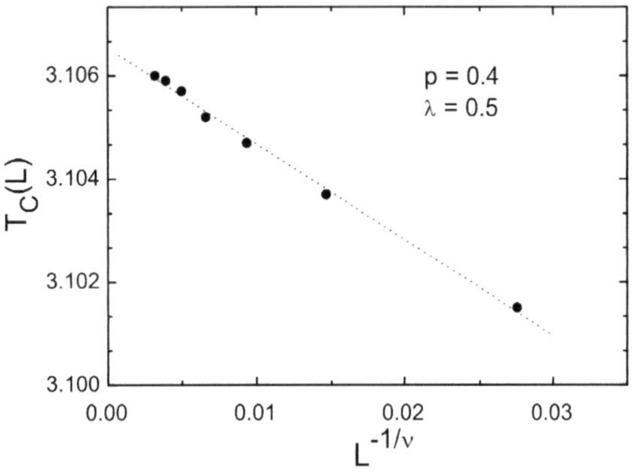

Figure 5. Size-dependent critical temperature $T_C(L)$vs. $L^{-1/\nu}$, for $\lambda = 0.5$ and $p = 0.4$.

Table 1. Critical exponents for three dilutions and $\lambda = 0.5$.

p	1	0.8	0.6	0.4
$1/\nu$	1.591(7)	1.68(7)	1.63(4)	1.56(3)
β/ν	0.4920(5)	0.49(1)	0.50(13)	0.496(5)
γ/ν	2.008(6)	2.06(3)	1.950(6)	2.001(6)

The average magnetization m and the magnetic susceptibility χ scale with the lattice size as:

$$m \sim a_m L^{-\beta/\nu}, \quad \chi \sim a_\chi L^{\gamma/\nu}, \tag{6}$$

here a_m and a_χ are non-universal amplitudes. From these power-laws we extracted the exponents β and γ plotting in logarithm scale the lattice size dependence of the susceptibility and average magnetization. The critical exponents obtained to $p = 0.4$, 0.6, 0.8 and $\lambda = 0.5$ are listed in Table I. The critical exponents oscillate without having an apparent correlation with the dilution. One can note that they are pretty close of those obtained for disordered Ising model [12]

4. Conclusion

We carried out Monte Carlo simulations for study the influence of bond dilution on the critical properties of the Ising Model applied for cubic lattice. We obtained thermodynamic parameters for $|\lambda| \leq 1$.

Satisfactory results are obtained using the algorithm of Wolff and showed that this technique is appropriated to treat the Mixed-bond problem. On the other hand, the Monte Carlo technique results gives very good agreement with the effective field theory data.

The critical behavior of the mixed-bond model is governed by the same universality class as the site-diluted model and pure Ising model.

ACKNOWLEDGMENT

N.O. Moreno is partially supported by FAPITEC-SE and CNPq Brazilian agencies. J. B. Santos Filho acknowledge financial support by CAPES.

[1] R. Folk, Y. Holovatch, T. Yavorskii: Physics Uspiekhi **173** (2003) 175.
[2] A. B. Harris: J. Phys. C **7**(1974) 1671.
[3] S. Wiseman and E. Domany, Phys. Rev. Lett. **81**, (1998) 22.

[4] H.G. Ballesteros, L.A. Fernández, V. Martín-Mayor, A. Muñoz Sudupe, G. Parisi, J.J. Ruiz-Lorenzo, Phys. Rev. B **61**, (2000) 3215.
[5] U. Wolff, Phys. Rev. Lett. **62** (1989) 361.
[6] M. N. Barber Finite-size scaling. In: Phase Transitions and Critical Phenomena, eds. C.Domb and J.Lebowitz, Academic, New York, 1983, vol.8
[7] A. M. Ferrenberg, D.P. Landau, Phys. Rev. B **44** (1991) 5081.
[8] J. T. Chayes, L. Hayes, D. S. Fisher, T. Spencer, Phys. Rev. Lett. **57** (1986) 2999.
[9] L. E. Zamora, G. A. Pérez Alcázar, A. W. Bohórquez, A. A. Rivera and J. A. Plascak, Phys. Rev. B **51** (1995) 9329.
[10] C Domb and M. S. Green, 1976. Phase Transitions and Critical Phenomena, ed. por C. Domb e M. S. Green, (Academic, London)
[11] K. Binder and D. W. Heermann, Monte Carlo Simulation in Statistical Physics. An Introduction, 3rd ed. (Springer, Berlin, 1997).
[12] H. G. Ballesteros, *et al*, Phys. Rev. B **58** (1998) 2740.

International Conference on Defects in Insulating Materials IOP Publishing
Journal of Physics: Conference Series **249** (2010) 012039 doi:10.1088/1742-6596/249/1/012039

Magnetic Properties of the Metamagnet Ising Model in a three-dimensional Lattice in a Random and Uniform Field

João B. dos Santos-Filho,[1] Douglas F. de Albuquerque[2] and N O Moreno[3]

[1] Departamento de Física, ICEX, UFMG, 1270-901, Belo Horizonte, Brazil

[2] Departamento de Matemática Aplicada, CCET, UFS, 49100-000, São Cristovão, SE, Brazil

[3]Departamento de Física, CCET, UFS, 49100-000, São Cristovão, SE, Brazil

E-mail: douglas@ufs.br

Abstract. The Monte Carlo technique is here used for study the behavior of metamagnet Ising model in presence of random field. The phase diagram is obtained by employing Glauber's algorithm in a cubic lattice of linear size L with values ranging from 16 to 42 and with periodic boundary conditions.

1. Introduction

Random fields and disordered magnetic systems have been a considerable source of research in recent years [1]. The random field Ising model (RFIM) has been one of the most interesting subject of research in Physics of Condensed Matter in the last fifteen years and it occupies prominence position when we deal with disordered systems [2, 3]. The specimen most common of disorder are represented by: i) disorder in the bonds and ii) randomness in the strength of the applied magnetic field.In this model the disorder depends on the applied external magnetic field and although RFIM has deserved many investigations from both experimental and theoretical points of view[4] and no conclusive result has been achieved for the understanding the nature of the phase transitions and critical behavior. On the other hand, questions as the lower critical dimension [5, 6] and the existence of a static phase transition have already been solved from the theoretical point of view, while questions as the existence of the tricritical point are still opened[7]. The relevance of RFIM is due to the fact that it is the simplest to describe the essential physics of various class of disordered systems, which includes: i) structural phase transitions in random alloys[8], ii) commensurate charge-density-wave systems with impurity pinning[9, 10], iii) binary fluid mixtures in random porous media [11], iv) melting of intercalates in layered compounds such as TiS_2[12], v) frustration introduced by the disorder in interacting many body systems, besides explaining several aspects of electronic transport in disordered insulators[13] and vi) systems near the metal-insulator transition[14, 15].

In the last years, the physics of the hysteresis, of the avalanche behavior and of the origin of self-organized criticality[15] have been modelled employing the non-equilibrium behavior of the RFIM. In particular, a new class of problems, such as self-generated glassy behabior, has been studied through the non-disordered model with infinitesimal random field [16]. Recently, random magnetic fields have been applied also in metamagnet systems like Ising model and importants results have been obtained [17, 18]. An ideal metamagnet crystal can be constructed overlapping identical layers of spin with ferromagnetic coupling between spins next neighbors of each layer and with antiferromagnetic coupling next neighbors spins of the adjacent layers. Another way to think a metamagnet crystal is to consider a cubic crystal with the couplings between the first neighbors as being of antiferromagnetic character and the couplings between the second neighbors as being ferromagnetic. In these systems, only the competition between the ferromagnetic and antiferromagnetic ordering are interesting. However, the application of a random and uniform magnetic field can yield the appearance of new phenomena and a richer critical behabior becomes possible and, in particular, in the present work we study the RFIM applied to a metamagnet system in a cubic lattice by employing Monte Carlo Method.

2. Model and Simulation

Metamagnet ideal can be considered as a set of spins with uniaxial anisotropy ferromagnetic interactions within each ($J_1 > 0$) and antiferromagnetic interaction ($J_2 < 0$). The hamiltonian model for spin-1/2 is given by:

$$\mathcal{H} = -\sum_{\langle i,j\rangle} J_1\sigma_i\sigma_j - \sum_{\langle i,j\rangle} J_2\sigma_i\sigma_k - \sum_{i=1}^{N}(h-h_i)\sigma_i\,, \tag{1}$$

where the first is executed on all pairs of spin nearest-neighbors on same plane and second sum run over all pairs of spin nearest-neighbor in parallel planes, h is the strength of the external uniform magnetic field and h_i is the random magnetic field which obeys the bimodal distribution given by:

$$P(h_i) = \frac{1}{2}[\delta(h_i - h_r) + \delta(h_i + h_r)]\,, \tag{2}$$

where h_r is the strength of the random field.

To study this system we employed Monte Carlo simulation technique [19] by using the algorithm of Glauber in a cubic lattice of linear size L with values ranging from 16 to 42 and with periodic boundary conditions. To reach the equilibrium state we take, for guarantee, at least 2×10^4 Monte Carlo steps (MCs) for all the lattice sites we studied and more 3×10^4 MCS to estimate the average values of the quantities of interest. In our work we consider one MCs equivalent L^3 trials for change the state of a spin of the lattice.

We calculated the sublattice magnetization per spin belong to the different planes by using

$$m_A = \left[\frac{2}{N} \left\langle \sum_{i \in A} \sigma_i \right\rangle \right], \tag{3}$$

$$m_B = \left[\frac{2}{M} \left\langle \sum_{i \in B} \sigma_i \right\rangle \right], \tag{4}$$

The transition lines of the phase diagram were obtained from the staggered magnetization, m_s and magnetization m. Thus we calculate $m_s = m_A - m_B$ and $m = m_A + m_B$ that are our parameters of order for antiferromagnetic and ferromagnetic phase, respectively. In the above equations $[\cdots]$ denotes the average over the disorder and $\langle \cdots \rangle$ denotes the thermal average.

3. Results and Conclusions

The Figure 1 illustrate the complete phase diagram, for two selected values of the random field h_r, showing the continuous and discontinuous transition lines separating the antiferromagnetic and paramagnetic phases. The tricritical point, which is indicated by an open square, joins these two lines. The points of the phase diagram are obtainded from the knowledge of the point of maximum of the curve for the susceptibility, see Figure 2. It was obtained of the results by fixing H and changing T. In the case of the discontinuous transition, we determined the magnetization curve as a function of the field, for a fixed value of temperature. This procedure is not an efficient to localize the tricritical point one, because it is difficult to distinguish a continuous from a discontinuous curve, especially near the tricritical point, but it gives an idea of the range of values of the field where the transition is of first order. In this work the location of the tricritical point was achieved through the disappearance of hysteresis [20]. For a fixed value of temperature, we drew the magnetization curves for increasing and decreasing values of the magnetic field.

On the other hand, Figure 3 shows the behavior of the magnetization m and the staggered magnetization m_s as function of the temperature for different values of the field external uniform.In this diagram we can observe as the random field affects the behavior of the system. The behavior of the magnetization when the external field varies for different values of the temperature is shown in Figure 4. For $T = 2$ one get a first-order transistion and for $T = 3$ and $T = 4$ a transistion of second order.

We can observe in Figure 5 the variation of the susceptibility with the external field for different temperatures. In Figure 6 we can see better as the random field affects the order of the system, in this Figure we kept the temperature and we calculate the magnetization with the variation of the field uniform for different values of random field.

In summary, the present Monte Carlo simulations for a metamegnet Ising model in a random and uniform field show that the phase diagram in the plane uniforme field

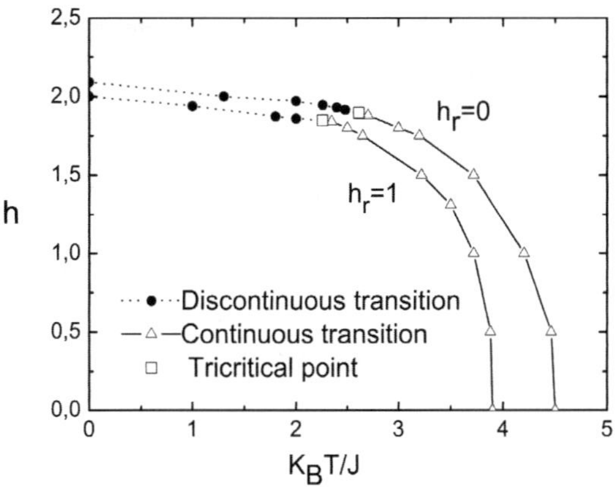

Figure 1: Phase diagram of metamagnet in a cubic lattice in the plane $t - h$ for $h_r = 0$ and $h_r = 1$.

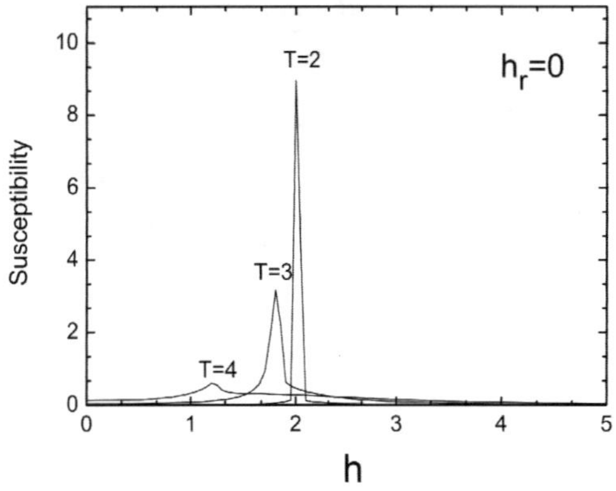

Figure 2: Susceptibility curve *versus* temperature, K_BT/J, for different values of external field.

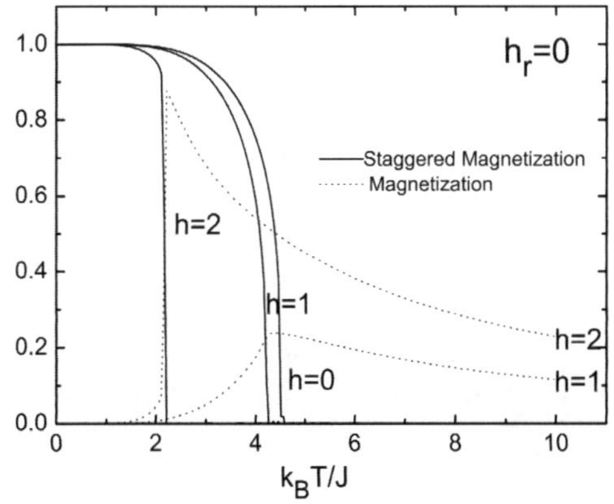

Figure 3: Curves of magnetization m and staggered magnetization m_s for different values of external field.

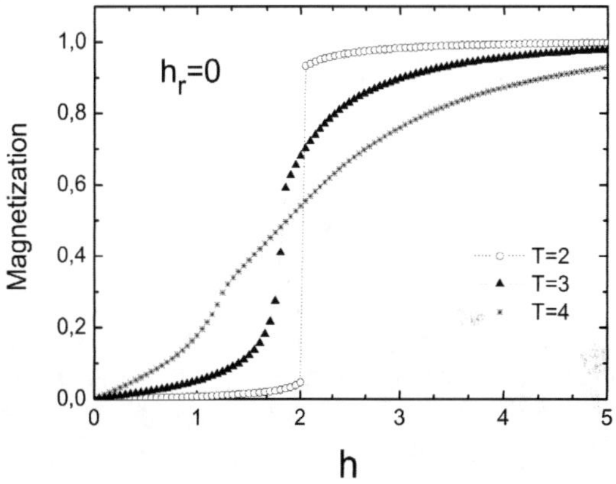

Figure 4: Susceptibility curves *versus* external field h for different temperatures.

h *versus* temperature present continuous and first-order transition lines separated by possible tricritical points.

3.1. Acknowledgments

Authors wishing to acknowledge partial support from CAPES, FAPITEC and CNPq Brazilian Researches.

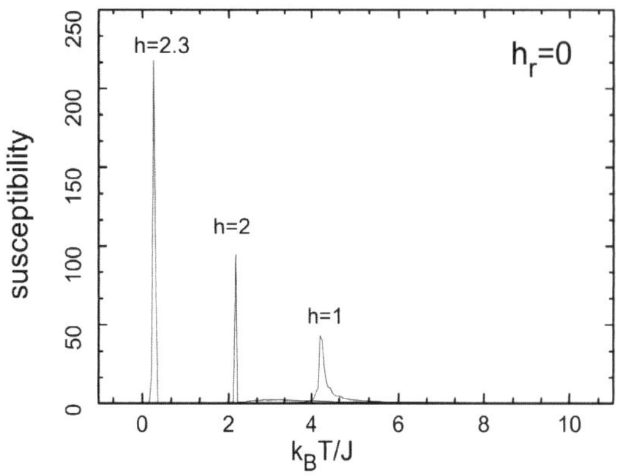

Figure 5: Curves of magnetization m versus external field for different values of temperature.

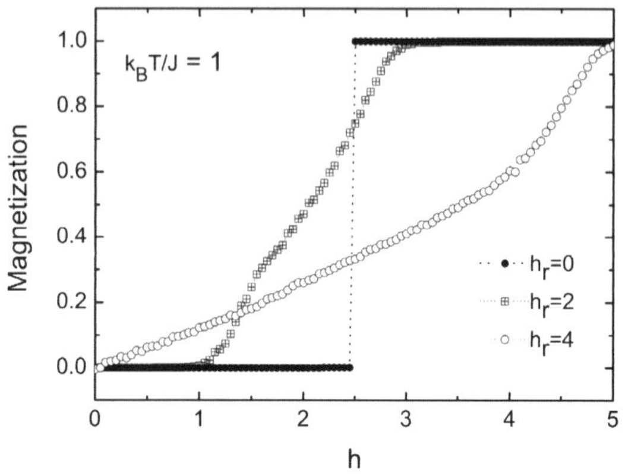

Figure 6: Curves of magnetization m versus external field for different values of random field h_r.

References

[1] Belanger D P 2000 *Braz. J. Phys.* **30** 682
[2] Belanger D P 1998 *Spin Glasses and Random Fields* (Singapore: edited by A. P. Young, World Scientific)
[3] Birgeneau R J 1998 *J. Magn. Magn. Mater.* **177** 1
[4] Shapir Y 1992 *Recent Progress in Random Magnets* (Singapore: World Scientific) p. 309
[5] Imbrie J Z 1984 *Phys. Rev. Lett.* **53** 1747
[6] Imry Y and Ma S K 1975 *Phys. Rev. Lett.* **35** 1399
[7] de Arruda A S, Figueiredo W and Sebastianes R M 1989 *Phys. Rev. B* **39** 4409
[8] Childress J R and Chien C L 1991 Phys. Rev. B **43** 8089
[9] Fisher D S 1983 Phys. Rev. Lett. **50** 1486
[10] Fisher D S 1985 Phys. Rev. B **31** 1396
[11] Maher J V, Goldburg W I, Pohlm D W and Lanz M 1984 Phys. Rev. Lett. **53** 60
[12] Suter R M, Shafer M W, Hornm P M and Dimon P 1982 Phys. Rev. B **26** 1495
[13] Efros A L and Shklovskii B L 1975 J. Phys. C **8** L49
[14] Kirkpatrick T R and Belitz D 1994 Phys. Rev. Lett. **73** 862
[15] Pastor A A and Dobrosavljević V 1999 Phys. Rev. Lett. **83** 4642
[16] Mezard M 1999 Physica A **265** 352
[17] Weizenmann A, Godoy M, de Arruda A S, de Albuquerque D F and Moreno N O 2007 Physica. B, Condensed Matter **398** 297
[18] dos Santos J B, de Albuquerque D F, Moreno N O and de Arruda A S 2007 Physica. B, Condensed Matter 294
[19] Landau D P and Binder K 2005 *A Guide to Monte Carlo Simulations in Statistical Physics* (Cambridge University Press)
[20] Landau D P 1976 Phys. Rev. B **31** 4054

Computer modelling of intrinsic defects and rare earth doping in KYF_4, K_2YF_5 and KY_3F_{10}

E. M. Maddock[1], R. A. Jackson[1] and M. E. G. Valerio[2]

[1] Lennard-Jones Laboratories, School of Physical and Geographical Sciences, Keele University, Keele, Staffordshire, ST5 5BG, UK.

[2] Physics Department, Federal University of Sergipe, Campus Universitario, 49100-000 São Cristovão, SE, Brazil.

r.a.jackson@chem.keele.ac.uk

Abstract. A computational study of KYF_4, KY_3F_{10} and K_2YF_5 is presented. Energy minimisation techniques have been used to obtain structural models of the materials and intrinsic defect calculations have been performed. Rare earth doping has been considered at all cation sites and solution energies are presented which show the preference of isovalent doping at the Y^{3+} site.

1. Introduction

Mixed metal fluorides of the KYF family have raised interest for the development of solid state lasers since they are able to accommodate rare earth ions at the non-centrosymmetric Y^{3+} sites. This paper presents computational studies of KYF_4, K_2YF_5 and KY_3F_{10}, each of which belongs to a different space group [1-3], however shows similar responses to both intrinsic defect formation and rare earth doping processes.

This work follows studies of other fluoride materials such as BaY_2F_8 [4] and LaF_3 [5] where interesting trends have been obtained which have rendered these materials as possible candidates for optical applications. Rare earth doping can be used to enhance optical properties of materials and therefore calculations can be performed to obtain the solution energies which represent the energy of the whole doping process including any charge compensatory mechanisms that may be required.

2. Methodology

Energy minimisation calculations were carried out using energy minimisation techniques within the GULP code [6]. Buckingham potentials with an additional electrostatic term were used to model the structures (1). The K-F potential was fitted to give good representations of the KYF materials, the Y-F and F-F potentials were obtained from previous work [4]. The Mott-Littleton approximation [7] was used to model defects with region I and region II having cut-off radii of 10 and 15Å respectively.

$$V(r) = Aexp\left(\frac{-r_{ij}}{\rho}\right) - Cr^{-6} + \frac{q_1 q_2}{r_{ij}} \quad (1)$$

Table 1. Potentials used for KYF material calculations

Interaction	A (eV)	ρ (Å)	C (eV Å^6)	k (eV Å^{-2})
$K_{core} - F_{shell}$	5500.0000	0.2635	11.00	-
$Y_{core} - F_{shell}$	1547.6209	0.3023	0.00	-
$F_{shell} - F_{shell}$	1127.7000	0.2753	15.83	-
F_{core} - F_{shell}	-	-	-	20.77

2.1. Solution Energies

Energies for the substitution of cations by trivalent rare earth ions were calculated and converted into solution energies using the expressions given below. Unbound solution energies were calculated using average vacancy energies while bound solution energies have been calculated by specifying the identity and position of charge compensatory vacancies.

i. Substitution at Y^+ site (expression common to all materials)

$$E_{sol} = M_Y - E_{latt}(MF_3) + E_{latt}(YF_3)$$

ii. Substitution at K^+ site

a. $2K^+$ vacancies as charge compensation (expression common to all materials)

$$E_{sol} = M_K^{\cdot\cdot} + 2V_K^{/} - E_{latt}(MF_3) + 3E_{latt}(KF)$$

b. Alternative charge compensation

i. KYF_4

$$E_{sol} = M_K^{\cdot\cdot} + V_Y^{///} + V_F^{\cdot} + E_{latt}(KYF_4) - E_{latt}(MF_3)$$

ii. KY_3F_{10}

$$E_{sol} = M_K^{\cdot\cdot} + 3V_Y^{///} + 7V_F^{\cdot} + E_{latt}(KY_3F_{10}) - E_{latt}(MF_3)$$

iii. K_2YF_5

$$E_{sol} = M_K^{\cdot\cdot} + V_Y^{///} + V_K^{/} + V_F^{\cdot} + E_{latt}(K_2YF_5) - E_{latt}(MF_3)$$

3. Results and Discussion

Comparison of the generated models for the KYF materials with experimental data shows a good correlation of parameters, with all modelled parameters within 4.5% of the experimental values.

Table 2. Comparison of calculated and experimental structural parameters of (a) KYF_4, (b) KY_3F_{10} and (c) K_2YF_5.

Parameter	Exp.	Calc.	Difference	Units	%
(a) KYF_4 [1]					
Volume	1729.623694	1717.199642	-12.424052	Å^3	-0.72
a/b	14.060000	13.952835	-0.107165	Å	-0.76
c	10.103000	10.185098	0.082098	Å	0.81
(b) KY_3F_{10} [2]					
Volume	1538.059079	1472.215745	-65.843334	Å^3	-4.28
a/b/c	11.543150	11.376024	-0.167126	Å	-1.45
(c) K_2YF_5 [3]					
Volume	517.823843	540.049545	22.225701	Å^3	4.29
a	10.791000	11.067484	0.276484	Å	2.56
b	6.607000	7.753236	0.146236	Å	2.21
c	7.263000	7.225580	-0.037420	Å	-0.52

Intrinsic defect calculations show that the formation of anion vacancies is more favourable than the formation of cation vacancies and in particular yttrium vacancies. For all the materials the energy required for the formation of a fluorine vacancy is between 4.5 - 5.5 eV, with inequivalent fluorine species showing slightly differing vacancy formation energies.

Table 3. Solution energies of rare earth doping in KYF_4

Site RE^{3+}	Y1 Charge compensation not required	Y2 Charge compensation not required	K1 2K vacancy charge compensation	K1 Y and F vacancy charge compensation
La	0.9997	0.5097	6.4401	10.0247
Ce	0.8660	0.4324	6.4556	10.0403
Pr	0.7311	0.3552	6.4638	10.0484
Nd	0.6462	0.3119	6.4825	10.0672
Sm	0.3009	0.0800	6.3484	9.9331
Eu	0.3501	0.1814	6.5200	10.1047
Gd	0.2267	0.1168	6.4800	10.0647
Tb	0.1008	0.0208	6.3972	9.9819
Dy	0.1119	0.0914	6.5051	10.0897
Ho	0.2183	0.1269	6.5453	10.1299
Er	0.0408	0.0408	6.3504	9.9350
Tm	-0.0072	0.0873	6.3583	9.9430
Yb	0.0915	0.0856	6.5449	10.1295
Lu	0.0126	0.1562	6.4065	9.9912

Potassium vacancy formation energies are in the region of 6.0 – 7.5 eV for all materials. In contrast, the energy required for the formation of a yttrium vacancy for the KYF materials is in the region of 47.5 – 51.0 eV. The significantly greater energy required to form a yttrium vacancy can be attributed to the greater structural destabilisation induced by the removal of a trivalent ion.

Anion Frenkel defect energies have been calculated for each of the materials and energies in the region of 0.5 – 2.0 eV have been obtained for all materials. When compared to the calculated Schottky energies of 10.6971 eV for K_2YF_5, 12.2911 eV for KYF_4 and 14.7897 eV for KY_3F_{10}, it can be seen that Frenkel defect formation is much more energetically favourable in these materials.

Table 4. Solution energies of rare earth doping in KY_3F_{10}

Site RE^{3+}	Y Charge compensation not required	K 2K vacancy charge compensation	K 3Y + 7F vacancy charge compensation
La	0.6759	6.9777	34.5358
Ce	0.5557	7.0701	34.6282
Pr	0.4412	7.1562	34.7143
Nd	0.3626	7.2388	34.7969
Sm	0.1293	7.2064	34.7645
Eu	0.1736	7.4791	35.0372
Gd	0.1290	7.4623	35.0204
Tb	0.0383	7.3959	34.9540
Dy	0.0955	7.5647	35.1228
Ho	0.0623	7.6625	35.2206
Er	0.0111	7.6207	35.1788
Tm	0.0992	7.6806	35.2387
Yb	0.0245	7.7445	35.3026
Lu	0.1664	7.8176	35.3757

The solution energy results for unbound defects in all materials show that rare earth doping is significantly more favourable at the yttrium site than the potassium site. This can be attributed to the greater similarity of ionic radii of the rare earth ions to the yttrium ion coupled with no charge compensatory measures being required leading to reduced structural destabilisation. Calculations involving rare earth doping at the potassium site revealed that charge compensation through two potassium vacancies was far more favourable than using the alternative charge compensatory mechanism for each material which involved the formation of yttrium vacancies.

Table 5. Solution energies of rare earth doping in K_2YF_5

Site RE^{3+}	Y Charge compensation not required	K1 2K vacancy charge compensation	K1 Y, K and 2F vacancy charge compensation
La	0.9685	5.9279	12.6677
Ce	0.8364	5.9331	12.6730
Pr	0.7029	5.9343	12.6741
Nd	0.6163	5.9438	12.6837
Sm	0.2871	5.8437	12.5836
Eu	0.3275	6.0012	12.7410
Gd	0.2180	5.9866	12.7265
Tb	0.0978	5.9061	12.6460
Dy	0.1114	6.0047	12.7445
Ho	0.1853	6.0215	12.7613
Er	0.0285	5.9678	12.7076
Tm	0.0047	6.0206	12.7604
Yb	0.0650	6.0168	12.7566
Lu	0.0238	6.0808	12.8207

Bound defect calculations have been carried out to determine whether, when doping at the potassium site, the specific position and identity of the charge compensatory vacancies influences the energetics of the doping process.

Results show that where two potassium vacancies are used as charge compensation the position and therefore identity of the vacancies influences the favourability of the doping process. Figure 5 shows that in K_2YF_5 there are clear energetic differences in response to the specific substitution and vacancy site combination. Despite this, it can be seen that Sm^{3+} and Tb^{3+} dopants are the most favourable irrespective of the particular combination of substitution and vacancy sites. A similar conclusion can be drawn from doping rare earth ions into KYF_4 as shown by Figure 4.

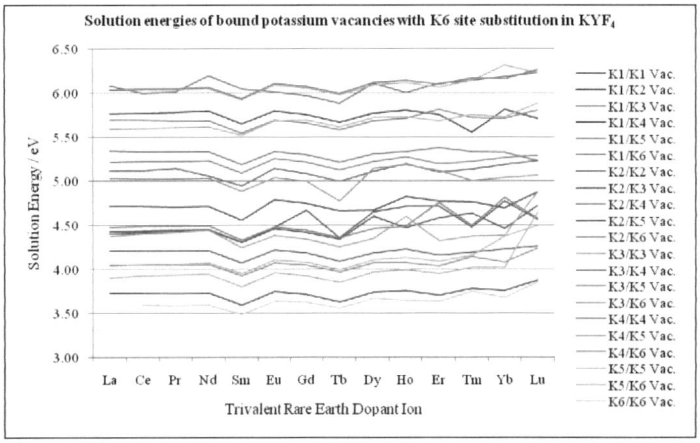

Figure 1. Bound solution energies of rare earth doping at a potassium site in KYF_4.

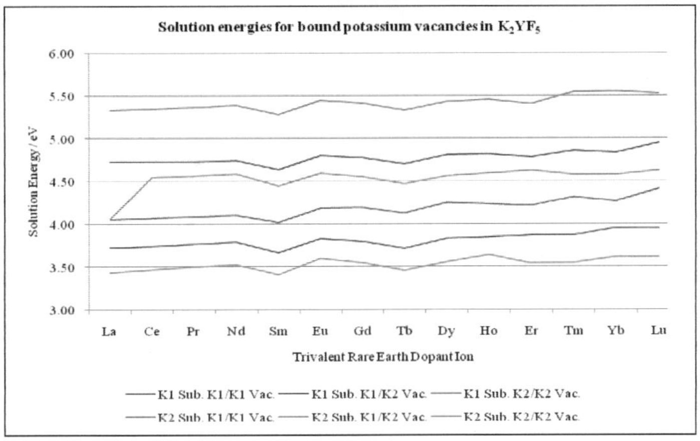

Figure 2. Bound solution energies of rare earth doping at the potassium sites in K_2YF_5.

4. Conclusion

Structural models for KYF_4, K_2YF_5 and KY_3F_{10} have been generated which show reasonable agreement with experimental data. It has been shown that for all materials the formation of anion Frenkel defects is significantly more favourable than the formation of Schottky defects which follows since, the formation of a Y^{3+} vacancy in these materials is particularly unfavourable.

Calculations involving the substitution of trivalent rare earth ions into the materials have shown that substitution at the Y^{3+} site is significantly more favourable. The similarity in ionic radii of the rare earth ions and the Y^{3+} ion allow isovalent substitution with minimal structural distortion. The energetic expense of substitution at the K^+ site can be attributed to the charge compensatory mechanism employed as shown by the bound defect calculations. In general however it can be noted that Sm^{3+} and Tb^{3+} dopants show particular interest as they exhibit lower solution energies.

5. Acknowledgements

The authors would like to thank iEPSAM at Keele University for financial support.

References

[1] Le Fur Y, Khaidukov NM, Aleonard S, Acta Crystallographica C**48** 978-982 (1992)
[2] Kharitonov YA, Gorbunov YA, Maximov BA, Kristallografiya **28** (5) 1031-1032 (1983)
[3] Grzechnik A, Crichton WA, Gesland J-Y, Solid State Sciences **5** 757-764 (2003)
[4] Amaral JB, Couto dos Santos MA, Valerio MEG, Jackson RA, Applied Physics B **81** (6) 841-846 (2005)
[5] Jackson RA, Valerio MEG, Amaral JB, Couto dos Santos MA, Maddock EM, Physica Status Solidi (c) **4** (3) 1185-1188 (2007)
[6] Gale JD, Journal of the Chemical Society Faraday Transactions **93** 629-637 (1997)
[7] Mott NF & Littleton MJ, Transactions of the Faraday Society **34** 485-499 (1938)

Quantum rings in a space with topological defects

A. L. Silva Netto and C. Furtado
DF-CCEN, Universidade Federal da Paraíba, Campus I, 58051-970 , João Pessoa, PB, Brazil

C. Chesman
DFTE, Universidade Federal da Rio Grande do Norte, Natal, 59078-970 RN, Brazil

Abstract. Topological defects generally they occur in a system when phase transitions take place. We have examples of arrangements with these defects both in Gravitation and Condensed Matter. Depending on how they are generated, this kind of defects make the space present curvature or torsion. There is a wide literature about this matter, much from a gauge field point of view [1, 2, 3], and probing different aspects [6] as the existence of Landau levels [4, 5] or quantum phases [7], for example. However, there is not an unique fundamental theory for describing defects in solids. In our work we follow the Geometric Thery of Defects (TGD) which has the work by Katanaev [8] as a good way of describing continuum media with topological defects. This last theoretical framework is our compass. Besides, we are interested in mesoscopic systems, an important subject right before the role played by nanodevices. In our work we study quantum rings in a space with a screw dislocation. This kind of space presents torsion. The confinement is modelled by a potential that describes different kinds of mesoscopic systems (quantum dots, antidots and quantum wires). We found some results showing the changes in the behaviour of properties as energy eigenvalues and magnetization.

1. Introduction

The study of rings of mesoscopic dimension in the presence of external fields exhibits a series of interesting phenomena in physics, for example Aharonov-Bohm (AB) [9, 10], quantum hall [11], persistent currents [12, 13] and Berry's phase [14]. This results has been exploited in various experiments [15, 16]. The simplest model for a ring-like device is a one-dimensional ($1D$) ring, in which the applied magnetic field can be represented by an AB flux tube. Such $1D$ models have played an important role in the understanding of quantum interference effects in rings.

Defects in solids modify properties as their melting point, stress and so on. However there is not a fundamental theory despite of several articles have been published about this matter. One of the best theories describing defects in solids is based upon Riemann-Cartan geometry in the seminal article by Katanaev [8]. In this formalism non-Euclidian metrics account for the boundary conditions imposed by defects in elastic media. The theory, in the continuum limit, describes the solid by a Riemann-Cartan manifold where curvature and torsion are associated to disclinations and dislocations, respectively, in the medium [17]. Besides, in this theoretical framework, the elastic deformations are incorporated in the metric of the manifold. We can assume that a medium without deformation is invariant by translations (spatial homogeneity) and by rotations (spatial isotropy) in any coordinates framework. In this preferential coordinates framework the medium is described for a euclidian metric as

$$ds^2 = \sum_{i=1}^{3} \left(dx^i\right)^2 \quad . \tag{1}$$

The metric in (1) is named cartesian. For this kind of space, if we move a vector along a closed path, the vector can return to its initial position. This movement along a closed path is named parallel transport [18]. If there is a defect, the parallel transport the vectors moved cannot be back to the same initial point from where they had left.

In this work we deal with dislocations which can be viewed using the Volterra process [19]. The existence of dislocations is related to the presence of torsion in the medium. Torsion tensor and Burgers vector are related by Burgers

$$b^i = \int\int_S dx^\mu \wedge dx^\nu \, T_{\mu\nu}{}^i \quad , \tag{2}$$

where $dx^\mu \wedge dx^\nu$ is the surface element and $T^i_{\mu\nu}$ is the torsion tensor. The integration region when Burgers vector points out in a certain direction x^3 corresponds to the surface generated by Volterra process, which before the dislocation appearing was formed by directions x^1 e x^2. By the expression (2) we see torsion tensor represents physically a superficial density of the Burgers vector [20]. In this work we study the quantum rings in the presence of topological defects. We use geometric theory of defects for describing two-dimensional ($2D$) quantum rings in the presence of a defect. We investigate situations where the defect is related to torsion. In addition we consider a potential that can describe different kinds of mesoscopic systems (quantum dots, wires or antidots) [16, 21, 22]. We apply this potential in a $2D$ ring and calculate their energy spectrum. Different kinds of insulators can be obtained from the values for effective mass through our data. It is shown that the energy spectrum depends on geometric parameters related to the topological defect.

2. Confining potential

Let be a cylindrical wire with a topological defect through his center. Our space has a screw dislocation, mathematically represented for this metric

$$ds^2 = d\rho^2 + (dz + \beta d\varphi)^2 + \rho^2 d\varphi^2, \tag{3}$$

There is a coupling between the z and φ coordinates, something intrisic to how screw dislocations are constructed. The strength of the Burgers vector and β are related by $b^z = 2\pi\beta$. In our approach the modelling of quantum rings is made with the potential $V_{TI}(\rho)$ proposed by Tan and Inkson, which describe some kinds of mesoscopic systems, e.g., dots, antidots and wires [16].

$$V_{TI}(\rho) = \frac{a_1}{\rho^2} + a_2\rho^2 - V_0, \tag{4}$$

with

$$\omega_0 \equiv \left(\frac{8a_2}{\mu}\right)^{1/2} \quad , \quad V_0 \equiv 2\left(a_1 a_2\right)^{1/2} \quad , \tag{5}$$

From the metric we obtain the hamiltonian and reach the following time independent Schrödinger equation (TISE)

$$\frac{1}{\rho}\frac{\partial}{\partial\rho}\left(\rho\frac{\partial\Psi}{\partial\rho}\right) + \frac{1}{\rho^2}\frac{\partial^2\Psi}{\partial\varphi^2} - 2\beta\frac{1}{\rho^2}\left(\frac{\partial}{\partial\varphi}\frac{\partial}{\partial z}\right)\Psi + \left(1 + \frac{\beta^2}{\rho^2}\right)\frac{\partial^2\Psi}{\partial z^2} - \frac{2\mu a_1}{\hbar^2}\frac{\Psi}{\rho^2} - \frac{2\mu a_2}{\hbar^2}\rho^2\Psi + 2\mu\frac{E - V_0}{\hbar^2}\Psi = 0 \tag{6}$$

Assuming the *ansatz* $\Psi(\rho,\varphi,z) = R(\rho)e^{im\varphi}e^{ikz}$, we obtain the radial equation

$$\frac{1}{\rho}\frac{d}{d\rho}\left(\rho\frac{dR}{d\rho}\right) - \frac{M^2}{\rho^2}R + \left(2\mu\frac{E-V_0}{\hbar^2} - k^2 - \frac{2\mu a_2}{\hbar^2}\rho^2\right)R = 0, \tag{7}$$

where $M \equiv (m - k\beta)^2 + \frac{2\mu a_1}{\hbar^2}$.
By the change $\xi = \sqrt{\frac{2\mu a_2}{\hbar^2}}\rho^2$, we also obtain

$$\xi\frac{d^2R}{d\xi^2} + \frac{dR}{d\xi} - \frac{M^2}{4\xi}R + \beta' R - \frac{\xi}{4}R = 0, \tag{8}$$

which has the solution

$$R(\xi) = e^{-\frac{\xi}{2}}\xi^{\frac{|M|}{2}}F\left(-\beta' + \frac{|M|+1}{2}, |M|+1, \xi\right) \tag{9}$$

Normalization condition impose the hypergeometric function $F\left(-\beta' + \frac{|M|+1}{2}, |M|+1, \xi\right)$ must be a n-degree polynomial [23] and this way

$$-\beta' + \frac{|M|+1}{2} = -n \quad , \tag{10}$$

what give us the following energy eigenvalues

$$E = \hbar\omega_0\left(n + \frac{|M|}{2} + \frac{1}{2}\right) + \frac{k^2\hbar^2}{2\mu} + V_0 \tag{11}$$

In the first parcel of (14) we see that, even without any external magnetic field, appears an amount similar to those when Landau levels are present. Here, we could say torsion (because of β) played a role like an effective magnetic (or Landau-like) contribution.

3. External Fields

The last result, where torsion plays a role similar to magnetic fields, motivate us to investigate how is the influence upon system here studied if an external field plus potential $V_{TI}(\rho)$ are present. Initilly we turn on an external magnetic field without confinement and find the following TISE

$$\frac{1}{\rho}\frac{\partial}{\partial\rho}\left(\rho\frac{\partial\Psi}{\partial\rho}\right) + \left[\frac{1}{\rho}\left(\frac{\partial}{\partial\varphi} - \beta\frac{\partial}{\partial z}\right) - \frac{ie}{\hbar}\frac{B\rho}{2}\right]^2\Psi + \frac{\partial^2\Psi}{\partial z^2} + \frac{2\mu E}{\hbar^2}\Psi = 0 \quad , \tag{12}$$

Its radial solution and energy eigenvalues are, respectively, given by

$$R(\xi) = e^{-\frac{\xi}{2}}\xi^{\frac{|m-k\beta|}{2}}F\left(-\beta' + \frac{|m-k\beta|+1}{2}, |m-k\beta|+1, \xi\right) \quad , \tag{13}$$

$$E = \hbar\omega_c\left[n + \frac{|m-k\beta|}{2} - \frac{m-k\beta}{2} + \frac{1}{2}\right] + \frac{k^2\hbar^2}{2\mu} \quad . \tag{14}$$

From the expression above, we find the following for magnetization

$$\mathcal{M} = -\frac{e\hbar}{\mu}\left[n + \frac{|m-k\beta|}{2} - \frac{m-k\beta}{2} + \frac{1}{2}\right] \quad . \tag{15}$$

Next, if we add the potential $V_{TI}(\rho)$ to the external magnetic field, we find the following TISE

$$\frac{1}{\rho}\frac{\partial}{\partial\rho}\left(\rho\frac{\partial\Psi}{\partial\rho}\right)+\left[\frac{1}{\rho}\left(\frac{\partial}{\partial\varphi}-\beta\frac{\partial}{\partial z}\right)-\frac{ie}{\hbar c}\frac{B\rho}{2}\right]^2\Psi+\frac{\partial^2\Psi}{\partial z^2}-$$
$$\left(\frac{2\mu a_1}{\hbar^2}\frac{1}{\rho^2}+\frac{2\mu a_2}{\hbar^2}\rho^2+\frac{2\mu V_0}{\hbar^2}\right)\Psi+\frac{2\mu E}{\hbar^2}\Psi=0 \quad . \tag{16}$$

Using the *ansatz* $\Psi(\rho,\varphi,z) = R(\rho)e^{im\varphi}e^{ikz}$ again and with the variable change $\xi = \lambda\rho^2$, we obtain

$$\xi\frac{d^2R}{d\xi^2}+\frac{dR}{d\xi}-\frac{M'^2}{4\xi}R+\beta' R-\frac{\xi}{4}R=0 \quad , \tag{17}$$

where $M'^2 \equiv (m-k\beta)^2-\frac{2\mu a_1}{\hbar^2}$, $\lambda^2 \equiv \frac{e^2B^2}{4\hbar^2c^2}+\frac{2\mu a_2}{\hbar^2}$ and $\beta' \equiv \frac{1}{4\lambda}\left(\frac{2\mu E}{\hbar^2}+\frac{eB}{\hbar c}(m-k\beta)-k^2-\frac{2\mu V_0}{\hbar^2}\right)$.

So, the radial solution and eigenvalues are, respectively, given by

$$R(\xi)=e^{-\frac{\xi}{2}}\xi^{\frac{|M'|}{2}}F\left(-\beta'+\frac{|M'|+1}{2},|M'|+1,\xi\right) \quad , \tag{18}$$

$$E=\hbar\omega\left[n+\frac{|M'|}{2}+\frac{1}{2}\right]-\frac{(m-k\beta)}{2}\hbar\omega_c+\frac{k^2\hbar^2}{2\mu}+V_0 \quad . \tag{19}$$

Here $\omega = \omega_0 + \omega_c$. We see there are two contributions to the Landau-like frequency ω: one because the potential V_{TI} and another related to the external magnetic field.

From (19) we find the following for magnetization

$$\mathcal{M}=-\frac{e\hbar}{\mu}\left[n+\frac{|M'|}{2}+\frac{1}{2}\right]+\frac{e\hbar}{\mu}\frac{(m-k\beta)}{2} \quad . \tag{20}$$

Looking at the last expressions, we see the torsion causes a shift in the quantum number m, what modifies the energy spectrum and how the magnetic field influence the physical properties we have approached.

4. Concluding remarks

Through this work we have found the torsion influences physical properties, namely, energy spectrum and magnetization. There is a shift on quantum number m and, mathematically, torsion has a behaviour as if it would be a kind of magnetic field. We can apply this formalism to different insulators, simply substituting the known values for effective mass in the literature. The torsion influence occurs by the presence of Burgers vector contribution which is related to β in our data. We also find that our system presents a quantization as if would be a Landau system: here, ω_0 coud play the role of cyclotron frequency.

Acknowledgments

This work was partially supported by PRONEX/FAPESQ-PB, CNPq and CAPES (PROCAD).

References

[1] Holz A 1988 *Class. Quantum Grav.* **5** 1259
Holz A 1992 *J. Phys.* A **25** L1
[2] Kleinert H 1990 *Gauge fields in condensed matter* vol 2 (Singapore: World Scientific)
[3] Kohler C 1998 *Phys. Lett.* A **237** 195
[4] Furtado C *et al* 1994 *Phys. Lett.* A **195** Issue 1 90
[5] Furtado C and Moraes F 1999 *Europhys. Lett.* **45** 279

[6] Moraes F 2000 *Braz. J. Phys.* **30** number 2 304
[7] Furtado C, Bezerra V B and Moraes F 2000 *Europhys. Lett.* **52** 1
[8] Katanaev M O and Volovich I V 1992 *Ann. Phys.* **216** 1
[9] Aharonov Y and Bohm D 1959 *Phys. Rev.* **115** 485
[10] Keyser U F 2002 *et al Semicond. Sci. Technol.* **17** L22
[11] Halperin B I 1982 *Phys. Rev.* B **25** 2185
[12] Tan W C and Inkson J C 1999 *Phys. Rev.* B **60** 5626
[13] Avishai Y, Hatsugai Y, and Kohmoto M 1993 *Phys. Rev.* B **47** 9501
[14] Chandrasekhar V *et al* 1991 *Phys. Rev. Lett.* **67** 3578
[15] Mally D, Chapelier C and Benoit A 1993 *Phys. Rev. Lett.* **70** 2020
[16] Tan W C and Inkson J C 1996 *Semicond. Sci. Technol.* **11** 1635
[17] Filgueiras C and Moraes F 2008 (To appear in *Annals of Physics*) doi:10.1016/j.aop.2008.08.002
Filgueiras C and Moraes F 2008 On the quantum dynamics of a point particle in conical space *Preprint* quant-ph/0809.1537 v1
[18] Landau L D and Lifshitz E M 1980 *Teoria do Campo* (Moscou: Mir)
[19] Puntigam R A and Soleng H H 1997 *Class. Quantum Grav.* **14** 1129
[20] Katanaev M O 2005 Geometric theory of defects *Preprint* cond-mat/0407469 v3
Katanaev M O 2005 Introduction to the Geometric Theory of Defects *Preprint* cond-mat/0502123 v1
[21] Silva Netto A L S, Chesman C and Furtado C 2008 *Phys. Lett.* A **372** 3894
[22] Furtado C, Rosas A and Azevedo S 2007 *Europhysics Letters* **79** number 5 57001
[23] Furtado C and Moraes F 2000 *J. Phys.* A*: Math. Gen.* **33**, 5513

Journal of Physics: Conference Series **249** (2010) 012042 doi:10.1088/1742-6596/249/1/012042

Intrinsic Defects in Strontium Aluminates studied via Computer Simulation Technique

M V dos S Rezende[1,3], R M Araujo[1], M E G Valerio[1] and R A Jackson[2]
[1]Physics Department, Federal University of Sergipe, Campus Universitário, 491000-000 São Cristovão-SE, Brazil
[2]School of Physical and Geographical Sciences,, Keele University, Keele, Staffordshire, UK ST5 5BG

E-mail: mvsrezende@gmail.com, mvalerio@ufs.br

Abstract. We present the results of computer simulation studies for all 10 different crystalline phases of the strontium aluminate family. After deriving a reliable potential model that reproduces all structures, the energetic process involving in creation of intrinsic defects were calculated, namely the Schottky, pseudo-Schottky, anti-Schottky, and Frenkel defects. The results indicated that the entire family of strontium aluminates can be divided in three groups according to the type of the predominant intrinsic defect formed, SrO pseudo-Schottky defect, oxygen Frenkel defect or Strontium Frenkel defect.

1. Introduction

Strontium aluminates have been studied because of its interesting structural and optical properties. Several of these compounds have long lasting phosphorescence for more than 1 h, when doped with trivalent and divalent rare-earth ions [1]. Results reported in the literature showed that $SrAl_2O_4$:Eu^{2+},Dy^{3+},B^{3+}, $Sr_4Al_{14}O_{25}$:Eu^{2+},Dy^{3+},B^{3+}[2], $SrAl_4O_7$:Eu^{2+},Dy^{3+}[3], $SrAl_{12}O_{19}$: Eu^{2+}, $Sr_2Al_6O_{11}$: Eu^{2+} [4], have a wide range of emission colors. Potential applications of these new phosphors are numerous, especially in the areas of safety improvement and energy saving (e.g., traffic signs, emergency signs, safety clothes, advertising, etc.).

Phosphorescence (or sometimes also confused with afterglow) refers to the light emission of an insulator that persists at room temperature after stopping excitation (usually UV irradiation). This delayed light emission is caused by the trapping of photo-generated electrons and/or holes at intrinsic or extrinsic defect sites of the material [5,6].

Computer simulation is a useful tool that can provide information on the relative energetic stability of different intrinsic and extrinsic defects, which, in turn, is a strong indication of the most probable one. As a consequence, this information can be used to explain the phosphorescence mechanisms. For example, the phosphorescence mechanism for $SrAl_2O_4$:Eu^{2+}, Dy^{3+}, B^{3+} proposed by Clabau et al. [1] admits that, under UV irradiation, electrons are promoted from the occupied 4f levels of Eu^{2+} to the empty 5d levels and from the top of the valence band to the unoccupied 4f levels of residual Eu^{3+}, a kind of charge transfer mechanism. The Eu^{2+} electrons promoted to the 5d levels, that supposedly are located in the conduction band of the solid, can be trapped at oxygen vacancy (V_O) defects located in the vicinity of the photogenerated Eu^{3+} cations. The accompanying holes created in the valence band

[3] To whom any correspondence should be addressed.

can be trapped at Sr or Al vacancy (V_{Sr} or V_{Al}) levels. This proposed mechanism is very dependent on the existence of different stable defects in the $SrAl_2O_4$ matrix, both from intrinsic and extrinsic origin and the existence of that defects are admitted by the authors without any further background.

The aim of the present work is to study all types of intrinsic defect, Frenkel, Schottky, pseudo-Schottky and anti-Schottky type defects, calculating the energetic of the production and stabilization of such defects in all strontium aluminates crystalline family.

Computer modelling methods employed in this work are based initially on well established approaches that use interatomic potentials to represent the interactions between the ions, coupled with energy minimization and Mott-Littleton methods to calculate the properties of 'pure' and the defective materials. In the first step, we obtained the potential parameters through the empirical fitting methods. In the second step, this potential set was used to model the intrinsic defects in all crystalline phases, ranging from Schottky, Pseudo-Schottky, Anti-Schottky, and Frenkel defects.

2. Structural information

The strontium aluminates compounds present different crystallographic forms depending of the ratio $SrO:Al_2O_3$. $Sr_4Al_{14}O_{25}$[7] and $Sr_2Al_6O_{11}$[8] exist in orthorhombic phase, while the $Sr_3Al_2O_6$[9] and $Sr_9Al_6O_{18}$[10] exists in cubic phase. The hexagonal phase is observed in $Sr_7Al_{12}O_{25}$[11] and $SrAl_{12}O_{19}$[12], and monoclinic phase is observed for $SrAl_4O_7$[13] and $Sr_{10}Al_6O_{19}$[14].

The $SrAl_2O_4$ exists in two crystallographic forms, and a reversible phase transition between the two occurs at 650°C (923K). At low temperature the monoclinic phase is observed [15] while at high temperature a hexagonal phase is observed [16]. There are two crystallographic non equivalent sites for Sr^{2+} in the monoclinic phase and three crystallographic non equivalent sites for Sr^{2+} in the hexagonal phase. The Sr^{2+} and rare earth ions are very similar in their ionic size (i.e., 1.21 and 0.86 to 0.96 Å, respectively) thus one would expect that the Sr^{2+} sites are the preferable sites for incorporation of the rare earth dopants.

Figures 1 and 2 show schematic representations of the $SrAl_2O_4$ phases. It can be seen that the Al tetrahedrons in hexagonal phase are more organized than in the monoclinic phase. The Al sites are almost perfect tetrahedral coordinated in the monoclinic phase while in the hexagonal phase the tetrahedral are slightly distorted. Thus the phase transition from monoclinic to hexagonal phase involves an organization of the special arrangement of the ions in the crystalline structure.

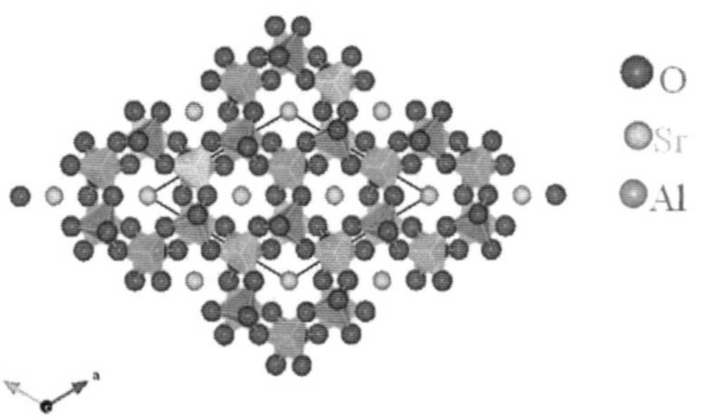

Figure 1 – hexagonal phase $SrAl_2O_4$.

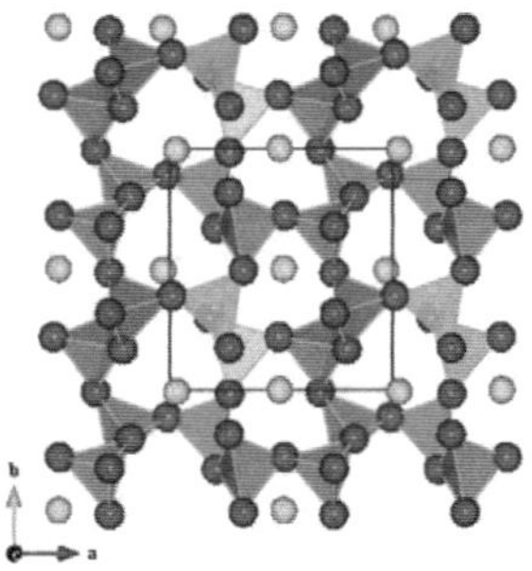

Figure 2 – monoclinic phase $SrAl_2O_4$.

3. Methodology

3.1. Derivation of potentials

The computational method used in the present work admits that any ionic material can be described in terms of ions interacting through effective potentials. Having specified these potentials, the lattice

Journal of Physics: Conference Series **249** (2010) 012042 doi:10.1088/1742-6596/249/1/012042

energy of the material is minimized by varying the structural parameters (atomic positions and unit cell constants). The interatomic potentials used in this paper were obtained using empirical fitting methods as embodied in the GULP code [17]. The Buckingham potentials supplemented by electrostatic interaction terms, was used here since it successfully describes ion interaction in a huge variety of materials, as below

$$V(r_{ij}) = A_{ij}\exp(-r_{ij}/\rho_{ij}) - C_{ij}r_{ij}^{-6} + q_i q_j / r_{ij} \tag{1}$$

here q_i , q_j are charges on ions i and j, and A, ρ, and C are parameters whose values have to be obtained for each ion pair. The short range term, described by the first two terms in equation (1), are important to mimic two important non-Coulombic interactions: i-the repulsion due to superposition of the electronic clouds of the pair of ions related to the Pauli exclusion principle, sometimes called as Pauli repulsion, and ii-attractive terms due to polarization of the ions and other quantum attractive terms like exchange interactions.

Other important aspect of the modelling strategy is the explicit inclusion of the O^{2-} polarisability via the Shell Model, as proposed by Dick and Overhasuser [18]. This model considers that the polarisable specie is divided between a core that concentrates the mass of the ion and a shell coupled by a mass less spring constant. The charge of the ion is divided between the core and the shell. This model adds two extra parameters per polarisable ion, the spring constant and the shell charge. The short range terms act between the O^{2-} shells or between O^{2-} shells and the Sr^{2+} or Al^{3+} ions. The cations were considered rigid ions in the present work.

The short range potential parameters plus the shell model parameters for the anion were obtained by manual fitting to the structural properties of all strontium aluminate crystalline phases, and the result of that can be seen in table 1. During the fitting procedure, only the potentials parameters for the Sr-O and Al-O interactions were varied. The O-O interaction was held fix following Catlow et al [19]. Table 2 gives the comparison between the experimental and calculated lattice parameters. The structural parameters were evaluated at temperatures where each material was measured, according to the reference quoted on the table. The temperature effect is included in the GULP package based on the harmonic approximation [20]. It can be seen that the potential parameters are quite successfully on reproducing all the crystalline structures within 3%.

Table 1. Potential parameters, shell model constants and ionic charges obtained during the fitting to the structure of all strontium aluminate phases.

Interacion	A(eV)	ρ(Å)	C(eV.Å^6)	Cutoffs(Å)
Sr_{core}-O_{Shell}	1449.0	0.3427	0.00	10.0
Al_{core}-O_{Shell}	1235.3	0.3109	0.00	10.0
O_{shell}-O_{Shell}	22764.0	0.1490	27.88	10.0
Spring				
O_{core}-O_{Shell}	70.0	-	-	-
Charge	Z_{Sr}	Z_{Al}	Z_{Ocore}	Z_{Oshell}
	2.00	3.00	0.900	-2.900

The lattice energy and the lattice energy per ion of each of the strontium aluminate phases were calculated at 0K and 300K and they are shown in table 3. These values can be used to estimate the relative stability of each of these phases and it is shown from the energetic point of view that the most stable phase would be $SrAl_{12}O_{19}$. Comparing the two $SrAl_2O_4$ phases, it is shown that there is practically no difference between their lattice energies indicating that these two phases may coexist. This result is consistent with the experimental finds by Montes et al [21] that reports the existence of these two phases in their samples.

Table 2. Comparison of calculated and experimental structures.

Aluminates	Ref.	Parameter	Experimental	Calculated	Diff.%
$Sr_4Al_{14}O_{25}$ (300K)	[7]	a(Å)	24.7451	25.103094	1.45
		b(Å)	8.4735	8.595316	1.44
		c(Å)	4.8808	4.933321	1.08
$Sr_2Al_6O_{11}$ (300K)	[8]	a(Å)	21.9140	22.231366	1.45
		b(Å)	4.8843	4.947937	1.30
		c(Å)	8.4039	8.508724	1.25
$Sr_3Al_2O_6$ (300K)	[9]	a(Å)	15.8556	15.949574	0.59
		b(Å)	15.8556	15.949574	0.59
		c(Å)	15.8556	15.949574	0.59
$Sr_9Al_6O_{18}$ (300K)	[10]	a(Å)	15.8476	15.949550	0.64
		b(Å)	15.8476	15.949550	0.64
		c(Å)	15.8476	15.949550	0.64
$Sr_7Al_{12}O_{25}$ (300K)	[11]	a(Å)	17.9100	18.355195	2.49
		b(Å)	17.9100	18.355195	2.49
		c(Å)	7.160000	6.931778	-3.19
$SrAl_{12}O_{19}$ (300K)	[12]	a(Å)	5.6581	5.640638	-0.31
		b(Å)	5.6581	5.640638	-0.31
		c(Å)	21.7776	21.917942	0.64
$SrAl_4O_7$ (300K)	[13]	a(Å)	13.0389	13.162647	0.95
		b(Å)	9.0113	9.160527	1.66
		c(Å)	5.5358	5.559149	0.42
		β/graus	106.7000	106.987196	0.27
$Sr_{10}Al_{16}O_{19}$ (300K)	[14]	a(Å)	34.5820	34.722108	0.41
		b(Å)	7.8460	7.910204	0.82
		c(Å)	15.7480	15.826944	0.50
		β/graus	103.6800	103.499167	-0.17
$SrAl_2O_4$ (monoclinic) (300K)	[15]	a(Å)	8.4470	8.442699	-0.05
		b(Å)	8.8160	9.033006	2.46
		c(Å)	5.1630	5.229848	1.29
		β/graus	93.4200	92.572269	-0.91
$SrAl_2O_4$ (hexagonal) (1073K)	[16]	a(Å)	8.9260	9.070489	1.62
		b(Å)	8.9260	9.070489	1.62
		c(Å)	8.4985	8.513597	0.18

Table 3. Lattice energies calculated at 0K and 300K and lattice energy per ion for each of the strontium aluminate phase.

Compoud	Lattice energy (eV)	Lattice Energy (eV) per íon	Lattice energy (eV)	Lattice Energy (eV) per íon
	0K		300K	
$SrAl_2O_4$(monoclinic)	-192.77	-27.54	-192.39	-27.48
$SrAl_2O_4$(hexagonal)	-192.77	-27.54	-192.37	-27.48
$SrAl_4O_7$	-350.28	-29.19	-349.52	-29.13
$Sr_4Al_{14}O_{25}$	-1245.10	-28.96	-1242.45	-28.89
$Sr_3Al_2O_6$	-262.06	-23.82	-261.61	-23.78
$Sr_{10}Al_6O_{19}$	-819.77	-23.42	-818.53	-23.39
$Sr_9Al_6O_{18}$	-786.19	-23.82	-784.82	-23.78
$SrAl_{12}O_{19}$	-983.4	-30.73	-981.15	-30.66
$Sr_2Al_6O_{11}$	-544.03	-28.63	-542.81	-28.57
$Sr_7Al_{12}O_{25}$	-1167.04	-26.52	-1162.31	-26.42

3.2. Defect calculations

Defects were modelled using the Mott–Littleton approximation [22], in which a spherical region of the lattice surrounding the defect (region I) is treated explicitly, with all interactions being considered, and more distant parts of the lattice (region II) are treated using a continuum approach. A further division of the region II in region II(a) and II(b) are necessary to accommodate the distortions of the lattice induced by the defect. In region II(a) all displacements of the ions are calculated considering an harmonic approximation. Region II(b) is treated as a continuum dielectric media since the influence of the defect is negligible. Consistent cut-off radii of 10 and 14 Å have been used for regions I, and II (a) respectively. All this approximations are embodied in the GULP code.

As for the intrinsic defect (Frenkel, Schottky and others) standard defect reactions have been devised and they are listed in table 4. Basic defect formation energies were calculated, and used to evalulate the energies of defect reactions involved in the intrinsic disorders.

Table 4. Defect reactions used to compute the energies of the intrinsic defects for all Sr-Al-O crystalline systems.

Defect type	reactions
Frenkel defect	$Sr_{Sr} \rightarrow V''_{Sr} + Sr_i^{\bullet\bullet}$
	$Al_{Al} \rightarrow V'''_{Al} + Al_i^{\bullet\bullet\bullet}$
	$O_O \rightarrow V_O^{\bullet\bullet} + O_i$
Schottky defect	$xSr_{Sr} + yAl_{Al} + zO_O \rightarrow xV''_{Sr} + yV'''_{Al} + zV_O^{\bullet\bullet} + Sr_xAl_yO_z$
Anti-Shottky	$Sr_xAl_yO_z \rightarrow xSr_i + yAl_i + zO_i$
Pseudo-Schottky	$Sr_{Sr} + O_O \rightarrow V''_{Sr} + V_O^{\bullet\bullet} + SrO$
	$2Al_{Al} + 3O_O \rightarrow 2V'''_{Al} + 3V_O^{\bullet\bullet} + Al_2O_3$

4. Results and discussions

Computer simulations are an ideal way to approach intrinsic disorder in these complex systems because each defect can be treated separately and systematically. In strontium aluminates, vacancy and interstitial formation energies are calculated for each of the possible species. From these quantities, the energies per defect for the various possible types of disorder may be obtained.

In Table 5 all formation energies for the vacancy defects are displayed. It should be noticed that all possible non equivalent sites were considered for each material. In table 6(a), the formation energies for Sr, Al or O interstitials are shown for all 9 crystalline structures. In table 6(b) the interstitial positions used to calculate the previous values are shown.

Using the reactions presented in table 4 with values in tables 3, 5 and 6(a), the Schottky, pseudo-Schottky, anti-Schottky and Frenkel ddefect energies were calculated and this values are given in table 7. It should be pointed that only the lowest defect energy in each case were considered among all possible combinations of the values in tables 5 and 6(a). Also, the values quoted in table 7 were normalized by the number of basic defects forming the intrinsic defect, e.g., the Frenkel energies were divided by 2, Al_2O_3 pseudo-Schottky energies were divided by 5, and so on. The values in table 7 were computed considering that the basic defects involved in a specific intrinsic defect are not bound meaning that each constituent of the defect are formed separately in the lattice.

Table 5. Formation energy of all vacancy defect at non equivalent site.

$Sr_{10}Al_6O_{19}$

Defect	Energy(eV)
V_{Sr1}	18.96
V_{Sr2}	18.46
V_{Sr3}	18.62
V_{Sr4}	18.90
V_{Sr5}	19.13
V_{Sr6}	18.79
V_{Sr7}	18.86
V_{Sr8}	18.67
V_{Sr9}	19.10
V_{Al1}	58.82
V_{Al2}	57.76
V_{Al3}	58.10
V_{Al4}	58.09
V_{Al5}	58.46
V_{Al6}	59.98
V_{O1}	20.90
V_{O2}	22.41
V_{O3}	21.32
V_{O4}	21.15
V_{O5}	21.25
V_{O6}	22.02
V_{O7}	21.54
V_{O8}	22.38
V_{O9}	22.07
V_{O10}	22.47
V_{O11}	21.70
V_{O12}	19.19
V_{O13}	21.10
V_{O14}	21.70
V_{O15}	21.81
V_{O16}	20.76
V_{O17}	21.41
V_{O18}	21.35
V_{O19}	20.88

$Sr_2Al_6O_{11}$

Defect	Energy(eV)
V_{Sr1}	19.17
V_{Sr2}	18.83
V_{Al1}	59.36
V_{Al2}	56.69
V_{Al3}	56.88
V_{Al4}	55.35
V_{Al5}	57.29
V_{O1}	22.32
V_{O2}	24.79
V_{O3}	22.14
V_{O4}	23.40
V_{O5}	22.22
V_{O6}	22.68
V_{O7}	24.25

$Sr_3Al_2O_6$

Defect	Energy(eV)
V_{Sr1}	19.29
V_{Sr2}	17.61
V_{Sr3}	18.00
V_{Sr4}	17.84
V_{Sr5}	18.80
V_{Sr6}	19.73
V_{Al1}	60.73
V_{Al2}	59.19
V_{O1}	23.37
V_{O2}	23.19
V_{O3}	23.00
V_{O4}	21.87
V_{O5}	21.89
V_{O6}	22.85

$Sr_4Al_{14}O_{25}$

Defect	Energy(eV)
V_{Sr1}	18.77
V_{Sr2}	19.40
V_{Al1}	55.63
V_{Al2}	55.49
V_{Al3}	55.44
V_{Al4}	56.90
V_{Al5}	59.33
V_{Al6}	56.65
V_{O1}	22.72
V_{O2}	23.50
V_{O3}	24.49
V_{O4}	26.89
V_{O5}	22.33
V_{O6}	22.48
V_{O7}	25.18
V_{O8}	23.06
V_{O9}	24.62

$SrAl_2O_4$(hex)

Defect	Energy(eV)
V_{Sr1}	19.38
V_{Sr2}	20.04
V_{Al1}	56.66
V_{Al2}	57.83
V_{O1}	23.39
V_{O2}	24.31
V_{O3}	23.00
V_{O4}	24.27

$SrAl_4O_7$

Defect	Energy(eV)
V_{Sr1}	19.78
V_{Al1}	56.32
V_{Al2}	55.68
V_{O1}	23.60
V_{O2}	23.36
V_{O3}	25.25
V_{O4}	24.12

$Sr_9Al_6O_{18}$

Defect	Energy(eV)
V_{Sr1}	19.29
V_{Sr2}	17.61
V_{Sr3}	18.00
V_{Sr4}	17.84
V_{Sr5}	18.80
V_{Sr6}	19.73
V_{Al1}	60.73
V_{Al2}	59.19
V_{O1}	23.37
V_{O2}	23.19
V_{O3}	23.00
V_{O4}	21.87
V_{O5}	21.89
V_{O6}	22.85

$SrAl_2O_4$(mono)

Defect	Energy(eV)
V_{Sr1}	19.48
V_{Sr2}	19.68
V_{Al1}	57.78
V_{Al2}	57.73
V_{Al3}	56.81
V_{Al4}	57.61
V_{O1}	23.54
V_{O2}	24.52
V_{O3}	24.61
V_{O4}	24.54
V_{O5}	24.17
V_{O6}	24.64
V_{O7}	23.61
V_{O8}	23.10

$SrAl_{12}O_{19}$

Defect	Energy(eV)
V_{Sr1}	19.41
V_{Al1}	54.70
V_{Al2}	54.84
V_{Al3}	54.18
V_{Al4}	54.49
V_{Al5}	58.28
V_{O1}	23.99
V_{O2}	22.75
V_{O3}	22.64
V_{O4}	24.68
V_{O5}	23.36

Table 6(a). Formation energies of the interstitial defects for all strontium aluminate systems.

Compoud	Defect Energy(eV)								
Def. type	$SrAl_2O_4$(m)	$SrAl_2O_4$(h)	$SrAl_4O_7$	$Sr_4Al_{14}O_{25}$	$Sr_3Al_2O_6$	$Sr_{10}Al_6O_{19}$	$Sr_9Al_6O_{18}$	$SrAl_{12}O_{19}$	$Sr_2Al_6O_{11}$
Sr_i	-12.09	-12.39	-12.15	-12.34	-14.92	-11.04	-10.00	-6.60	-11.74
Al_i	-44.31	-44.25	-42.80	-39.59	-45.25	-45.61	-45.39	-43.96	-43.04
O_i	-14.43	-15.29	-18.52	-14.86	-10.53	-15.18	-14.60	-15.90	-14.48

Table 6(b): Interstitial positions used for calculating the interstitial formation energies.

	Intersticial position(fractional)		
Compoud	x	y	z
$SrAl_2O_4$(monoclinic)	0.3680	0.1580	0.2945
$SrAl_2O_4$(hexagonal)	0.3320	0.3370	1.0000
$SrAl_4O_7$	0.0000	0.4050	0.1250
$Sr_4Al_{14}O_{25}$	0.2500	0.5000	0.7965
$Sr_3Al_2O_6$	0.1145	0.8705	0.7437
$Sr_{10}Al_6O_{19}$	1.0000	0.9810	0.5671
$Sr_9Al_6O_{18}$	0.2500	0.0000	0.0000
$SrAl_{12}O_{19}$	0.3333	0.1667	0.1250
$Sr_2Al_6O_{11}$	0.1804	0.0604	0.0000

Table 7. Solution energies of the intrinsic defects for the Sr-Al-O systems.

Compoud	Defect Energy(eV)								
Defect type	$SrAl_2O_4$(h)	$SrAl_2O_4$(m)	$SrAl_4O_7$	$Sr_4Al_{14}O_{25}$	$Sr_3Al_2O_6$	$Sr_{10}Al_6O_{19}$	$Sr_9Al_6O_{18}$	$SrAl_{12}O_{19}$	$Sr_2Al_6O_{11}$
Frenkel									
Sr	**3.70**	**3.50**	3.82	**3.22**	**1.35**	3.71	3.81	6.41	3.55
Al	6.25	6.21	6.44	7.93	6.97	6.08	6.90	5.11	6.16
O	4.34	3.86	**2.42**	3.74	5.67	2.01	3.64	**3.37**	3.83
Pseudo-Schottky									
SrO	4.08	3.98	4.36	3.34	2.53	**1.61**	**2.53**	3.81	**3.27**
Al_2O_3	4.94	4.82	4.64	3.93	5.15	2.97	5.15	3.65	3.81
Schottky	4.68	4.56	4.65	3.82	3.67	2.17	3.67	3.63	3.65
Anti-Schottky	4.91	4.39	7.74	6.28	5.78	4.21	4.88	4.60	5.42

Results presented in table 7 reveals that the 9 crystalline systems can be divided in 3 groups according to the preferred intrinsic disorder. Group I contains both phases of $SrAl_2O_4$, $Sr_4Al_{14}O_{25}$ and $Sr_3Al_2O_6$ where the Sr Frenkel disorder gives the lowest energies. In group II are the systems where the O Frenkel defect is more likely to occur and this group is formed by the $SrAl_4O_7$ and $SrAl_{12}O_{19}$. Finally, $Sr_{10}Al_5O_{19}$, $Sr_9Al_6O_{18}$ and $Sr_2Al_6O_{11}$ form group III, where the pair of Sr and O vacancies (pseudo-Schottky type defect) are the most favourable defect.

For the $SrAl_{12}O_{19}$ case, our predictions agree with the results obtained by Jae-Gwan Park and Cormack [23]. Experimental results by Beauger et al [24] showed that during the synthesis of $SrAl_2O_4$, the $Sr_3Al_2O_6$ is also formed as a marginal phase. The authors attributed that to the presence of strontium vacancies (V_{Sr}), and hence oxygen vacancies (V_O), in the main $SrAl_2O_4$ phase leading to a stoichiometric deviation from the $SrAl_2O_4$ phase forming a region rich with Sr and O that can give rise to the undesirable $Sr_3Al_2O_6$ phase. They also showed that their thermoluminescence (TL) measurements carried out for the stoichiometric and nonstoichiometric samples of $SrAl_2O_4$: Eu^{2+} [24] indicate that Sr and O vacancies could be present. This agree with the new phosphorescence mechanism proposed of Clabau et al. proposed for $SrAl_2O_4$:Eu^{2+}, Dy^{3+}, B^{3+}[1].

5. Conclusions

A new set of potential parameters were fitted in order to reproduce all the experimental crystalline structure available for the Sr-Al-O compounds and the potential parameters were able to reasonably reproduce all the crystalline structures within 3%. This potential set was then used to model the intrinsic defects of the strontium aluminates compounds. According to the obtained energy, we found that in $Sr_2Al_6O_{11}$, $Sr_9Al_6O_{18}$ and $Sr_{10}Al_6O_{19}$ SrO pseudo-Schottky defect is more likely to occur. In $SrAl_4O_7$ and $SrAl_{12}O_{19}$ the lowest energy was obtained for the oxygen Frenkel defect. On the other hand, in $SrAl_2O_4$ (monoclinic), $SrAl_2O_4$ (hexagonal), $Sr_3Al_2O_6$, and $Sr_4Al_{14}O_{25}$ the Strontium Frenkel defect is the preferred intrinsic defect.

Acknowledgements

The authors are grateful to CNPq, CAPES and FINEP for financial support.

Reference

[1] Clabau F, Rocquefelte X, Jobic S, Deniard S, Whangbo M-H, Garcia A, Mercier T Le 2005 *Chem. Mater.* **17** 3904-3912.

[2] Lin Y, Tang Z, Zhang Z 2001 *Materials Letters* **18** 5114.

[3] Chang C, Mao D, Shen J, Feng C 2003 *J. Alloys Compd.* **348** 224.

[4] Zhong R, Zhang J, Zhang X, Lu S, Wang X-J 2006 *Journal of Luminescence* **119–120** 327-331.

[5] Shionoya S, Yen W 1999 *Phosphor Handbook*, CRC Press, New York.

[6] McKeever S 1988 *Thermoluminescence of Solids*, Cambridge Press University, Cambridge.

[7] Wang D, Wang M-Q, Lu G-G 1999 *Journal of Materials Science* **34** 4959-4964.

[8] Takeda T, Takahashi T, Takeda K, Takahashi T, Uheda K, Takizawa K, Takeda H 2004 *Journal of the Japan Society of Powder Metallurgy* **49** 1128-1133.

[9] Chakoumakos B C, Lager G A, Fernandez-Baca J A 1992 *Acta Crystallographica C* **48** 414-419.

[10] Alonso J A, Rasines I, Soubeyroux J L 1990 *Inorganic Chemistry* **29** 4768-4771.

[11] Nevskii N N, Glasser L D, Ilyukhin V V, Belov N V 1979 *Kristallografiya* **24** 161-166.

[12] Park J-G, Cormack A N 1996 *Journal of Solid State Chemistry* **121** 278-290.

[13] Lindop A J, Goodwin D W 1972 *Acta Crystallographica* **28** 2625-2526.

[14] Kahlenberg V 2003 *Materials Research Bulletin* **37** 715-726.

[15] A. R. Schulze, Hk. Mueller-Buschbaum, Zeitschrift fuer Anorganische und Allgemeine Chemie 475 (1981) 205-210.

[16] Fukuda K, Fukushima K 2005 *Journal of Solid State Chemistry* **178** 2709-2714.

[17] Gale J D 1997 *J. Chem. Soc., Faraday Trans.* **93** 629.

[18] Dick B J, Overhauser A W 1958 *Phys. Rev.* **112** 90.

[19] Sandes M J, Leslie M, Catlow C R A 1984 *J. Chem. Soc. Chem. Commun* 1271.

[20] LeSar R, Najafabati R, Srolovitz D J 1991 *J. Chem. Phys.* **94** 5090–5097.

[21] Montes P J R, Valerio M E G, Azevedo G M 2008 *Nuclear Instruments and Methods in Physics Research B* **266** 2923-2927.

[22] Mott N F, Littleton M J 1938 *Trans. Farady Soc.* **34** 485.

[23] Park J-G, Cormack N A 2000 *Korean J. Crystallography* **11** 176-181.

[24] Beauger, *Thesis C, Université de Nice* 1999 Nice, France.

Journal of Physics: Conference Series **249** (2010) 012043 doi:10.1088/1742-6596/249/1/012043

Growth and characterization of single crystal fibers of Nd^{3+}:$NaLa(WO_4)_2$

J R de Moraes[1], S L Baldochi[1], S Ganschow[2], A M E Santo[3] and A A Martin[3]

[1]Instituto de Pesquisas Energéticas e Nucleares, IPEN – CNEN/SP
Av. Lineu Prestes, 2242 – CEP 05508-000 – São Paulo/SP – Brazil
[2]Institute for Crystal Growth (Institut für Kristallzüchtung)
Max-Born-Strasse 2, D-12489 Berlin – Germany
[3]Instituto de Pesquisa & Desenvolvimento, Universidade do Vale do Paraíba-UNIVAP
Av. Shishima Hifumi, 2911 – CEP 12244-000 – São José dos Campos/SP – Brazil

E-mail: baldochi@ipen.br

Abstract. Single crystal fibers of pure and Nd^{3+}-doped $NaLa(WO_4)_2$ were grown by the micro-pulling down method (μPD) under air atmosphere. The starting materials were synthesized through solid state reaction. Inductive and resistive configurations of μPD systems were tested. The control of the growth process was better achieved through the resistive mode than in the inductive one. Optical and structural properties are compatible with the data already reported in the literature for bulk crystals of this host.

1. Introduction

Recently, there has been an increasing interest in the production of single-crystalline fibers of many different compounds. Their unique properties make their use in a variety of optical and electronic devices possible. Additionally, the growth of single crystalline fibers is fast and relatively low-cost compared to others crystal growth techniques from the melt, making their production a very suitable tool in the search for new materials.

In these last decades several works were reported regarding the properties of alkali rare earth tungstates - $A(RE)(WO_4)_2$, where A= K, Na, Li; and (RE) = rare earth elements. They are promising materials for optical applications due to their excellent luminescence properties, particularly as laser hosts, when doped with optically active rare earth ions [1]. In special, Nd^{3+} was the first of the trivalent rare earth ions to be used in a laser, and remains by far the most important element in this group. Stimulated emission has been obtained with this ion incorporated in at least 100 different host materials, and a higher power level has been obtained from Nd lasers than from any other four-level material [2]. Recent spectroscopic studies of Nd^{3+}:$NaLa(WO_4)_2$ crystals have shown high luminescent quantum efficiency, moderate emission cross-section and FWHM large emission in the 1.06 and 1.35μm regions, making this material a promising active host for tunable solid state laser [3-4].

Many reports about Czochralski growth (CZ) of $Na(RE)(WO_4)_2$ and TSSG of $K(RE)(WO_4)_2$ crystals have already been published. However, very little investigations were performed on single crystal fiber growth of these materials. In fact, only one report was found on fiber growth of $NaGd(WO_4)_2$ and $NaY(WO_4)_2$ [5].

In this work, the preparation of pure and Nd^{3+}-doped $NaLa(WO_4)_2$ (NLW) compounds through solid state reaction and the growth process of single crystal fibers through the micro-pulling down method (μPD) was studied. $NaLa(WO_4)_2$ crystals are isostructural with scheelite tetragonal structure ($CaWO_4$ – space group C^6_{4h}-I^4_1/a). They melt congruently at 1253°C [6]. In the μPD method, the fiber growth stability is controlled through the capillarity effect where the meniscus shape must be carefully controled so that an uniform fiber pulling process occurs [7]. Details about μPD method can be found in the refs. [8,9,10].

2. Experimental procedure

Starting materials for fiber growth experiments were synthesized in a kanthal resistive furnace equipped with a temperature controller model 2416 from Eurotherm. The synthesis process was realized through solid state reaction as described in the literature by Serrano *et al.* [11] which is summarized as follows.

The initial chemicals of analytical grade – La_2O_3, Nd_2O_3, WO_3 and Na_2CO_3 – were carefully mixed in an agate mortar, in the appropriate molar ratio in accordance with the reaction

$$Na_2CO_3 + (1-x)\ La_2O_3 + x\ Nd_2O_3 + 4\ WO_3 \rightarrow 2\ NaLa_{1-x}Nd_x(WO_4)_2 + CO_2\uparrow \quad (x = 0, 0.01, 0.02) \quad (1)$$

to obtain $NaLa_{1-x}Nd_x(WO_4)_2$ compounds. The powder mixture was annealed in two steps under air atmosphere. The treatment of the material was performed with alumina or platinum crucibles. Firstly, the powder was heated at a rate of 50°C/h to 750°C and treated in this temperature for a period of 18h. After cooled down to room temperature the product of reaction was mixed in an agata mortar and submitted to a second heating step at a rate of 100°C/h to 850°C and held at this temperature for 24h.

X-ray diffraction (XRD) analyses of the synthesized tungstates were carried out with a Bruker AXS A8 Advance diffractometer with a Cu cathode (λ = 1.054Å). The data were collected using a Ni filtered Cu-target tube at room temperature in the 2θ range of 15–80°, step 0.02°/5s. XRD patterns of pure and doped compounds were compared and lattice parameters were determined through the Rietveld method using the DBWS software [12].

Single crystal fibers growth experiments were performed in a commercial resistive μPD system from Linn High Therm and in a homemade inductive μPD system equipped with an induction Versa Power (5KW 10-50KHz) InductoTherm frequency converter. The visual control of the experiments was performed with a stereo microscope. The starting materials were manually pressed into cylindrical pellets of 3.0mm of diameter and 10.0mm in thickness and placed in Pt crucibles in both cases.

Micro-EDX analyses were performed in order to characterize the Nd^{3+} incorporation. The measurements were taken longitudinally using a fluorescence spectrometer μEDX 1300, Shimadzu Inc. The scanning was done with steps of 1.0mm and count time of 100s per point. Absorption spectroscopy was performed at room temperature in the VIS and near IR range (450-950nm) with a CARY-17 Spectrophotometer, Conversion *OLIS Inc.*

3. Results and discussion

3.1. Starting material synthesis and characterization

Evaluation of X-ray powder diffraction reveals that the obtained polycrystalline compounds, when annealed in an alumina or platinum crucible, are formed by a main phase $NaLa(WO_4)_2$ (84wt%) and a secondary phase $Na_2W_4O_{13}$ (16wt%). A non-stoichiometric reaction is probably the reason of the secondary phase formation.

The high temperature may result in evaporation of one or more components resulting non-stoichiometric reactions. Analyzing the synthesis equation (1), the formation of $Na_2W_4O_{13}$ phase can be only explained if we consider a lower concentration of La_2O_3, during the processing of the samples, stimulating the occurrence of a secondary reaction with Na_2CO_3 + $4WO_3$ resulting in $Na_2W_4O_{13}$ compound. The presence of this secondary compound was not detected in grown fibers.

Figure 1 shows the XRD patterns, analyzed by DBWS software, of a pure synthesized powder identified as the scheelite tetragonal NLW (main phase).

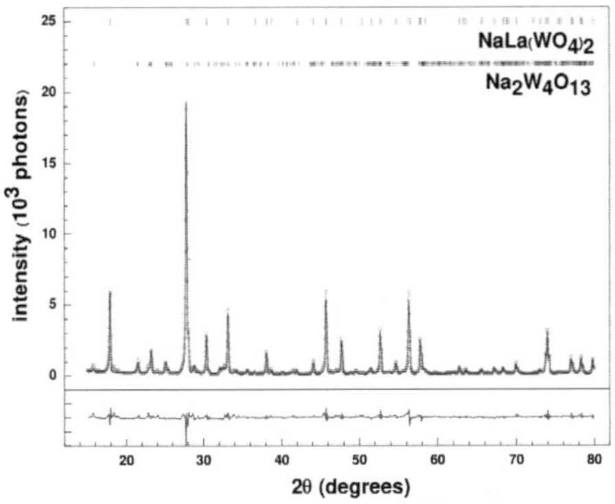

Figure 1. X-ray powder diffraction pattern of solid state reaction synthesized NLW.

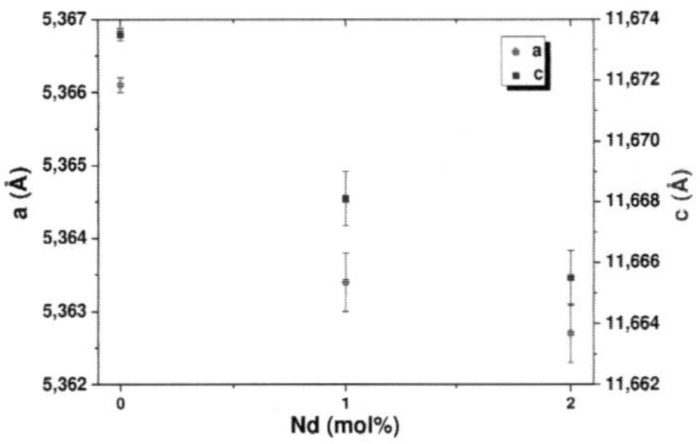

Figure 2. Lattice parameters variations as a function of Nd concentration.

Lattice parameters were obtained for all NLW compounds as shown in table 1. The decrease of the cell parameters due to the La substitution by Nd ions can be observed in figure 2. This result is justified since ionic radii of Nd is smaller (109.0pm) than of La^{3+} (118.0pm) and Na^{+}(116.0pm) [13] for octahedral coordination [14] as in this case. The calculated values are comparable with those observed in bulk crystals [15].

Table 1. Lattice parameters of pure and Nd-doped NLW powders.

compound	*a* (Å)	*c* (Å)	volume (Å^3)
bulk crystal [15]	5.349	11.628	332.7
$NaLa(WO_4)_2$	5.3661(1)	11.6735(2)	336.13(1)
$NaLa_{0,99}Nd_{0,01}(WO_4)_2$	5.3634(4)	11.6681(9)	335.65(5)
$NaLa_{0,98}Nd_{0,02}(WO_4)_2$	5.3627(4)	11.6655(9)	335.48(5)

3.2. Crystal fiber growth through the μPD method

In order to obtain seeds, fiber growth tests in an inductive μPD system were first performed using as starting materials pieces of CZ crystals of Nd:NLW. The initial seed was an arbitrary oriented piece of crystal grown through the CZ method. The growth conditions are summarized in table 2.

Table 2. Growth conditions of Nd-doped NLW in the inductive μPD system.

crucible (cylindrical)	Pt (3.0 cm^3)
nozzle diameter	0.6mm
pulling rate	0.3-0.5mm/min
atmosphere	air

Power control on this system was observed to be very critical at the start of the growth process. Variations of about 0,1kW in the radio frequency system result in significant changes in the growth interface as the liquid quickly flows to the exterior side of crucible. An appropriate control was only obtained with power variations of about 0,05kW which is very close to the limit of this equipment.

Such small variations make it possible to obtain controlled concave and convex liquid drops in the nozzle facilitating the seeding process. Once a stable meniscus was formed, Nd-doped NLW fibers with ~1.0mm in diameter and 10mm in length were obtained. All of them are transparent and uniform as showed in the figures 3.

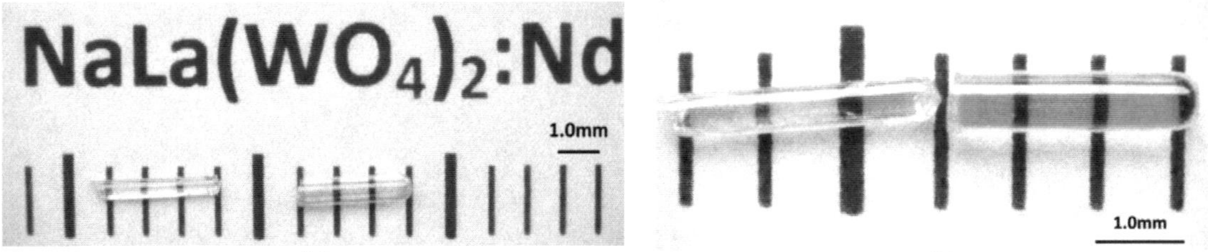

(a) (b)

Figure 3. (a) As-grown Nd:NLW fiber obtained through inductive μPD and (b) its optical zoom.

Growth experiments were also performed in a μPD resistive mode. In this system there are two heater components (main and after heater – figure 4) that can be fitted to provide an ideal temperature gradient for stable crystallization conditions. The control of the process was better achieved through this system than in the inductive one. By adjusting the distances between the main and after heaters and also between the after heater spirals a better temperature control was obtained. Fiber growth parameters are summarized in table 3.

Table 3. Growth conditions of pure and Nd-doped NLW in the resistive μPD system.

crucible	Pt (15.0 x 4.0 x 1.0mm^3)
nozzle diameter	0.8 and 1.0mm
pulling rate	0.2-0.4mm/min
atmosphere	air

In the growth process a constant and uniform meniscus, which is essential to produce regular fibers in diameter, was obtained through the temperature control of the crucible and after heater. Once a stable flat crystallization interface was achieved, no more changes of the pulling rate and main and after heater temperatures were necessary. The dimensions of the growth zone were measured through video microscope images during the crystal growth process as shown in figure 5. The meniscus height is given by the following empiric equation [16]:

$$H_{max} \approx \frac{D_{fiber}}{2} = R_{fiber} \qquad (2)$$

The control of the temperature of the after heater (Pt wire – figure 4) allowed the pulling of free crack fibers of all the grown compositions. Transparent and uniform fibers with 0.8mm in diameter and 15mm in length were pulled (figure 6-8). Doped fibers showed transparent purple color.

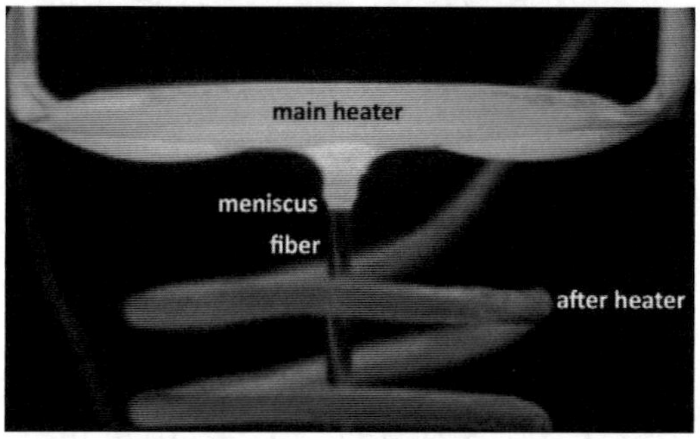

Figure 4. Nd-doped $NaLa(WO_4)_2$ single-crystalline fiber during growth process. *Note:* 0.8mm diameter of the crystal as scale reference.

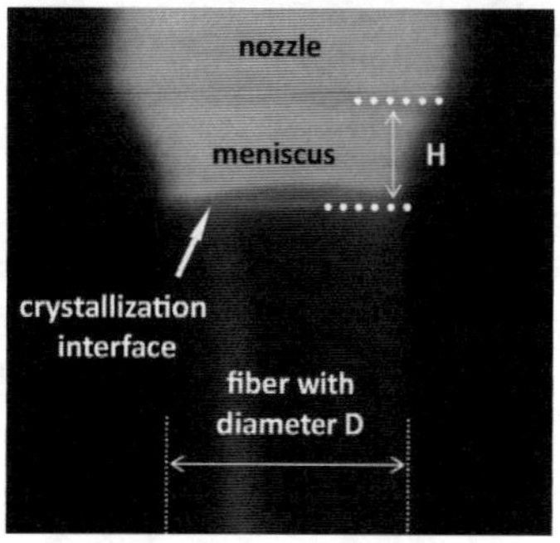

Figure 5. Meniscus shape during Nd-doped $NaLa(WO_4)_2$ single-crystalline fiber growth process.

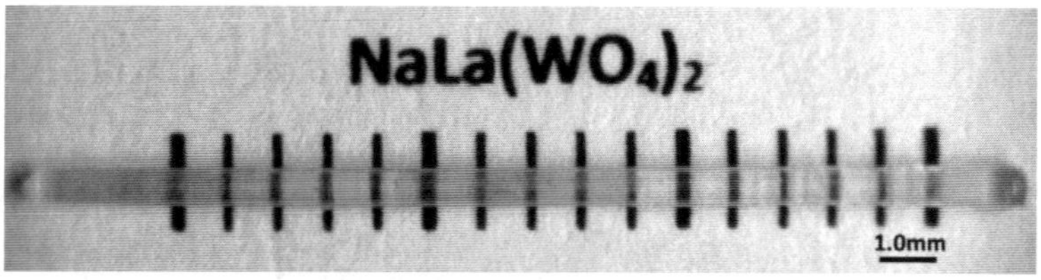

Figure 6. Stoichiometric $NaLa(WO_4)_2$ single-crystalline fiber.

Figure 7. As grown $NaLa_{0,99}Nd_{0,01}(WO_4)_2$ single-crystalline fiber.

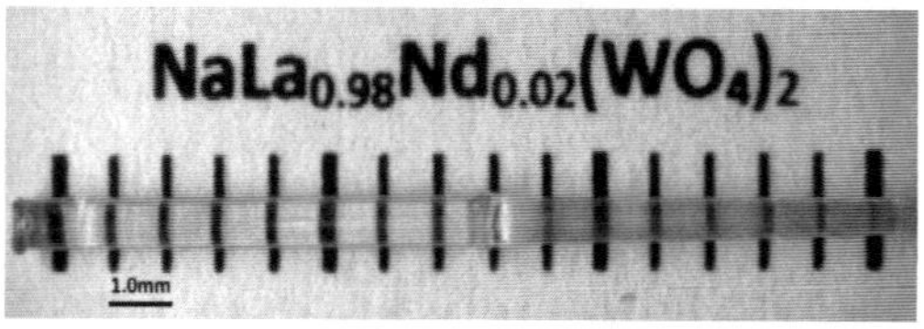

Figure 8. As grown $NaLa_{0,98}Nd_{0,02}(WO_4)_2$ single-crystalline fiber.

3.3. Fiber characterization

The main fluorescence lines of La and Nd elements are very close, which difficultates the determination of the real dopant concentration in NLW crystals through the μEDX technique. Due to this, the data was obtained taking into account the quantification of La atoms with Na atoms and the $(WO_4)_2$ group fixed, and using the Nd element as measurement balance. It was also considered that the Nd ions occupy La sites due to their similar valence and ionic radii, as discussed in the 3.1 section of this work. The average concentrations in the fibers are summarized in the table 4.

Table 4. Average concentration of Nd ion in the NLW fibers.

nominal concentration (mol%)	measured concentration (mol%)
1.0	1.58(28)
2.0	1.98(19)

The absorption spectrum (figure 9) showed absorption bands of the neodymium trivalent cation (Nd^{3+}) in the NLW host (in the single-crystalline fiber shape) in accordance with data already reported at literature for bulk crystals [4]. All of the observed transitions were identified.

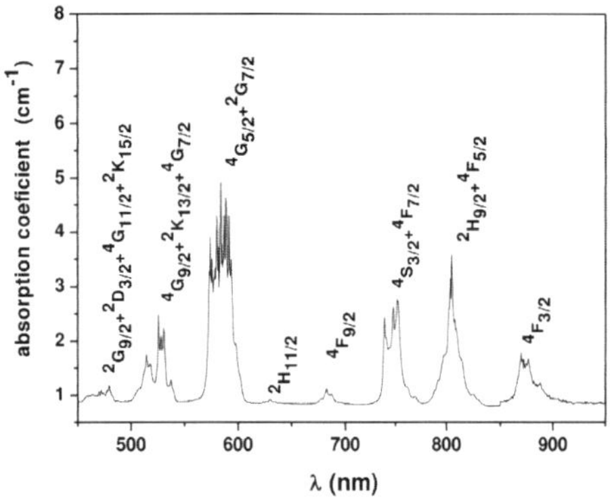

Figure 9. Absorption spectrum at room temperature in the VIS and near IR range of Nd^{3+}-doped NLW host.

4. Conclusions

The ideal conditions to obtain single-crystalline fibers of pure and Nd-doped $NaLa(WO_4)_2$ compounds with good optical quality for optical properties evaluation were established. The control of the growth process in the μPD system was better achieved through the resistive mode than in the inductive one. Optical properties are compatible with the data already reported in the literature for bulk crystals of this host. In addition, the results show that the incorporation of neodymium trivalent cation (Nd^{3+}) by the NLW host is relatively easy.

Acknowledgments

The authors acknowledge CAPES/DAAD, FAPESP and CNPq for their financial support.

References

[1] Kaminskii A A 1996 *Crystalline laser: physical processes and operating scheme* (Boca Raton, FL: CRC Press)
[2] Koechner W 2006 *Solid state laser engineering* (New York, NY:Springer)
[3] Wei Y, Tu C, Wang H, Yang F, Jia G, You Z, Lu X, Li J, Zhu Z and Wang Y 2007 *Opt. Mater.* **29** 1653
[4] Garcia-Cortés A, Cascales C, Andrés A, Zaldo C, Zharikov E, Subbotin K A, Bjurshagen S, Pasiskevicius V and Rico M 2007 *IEEE J. Quant. Electron.* **43** (2) 157
[5] Terada Y, Shimamura K and Fukuda T 1998 *J. Alloys. Comp.* **275-277** 697
[6] Rode E Ya, Karpov V N and Ivanova M M 1971 *Russ. J. Inorg. Chem.* **16** (6) 905
[7] Santo A M E, Librantz A F H, Gomes L, Pizani P S, Ranieri I M, Vieira Jr N D and Baldochi S L 2006 *J. Crystal Growth* **292** 149
[8] Yoon D H, and Fukuda T 1994 *J. Crystal Growth* **144** 207
[9] Fukuda T, Rudolph P *et al* (eds) 2004 *Fiber crystal growth from the melt* (Berlin:Springer)
[10] Fukuda T and Chani V I (eds) 2007 *Shaped crystals: growth by micro-pulling-down technique* (Berlin:Springer)
[11] Serrano M D, Esteban-Betegón F and Zaldo C 2004 *J. Crystal Growth* **275** e819
[12] R. A. Young, A. C. Larson and C. O. Paiva-Santos, User´s Guide to Program DBWS-9807a for Rietveld Analysis of X-ray and Neutron Powder Diffraction Patters (Institute of Technology, Atlanta 1999).
[13] Shannon R D and Prewitt C T 1969 *Acta Cryst.* B **25** 925
[14] Méndez-Blas A, Rico M, Volkov V, Cascales C, Zaldo C, Coya C, Kling A and Alves L C 2004 *J. Phys.: Condens. Matter* **16** 2139
[15] Li H, Hong G and Yue S 1990 J. Rare Earth **8** 37 41
[16] Karek A, Lebbou K, Diaf M, Brenier A and Boulon G 2007 *Mat. Res. Bull.* **42** 532-543

Journal of Physics: Conference Series **249** (2010) 012044 doi:10.1088/1742-6596/249/1/012044

GROWTH AND CHARACTERIZATION OF $LiYF_4$: Er^{+3} FIBERS

F R Silva[1], A Mirage[1], A M E Santo[2], A A Martin[2], S L Baldochi[1]

[1]Instituto de Pesquisas Energéticas e Nucleares – IPEN-CNEN/SP – Brazil
[2]Instituto de Pesquisa e Desenvolvimento IP&D/UNIVAP - Brazil

Contact address: fersilva@ipen.br

Abstract. The growth of single crystal fibers of $LiYF_4$ (YLF) doped with Er^{3+} was studied through the resistive micro-pulling-down (μ-PD) technique. The growth chamber was modified to achieve vacuum of approximately 10^{-7}torr. Single crystal fibers of YLF: Er (1, 10 and 20 mol %), with 0.7 mm of diameter and up to 120mm of length were grown. The use of a static atmosphere was found to be inappropriate for elimination of spurious contamination from growth environment for this experimental setup.

1. Introduction

Laser systems are attractive tools for many applications, as for example environmental monitoring, medicine, industry and basic research. Hence, laser action in solid state systems has been investigated using a wide variety of materials in different conditions and new laser systems are under continuous development as a consequence of the demand created by new applications. The need of compact laser sources led, for example, to the investigation of laser crystals in the fiber geometry. The light-guiding properties of fibers can be combined with the conversion properties of bulk single-crystals as active laser media [1]. Moreover, the fiber geometry improves the output power on these systems because it removes heat efficiently due to the short distance between the pumping region and environment [2].

Yttrium lithium fluoride crystals ($LiYF_4$ or YLF) doped with rare earth elements present laser emission in a wide wavelength range. This fluoride material is a well-known laser host in the shape of bulk crystals. Nevertheless, there are not many reports about its growth as fibers, particularly when doped with Er^{+3}. Undoped and Nd^{3+} doped YLF fibers were grown by Santo et al. [1, 3] from a composition slightly enriched with LiF. A peritetic transient preceding the growth of the crystalline YLF: Nd was reported.

In this work we report the preparation and characterization of YLF-doped Er^{+3} fibers grown by the micro-pulling down technique [2]. YLF:Er fibers were obtained without the formation of an initial peritetic transient.

2. Experimental procedure

Starting compounds (YF_3, ErF_3) were obtained through hidrofluorination of commercial oxides [4]. The LiF compound was pre-purified through the zone melting process. Pure YLF single crystals also obtained through zone melting were used as starting material. The Er-doped YLF was prepared through melting of the components under protective reactive atmosphere (HF). Five mol percent of LiF excess was used in the initial charge, since previous works have shown that the best results to the growth of YLF fibers, in a micro-pulling down system, were achieved when such LiF excess was introduced [1].

Previously to the growth process, the quartz growth chamber was first heated under vacuum of 10^{-3}torr and back-filled with ultra-pure argon. Subsequently, the chamber was evacuated up to a pressure of 10^{-7}torr, which was sustained for at least 24 hours and was back-filled with a mixed atmosphere of CF_4 plus Ar for the fiber growth. The growth experiments were performed under a static atmosphere of this mixture in a modified commercial micro-pulling down system, in resistive mode. Platinum handmade crucibles were used; the method to fabricate the crucible was adapted in order to achieve reproducibility [5].

The fibers were analyzed through an Infinity video-microscope of inspection; model InfiniVar CFM-2TM, for evaluation of macroscopic defects and segregation. Micro-X-ray Fluorescence Spectrometry (micro-EDX) was carried out by Shimadzu spectrometer, model micro-EDX-1300, using a Si(Li) detector operated at 50kV and 50μA. The measurements were performed with a scan rate of 100s per point (live time) using an incident beam diameter of 50μm, to determine the dopant distribution. In the micro-EDX technique the Li and F quantities are not measurable; therefore the values of these ions were fixed according to the stoichiometric parameters.

Diffraction patterns were obtained, by pulverizing the single crystals fibers. The measurements were performed on a powder X-ray diffractometer Bruker-AXS, model-D8, at room temperature, in a 2θ interval ranging from 20 to 80 degrees in steps of 0.05 degree. An Scanning Electron Microscope (SEM) with incorporated Energy Dispersive X-ray (EDS), OXFORD model LEO 440, was used to determine the formation of defects and to investigate the chemical composition of the fibers. The visible and the infrared absorption spectra were collected in a Cary 17D-Olis modified spectrometer and in a Thermo-Nicholet – 6750 FTIR spectrometer, respectively, to verify the incorporation of Er^{3+} and the eventual occurrence of water contamination.

3. Results and discussion

The single crystal fibers were grown with lengths of up to 120mm, and with large transparent regions of constant diameters (~0.7mm). Erbium doped YLF: fibers with 1, 10 and 20mol% concentration are presented in figure 1.

An uniform YLF phase was directly crystallized at the seeding process, differently from previous works where a peritetic transient preceding the growth of pure and doped crystalline YLF was reported [1, 3]. In the previous works the fibers were grown under gas flux and on the present one the growth was done under a static atmosphere after a high vacuum treatment.

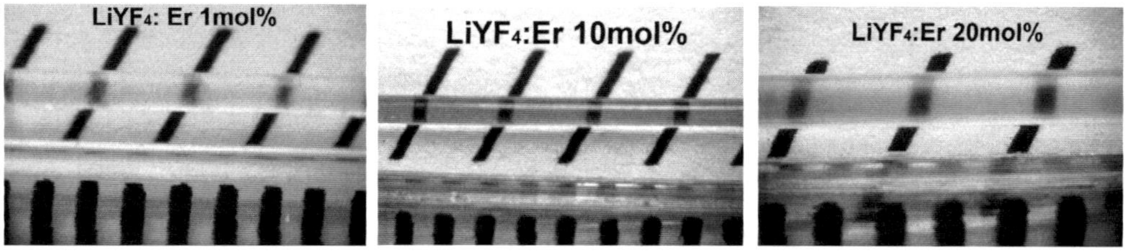

Figure 1. Optical characterization $LiYF_4$: Er^{3+} fibers with concentrations of 1, 10 and 20 mol%.

Pastor *et al.* [6] reported that YLF crystals present congruent melting behavior in highly pure atmosphere conditions. The above results indicate that the high vacuum treatment preceding the growth was probably more effective than the gas flux treatment to eliminate spurious impurities and consequently the initial peritetic transient.

A single phase identified as YLF: Er was observed in the X-ray power diffractogram showed on figure 2. The micrographs obtained through SEM substantiated the existence of only one phase (figure 3). Nonetheless, micro-inclusions (P1) were observed within some of the fibers, as well as an external thin superficial layer (P3).

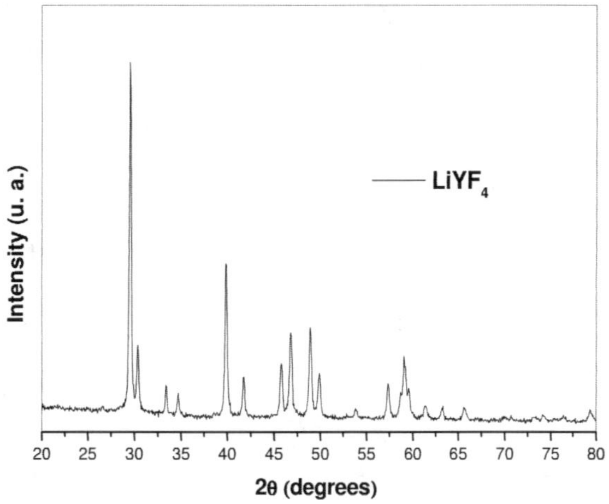

Figure 2. X-ray powder diffractogram of $LiYF_4$: Er^{3+} 10mol% fiber.

It is important to mention that by reason of a mechanical problem in the pulling system used, which resulted in significant diameter variations in a few of the fibers, adjustments of some of the growth conditions were occasionally required, such as temperature gradient and pulling rate. In a few cases the apparatus misalignment problem resulted also in oxygen contamination because of bellows connections detachment. Although oxygen presence was detected though SEM/EDS, it was not possible to determine the exact concentrations of this impurity. The radial Erbium incorporation was found to be almost constant (Table 1).

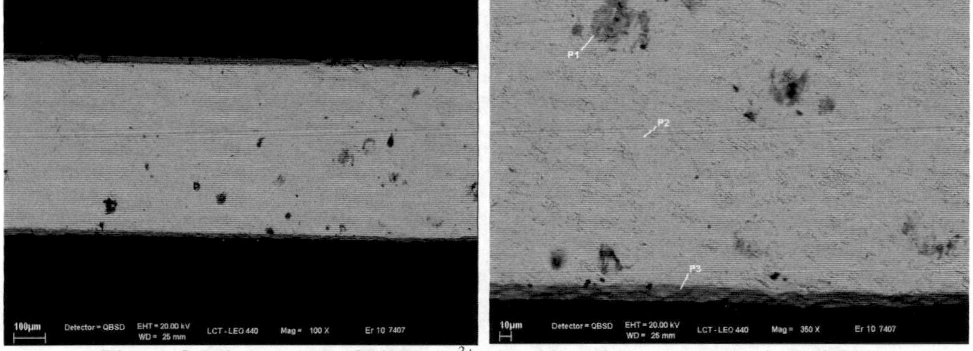

Figure 3. Micrographs of $LiYF_4$ Er^{3+} 10mol% and the measure points of EDS

Table 1 . Concentrations of Er^{+3} in the following points of interest: P1 (micro-inclusion), P2 (center of the fiber), P3 (superficial layer) determined by EDS on SEM measurements.

Element	P1 (inclusion)	P2 (center)	P3 (surface)
	% molar	% molar	% molar
Er	13.5	13.5	12.9
O	detected	not detected	detected

Erbium incorporation along the fibers was determined through micro-EDX to be approximately constant (figure 4). Measured concentrations were close to the nominal values of the starting materials with 1 and 20 mol% of Er. Nonetheless, a higher concentration (11 mol% of Er^{3+}) was found in the sample corresponding to the nominal concentration of 10 mol%.

The dopant fluctuation observed in these micro-EDX measurements can be attributed to experimental error or to small diameter variations at some stage in the pulling process. Small mechanical instabilities during fibers pulling may result in unstable solid-liquid interfaces. The growth process of the 20mol% Er fiber was particularly mechanically disturbed, and consequently resulted in the sharpest oscillations. The growth process may be further disturbed by crucible imperfections, which may have also contributed to the dopant fluctuation observed. Nevertheless, only 5mm samples are necessary for the laser action tests and spectroscopic characterization, therefore uniform dopant samples could be selected from the grown fibers.

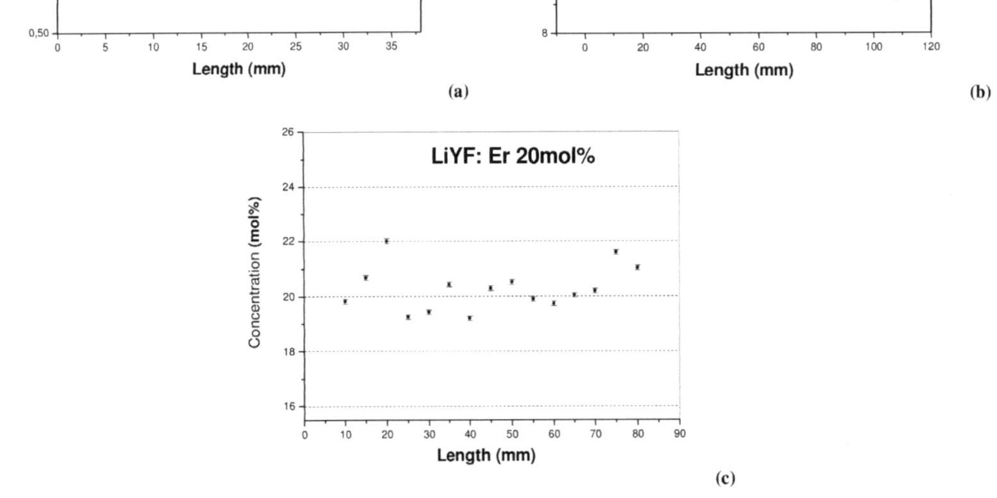

Figure 4. Distribution of Er^{3+} in $LiYF_4$ fibers measured by micro-EDX technique.

The characteristic absorption bands of Er^{3+} were observed in the measured visible absorption spectrum (figure 3a and 3b). A broad absorption band in 3000-3600cm^{-1} region was observed (figure 3c) in the IR, which can be associated to OH^- incorporation resulting from moisture and oxygen contamination during the growth process [7].

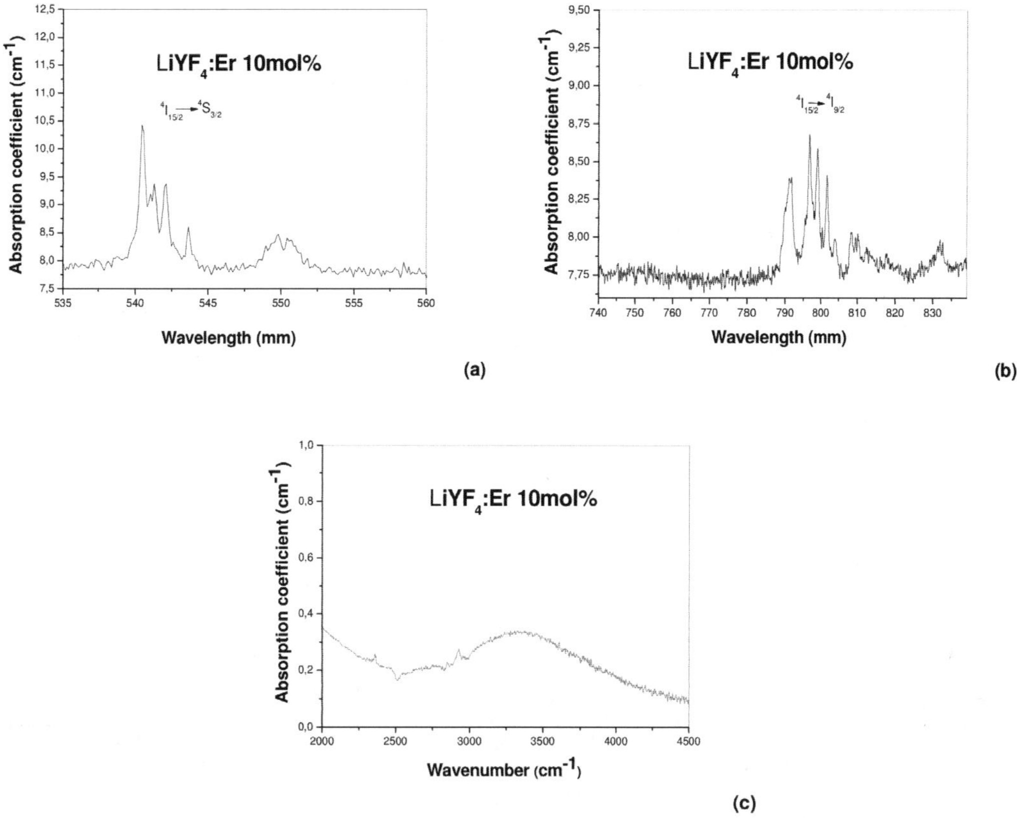

Figure 5. Visible and infrared absorption spectra of $LiYF_4$: Er^{3+} 10mol%

4. Conclusion

The micro-pulling down method was determined to be suitable to grow transparent $LiYF_4$: Er^{3+} fibers. The high vacuum treatment preceding the growth process was able to eliminate the peritetic transient, though the static atmosphere did not prevent oxygen contamination caused by mechanical problems. Additional studies are underway to eliminate mechanical instabilities observed in our system which resulted in fluctuations of the dopant concentration along the fibers.

Acknowledgments

The authors acknowledge CNPq and FAPESP for financial support.

References

[1] Santo A M E, Ranieri I M, Brito G E S, Epelbaum B M, Morato S P, Vieira N D, Baldochi S L, 2005 *J. Crystal Growth*, **275** 528–33.

[2] Fukuda T,. Rudolph P and Uda S 2004 *Fiber Crystal Growth from the Melt.* (Berlin: Springer) p.1-40.

[3] Santo A M E, Librantz A F H, Gomes L, Pizani P S, Ranieri I M, Vieira N D, Baldochi S L 2006 *J. Crystal Growth,* **292** 149–154.

[4] Abell J S, Harris I R, Cockayne B, Plant J G 1976 *J. Mat. Science* **11** 1807-16.

[5] Silva F R, "Crescimento de fibras de $LiYF_4$ dopadas com Nd^{3+} e Er^{3+} para aplicações em lasers de estado sólido" (Master in Science thesis), IPEN-University of São Paulo, 2008.

[6] Pastor R O, Robinson M, Akutawaga W M, 1975 *Mat. Res. Bulletin,* **10** 501-9.

[7] Ranieri I M, Baldochi S L, Santo A M E, Gomes L, Courrol L C, Tarelho L V G, deRossi W, Berretta J R, Costa F E, Nogueira G E C, Wetter N U, Zezell D M, Vieira N D and Morato S P 1996 *J. Crystal Growth,* **166** 423-8.

A Structural Study of Delafossite-type $CuInO_2$ Thin Films

Alan V. Chadwick[1], Aran N. Blacklocks[1], Aline Rougier[2] and Cedric Yaicle[2]

[1] Functional Materials Group, School of Physical Sciences
University of Kent, Canterbury, Kent CT 7NH, UK

[2] Laboratoire de Réactivité et de Chimie des Solides, Université de Picardie Jules Verne, Amiens, France

a.v.chadwick@kent.ac.uk

Abstract. This contribution reports a study of the preparation of pure and doped delafossite-type material $CuInO_2$ using pulsed laser deposition. A series of films have been prepared at different oxygen partial pressures. X-ray absorption spectroscopy (XAS) has been used to characterise the microstructure of the films. The results show that the structure of the films is extremely sensitive to the preparation conditions.

1. Introduction

Transparent conducting oxides (TCO) are an important group of materials used in optoelectronics that combine low electrical resistance (typically $<10^{-3}$ Ωcm) with high optical transparency (>80%) in the visible range of the electromagnetic spectrum [1]. Optically transparent oxides tend to be electrical insulators as a result of their large electronic band gap of ≥3.1eV, corresponding to the energy of a 400nm violet photon. Therefore visible photons with an energy ranging from ~1.8eV (700nm) to 3.1eV (400nm) cannot excite electrons from the valence band into the conduction band and as a result the light is transmitted. However, TCOs are semiconductor materials and the visible photons have enough energy to excite electrons from donor levels to the conduction band (n-type TCO) or holes from the acceptor level into the valence band (p-type TCO). These acceptor or donor levels within the TCO band gap are created by introducing defects through non-stoichiometry and/or doping with appropriate impurity ions in a controlled manner.

The first observation of TCO materials was CdO in 1951 in which it was observed that electrical conductivity and optical transparency could co-exist [2,3]. Similar behaviour was also reported in In_2O_3 a few years later in 1956 [4] and applications were found for heated windows. A concentrated research effort in the following decade led to the discovery of films of doped SnO_2 and In_2O_3:Sn ($In_{2-x}Sn_xO_3$; also known as ITO) displaying excellent electrical and optical properties [5,6]. These films were primarily used at the time to reduce heat losses from sodium lamps. At around the same time TCO properties were discovered in ZnO which led to use as part of light emitting diodes (LED) [7,8]. The most common n-type TCOs have remained virtually unchanged over the years with the principal constituents remaining the simple binary oxides SnO_2, In_2O_3, ZnO and CdO. More recent work has begun to explore new binary oxide combinations, and even ternary oxides such as Cd_2SnO_4, Zn_2SnO_4, $MgIn_2O_4$, $CdSb_2O_6$:Y, $ZnSnO_3$, $GaInO_3$, $Zn_2In_2O_5$, and $In_4Sn_3O_{12}$ to name but a few of the prominent examples [9,10].

Although TCOs have many applications as mentioned above, very little work has been done on active device fabrication using TCOs. This is because all of the examples mentioned so far have been n-type semiconductors, and this monopolarity has restricted the use of TCOs to narrow areas such as transparent electrodes within optoelectronic devices. The corresponding p-type TCOs, essential for junction devices were not discovered in thin film form until 1997 when p-type conductivity was reported in highly transparent $CuAlO_{2+x}$ [11]. The discovery of p-type transparent conductive oxides opens up a range of novel applications termed "Transparent Electronics" or "Invisible Electronics" [12] where the combination of the two type of TCO in the form of a p-n junction could lead "functional" window that transmits visible light, yet generates electricity in response to the absorption of ultraviolet photons [11].

Materials with the delafossite-type structure $M^IM^{III}O_2$ (M^I = monovalent ions, Cu^+ and Ag^+; M^{III} = trivalent ions, Al^{3+}, Ga^{3+}, In^{3+}, Cr^{3+}, Fe^{3+}, Co^{3+} etc.) have been chosen as candidates for p-type TCO materials.

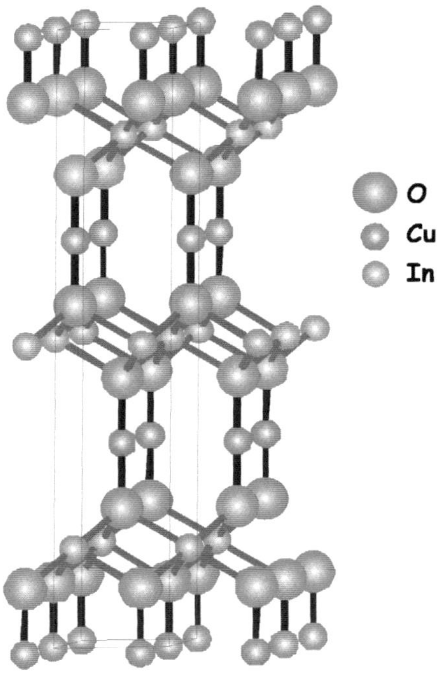

Figure 1 Crystal structure of delafossite-type ($M^IM^{III}O_2$) copper indium oxide

The delafossite structure is shown above in Figure can be seen to have alternate stacking of an O–M^I–O dumbbell layer and an $M^{III}O_6$ edge sharing octahedral layer. The oxide ions form a "pseudo-tetrahedral" coordination ($M^{III}{}_3M^IO$) [13] allowing a reduction in the non-bonding nature of the oxide ions and the delocalisation of holes at the valence band edge. The layered structure involving the M^IO_2 dumbbell effectively reduces the level of cross-linking of M^I ions allowing the enlargement of the band gap [10].

Since the first discovery of p-type TCO films in 1997 [11] a number of promising p-type materials have been found, however at present these display conductivity over an order of magnitude less than

Journal of Physics: Conference Series **249** (2010) 012045 doi:10.1088/1742-6596/249/1/012045

n-type materials. One of the classes of material showing the most promise to fulfil the role of a p-type TCO is that of the Cu based delafossite type oxides with examples including $CuAlO_2$ [14], $CuGaO_2$ [15], $CuScO_2$ [16], $CuFeO_2$ [17] and $CuInO_2$ [18,19]. With p-type TCOs now established, more attention has been focussed on the possibility of realising all TCO p-n junctions. However, it has also been shown that in the fabrication of p-n heterojunctions that lattice defects present in the interface or grain boundaries between the different n- and p-type components can degrade the electrical or optical performance of the devices [15]. It was then suggested that improved lattice matching of the p- and n-type TCO components in the form of an all delafossite p-n junction would improve active device fabrication. Until this point these materials had predominantly presented p-type conductivity, and the first delafossite type TCO with n-type conductivity, $AgInO_2$, was discovered in 1998 [20]. It was then suggested that all delafossite p–n heterojunctions could therefore be realised [21] involving a Cu based p-type delafossite TCO, and the n-type $AgInO_2$. Despite the all delafossite TCO p-n heterojunction now being a possibility, a true homojunction with both n- and p-type TCOs formed from the same material would be preferable to give the best lattice matching [22,23]. Although $AgInO_2$ was the first delafossite type TCO to be donor doped to give n-type conductivity, acceptor doping has been unsuccessful in producing p-type conductivity in $AgInO_2$ [24]. Recent work, however, has shown that $CuInO_2$ is unique in that bipolar doping has been reported [14]. Acceptor doping with Ca^{2+} can increase the p-type character, and donor doping with Sn^{4+} can lead to n-type conductivity. The discovery of bipolarity led soon after to the first fabrication of an all $CuInO_2$ based p-n homojunction [25] although it can be noted that improvement is still sought in terms of film quality and conductivity levels.

The discovery of both p-type and n-type conductivity in doped $CuInO_2$ has led to use in TCO p-n homojunctions [25]. However, the levels of conductivity seen in the individual components are still several orders of magnitude lower than in the leading n-type TCOs such as ITO, and improvements must be made if the field of "transparent electronics" is to be fully realised. The defect chemistry of the bipolar doped $CuInO_2$ system (Ca^{2+} for p-type; Sn^{4+} for n-type) is currently not well understood, with very little in the literature regarding point defects such as the doping site preference or any detailed mechanism for the bipolar semiconductivity [26]. The work reported here is an experimental analysis of the local defect structure of thin films of undoped and doped (Sn, Ca or Eu) $CuInO_2$ using x-ray absorption spectroscopy (XAS) at the host Cu K- and In K-edges along with the dopant Ca K-, Sn K- and Eu L_{III}-edges. The specific objective of the work was to reveal the role of oxygen partial pressure during film preparation on the average structure of the deposited $CuInO_2$ film. The second objective was to provide experimental evidence for the dopant site preferences for the Ca, Sn and Eu dopants and to observe the effects of increased doping levels (from 2 to 10 atom%) on the bulk structure.

2. Experimental

2.1. Materials

A range of undoped and doped (Ca, Sn or Eu) thin film delafossite-like '$CuInO_2$' samples were studied. These were prepared from a target of $Cu_2In_2O_5$ (and additional dopant targets) by pulsed laser deposition (PLD) using a KrF excimer laser beam (Lambda Physik Compex 102, λ = 248 nm) with a laser fluence of 1–2 J cm^{-2}. The substrates were polished sapphire. The base pressure in the chamber was in the order of 1×10^{-6} mbar whereas the oxygen pressure varied from 5×10^{-4} to 2×10^{-2} mbar. The films were characterised by X-ray diffraction (XRD), transmission electron microscopy (TEM) and Rutherford back-scattering (RBS). Further details of the preparation and characterisation of these films can be found in reference [27].

2.2. X-ray absorption measurements

Sn, In and Cu K-edge XAS were recorded on station 9.3, with Ca K, Cu K and Eu L_{III} edge XAS being recorded on station 7.1, both utilising double crystal Si(111) monochromators and harmonic rejection set to 50%. The measurements were all made at room temperature and collected in fluorescence mode using a 9-element solid state detector with high count rate electronics.

During these experiments problems were encountered due to diffraction peak contamination. These can arise where the incident photon wavelength satisfies a Bragg condition for a set of planes in the film or its substrate. The dominant effect in this case occurs when the Bragg reflected beam enters the fluorescence detector and I_f is increased. In this case the apparent absorption coefficient appears bigger than it actually is. When the diffracted beam enters the I_0 chamber the effect is the opposite and a decrease in absorption coefficient is observed. Several methods have been employed to work around this problem, including spinning or wobbling the sample, using energy discriminating detectors, multiple detector arrays, masking and dodging [28-31]. The method chosen was to rotate the samples during data collection. When the sample is spinning, the diffraction condition is met for only a brief time compared to the overall data averaging time. Hence, the amplitude of the diffraction feature in the XAS is significantly reduced, often to a level below visual detection. Although the contaminating peak has not been completely removed, the amplitude is reduced such that the fitting and removal of a cubic-spline curve is adequate in eliminating any residual non-oscillatory contribution. The spinner was mounted onto the beam line between the ion chambers I_0 and I_t and set to an angle of 45° with respect to both the direction of the incident beam and the fluorescence detector. The rotation speed was a few hundred rpm.

Samples of Cu_2O, CuO, In_2O_3, SnO_2, Eu_2O_3 and $CaCO_3$ were mixed with PVP and pressed into pellets with a 13mm IR press for use as standards. The amount of material used in each pellet was adjusted to give an absorption of about $\mu d=1$.

The raw data were converted to an energy scale with the Daresbury programme EXCALIB. Background subtraction for both X-ray absorption near edge structure (XANES) and X-ray absorption fine structure (EXAFS) work was carried out using VIPER [32,33]. EXAFS analysis was performed using EXCURV98 [34,35]. The phase shifts were calculated *ab initio* from Hedin-Lundqvist exchange potentials and Von Barth ground state potentials. Theoretical fits to the k^3-weighted EXAFS were produced by least-squares refining E_f (an energy offset), and for each shell RD (absorber-scatter distance) and $2\sigma^2$ (Debye-Waller factor representing both static and thermal disorder). The coordination number can also be refined, but this option was not utilised in this work. The quality of the fit is measured by an R-factor [35] and the errors in RD are approximately ±0.02Å and approximately ±20% in $2\sigma^2$ and CN.

3. Results and discussion

XRD peaks that indexed to delafossite-type $CuInO_2$ were seen in all samples, with preferential orientation mostly for the (012) plane. The presence of In_2O_3 is also seen in the sample prepared at the low partial pressure of 5×10^{-4} mbar. The TEM micrographs showed that the films also contained some amorphous material.

As mentioned in experimental section 2.2, it was found that the fluorescence XAS signal detected was contaminated with Bragg peaks in the higher K region, which may be caused by either the films being epitaxial, or more likely as a result of diffraction from the Al_2O_3 substrate. Unfortunately, with the films being relatively thin, simply changing the angle away from the preferred 45° angle relative to the beam and detector also results in a large reduction in fluorescence intensity observed for the sample therefore making this method largely unsuitable for removing the interference. However, as seen in Figure 2 the effect of spinning the sample was very effective in removing the interference at high *k* values.

The Cu K-edge XANES spectra for undoped delafossite-type $CuInO_2$ films prepared under varying oxygen partial pressures is shown below in Figure 3. The average oxidation state of the Cu site is difficult to quantify by XANES with the shift for Cu^0 and Cu^I being very similar. An added

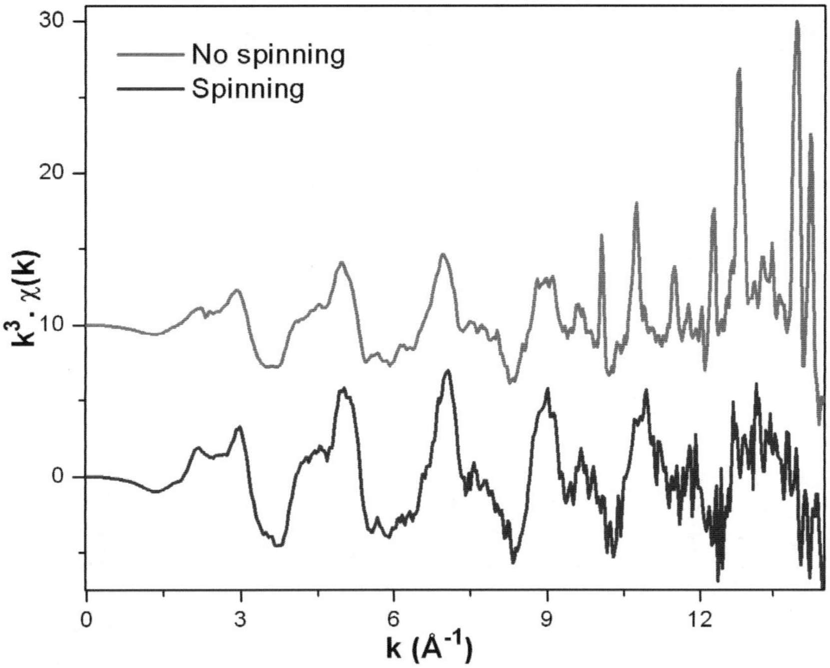

Figure 2 Example of Bragg peak reduction at the Cu K edge by sample rotation

complication is that the change in environment can also have an effect on the edge position. The films prepared at 5×10^{-4} and 2×10^{-3} mbar appear to have an oxidation state of around +1, with the 1×10^{-2} mbar film appearing to be very slightly above +1. The 2×10^{-2} mbar film appears to be further oxidised to around Cu^{II}, although again the exact oxidation state can not be quantified. The delafossite type $CuInO_2$ has oxidation states of Cu^{I} and In^{III} [19], and therefore the only films that are likely to present the delafossite phase are the 5×10^{-4} to 1×10^{-2} mbar prepared samples, although it has been suggested that the material can be oxidised to $CuInO_{2.5}$ with copper present as Cu^{2+} [36]. The Cu K-edge XANES spectra for the 10% Ca doped samples appeared very similar to those of the undoped '$CuInO_2$' films, suggesting that the oxygen pressure of the film preparation is very important to the deposited sample, and that the presence of the dopant has little effect on the overall structure. This was supported by the Cu K-edge XANES spectra for the 10% Sn doped '$CuInO_2$' films. Again the samples prepared between 5×10^{-4} and 1×10^{-2} mbar appear to have copper oxidation states of roughly Cu^{+} and are very similar to the equivalent undoped films.

The In K-edge XANES spectra for the undoped '$CuInO_2$' films showed that in all cases there appeared to be no deviation away from the expected In^{3+} oxidation state, showing that any additional oxidation taking place is compensated by increasing the oxidation state of the copper from Cu^{+} towards Cu^{2+} as expected [36,37]. The Eu L_{III}-edge XANES spectra showed that the Eu appears to be present as Eu^{3+} at all doping concentrations from 2 to 10% with the Eu L_{III}-edge position appearing in an identical position to that of Eu_2O_3. Sn K- and Ca K-edge data were also recorded, however the poor quality data did not allow for quantitative or qualitative XANES information to be extracted. This is the result mostly from the weak/noisy signal leading to large data collection times, which when coupled with the lack of a monitor spectrum meant the spectra could not be accurately calibrated.

The Cu K-edge EXAFS and Fourier transform of the '$CuInO_2$' film deposited with an oxygen pressure of 2×10^{-2} mbar is shown in Figiure 4 with the best fit parameters displayed in Table 1. The data has only been fitted to three shells as there is very little contribution visible in the EXAFS and

Fourier transforms above 3.5 to 4 Å. The first point to note is the high level of disorder seen in the sample

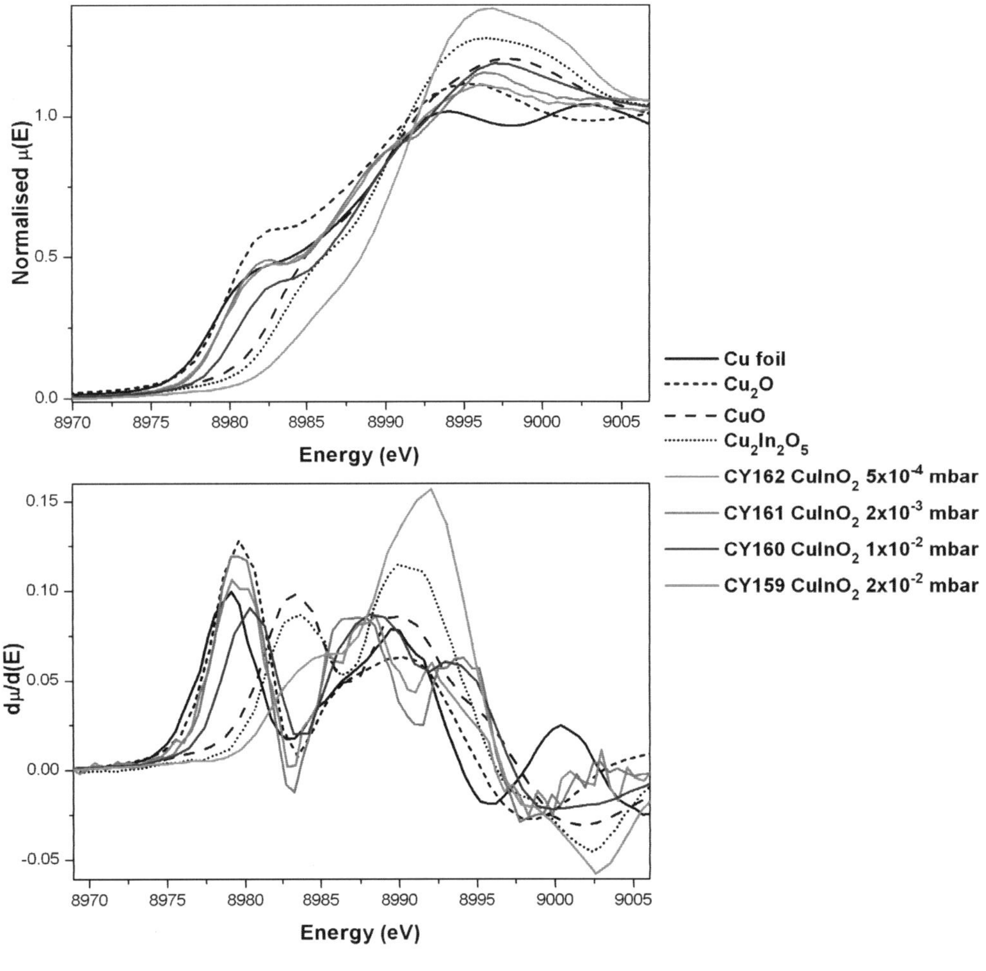

Figure 3 Cu K edge XANES (top) and derivative (bottom) for delafossite-type $CuInO_2$ films prepared under different pO_2

through the high Debye-Waller factors for the Cu-Cu and Cu-In correlations. These values could be expected to be higher than normally seen in other compounds due to the Cu movement perpendicular to the O-Cu-O bond as in Cu_2O. Similarly high values of ~0.04Å^2 have been reported [38] for the $2\sigma^2$ values for the Cu-Cu correlation in $CuGaO_2$. The data does not fit well to the delafossite-type $CuInO_2$ model, with the Cu-Cu correlation unable to settle at a distance close to the 3.292Å expected. The Cu-O distance of 1.938Å would also suggest an increased oxidation state from Cu^+ towards Cu^{2+} as seen in the XANES and similar to the Cu^{II}-O distance of 1.956Å in CuO. The low Debye-Waller factor for the Cu-O shell would also suggest an increase in coordination number (CN and $2\sigma^2$ are highly correlated) as would be expected in comparison to the Cu^{II} standards CuO and $Cu_2In_2O_5$ which show a first shell coordination number of four compared to two for the linear CuO_2 geometry expected in Cu_2O and $CuInO_2$. The structure of $CuInO_{2+\delta}$ appears to be mistakenly reported in the literature, with the delafossite type $CuInO_{2.66}$ structure [36] reported in the ICSD [39] claiming an additional 18 O atoms surround each Cu atom at a distance of 1.936Å which would not provide the correct stoichiometry. The possibility of the 2×10^{-2} mbar film displaying the $CuInO_{2+\delta}$ structure has therefore not been modelled due to the uncertainty in structure, and the apparent high level of disorder in the sample.

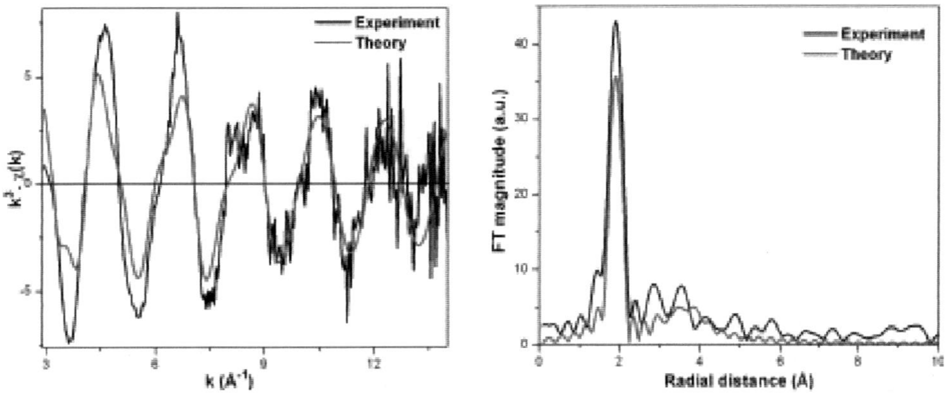

Figure 4 Cu K edge EXAFS (left) and Fourier transform (right) for '$CuInO_2$' film prepared at pO_2 2×10^{-2} mbar

$CuInO_2$ crystallographic data [19]			EXAFS					
Atom	CN	RD (Å)	Atom	CN	RD (Å)	$2\sigma^2$ (Å^2)	E_f(eV)	R (%)
O	2	1.845	O	2	1.938	0.001	0.18	50.77
Cu	6	3.292	Cu	6	3.351	0.048		
In	6	3.466	In	6	3.501	0.045		

Table 1 Best-fit parameters for Cu K edge EXAFS of '$CuInO_2$' made at pO_2 2×10^{-2} mbar

The Cu K-edge EXAFS and best-fit parameters for the pO_2 1×10^{-2} mbar film are displayed in Figure 5 and Table 2, respectively.

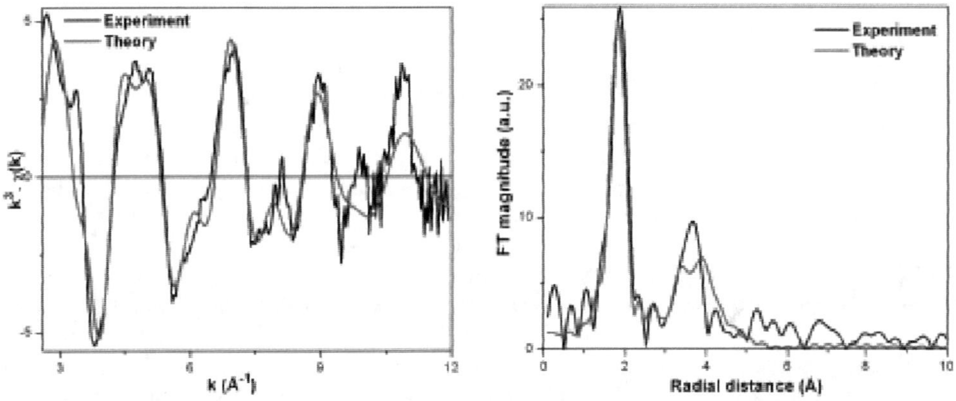

Figure 5 Cu K edge EXAFS (left) and Fourier transform (right) for '$CuInO_2$' film prepared at pO_2 1×10^{-2} mbar

$CuInO_2$ crystallographic data [19]			EXAFS					
Atom	CN	RD (Å)	Atom	CN	RD (Å)	$2\sigma^2$ ($Å^2$)	E_f (eV)	R (%)
O	2	1.845	O	2	1.861	0.010		
Cu	6	3.292	Cu	6	3.287	0.046	-6.63	39.08
In	6	3.466	In	6	3.479	0.045		

Table 2 Best fit parameters for Cu K edge EXAFS of '$CuInO_2$' film prepared at pO_2 1×10^{-2} mbar

The EXAFS refinement fits much better to the delafossite-type $CuInO_2$ model (R=39%) with excellent agreement in the radial distances well within the expected ±0.02Å. The Debye-Waller factors for the Cu-Cu and Cu-In shells remain high however, which may be as a result of the expected disorder in the sample due to the movement around the O-Cu-O linear bond, or a suggestion that the sample is fairly disordered overall.

The Cu K-edge EXAFS for the films prepared at oxygen pressures of 2×10^{-3} mbar and 5×10^{-4} mbar gave very poor fits to the delafossite structure. The Cu-Cu distances did not settle close to the 3.29Å distance expected, and falling a large distance outside the expected ±0.02Å error. Additionally the Debye-Waller factors were very large suggesting that the Cu environment in this sample is not correctly modelled by a Cu site in delafossite type $CuInO_2$. The most striking observation, however, was the presence of the peaks at around 2.5Å and 4.5-5Å in the sample prepared at 5×10^{-4} mbar which shows that there is Cu metal present in the sample. It would appear that the low oxygen pressure is not high enough to allow all of the copper to form in an oxidised form, resulting in a mixed structure with both Cu^0 and Cu^+ present.

The Cu K-edge EXAFS for the various doped films confirmed the picture that emerged from the XANES data. The spectra were not affected by the doping and reflected the spectra of the undoped films prepared at the same oxygen partial pressures.

The In K-edge EXAFS of the undoped '$CuInO_2$' film prepared at the oxygen pressure of $2x10^{-2}$ mbar confirmed the picture that emerged from the Cu K-edge EXAFS for this film; i.e. the local environment is not the delafossite-type $CuInO_2$ structure. The difference is again the copper sites, with the In-Cu shell unable to refine to the distance expected of 3.466Å and the excessively large $2\sigma^2$ value suggesting that the correlation does not really exist.

The In K-edge EXAFS and Fourier transforms for the undoped film prepared with a pO_2 of 1×10^{-2} mbar is displayed in Figure 6 with the best-fit parameters shown in Table 3. The displayed results show an excellent agreement (R=39.78%) to the expected In environment in delafossite $CuInO_2$, and also to the Cu-In distances provided by the Cu K-edge EXAFS of the same sample within the expected ±0.02Å error. The Debye-Waller factors are also reasonable and show the expected increase for the outer shells. . They also appear to show that there is a significantly lower level of disorder presented in the average In environment compared to Cu environment. This would be consistent with the Cu K- and In K-edge XANES which showed variation in valency and geometry around the Cu with changing pO_2, whilst the In site remained consistent.

The refinement of the In K-edge EXAFS results for the undoped '$CuInO_2$' film prepared at pO_2 2×10^{-3} mbar appears to show the local structure of the indium fits well to the delafossite-type $CuInO_2$ model (R=30.74%) with good agreement in radial distances and reasonable $2\sigma^2$ values suggesting the local structure displays relatively low value of disorder compared to the Cu site. This however, is in stark contrast to the Cu K-edge EXAFS for the same sample which suggests the average Cu environment is not representative of the $CuInO_2$ structure. The combination of the results for the two absorption edges would suggest that the film may not present a single phase of $CuInO_2$, with perhaps an additional Cu containing phase a possibility to explain the disorder seen in the average local structure of the copper. To confirm this other techniques would need to be employed such as high resolution XRD to provide a model for further EXAFS refinement.

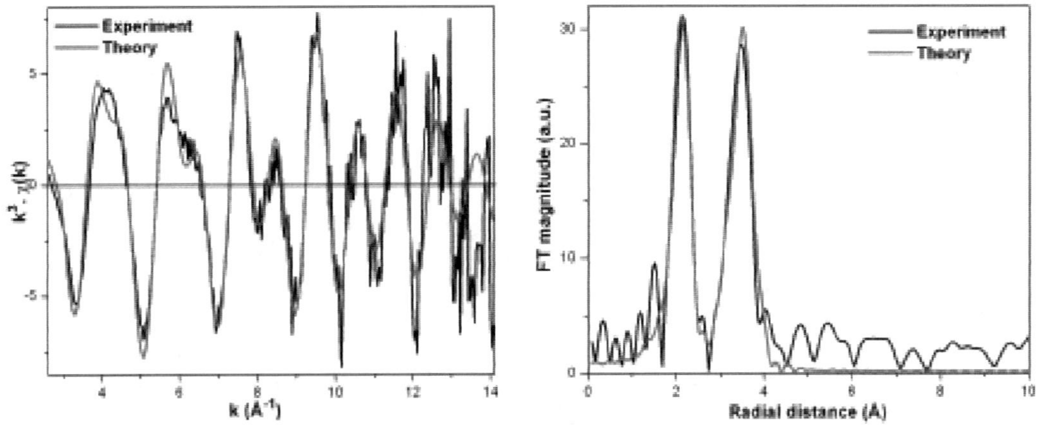

Figure 6 In K edge EXAFS (left) and Fourier transform (right) for '$CuInO_2$' film prepared at pO_2 1×10^{-2} mbar

$CuInO_2$ crystallographic data [19]			EXAFS					
Atom	CN	RD (Å)	Atom	CN	RD (Å)	$2\sigma^2$ ($Å^2$)	E_f(eV)	R (%)
O	6	2.173	O	6	2.153	0.013		
In	6	3.292	In	6	3.302	0.015	-5.36	39.78
Cu	6	3.466	Cu	6	3.466	0.032		

Table 3 Best fit parameters for In K edge EXAFS of '$CuInO_2$' film prepared at pO_2 1×10^{-2} mbar

The results of the In K-edge EXAFS of the undoped '$CuInO_2$' film prepared under an oxygen partial pressure of 5×10^{-4} mbar did not fit well to the $CuInO_2$ model, with variation in the In-In and the In-Cu distances in particular well outside of the ±0.02Å error, with the Cu shell refining to 3.734Å – a value around 0.27Å higher than expected. The average local environment of the indium here actually appears to be In_2O_3 which is supported by XRD data.. The sample appears therefore to be a mixed structure containing In_2O_3, Cu metal and some oxidised Cu component, although only very weak signals are seen in the XRD pattern for the other components suggesting there may be a large amount of amorphous material or very small nanocrystalline regions present.

It has been assumed experimentally and suggested by computer simulation studies in the literature [26] that for doping of Sn^{4+} and Ca^{2+} into the $CuInO_2$ structure, the dopant is incorporated by substitution onto an In^{3+} site. In a similar fashion, it is expected that the Eu^{3+} will be incorporated onto an In^{3+} site. The Eu L_{III}-edge EXAFS data gave a poor fit for the substitution onto a Cu site in all samples (2 to 10% Eu doping levels), with poor agreement to the radial distances expected for the $CuInO_2$ model and unrealistic and unacceptable values of −37 to −40eV for the Fermi energy (E_f). In comparison, the data fitted much better with a model of Eu on an In site, with much more sensible values for E_f. There also appears to be a local expansion around the Eu compared to undoped model which would be expected from the relative ionic radii (In^{3+} = 0.94Å, Eu^{3+} = 1.09Å [40]). Unfortunately the Ca and Sn data were not of sufficient quality to identify the dopant site.

4. Conclusions

The combination of Cu K- and In K-edge EXAFS and XANES spectra show that the structure of films deposited by PLD is very sensitive to the partial pressure of oxygen used during deposition as suggested in the literature [18,26,27,36]. This is also seen to a lesser extent in tour preliminary XRD data in which the main effect seen in the longer range order is a change in the preferential orientation. Below a pressure of 5×10^{-4} mbar, no $CuInO_2$ is seen with a mixture of copper and indium oxides formed.

The only undoped sample that fits well to the delafossite-type $CuInO_2$ EXAFS model for both Cu and In sites is that prepared at 1×10^{-2} mbar. This is a similar value suggested for the optimal growth of $CuInO_2$ by several other groups using PLD [18,41,42]. The XRD patterns for the 2×10^{-3} and 1×10^{-2} mbar oxygen films appear to be very similar, but the EXAFS reveals that although the average indium environment is delafossite $CuInO_2$ that the average copper environment deviates from $CuInO_2$ in the 2×10^{-3} mbar film. As seen in previous studies [18,42] this could be evidence of partial decomposition of the ternary $CuInO_2$ structure into binary oxides (e.g. Cu_2O or In_2O_3) when the films are prepared under lower oxygen pressure. This pattern is further highlighted at the lowest oxygen pressure of 5×10^{-4} mbar in which the average indium environment is now mostly In_2O_3 and the average Cu environment displays a mixture of both metallic and oxidised Cu species.

HRTEM measurements on these films reveal that regions of both crystalline and amorphous material are visible within the grains [27] with region sizes as small as ~5nm. This could go some way to explaining the higher than expected Debye-Waller factor values observed for the Cu-Cu and Cu-In correlations in EXAFS, in which the ***average local structure*** of all Cu atoms in the sample is observed. A mixture of amorphous and crystalline material would create an increase in disorder in the average structure in the sample visible by an attenuation in the outer shell correlations in the EXAFS. An additional factor could be the minor decreases in coordination number/increases in Debye-Waller factor expected for small nano-particles [43] although it can be stressed that the size-effect will be minor compared to the effect of an amorphous/crystalline mixture.

For the doped (Ca, Sn or Eu) '$CuInO_2$' films, the bulk structure is very similar to that of the undoped films, only demonstrating the delafossite-type $CuInO_2$ structure when prepared under a partial oxygen pressure of 1×10^{-2} mbar. The dopant site identification was less successful with large background interference from the In L-edges rendering the Ca K edge data unreliable, and the energy range of the solid state fluorescence detector reducing the efficiency of the Sn K-edge data. However, the Eu was successfully identified as being incorporated as Eu^{3+} on an In^{3+} site in agreement with the expected site preference for both Ca^{2+} and Sn^{4+} [26]. The $CuInO_2$ matrix appears to become less ordered with increasing dopant concentration in agreement with the long range order seen by XRD. This pattern is also observed in the dopant site in the Eu L_{III}-edge EXAFS in which a clear attenuation in the outer shell intensities can be seen.

In summary, the conclusions for the study are:

- Delafossite-type $CuInO_2$ formed at pO_2 1×10^{-2} mbar
- Films prepared at $pO_2 \leq 2\times10^{-3}$ mbar do not form phase-pure delafossite $CuInO_2$, decomposing into a mixture including $CuInO_2$, binary oxides such as Cu_2O and In_2O_3 as well as some Cu metal formation at pO_2 5×10^{-4} mbar
- At pO_2 2×10^{-2} mbar the copper is oxidised to Cu^{2+} and the film has become less ordered, no longer appearing to display the delafossite-type structure
- All films display a mixture of crystalline and amorphous material
- Eu is doped as Eu^{3+} on an In^{3+} site in delafossite type $CuInO_2$ at pO_2 1×10^{-2} mbar
- Higher doping concentration leads to a loss of film crystallinity observed by both EXAFS and XRD

Acknowledgements
We acknowledge the support from EU Interreg3 for the grant under which this work was undertaken. We are very grateful for help from staff at the Daresbury SRS that has facilitated the work , especially Ian Harvey and Steve Fiddy.

References

[1] Minami, T., Semicond. Sci. Technol., **20**, (2005) S35-S44.
[2] Hiesinger, L. and König, H., *Festschrift W.C. Heraeus*. (Hanau, 1951).
[3] Helwig, G., Z. Physik, **132**, (1952) 621-642.
[4] Thelen, A. and König, H., Naturwissenschaften, **13**, (1956) 297-298.
[5] Groth, R. and Kauer, E., Philips Tech. Rev., **26**, (1965) 105-111.
[6] Groth, R., Phy. Stat. Sol., **14**, (1966) 69-75.
[7] Drapak, I. T., Semiconductors, **2**, (1968) 624.
[8] Ozgur, U., Alivov, Y. I., Liu, C., Teke, A., Reshchikov, M. A., Dogan, S., Avrutin, V., Cho, S. J. and Morkoc, H., J. Appl. Phys., **98**, (2005) 041301.
[9] Dali, S. E., Jayachandran, M. and Chockalingam, M. J., J. Mater. Sci. Lett., **18**, (1999) 915-917.
[10] Minami, T., Kakumu, T., Shimokawa, K. and Takata, S., Thin Solid Films, **317**, (1998) 318-321.
[11] Kawazoe, H., Yasukawa, M., Hyodo, H., Kurita, M., Yanagi, H. and Hosono, H., Nature, **389**, (1997) 939-942.
[12] Thomas, G., Nature, **389**, (1997) 907-908.
[13] Kawazoe, H., Yanagi, H., Ueda, K. and Hosono, H., MRS Bulletin, **25**, (2000) 28-36.
[14] Yanagi, H., Inoue, S., Ueda, K., Kawazoe, H., Hosono, H. and Hamada, N., J. Appl. Phys., **88**, (2000) 4159-4163.
[15] Ueda, K., Hase, T., Yanagi, H., Kawazoe, H., Hosono, H., Ohta, H., Orita, M. and Hirano, M., J. Appl. Phys., **89**, (2001) 1790-1793.
[16] Duan, N., Sleight, A. W., Jayaraj, M. K. and Tate, J., Appl. Phys. Lett., **77**, (2000) 1325-1326.
[17] Barnabe, A., Mugnier, E., Presmanes, L. and Tailhades, P., Mater. Lett., **60**, (2006) 3468-3470.
[18] Yanagi, H., Hase, T., Ibuki, S., Ueda, K. and Hosono, H., Appl. Phys. Lett., **78**, (2001) 1583-1585.
[19] Shimode, M., Sasaki, M. and Mukaida, K., J. Solid State Chem., **151**, (2000) 16-20.
[20] Otabe, T., Ueda, K., Kudoh, A., Hosono, H. and Kawazoe, H., Appl. Phys. Lett., **72**, (1998) 1036-1038.
[21] Ibuki, S., Yanagi, H., Ueda, K., Kawazoe, H. and Hosono, H., J. Appl. Phys., **88**, (2000) 3067-3069.
[22] Ohta, H., Nomura, K., Hiramatsu, H., Ueda, K., Kamiya, T., Hirano, M. and Hosono, H., Solid-State Electron., **47**, (2003) 2261-2267.
[23] Hosono, H., Ohta, H., Orita, M., Ueda, K. and Hirano, M., Vacuum, **66**, (2002) 419-425.
[24] Marquardt, M. A., Ashmore, N. A. and Cann, D. P., Thin Solid Films, **496**, (2006) 146-156.
[25] Yanagi, H., Ueda, K., Ohta, H., Orita, M., Hirano, M. and Hosono, H., Solid State Commun., **121**, (2001) 15-17.
[26] Liu, L., Bai, K., Gong, H. and Wu, P., Phys. Rev. B, **72**, (2005) 125204.
[27] Yaicle, C., Blacklocks, A. N., Chadwick, A. V., Perriere, J. and Rougier, A., Appl. Surf. Sci., **254**, (2007) 1343-1346.
[28] Harris, V. G., Rev. Sci. Instrum., **68**, (1997) 23-29.
[29] Heald, S. M., Chambers, S. A. and Droubay, T., Phys. Scr., **T115**, (2005) 597-599.
[30] Erbil, A., Cargill Iii, G. S., Frahm, R. and Boehme, R. F., Phys. Rev. B, **37**, (1988) 2450-2464.

[31] Tourillon, G., Dartyge, E., Fontaine, A., Lemonnier, M. and Bartol, F., Phys. Lett. A, **121**, (1987) 251-257.
[32] Klementev, K. V., J. Phys. D: Appl. Phys., **34**, (2001) 209-217.
[33] www.desy.de/~klmn/viper.html
[34] Gurman, S. J., Binsted, N. and Ross, I., J. Phys. C: Solid State Phys., **17**, (1984) 143-151.
[35] Binsted, N., EXCURV98: CCLRC Daresbury Laboratory computer program (1998).
[36] Garlea, O., Bordet, P., Darie, C., Isnard, O. and Ballou, R., J. Phys.: Condens. Matter, **16**, (2004) S811-S816.
[37] Park, S. and Keszler, D. A., J. Solid State Chem., **173**, (2003) 355-358.
[38] Pellicer-Porres, J., Segura, A., Ferrer-Roca, C., Martinez-Garcia, D., Sans, J. A., Martinez, E., Itie, J. P., Polian, A., Baudelet, F., Munoz, A., Rodriguez-Hernandez, P. and Munsch, P., Phys. Rev. B, **69**, (2004) 024109.
[39] Fletcher, D. A., McMeeking, R. F. and Parkin, D., J. Chem. Inf. Comput. Sci., **36**, (1996) 746-749.
[40] http://www.webelements.com/.
[41] Teplin, C. W., Kaydanova, T., Young, D. L., Perkins, J. D., Ginley, D. S., Ode, A. and Readey, D. W., Appl. Phys. Lett., **85**, (2004) 3789-3791.
[42] Makhova, L., Wett, D., Lorenz, M. and Konovalov, I., Physica Status Solidi a-Applications and Materials Science, **203**, (2006) 2861-2866.
[43] Savin, S. L. P., Blacklocks, A. N., Berko, A., Edwards, W. and Chadwick, A. V., C. R. Chimie, **11,** (2008) .948-963.

GROWTH AND CHARACTERIZATION OF THE $LiGd_{0.25}Lu_{0.75}F_4$ CRYSTAL DOPED WITH NEODYMIUM

I A Santos[1], R Bertram[2], L Gomes[1], S L Baldochi[1], I M Ranieri[1]
[1] Instituto de Pesquisas Energéticas e Nucleares - IPEN-CNEN/SP, CP 11049, 05422-970, São Paulo, SP, Brazil.
[2] Leibniz Institute for Crystal Growth (IKZ), Max-Born-Str. 2, 12489 Berlin, Germany

E-mail: iranieri@ipen.br

Abstract. Mixed crystals based on scheelites type $LiLnF_4$ (Ln = lanthanides, Y) are an alternative in the development materials with better structural features to be doped with light rare earth ions. In this work a $LiGd_{0.232}Lu_{0.75}Nd_{0.018}F_4$ crystal was successfully grown using the Czochralski method in a commercial system with automated diameter control, under a mixed atmosphere of argon and CF_4. It was demonstrated that G25L75LF:Nd is an ideal solid solution that can be obtained with good optical quality from its stoichiometric composition. The spectroscopic studies showed that Nd presents larger bandwidth at 792 and 797 nm, which can be suitable to obtain a laser medium for mode-locking purposes.

1. Introduction

Crystals with the general formula $LiLnF_4$ or LnLF are isostructural to $LiYF_4$, which has the scheelite structure with space group $I4_1/a$. These crystals have high chemical stability with good mechanical and thermal properties and, in general, present high solubility to the incorporation of rare earth ions. In addition, as in the case of many fluoride crystals, they present a large transmission range from the ultraviolet (UV) to the infrared (IR), making them suitable for laser operation.

$LiLuF_4$ (LLF) has a congruent melting behavior, and its crystals present good optical quality. Since Lu^{3+} effective radius is 0.977Å for coordination 8 [1], the smallest lanthanide, this matrix is the most compact in the $LiLnF_4$ family (Ln = Eu up to Lu) with lattice parameters: a = 5.124 (5) Å and c = 10.54 (1) Å [2]. This crystal presents laser action when doped with many lanthanides ions: cerium [3,4], holmium [5], thulium and praseodymium [6], for example. LLF:Nd^{3+} crystals have also been investigated because of ultraviolet optical transitions resultant from $4f^{n-1}5d$ levels [7, 8].

The $LiGdF_4$ (GLF) compound has a strong incongruent melting behavior and, consequently, the growth of optical quality crystals from the melt is more difficult. Laser action has been reported for GLF:Pr and GLF:Nd, although laser performance was limited by the poor quality of the samples [9, 10].

The growth of mixed crystals proved to be a viable alternative in order to obtain crystals with better quality or new properties [11, 12]. Therefore, new laser materials were achieved: Nd:$LiLu_{0.75}Y_{0.25}F_4$ crystals exhibited a broader emission band compared to Nd:YLF [13], and $LiGd_{0.50}Y_{0.473}Nd_{0.027}F_4$ crystals could be grown with good optical quality [14]. The present paper is the first report of a solid solution crystal grown from the $LiGdF_4$-$LiLuF_4$ system. The $LiGd_{0.232}Lu_{0.75}Nd_{0.018}F_4$ (G25L75LF:Nd) crystal was obtained from its stoichiometric composition, and presented a nearly congruent melting behavior. Characterization regarding its optical properties, composition and lattice parameters was performed. Optical characterization was focused on absorption; the composition of the crystal was determined using the ICP – OES technique and the lattice parameters of the crystal were calculated from X-ray diffraction patterns.

2. Experimental Setup

The starting materials were prepared from commercially available GdF_3 and LuF_3 (AC Materials, 99.99%). Commercial LiF (Aldrich, 99.9%), was previously purified using the zone melting technique. The compounds were mixed in the $LiGd_{0.232}Lu_{0.75}Nd_{0.018}F_4$ composition and placed into an open platinum boat, which was inserted in a sealed platinum reactor. The material was synthesized in a reactive atmosphere of HF + Ar.

The single crystal was grown through the Czochralski technique under a high purity CF_4 + Ar atmosphere, in a commercial system with automatic diameter control. The crystal was grown with a growth rate of 1 mm/h and a rotation rate of 15 rpm for a [1 0 0]-oriented boule; a seed of $LiLuF_4$ was used to achieve the crystallization process.

X-ray powder diffraction (XRD) analyses were performed at room temperature in a PANalytical MPD 1980 diffractometer. The diffraction patterns were taken in 2θ range from 18 to 70° in step scan mode, with steps of 0.02°, 1 s per point, using CuKα radiation and operating at 40 kV and 40 mA. Three samples were obtained from the top, middle and bottom of the crystal to determine the lattice parameter variation over the length of the boule. The diffraction patterns where treated with the Rietveld Method [15] using the GSAS program to calculate lattice parameters [16]. The rare-earth concentrations were determined by inductive coupled plasma with optical emission spectrometry (ICP-OES, IRIS Thermo Electron, USA). Absorption measurements were performed on a Cary 17D-Olis spectrometer and on a Thermo-Nicholet – 6750 spectrometer for the near and far-infrared regions respectively. Luminescence lifetime measurements were performed at room temperature, the samples were excited by pulsed laser radiation generated by a tunable optical parametric oscillator-infrared (OPO-IR) pumped by a second harmonic of a Q-switched Nd:YAG laser (Brilliant B from Quantel, France). Laser pulses (width = 4ns, frequency = 10 Hz and wavelength = 797 nm) were used to excite the $^4F_{5/2}$ excited state of Nd^{3+}. Luminescence lifetime was measured using a digital oscilloscope of 100 MS s^{-1} model TDS from Tektronix interfaced to a microcomputer.

3. Results and discussion

The G25L75LF:Nd single crystal, with a length of 65 mm, a diameter of 20 mm and 65g, is shown in Fig. 1. The numbers on the figure identify the positions where the samples were extracted from to perform X-ray diffraction analyses.

The concentration determined by ICP-OES is shown in Table 1. The data show a wide variation in Gd concentration throughout the crystal. This variation is mostly due to the fact that this solid solution presents its liquidus line in a higher temperature than its solidus line, giving rise to segregation. In addition, the ionic radius of Gd is smaller than that of Nd, inhibiting the incorporation of the latter in favor of Gd ions.

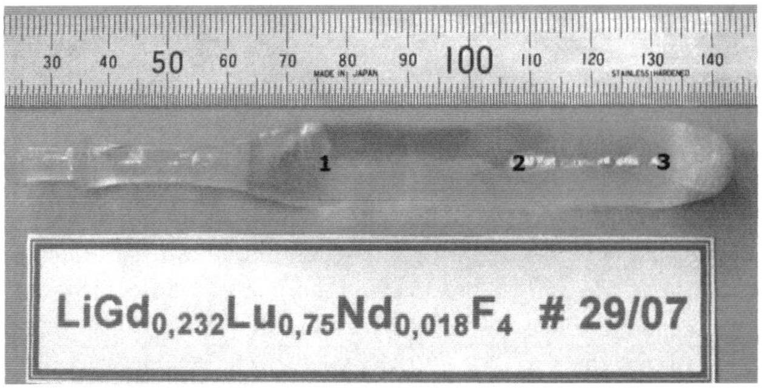

Fig. 1. The G25L75LF:Nd crystal grown using the Czochralski method.

Table 1. Rare earth concentrations determined by ICP-OES.

	Concentration (At. %)		
Solidified fraction	Gd	Lu	Nd
0.19	16.68(6)	82.82(4)	0.498(2)
0.28	17.50(5)	81.95(3)	0.545(3)
0.36	18.54(9)	80.85(6)	0.608(3)
0.45	19.51(9)	79.81(7)	0.673(4)
0.54	20.97(6)	78.26(5)	0.773(2)
0.64	22.91(4)	76.16(3)	0.935(3)

Rare earth segregation coefficients were estimated using the normal solidification equation, represented by Scheel's law (Equation 1).

$$C_S = C_0 k(1-g)^{(k-1)} \qquad (1)$$

Where C_s is the dopant concentration in the crystal; C_0 is the initial dopant concentration in the melt; k is the segregation coefficient and g is the solidified fraction. It was determined that Nd segregation coefficient is 0.22(1) (Fig. 2). This value is smaller than expected, since 0.31 and 0.40 are Nd segregation coefficients in $LiLuF_4$ and $LiGdF_4$, respectively. The same determination was made for Gd (k = 0.61) and Lu (k = 1.10) ions. These values reflect the variation of these constituents along the crystal.

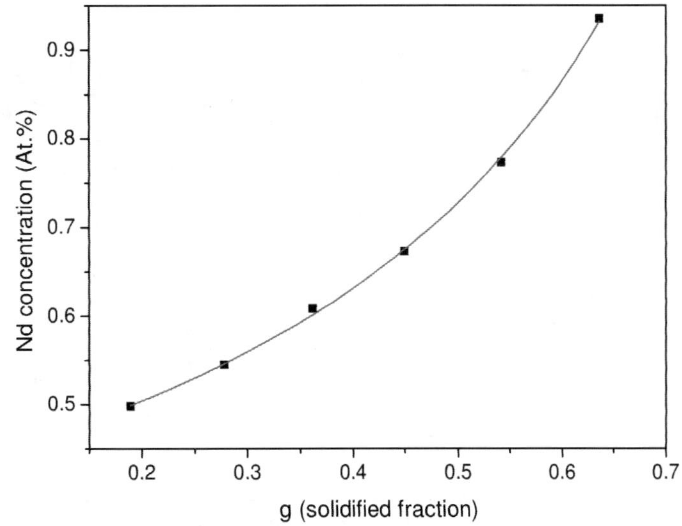

Fig. 2. Distribution of neodymium ions along the G25L75LF:Nd crystal.

The X-ray diffraction patterns obtained for the three assigned samples presented only a single phase, as expected for a single crystal (Fig.3). All reflections from the $LiGd_{1-x-y}Lu_xNd_yF_4$ phase can be indexed with the scheelite structure ($I4_1/a$, Z=4). The additional peaks are due to Si, which was added as an internal standard.

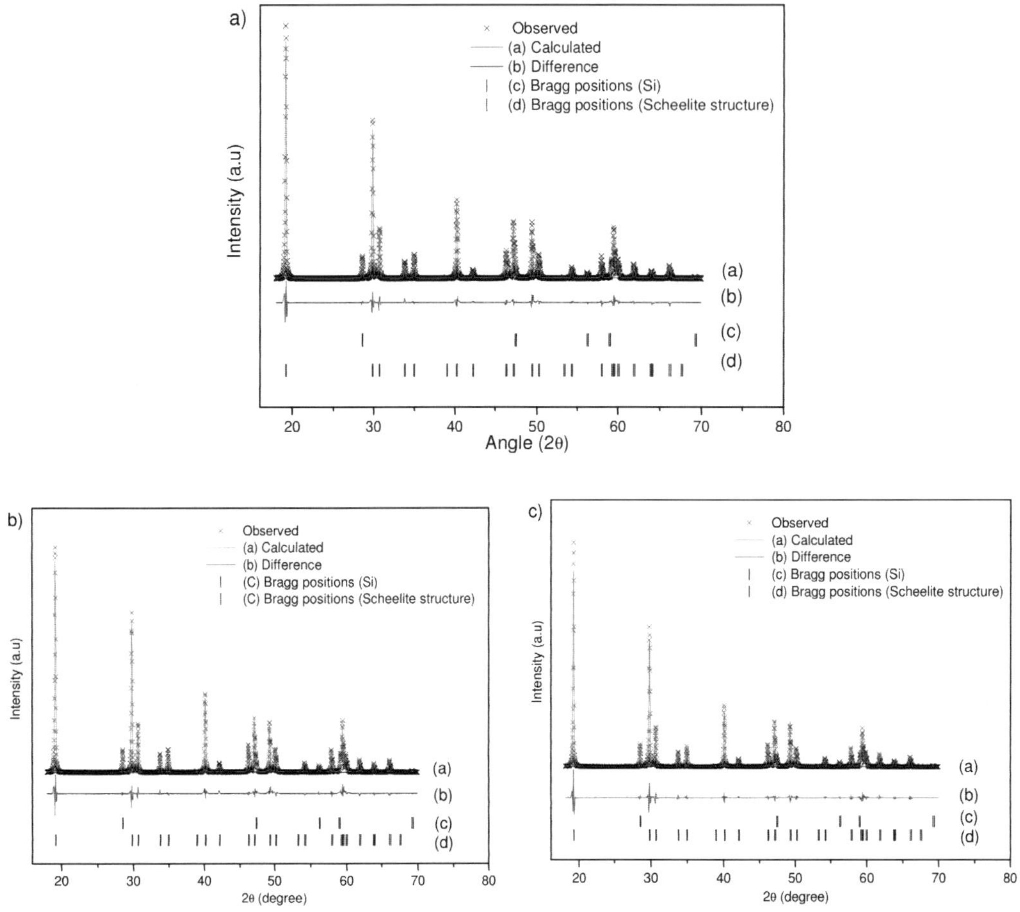

Fig. 3. X-ray diffraction patterns of three samples taken from the grown crystal and their respective fitted curves. a) $LiGd_{0.167}Lu_{0.828}Nd_{0.005}F_4$ (sample 1); b) $LiGd_{0.195}Lu_{0.798}Nd_{0.007}F_4$ (sample 3); c) Sample $LiGd_{0.229}Lu_{0.762}Nd_{0.009}F_4$ (sample 6).

Table 2. Lattice parameters calculated from X-ray diffraction patterns.

Lattice parameter: G25L75LF:Nd						
Sample	a (Å)	c (Å)	Vol. (Å^3)	Rp (%)	Rwp (%)	Goodness of fit
1	5.1416(1)	10.6183(2)	280.707(8)	12.4	16.5	1.90
2	5.1442(1)	10.6314(2)	281.339(7)	10.6	14.4	1.69
3	5.1476(1)	10.6503(2)	282.210(8)	11.4	15.0	1.77

The lattice parameters calculated from X-ray diffraction patterns using the Rietveld Method [15] are shown in Table 2. The increase in lattice parameters along the crystal is mainly due to Gd incorporation during the process of crystallization. The data presented in Table 1 show that Gd concentration in the beginning of the process is 16.68 At.% and it rose to 22.91 At.% at the last solidified portion. Since the ionic radius of Gd^{3+} is 7.8% larger than that of Lu^{3+} [1], lattice parameters variation is strongly related to Gd content in the $LiGd_{1-x}Lu_xF_4$ matrix.

The infrared spectrum revealed no contaminants related to OH^- or $Me(OH)_2$ (Me = divalent metals). The absorption bands in 4200 – 3700 cm^{-1} and 2300 – 1800 cm^{-1} are due to Nd transitions. Residual CO_2 present in the measurement chamber was detected by the absorption band in 2345 cm^{-1} (Fig. 3). Thus, the CF_4 + Ar atmosphere used during growth was suitable for obtaining a crystal free of contamination.

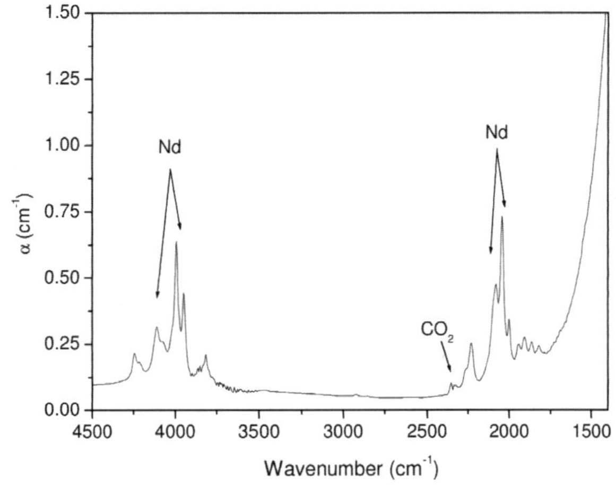

Fig. 3. Absorption coefficient in the infrared of a G25L75LF:Nd sample 66.5 mm thick.

Absorption spectra were obtained in polarizations π (E//c) and σ (E⊥c) (Fig. 3). It can be observed that emission in the π polarization is more efficient and, in general, the most commonly used to obtain laser action. This mechanism occurs when Nd^{3+} levels $^4F_{5/2}$ and $^2H_{3/2}$ are pumped by a diode laser. There is a nonradiative decay to the upper laser transition $^4F_{3/2}$ that decays radiatively to level $^4I_{1/2}$, generating laser action around 1.047 and 1.053 nm.

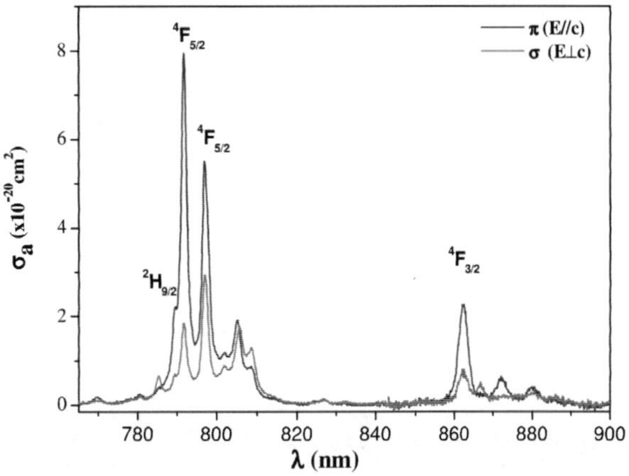

Fig. 3. π and σ polarization absorption spectra from a sample with 0.5 mol% Nd.

Spectroscopic studies showed that the G25L75LF:Nd crystal presents intermediate properties when compared to LLF:Nd and GLF:Nd (Table 3). The absorption bandwidth for transitions $^4F_{5/2}$, $^2H_{3/2}$ and $^4F_{3/2}$ were obtained through multi-Lorentzian fittings. It can be observed that bandwidths increased proportionally to Gd content, and there is also some shifting to larger wavelengths. These properties reveal how gadolinium influences spectroscopic properties in this crystal. The lifetime calculated for the $^4F_{3/2}$ transition decreased proportionally to Nd content, given that the interactions between Nd ions increase with concentration. This phenomenon is known as concentration quenching [17] and it takes into account the energy transfer from excited Nd^{3+} ions to centers Nd^{3+} - Nd^{3+} ion pairs through a dipole-dipole interaction. Since this process is competitive with luminescence radiation, it results in a decrease in luminescence efficiency and in luminescence lifetime (Table 3).

Table 3. Measured $^4F_{5/2}$ level lifetime and absorption bandwidth for the samples studied.

Sample	Lifetime (μs)	λ_{Peak} (nm)	bandwidth (nm)
$LiLu_{0.992}Nd_{0.008}F_4$	-	791.67	1.56
		796.87	0.93
		797.59	1.46
		862.27	2.55
$LiGd_{0.167}Lu_{0.828}Nd_{0.005}F_4$	511(1)	791.75	1.66
		796.83	1.11
		797.57	1.90
		862.28	2.68
$LiGd_{0.195}Lu_{0.798}Nd_{0.007}F_4$	496(1)	791.78	1.93
		796.84	1.24
		797.62	1.84
		862.29	2.63
$LiGd_{0.229}Lu_{0.762}Nd_{0.009}F_4$	480(1)	791.81	2.26
		796.84	1.29
		797.62	1.94
		862.33	2.72
$LiGd_{0.985}Nd_{0.015}F_4$	480 (5)	792.27	2.47
		796.72	1.20
		797.75	2.27
		862.91	2.67

4. Conclusions

It was demonstrated that G25L75LF:Nd is a solid solution that can be obtained with good optical quality from its stoichiometric composition. It presents accentuated Gd segregation that is typical of solid solutions with a lens-type phase diagram. Of the two ions to be incorporated into the crystal during the growth process, Gd^{3+} and Nd^{3+}, gadolinium proved more effective, since it has a smaller ionic radio than that of the neodymium ion. A mean Gd segregation coefficient of 0.22 was determined to the G25L75LF crystal. Nevertheless, the spectroscopic studies showed that due to the larger bandwidth of Nd at 792 and 797 nm in this matrix, it can be a suitable crystal to obtain a laser medium for mode-locking purposes, with advantages over Nd:YLF and Nd:LuLiF.

Acknowledgements

The authors thank FAPESP (05/57580-2), CAPES and CNPq for their financial support.

References

[1] Shannon R D 1976 *Acta Crystallogr.* A **32** 751
[2] Keller C and Schmutz H 1965 J. Inorg. Chem. **27** 900
[3] Johnson K S, Pask H M, Wtihford M J, Coutts D W 2005 *Opt. Commun.* **252** 132
[4] Ranieri I M, Shimamura K, Nakano K, Fujita T, Liu Z, Sarakura N, Fukuda T 2000 *J. Crystal Growth* **217** 151
[5] Bensalah A, Shimamura K, Sudesh V, Sato H, Ito K, Fukuda T 2001 *J. Crystal Growth*, **223** 539
[6] Cornacchia F, Richter A, Heumannie E, Huber G, Parised D, Tonelli M 2007 *Opt. Express* **15** 992
[7] Librantz A F H, Gomes L, Tarelho L G V, Ranieri I M 2004. *Appl. Phys.* **95** 1681
[8] Librantz A F H, Gomes L, Baldochi S L, Ranieri I M, Brito G E 2006 *J. Lumin.* **121** 137
[9] Sandrock T, Danger T, Heumann R, Huber G, Chai B H T 1994 Appl. Phys. B 58 149
[10] Zhang X X, Bass M, Villaverde A B, Lefaucheur J, Pham A, Chai B H T 1993 Appl. Phys. Lett. 62 1197
[11] Ranieri I M, Morato S P, Courrol L C, Shihomatsu H M, Bressiani A H A, Moraes N M P 2000 *J. Crystal Growth* **209/4** 906
[12] Ranieri I M, Shimamura K, Nakano K, Fujita T, Courrol L C, Morato S P, Fukuda T 2000 *J. Crystal Growth* **217** 145
[13] Maldonado E P, Barbosa E A, Courrol L C, Wetter N U, Ranieri I M, Morato S P, Vieira Jr. N D 2001 *Optical Engineering* **40** 1573
[14] Wetter N U, Matos P S F, Ranieri I M, Courrol L C, Morato, S P 2002 *Optics Communications* **204** 311
[15] Rietveld H M 1969 *J. Appl. Crystallogr.* **2** 65
[16] Larson A C and Dreele R B V 2007 General structure analysis system (GSAS), *Technical Report LAUR*, Los Alamos National Laboratory 86
[17] Zhang X X, Villaverde A B, Bass M, Chai B H T, Weidner H, Ramotar R I, Peale R E 1993 *J. Appl. Phys.* **2** 74

Synthesis and characterization of KY_3F_{10}:Yb:Nd:Tm crystals

H M Silva[1], L C Courrol[2], L Gomes[1], R Bertram[3], S L Baldochi[1], I M Ranieri[1]

[1] Instituto de Pesquisas Energéticas e Nucleares IPEN/CNEN – SP
Av. Prof. Lineu Prestes, 2242- CEP:05508000 - São Paulo-SP - Brasil
[2] Universidade Federal de São Paulo
Rua Prof. Artur Riedel, 275- CEP: 09972-270 - Diadema – SP - Brasil
[3] Institut für Kristallzüchtung (IKZ)
Max-Born-Strasse 2, D-12489, Berlim – Germany

E-mail: iranieri@ipen.br

Abstract . Energy transfer processes that generate thulium blue and ultraviolet emissions by upconversion were studied in KY_3F_{10}:Tm (KY3F:Tm), using Nd^{3+} and Yb^{3+} as sensitizers of Tm^{3+}. The upconversion mechanisms were determined by exciting into the Nd and Tm absorption bands with a diode laser at 797nm. It was observed that the intensity of the Tm^{3+} blue and ultraviolet emissions in 484, 453, 366 and 350 nm were very dependent of the Yb^{3+} concentration in the samples, confirming that the main energy transfer processes involve Nd^{3+} and Yb^{3+} ions. KY_3F_{10}:Nd (1.3 mol%):Tm (0.5 mol%) crystals codoped with 5, 10, 20 and 30 mol% Yb were prepared by slow cooling from the melt, to establish the optimal Yb concentration that maximizes the $Nd^{3+}\rightarrow Yb^{3+}\rightarrow Tm^{3+}$ energy transfer.

1.Introduction

Infrared to visible frequency upconversion (UC) in rare-earth (RE) doped materials has been extensively investigated due to its potential applications in UC lasers, high-density memories and solid-state color displays. Fluoride compounds in particular have been receiving special attention for UC applications owing to their low phonon energy, which means that nonradiative quenching of emission is limited and then higher efficiencies can be reached [1]. Also, they have good mechanical and thermal properties, lower melting temperatures than the oxides and are less hygroscopic than bromides, sulfides and chlorides. Some rare earth elements such as Pr^{3+}, Nd^{3+}, Ho^{3+}, Er^{3+} and Tm^{3+} are especially suitable for UC because these ions exhibit long-lived high excited $4f^n$ states that lead to strong green, blue or UV luminescence [2]. Furthermore, it is already well-known that the use of Yb^{3+} as a sensitizer enhances the efficiency of the process in one or two orders of magnitude [3,4].

The KY_3F_{10} (KY3F) is a cubic optically isotropic material with a 2x2x2 superstructure of fluorite, space group Fm-3m (Z=8), and melts congruently at 990 °C [5]. This feature allows KY3F to be grown either as bulk crystals through the Czochralski method or as fibers through the micro-pulling-down method [6]. These crystals have low phonon energy ($400cm^{-1}$) and the yttrium ion occupies sites with tetragonal symmetry (C_{4v}) [7, 8]. Previous studies have shown that this matrix exhibits high blue emission efficiency when KY3F:Tm is codoped with Yb and pumped with 975 nm-laser diode [1]. Blue emission enhancement of up to two orders of magnitude was also achieved, when Nd^{3+} was added as a second sensitizer in $LiYF_4$:Yb:Tm (YLF:Yb:Tm) crystals [9, 10, 11] and glasses [12, 13, 14]. In these materials, the population of the Tm ions in the $^1G_4$ emitting level was obtained by multi-ion interaction involving ground state absorption of pumped photons around 800 nm by the $Nd^{3+}(^4I_{9/2}\rightarrow{}^2H_{9/2}, {}^4F_{5/2})$ and $Tm^{3+}(^3H_6 \rightarrow {}^3F_4)$ ions, followed by energy-transfer processes involving the Nd^{3+}–Yb^{3+} and Yb^{3+}–Tm^{3+} pairs.

In this work the effect of Nd used as a second sensitizer to pump Tm ions in KY3F:Tm:Yb:Nd crystalline samples has been studied. $Nd^{3+}(^4I_{9/2}\rightarrow{}^2H_{9/2}, {}^4F_{5/2})$ and $Tm^{3+}(^3H_6 \rightarrow {}^3F_4)$ ions were pumped by a 797nm-laser diode and the energy-transfer processes involved in the upconversion mechanism were determined. Monocrystalline samples doped with 0.5 mol% Tm, 1.3 mol% Nd and diverse concentrations of Yb (5, 10, 20 and 30 mol%) were synthesized through slow cooling from the melt under a HF+Ar atmosphere.

2. Experimental Setup

The rare earth fluorides were synthesized from pure oxide powders (Alpha-Johnson Matthey or Aldrich, 99.99%) through hydrofluorination at high temperature (around 800ºC) under an HF + Argon atmosphere. The powders were contained in a cylindrical platinum boat, which was inserted in a sealed platinum tube. The transformation from oxide to fluoride occurs as follows: $RE_2O_3 + 6\ HF \leftrightarrow 2\ REF_3 + 3\ H_2O$ (RE=Y, Tm, Nd or Yb).

The bars of KY3F:Tm (0,5mol%) and KY3F:Tm (0,5mol%): Nd (0 or 1,3mol%) codoped with 0, 5, 10, 20 and 30 mol% Yb were synthesized using the same system and atmosphere. Fluorides react strongly with oxygen and moisture, when these impurities are present during the synthesis process, oxygen complexes are formed in the crystals. They are responsible for scattering centers that greatly influence the optical properties of the crystals causing losses in the laser media. The use of hydrofluoric acid as reactive atmosphere prevents the degradation of the materials. Mixtures of the rare earths fluorides and KF powder (99% -Johnson Mattey) were melted in an open platinum boat and slowly cooled with rates of 7.2, 10, 15 or 25°C/h. Resulting material always presented transparent crystalline regions. Powder X-ray diffraction (XRD) measurements were carried out in the 2θ range of 12.5-72.5^0 on a Bruker AXS model A8 Advanced difratometer, in the step scan mode, with 0.02°/step and step time of 5s. The lattice parameters were calculated using the LCLSQ program [15]. The rare earth concentrations in the samples were determined via inductively coupled plasma (ICP-OES). Absorption spectroscopy was carried out on a Cary 17-D Varian. An experimental arrange consisting of a chopper, a 50cm-monochromator from Spex, and photomultiplier with an extended S-20 cathode were used to obtain the emission spectra excited by a diode laser from SLD emitting around 797 nm. The signal was amplified with a lock-in and finally processed by a computer.

3. Results and Discussions

KY3F crystalline samples were successfully obtained directly from the synthesized bars. X-ray diffraction analysis demonstrated that only the desired phase is present in all samples. The triply codoped samples X-ray patterns are presented in Figure 1 and the lattice parameters are shown in table 1.

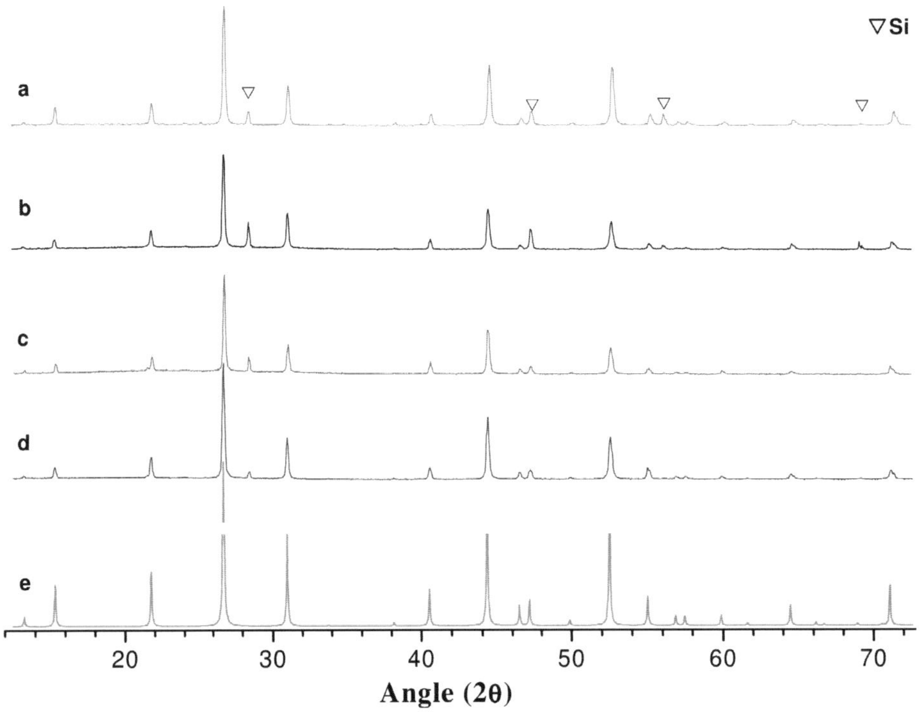

Figure 1. Powder diffraction patterns of KY3F doped with 0.5 mol% Tm, 1.3mol% Nd and different concentrations of Yb: a) 30mol% Yb; b) 20mol% Yb; c) 10mol% Yb; d) 5 mol% Yb; and e) undoped KY_3F_{10} according the ICSD pattern Nr. 155135.

Table 1. RE concentrations and calculated lattice parameters

Nominal concentration (mol %)	Measured concentrations (mol%)			Lattice parameters (Å)
	Tm^{3+}	Nd^{3+}	Yb^{3+}	
KY3F:Tm(0.5 mol%):Nd(1.3mol%)	0.47 (1)	1.26 (1)		11.545(1)
KY3F:Tm(0.5 mol%): Yb(10mol%)	0.48 (1)		9.48 (2)	11.523(4)
KY3F:Tm(0.5 mol%):Nd(1.3mol%): Yb(5mol%)	0.493 (4)	1.180 (1)	5.41 (8)	11.546 (2)
KY3F:Tm(0.5 mol%):Nd(1.3mol%): Yb(10mol%)	0.492 (5)	0.871 (4)	9.30 (1)	11.539 (2)
KY3F:Tm(0.5 mol%):Nd(1.3mol%): Yb(20mol%)	0.481 (1)	0.909 (1)	18.49 (1)	11.5271(1)
KY3F:Tm(0.5 mol%):Nd(1.3mol%): Yb(30mol%)	0.493 (2)	1.520 (1)	29.60 (2)	11.5052(9)

The concentrations of the samples presented in Table 1, show that measured concentrations are very close to the nominal values for the heavy rare earths. Different Nd concentrations could be obtained along the bars due to its segregation coefficient of approximately 0.7.

The KY3F:Tm:Nd sample present an absorption band with the most intense peak at 802 nm, due to transitions $^4I_{9/2} \rightarrow {}^4F_{5/2}$ of Nd^{3+}. In the same region, other four bands peaked at 773 and 778.5, 796 and

Journal of Physics: Conference Series **249** (2010) 012047 doi:10.1088/1742-6596/249/1/012047

802 nm correspond to transitions $^3H_6 \rightarrow {}^3H_4$ of Tm^{3+} (Figure 2). The excitation wavelength is highlighted in the figure. No blue emission was detectable when a sample of KY3F:Tm was excited at 797 nm; KY3F:Nd:Tm and KY3F:Yb:Tm samples presented smaller emissions when compared to KY3F:Yb:Nd:Tm (Figure 3).

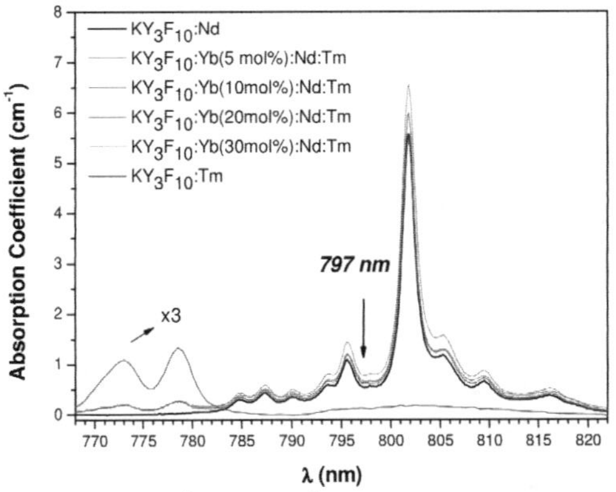

Figure 2. Nd^{3+} and Tm^{3+} absorption bands in the pumping region.

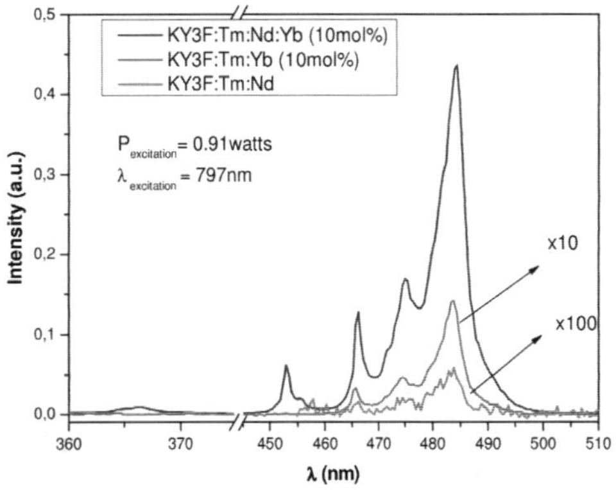

Figure 3. Blue emission of KY3F:Tm codoped with Yb and/or Nd

Figure 4. Luminescence of KY3F:Yb(10mol%):Nd:Tm crystal under 797 nm excitation.

In KY3F:Yb:Nd:Tm samples strong blue emission was always observed under 797 nm excitation (Figure 4); the emission bands peaked at 453nm and 484nm correspond to the $^1D_2 \rightarrow {}^3F_4$ and $^1G_4 \rightarrow {}^3H_6$ transitions of Tm^{3+} respectively [16].

The study of the dependence of the blue luminescence intensity upon the Yb^{3+} concentration revealed that the blue emission was dependent of the ytterbium concentration (Figure 5). The emission intensity of the 484 nm band grows considerably when the Yb concentrations changes from 5 to 10 mol% and then decreases for larger concentrations. The intensity of the 453 nm emission band is proportional to the concentration of Yb up to 20 mol%, and probably tends to saturate with increasing ytterbium concentration (Figure 6). It was concluded that the energy transfer between Nd^{3+} and Yb^{3+} is the main mechanism in the process to provide Tm upconversion excitation.

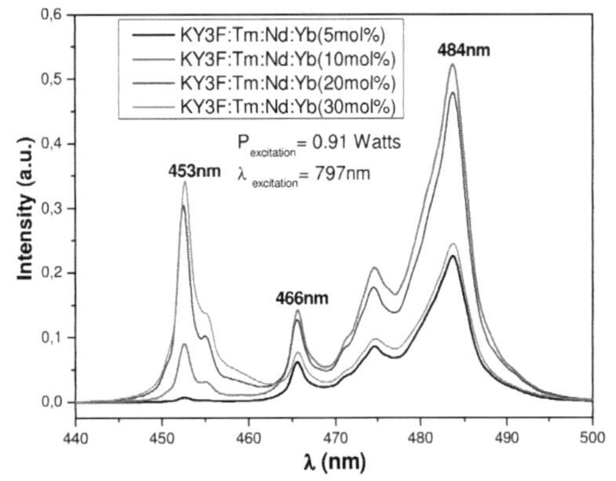

Figure 5. Blue emission in KY3F:Tm:Nd:Yb

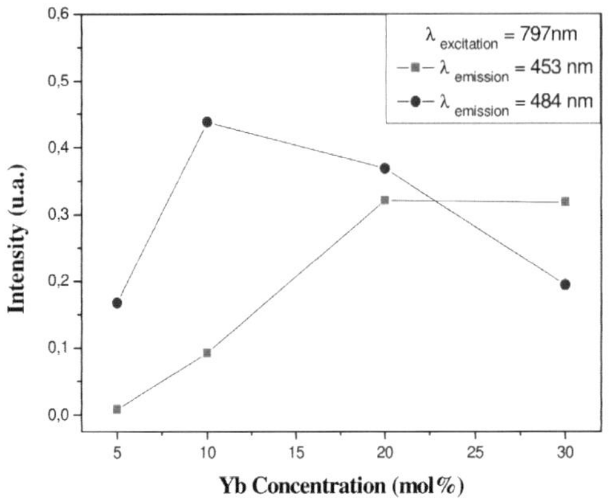

Figure 6. Blue emission intensity related to Yb concentration in KY3F:Tm:Nd:Yb.

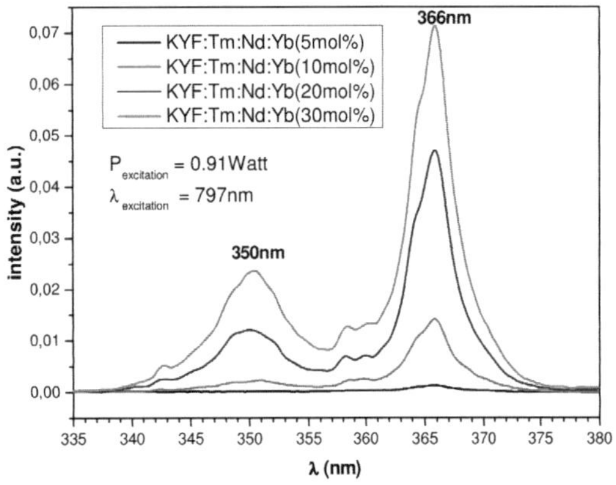

Figure 7. UV emission in KY3F:Tm:Nd:Yb

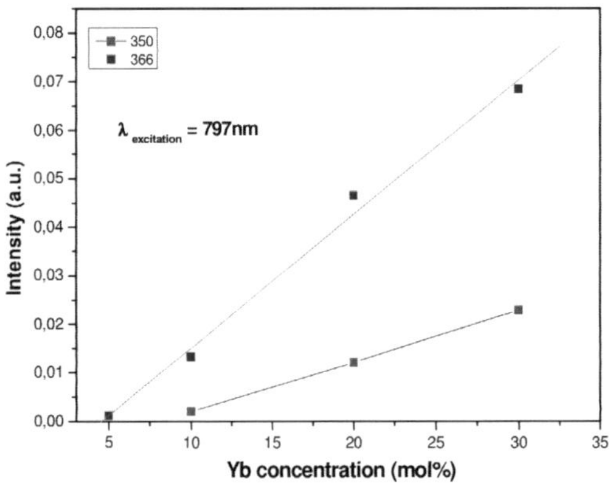

Figure 8. Yb^{3+} concentration dependence of the UV emission in KY3F:Tm:Nd:Yb

All tripled doped samples exhibited emission bands around 350 nm and 366 nm involving Tm^{3+} ions in $^3H_4$ and $^1D_2$ levels respectively (Figure 7). The intensities of these peaks are proportional to Yb concentration (Figure 8).

From the absorption and emission spectra of doped and codoped samples, as well as earlier YLF spectroscopy studies [9,10,11, 17], were constructed the energy level schemes which show the most probable mechanisms of energy transfer (Figure 9). The emission around 484nm including bands with smaller wavelength (460 and 470nm) are originated from $^1G_4 \rightarrow {^3H_6}$ transitions, and two pump photons participate in the upconversion excitation mechanism as they excite both Nd^{3+} and Tm^{3+} through ground-state absorption in the follow sequence:

(a) Tm^{3+}: $^3H_6$ + one photon $\rightarrow$ Tm^{3+}: $^3H_4$
Nd^{3+}: $^4I_{9/2}$ + one photon $\rightarrow$ Nd^{3+}: $^4F_{3/2}$
(b) Nd^{3+}: $^4F_{3/2}$ + Yb^{3+}: $^2F_{7/2}$ $\rightarrow$ Nd^{3+}: $^4I_{9/2}$ + Yb^{3+}: $^2F_{5/2}$
Yb^{3+}: $^2F_{5/2}$ + Tm^{3+}: $^3H_4$ $\rightarrow$ Yb^{3+}: $^2F_{7/2}$ + Tm^{3+}: $^1G_4$

Once its $^1G_4$ level is populated, the Tm^{3+} ion absorbs the energy from another excited Yb^{3+} ion and has its $^1D_2$ level populated:

(a) + (b)

$$\text{(c) } Tm^{3+}{:}^1G_4 + Yb^{3+}{:}^2F_{5/2} \rightarrow Tm^{3+}{:}^1D_2 + Yb^{3+}{:}^2F_{7/2}$$

If the decay comes from $^1D_2$ to $^1H_6$ ground state, Tm^{3+} emits at 350 and 366nm. If the transition is from $^1D_2$ to $3F_4$ level, the 453nm emission occurs.

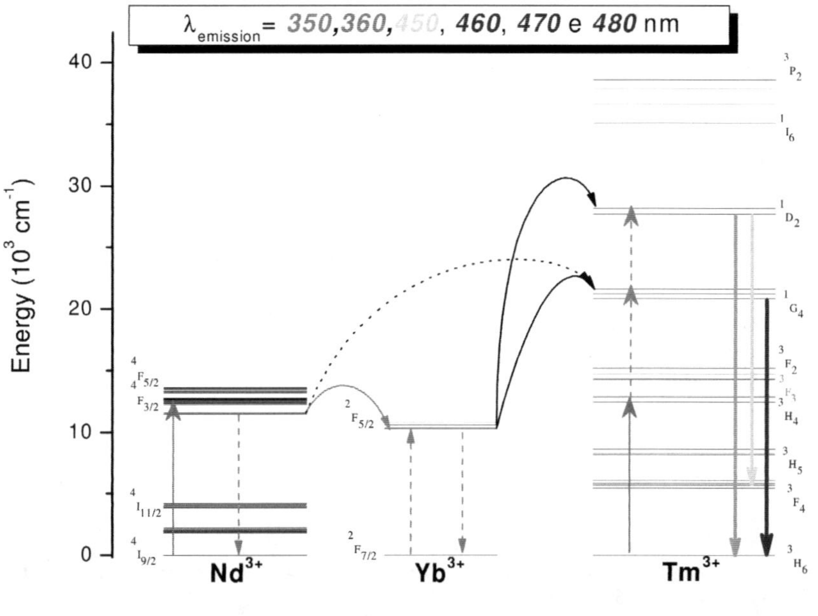

Figure 9. Energy levels scheme and energy transfer mechanisms of the Yb/Tm/Nd system related to the 484 nm emission.

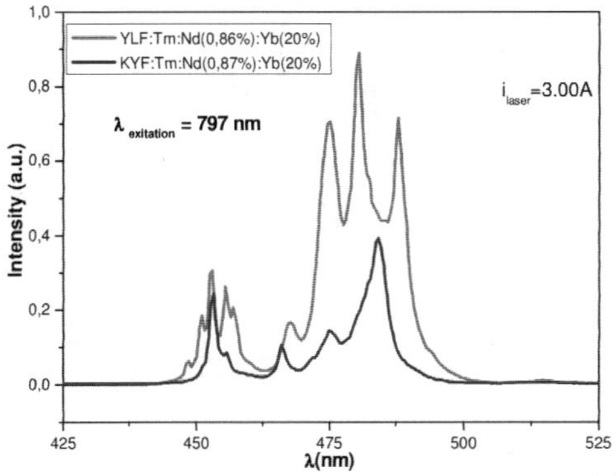

Figure 10. KY3F and YLF spectra – both pumping at 797 nm

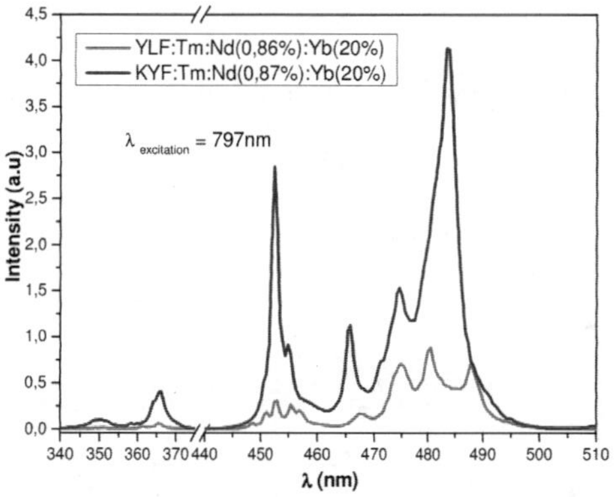

Figure 11. KY3F and YLF spectra supposing KY3F pumped at 802 nm

Comparison between $LiYF_4$ (YLF) and KY3F crystals results in a higher blue emission in YLF crystals when compared with KY3F, both excited at 797nm (Figure 10). Conversely, taking into account a resonant absorption at 802 nm for the KY3F, a correction can be made resulting in a 4-fold enhancement on the blue emission when compared to YLF (Figure 11).

4. Conclusions

KY3F samples of good quality were obtained using a simple synthesis method consisted by the slow cooling of the liquid charges. This method reduced the time to produce samples with sufficient transparency and free of macroscopic defects. Analysis of the Tm^{3+} blue emission for KY3F:Yb:Tm:Nd crystals indicates that Nd^{3+} ions contribute significantly to enhance the population of $^1G_4$ excited level. It was demonstrated that energy transfer between Nd^{3+} and Yb^{3+} is the main mechanism of the upconversion process when these crystals are pumped with 797 nm. The blue-emission intensity was enhanced around 30 times when compared to a KY3F:Yb:Tm crystal.

5. Acknowledgements

The authors thank **CNPq** for financial support.

6. References

[1] Rapaport A, Milliez J, Szipocs F, Bass M, Cassanho A and Jenssen H 2004 *Applied Optics* **43**(35) 6477-80.
[2] Joubert M-F 1999 *Opt. Mat.* **11** 181 – 203.
[3] Auzel F, 1969 *Ann. Telecom.* (Paris) **24**(9-10) 363-76.
[4] Geusic J E, Ostermayer F W, Marcos H M, Van Uitert L G and van der Ziel J P , 1958-1960 *J. Appl. Phys.* **42**(5).
[5] Grzechnik A, Nuss J, Friese K, Gesland J Y and Jansen M 2002 *Z. Kristallogr.* **217** 460-3
[6] Yoshikawa A, Kamada K, Nikl M, Aoki K, Sato H, J Pejchal and Fukuda T 2005 *Journal of Crystal Growth* **285**(4) 445-9.
[7] Porcher P and Caro P J 1976 *Chem. Phys.* **65**(1) 89.
[8] Porcher P and Caro P J 1978 *Chem. Phys* **68**, 4183–7.
[9] Courrol L C, Tarelho L V G, Baldochi S L, Gomes L and Vieira Jr N D 2005 *J. Applyed Physics* **98** 13504.
[10] Ranieri I M, Courrol L C, Carvalho A F, Gomes L and Baldochi S L 2007 *4th Brasilian mrs Meeting* **42** 2309-13.
[11] Courrol L C, Tarelho L V G, Baldochi S L, Gomes L and Vieira Jr 2007 *Journal of Luminescence* **122-123** 474-7.
[12] Qiu J, Mukai A, Makishima A and Kawamoto Y 2002 *J. Physics Condensed Matter* **14** 13827–34.
[13] Gouveia-Neto A S, Costa E B, Santos P V, Bueno L A and Ribeiro S J L 2003 *Journal of Applied Physics* **94**(9) 5678-81.
[14] Chen D, Wang Y, YuY, Liu F, Huang P 2007 *Optics Letters* **32**(21) 3068-70.
[15] Burnham C 1991 *Least squares refinement of crystallographic lattice parameters (LCLSQ)*, version 8.4, Harvard University.
[16] Diaf M, Braud A, Labbé C, Doualan J L , Girard S, Margerie J, Moncorgé R and Thuau M, 1999 *Can. J. Phys.* **77** 693–7.
[17] Zhang X X, Hong P, Bass M and Chai B H T 1995 *Physical Review B* **51** (14) 9298-9301.

Structure and IRR spectra of copper-exchanged soda-lime silica glass

M Suszynska[1,3], M Maczka[1], E Bukowska[1] and K J Berg[2]

[1] Institute of Low Temperature and Structure Research, PAS, Okolna Street 2, 50-950 Wroclaw, Poland
[2] Physics Department, M Luther University, Friedemann-Bach Square 6, 06-108 Halle/Saale, Germany

suszyn@int.pan.wroc.pl

Abstract. Multi-component soda lime silica glass (SLSG) specimens, with composition from the miscibility gap in the SiO_2-Na_2O system, were investigated after the substitution of copper for sodium and after thermal treatment of the exchanged samples in gaseous hydrogen. The behaviour of the obtained composite materials was studied by infrared reflection spectroscopy (IRRS). Some optical absorption micro-spectrophotometry (OAMS) measurements and X-ray-diffraction (XRD) analyses complemented these investigations. It has been stated that after exchange in the CuCl molten bath, the dopant was present in the glassy matrix in the form of cuprous and cupric ions. The cuprous ions participate in the formation of semiconducting Cu_2O nanoparticles. Hydrogenation of the exchanged specimens leads to the formation of metallic copper atoms which due to the van der Waals interactions precipitate in the form of colloidal copper nanoparticles. Changes of the matrix morphology are mainly related with the presence of cupric ions which behave as glass forming cations and such as these participate in the formation of mixed copper-sodium silicates. The effectiveness of these processes is substantially dependent on the exchange and hydrogenation parameters.

1. Introduction

Ion exchange between an oxide glass and a molten salt is used for preparation of composite materials for a variety of applications, such as chemical strengthening of glasses [1], formation of gradient index materials [2], and fabrication of passive integral optical devices [3, 4]. Although the formation of copper-colloids in these glasses is well documented [5, 6], the ion exchange process, in which a part of the univalent cations in a silica glass is replaced by other univalent cations from a molten salt, has been conducted to date without paying much attention to the alteration of the glass structure. It has been shown recently [7], that some structural changes occur during the exchange of soda lime silica glass and lead to distinct modifications of the phase separated occlusions already present in not-exchanged specimens, cf. figure 3 in [7].

The purpose of the work described was to study the interrelations between the exchange conditions, copper valence state, optical properties and the structure of copper-doped multi-component soda-lime silica (SLS) glass. The local structure of the glass network was investigated by infra-red-reflectance

[3] To whom any correspondence should be addressed.

spectroscopy (IRRS). Measurements of optical absorption micro-spectrophotometry (OAMS) and powder-X ray-diffraction (XRD) were made to obtain supplementary data.

2. Experimental part: specimens and measurements

A multicomponent soda-lime silica glass (SLSG), produced according to the Fourcault procedure (Torgau-factory, Germany), was used. The content of the main components (SiO_2/73 mol % and Na_2O/13 mol %) corresponds to the miscibility gap in this system [8].

Doped samples have been prepared by the ion exchange method, and to the first time for this type of glasses molten CuCl salt was used at the exchange temperature of 903 K for times between a few min. and 168 h. To avoid the formation of a metal lustre, samples were exchanged in the air atmosphere. Under these conditions, the dopant enters the matrix as cuprous and cupric ions, and the equilibrium between both valence states depends in a quite complex way on the temperature and time of exchange. To transform the copper ions into atoms and metallic nanoparticles, it was necessary to anneal the exchanged samples in gaseous hydrogen at temperatures ≥ 573 K; for the purposes of the present work 5 h at 773 K have been chosen. Not-exchanged specimens were used as a reference material.

According to the previously obtained EDX data, see figure 1 in [9], the copper concentration-profile in the SLSG specimens has a rectangular shape with a constant value of the concentration from the sample surface up to the end of the penetration depth. For each temperature, this depth increases with the increase of the exchange time. The sodium concentration-profile gives the degree of the Cu/Na-exchange. It has been stated that at the surface about 90 % of sodium has been exchanged independently of the exchange time.

To determine the local atomic arrangement around the copper ions and copper nanoparticles in the exchanged and hydrogenated specimens, respectively, infra-red reflectivity (IRR) measurements have been performed. The IRR spectra were measured with a BIORAD 575 C FT-IR spectrofotometer in the 1300-400 cm^{-1} range with a spectral resolution of 2 cm^{-1}. The measurements were performed at near normal incidence of 10° using a fixed-angle specular reflectance accessory.

Optical absorption micro-spectrophotometry (OAMS) allowed to measure the extinction spectra of doped samples (nominal thickness was about 25 μm) for a fixed value of wavelength (λ) as function of the penetration depth in steps equal to 0,5-1,0 μm [10]. Optical profiles of the copper incorporation into the SLSG samples were determined by using the ARL EMX SM 32 electron-probe microanalyzer.

The powder X-ray-diffraction (XRD) data were collected with the aim of a STOE powder X-ray diffractometer using the (Cu-K_α) radiation.

3. Results and discussion

3.1. The SiO_4 tetrahedron vibrational unity

The soda lime silica glasses are Si-O networks in which the SiO_4 tetrahedra are joined together by the oxygen atoms located at the vertices. These SiO_4 tetrahedral connections become modified by incorporation of the Na, K, Ca, Mg, Al and Fe ions (present in the SLSG specimens) which alter the polymerization degree of the glass framework. Also the dopant ions and/or particles affect the polymerization degree of the matrix. Five structural units (Q^0-Q^4) correspond to the SiO_4 tetrahedra with different number (4-0) of the bridging oxygen atoms [6]. The IRR spectra obtained for the exchanged and hydrogenated specimens reflect the modifications of the SiO_4 tetrahedral vibrational units. Mostly, the intensity, line-width and spectral position of the associated reflection bands are considered in terms of the symmetric stretching modes; this description is essentially the same as that presented in terms of the modes originating from bending of the SiO_4 units [11-13].

3.2. IRR spectra

Journal of Physics: Conference Series **249** (2010) 012048 doi:10.1088/1742-6596/249/1/012048

Figure 1 shows representative IRR spectra detected for not exchanged (curve a), copper exchanged (curve b) and hydrogenated (curve c) SLSG samples for which the exchange time at 903 K was equal 168 h. For shorter exchange times, the results were qualitatively similar.

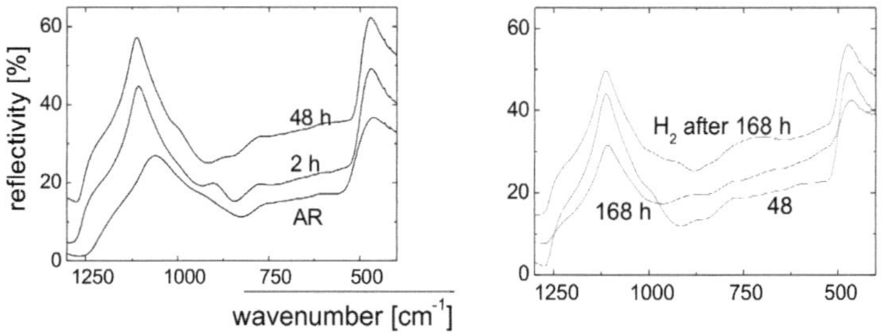

Figure 1. IRR spectra of SLSG specimens exchanged at 903 K for different times and hydrogenated after 168 h; results for not exchanged sample are also presented.

For not-exchanged samples the strongest bands are located at 1060 cm^{-1} and 462 cm^{-1}, and are related with the Q^3 stretching and bending vibrational modes, respectively. The shoulders, detected on the high and low vibration side of the 1060 cm^{-1} band, correspond to the Q^4 (1150 cm^{-1}) and Q^2 (930 cm^{-1}) units, respectively.

After the ion-exchange, the width of the main bands decreases (in comparison with not-exchanged specimens), suggesting that the glass framework became more ordered. Moreover, the frequency of the stretching Q^3 band strongly increases (1060 $\rightarrow$ 1109 cm^{-1}), the frequency of the bending Q^3 band also increases but not so much (462 $\rightarrow$ 473 cm^{-1}), the frequency of Q^2 slightly decreases, and the intensities of Q^4 and Q^2 remain nearly unchanged. This behavior could be explained in terms of some de-polymerization which occurs in the glass framework, but in conflict with this suggestion is the detected increase of the stretching Q^3 frequency. On the other hand, assuming that a part of the stretching Q^3 units disappears giving a mixture of (Q^4 + Q^2), the frequency of the stretching Q^3 band would increase and this increase should be accompanied by a decrease of the bending Q^3 frequency. Also this supposition is in conflict with the experimental observation. Moreover, the expected intensity increase of the Q^4 and Q^2 bands has not been detected. Taking into account the experimental data and all the mentioned incompatibilities one has to consider that the results are consequences of interactions between the dopant ions and the oxygen atoms from the silica framework.

Hydrogenation, performed after the ion-exchange, leads to a slight increase of the stretching Q^3 frequency (1109 $\rightarrow$ 1115 cm^{-1}), suggesting further decrease of the corresponding (Si-O) bond length, an increase of the interaction force and a slight decrease of the polymerization degree. The appearance of a new band at about 1007 cm^{-1} has been tentatively ascribed to the formation of some silicates, the structure of which comprises chains of connected SiO_4 tetrahedra [6]. The absence of bands characteristic of the Q^0 and Q^1 units indicates that clustering is limited to two cations per one SiO_4 tetrahedron, at least below the di-silicate composition.

The XRD-data (figure 2), have been accepted as confirmation of this hypothesis although a proper identification of the newly created phase was unsuccessful on the basis of the present data. Probably

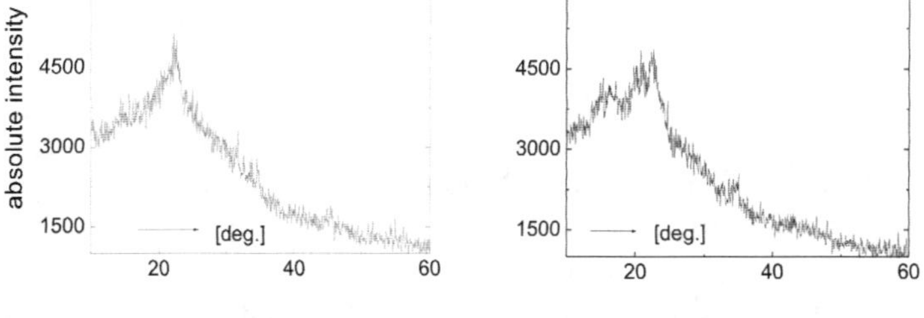

Figure 2. The XRD spectra of not-exchanged and samples hydro genated after 168 h of exchange (left & right hand side, respectively)

the number of the corresponding particles per unit area of the sample surface is very small and they are amorphous at this stage of the sample preparation.

3.3. Structural interpretation of the spectral features

The described results have shown that more complex structural modifications of the matrix must be considered to account for the behavior of the copper-doped soda-lime silica glass.

First of all, one has to stress that the mentioned changes of the glass morphology are characteristic only for specimens exchanged at temperatures above the glass-transformation temperature [7]. Under these conditions, the concentration of cupric ions is distinctly larger than that of the cuprous ions. Moreover, according to the optical microspectrophotometry the divalent copper ions are still present in the hydrogenated specimens. Figure 3 exemplifies the results of the OAMS after 168 h of exchange.

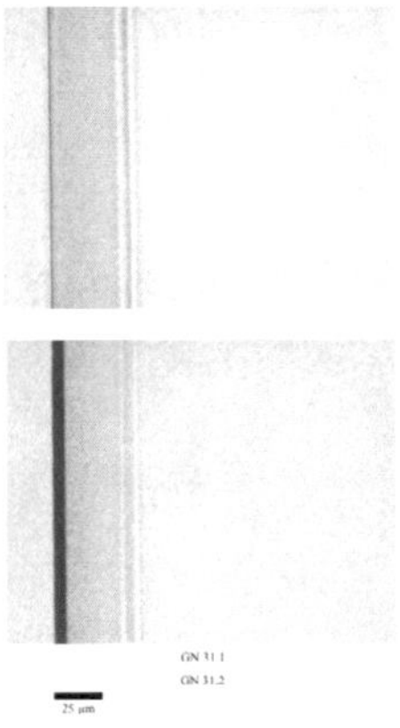

Figure 3 . The optical microspectrophotometry figures characteristic of SLSG specimens exchanged for 72 h (upper part) and hydrogenated afterwards (lower part).

The dark zone, which in the hydrogenated sample corresponds to the copper colloids, is followed by other colored zones related with the presence of cupric ions probably in the form of some silicates, see Section 3.2. The narrow and sharp zones in the "optical profiles" are compatible with the copper concentration profiles [9], and are characteristic of transitions between the compression and tension zones in exchanged and thermally treated specimens.

With respect to the IRRS results, substitution of sodium ions in the SLS glasses by other metal ions should affect the frequencies of both Q^3 modes (stretching and bonding) by shifting their frequencies towards lower or higher values. These changes are consequences of the altered glass framework where the number of the bridging Si-O-Si bonds, represented by the Q^2, Q^3, and Q^4 modes, and the bonding strength between the matrix oxygens and the dopant ions were changed.

While the cuprous ions are comparable in size with the exchanged sodium, the cupric ions are distinctly smaller and each divalent copper replaces two univalent sodium ions. The matrix becomes more tight and the shorter (Si-O) bonds could be responsible for higher frequencies of the related Q^3 units. For changes induced only by the structure (morphology), the frequencies of the stretching and bending Q^3 modes should be shifted in opposite directions. This means that for increasing number of the bridging-oxygens the frequency of the stretching band (1050-1100 cm^{-1}) should increase while the frequency of the bending Q^3 (450 cm^{-1}) should decrease. And for decreasing number of the bridging oxygens, the opposite effect should be observed. The bonding strengths, characteristic of both Q^3 modes, should be affected approximately in the same way. In the described case, the intensities of the Q^2, Q^3, and Q^4 bands are relatively small, and the frequencies of both Q^3 modes are shifted towards higher values. Although, this behavior could be explained in terms of the formation of a more tight glass network, it is also possible to consider that the divalent copper ions are not only located at positions anticipated for the glass-modifiers but are also able to enter the matrix as the glass-formers and form new silica phases. This possibility was, among others, considered for the phosphate glasses where the (Cu-O) bonds are relatively strong and have a large covalent contribution (similar to the case of Cu in SLSG). The glass-forming position for divalent copper was also evidenced for the silica glasses on the basis of some optical and magnetic characteristics [15-17]. The detection of a new IRR band (1007 cm^{-1}) in the hydrogenated SLSG specimens strongly confirms these suppositions.

4. Summary

Combined studies of the optical and structural characteristics of copper-doped SLSG specimens show that:

- Distinct morphology changes of the glassy matrix occur due to the joint effect of divalent copper and the high temperature treatment;
- These changes are related with the formation of strong chemical bonds between the dopant and the non-bridging oxygens of the matrix;
- The Cu-O interactions, induce the decrease of the Si-O bond lengths and result in an increase of the compactness of the glass framework;
- Some mixed sodium/copper silica phases could be formed in the exchanged SLS glass, and suitably sized voids in the hydrogenated specimens accommodate the copper nanoparticles;
- The described phenomena suggest that the divalent copper ions in the ion exchanged glasses not only replace sodium ions (one for two) but are also responsible for the alteration of the glass network;
- These preliminary investigations deserve further studies, and more work is needed to understand the role of the divalent cations in perturbing the network of the soda lime silica glasses.

Acknowledgments

The authors like to express their thanks to Dr. B. Macalik and Dr. J. Kowalczyk for preparation of samples, and to Prof. J. Janczak for valuable discussions of the XRD data.

References

[1]Suszynska M, Szmida M and Grau P, 2001 *Mater. Sci. Eng.* A **319-321** 702
[2]Poszner T, Schreiter G and Mueller R 1991 *J. Appl. Phys.* **70** 1966
[3]Shibaoka K and Suzuki K 1989 *J. Non-Crystal. Solids* **112** 364
[4]Flytzanis C F, Hache F, Klein M C, Ricard D and Roussignol Ph 1991 *Nonlinear Optics in Composite Materials (Progress in Optics vol 29)* ed E Wolf (New York: Elsevier)
[5]Doremus R, Kao S C and Garcia R 1992 *Appl. Optics* **31** 5773
[6]Colomban P and Schreiber H D 2005 *J. Raman Spec.* **36** 884
[7]Suszynska M, Krajczyk L, Macalik B, Capelletti R, Mora C and Gubanski A 2005 *phys. stat. sol.* c **2** 612
[8]Porai – Koshits E A and Averjanow V J 1968 *J Non-Cryst. Solids* **1** 29
[9]Macalik B, Krajczyk L, Okal J, Morawska-Kowal T, Nierzewski K D and Suszynska M 2002 *Radiat. Eff. Defects Solids* **157** 887
[10]Berg K J, Capelletti R, Krajczyk L and Suszynska M 1996 *IEEE Dielectrics and Electrical Insulators Society* ed.'s X Zhongfu and Z Hongyan, p. 378
[11]Sharof N A, Condrate R A and Ahmed A A 1991 *Mater. Lett.* **11** 115
[12]Kirk E T 1998 *Phys. Rev.* B **38** 1255
[13]Liem N Q, Thanh N T and Colomban Ph 2002 *J. Raman Spectroscopy* **33** 287
[14]Sitte H 1982 Zue Faerbung von Natrium - Silica Glaesern durch Kupfer - Natrium Ionenaustausch (Thesis at the M. Luther University) Halle/Saale, Germany
[15]Singh S P and Kumar 1995 *J. Mater. Sci.* **30** 2999
[16]Ezikov V I and Romanova D V 2002 *Glass Phys. Chem.* **28** 33

Journal of Physics: Conference Series **249** (2010) 012049 doi:10.1088/1742-6596/249/1/012049

Production and thermal stability of pure and Cr^{3+} -doped hydroxyapatite

T S de Araujo[1], S O de Souza[1], E M B de Sousa[2], M S Araújo[1]
[1]Physics Department, Federal University of Sergipe, São Cristóvão, SE 49100-000, Brazil
[2]Center of Development of the Nuclear Technology – CDTN/CNEN, Belo Horizonte, MG 30123-970, Brazil

E-mail: tatiana.araujo@cefetse.edu.br

Abstract. Hydroxyapatite (HAP) have been used as starting material for biomedical applications. The pure and Cr^{3+} -doped hydroxyapatite were prepared by chemical precipitation reactions at 100, 500 e 800 °C in order to investigate the thermal stability of these materials. The characterization of the thermal behavior of this phosphate, especially on the structural changes with heating, is very important for production of sunscreens The powders were characterized using chemical analysis: X-ray diffraction (XRD) and thermal analysis. The present study was successful in the preparation of pure hydroxyapatite and chromium substituted hydroxyapatites with good thermal stability and nanoparticles formation.

1. Introduction

In the last decade, synthetic hydroxyapatite (HAP) has been widely used in various clinical applications due to its compositional similarity to the mineral phase present in hard tissue of human body [1, 2]. Lately it has been studied for applications as active ingredients of inorganic sunscreens [3, 4]. The characterization of the thermal behavior of this phosphate, especially on the structural changes with heating, is very important for several applications. So it is very significant to reveal the thermal stability details of hydroxyapatite.

In the literature are found many methods to synthesize HAP crystals with different characteristics such as hydrolysis of other calcium phosphates, solid-state reactions [5] and various techniques of wet chemical synthesis [6, 7], etc. In this work the hydroxyapatite was prepared through aqueous precipitation processing. Unlike the synthesis for thermal processes, this method produces larger amounts of the material, besides being one of the most economical processes, simple with possibility of industrial application [8, 9]. The precipitation method makes possible the obtaining of homogeneous particles and the control of the concentration of the reactants [3, 9, 10].

Hydroxyapatite powders can be precipitated from conventional wet chemical methods and decomposed into a mixture of HAP and tricalcium phosphate, $Ca_3(PO_4)_2$ (TCP) by thermal treatment above 700 °C. This allows a direct processing of biphasic calcium phosphate ceramics HAP/TCP without the step of powder blending. The main interest in the last years is not only controlling the stoichiometry of synthetic HAP but also controlling the crystal size [11].

Considering calcium-deficient HAP change with thermal treatment to stoichiometric HAP and TCP according to the following equation:

$Ca_{10-x}(HPO_4)_x(PO_4)_{6-x}(OH)_{2-x} \rightarrow (1-x)Ca_{10}(PO_4)_6(OH)_2 + 3x\beta\text{-}Ca_3(PO_4)_2 + xH_2O$ [11]

The Cr^{3+} is a metal non toxicant that can be incorporate in HAP in order to produce fluorescence. For the measures of the fluorescence of the ions of Cr^{3+} in HAP is possible to identify the levels of HAP that are being incorporate to the human bone. The low toxicity of Cr^{3+} is partially due to the fact that many Cr(III) complexes are poorly soluble at physiological pH values and are not easily transported through the cell membranes. Complexes which are not able to penetrate into the cells may not induce toxic responses [2]. Besides Cr^{3+} is a metal that may generates optical absorption bands in the UV range when incorporated in a crystalline matrix. The biocompatibility HAP with the optical activity of the Cr^{3+} turn the HAP very interesting to be investigated as active ingredient for inorganic sunscreens.
The objective of this work was to know the thermal behavior of pure and Cr^{3+} -doped HAP. To study the thermal stability of HAP is important in the determination of the great conditions of temperature for calcination of the obtained precipitate powders, as well as to know the thermal behavior of pure HAP and doped with Cr^{3+}. For example, producing pure and Cr^{3+} -doped HAP in a temperature of appropriate calcination the formation of species reactivates they can be avoided, and that will guarantee the success of your application in sunscreens.

2. Experimental procedure

Nanocrystalline HAP was prepared from an aqueous precipitation reaction using mixture of $Ca(NO_3)_2.4H_2O$ with $Cr(NO_3)_3.9H_2O$ and $(NH_4)_2HPO_4$. The reaction can be written as follows [2, 3]:

$$10Ca(NO_3)_2.4H_2O + 6(NH_4)_2HPO_4 + 8NH_4OH \rightarrow Ca_{10}(PO_4)_6(OH)_2 + 20NH_4NO_3 + 6H_2O \quad (1)$$

The aqueous solution of $Ca(NO_3)_2.4H_2O$ was slowly added to $(NH_4)_2HPO_4$ at a rate of 1.5 ml/min, keeping the pH in 10.4. The pH was adjusted during the mixing using a solution of NH_4OH under a constant stirring condition at room temperature. After total addition of the $Ca(NO_3)_2.4H_2O$ with 0,01 mol/L $Cr(NO_3)_3.9H_2O$ solution, the suspension was matured during 30 min. Then, it was filtered and washed. The resulting precipitate received thermal treatment at 100 °C, 500 °C and 800 °C for 1 hour. It was varied the temperature of thermal treatment of the samples in order to verify the effects of the temperature in the thermal stability of the materials.
In this study the crystal behavior of the 100, 500 and 800 °C treated pure and Cr^{3+} -doped HAP powders were characterized by X-ray powder diffraction (XRD) and the thermal behavior was investigated by Differential thermal analysis (TG-DTA).
The XRD data were obtained on an X-ray difractometer (Rigaku DMAX2000) with CuKα radiation at 40 kV/40 mA with a spinning sample holder and a monochromator detector. The thermal decomposition of the samples was studied from 10 to 900 °C by TG, DTG and DTA using a TG-DTA apparatus of TA – Instruments, model SDT 2960, at a heating rate 10°C min^{-1}, synthetic air atmosphere, alumina crucibles and sample mass of approximately 9 to 13 mg.

3. Results and discussions

The XRD patterns of the pure HAP samples that received different thermal treatment are shown in figure 1. The standard HAP and CaO XRD patterns are also shown in the bottom part of the figures. The thermal treatment at 100 °C was sufficient to obtain the precipitate of HAP, however, broad diffraction peaks are observed indicating low crystallinity of the powders. It is possible to observe an increase of the crystallinity after calcination at 500 °C and a mixture of crystalline structures of HAP and calcium oxide (CaO) after calcination at 800°C. As this new phase is biocompatible [12], the material can be used in sunscreen. That shows that an extra amount of Ca^{2+} stayed in the mixture and reacted with the oxygen (O_2) of the environment. Raynald et al [13] it observed the presence of a small amount of CaO and great amount of TCP in the HAP matrix when preparing HAP by aqueous precipitation method of $(NH_4)_2HPO_4$ with $Ca(NO_3)_2$ calcined at 1250 °C. In the literature the decomposition of HAP in temperatures above 700 °C is frequently found during the production of ceramic HAP [11, 13-15].

Journal of Physics: Conference Series **249** (2010) 012049 doi:10.1088/1742-6596/249/1/012049

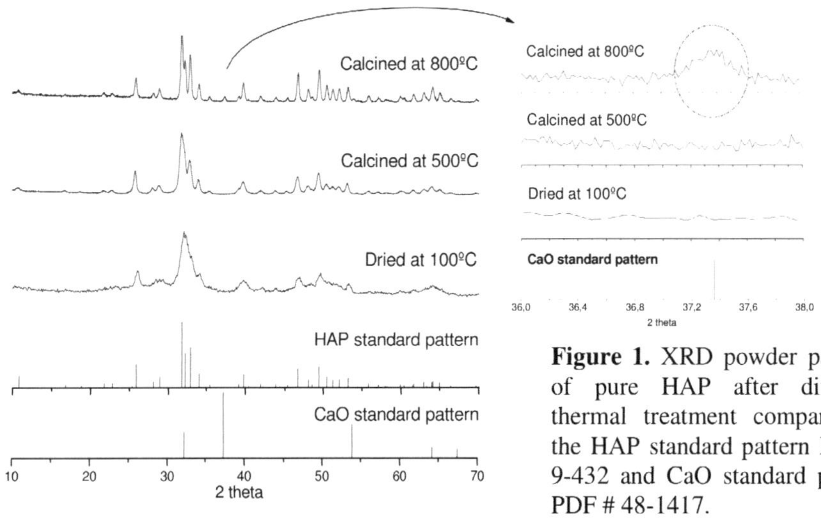

Figure 1. XRD powder patterns of pure HAP after different thermal treatment compared to the HAP standard pattern PDF # 9-432 and CaO standard pattern PDF # 48-1417.

Figure 2 compares the XRD patterns of the Cr^{3+} -doped HAP after different treatments of calcinations. All the patterns show only the peaks characteristic of HAP.
As shown in figure 2 the Cr^{3+} -doped HAP treated to 100 °C it presented an amorphous structure with only two picks crystalline of HAP. For the precipitate calcined at 800 °C there is the decomposition of the apatite to form TCP that is biocompatible.
Doped HAP show a significant correlation between the crystallinity and the thermal treatment. The crystallinity of the material decreases with the incorporation of the dopant. The correlation between crystallinity and thermal treatment is evident for pure and doped samples. That is more noted for dried at 100°C samples; the difractograms of pure and doped calcined 500°C samples are very similar. The results of XRD patterns are compatible with the results of Wakamura et al [16].
The best definition of the powder XRD peaks as function of the calcination temperature is also related with the increase in the crystallite size. Comparing the pure sample and doped calcined 500 °C significant difference it is not observed as the crystallite size of the two samples. The table 1 presents the medium crystallite sizes calculated from Scherrer equation. The effect of the thermal treatment is well known from materials engineering studies. The results of crystallite size are very similar for pure and doped samples. The results indicate that a temperature increase favors the growth of the crystallites. Besides, the low values of crystallite sizes (<40 nm) indicates the formation of nanoparticles.

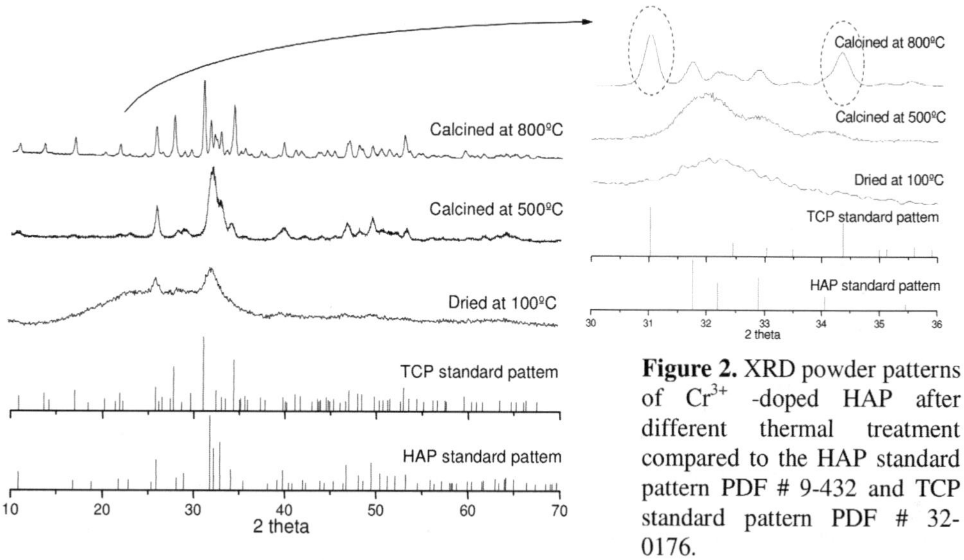

Figure 2. XRD powder patterns of Cr^{3+} -doped HAP after different thermal treatment compared to the HAP standard pattern PDF # 9-432 and TCP standard pattern PDF # 32-0176.

Table 1. Medium size crystallite obtained via XRD measurements using Scherrer's equation to HAP after different thermal treatment

Temperature	Crystallite size ± 1 (nm)
Pure HAP	
100	20
500	25
800	34
Cr^{3+} -doped HAP	
100	16
500	23
800	35

TG, DTG and DTA curves of pure HAP calcined at 500 ºC are shown in figure 3 and 4. The DTA curve shows a small exothermic peak in 191 ºC, attributed to HAP of low crystallinity (Figure 4). In fact the powder XRD patterns showed formation of little crystalline HAP at the temperature of 100 ºC. In fact the powder XRD patterns showed formation of little crystalline HAP at the temperature of 100 ºC. The TG curve indicate a mass variation of 2.5% up to 800ºC. The weight loss under 750 ºC is attributed to the elimination of H_2O (Figure 3). As expected there are not peaks indicating phase transformation of the material in the temperature of 500 ºC. The studies of XRD patterns showed that in 800 ºC there is decomposition of the apatite into a mixture of HAP/CaO. Therefore, the mass loss starting from 770 ºC refers to the decomposition of HAP in CaO.

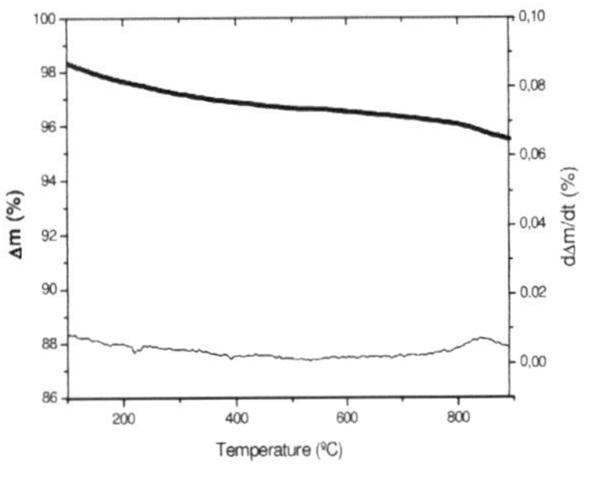

Figure 3. TG (bold line) and DTG of pure hydroxyapatite powder calcined at 500 °C.

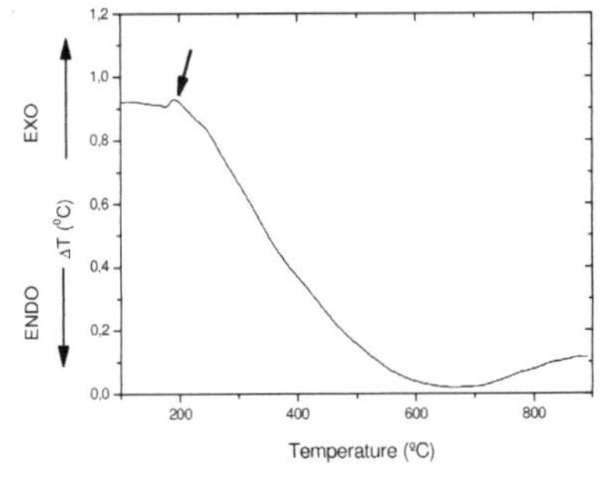

Figure 4. DTA of pure hydroxyapatite powder calcined at 500 ° C.

In the figures 5 and 6 we can observe the thermal behavior of pure HAP dried at 100 °C. The results show an increase in the mass loss around 248 °C with relationship to the roasted sample (Figure 5). The existence of the exothermic process in approximately 230 °C probably indicates a crystallization of HAP starting from the amorphous phases (Figure 6). This result is in agreement with the analyses of XRD patterns that demonstrated a better crystallinity degree for samples that were calcined 500 °C than those that received thermal treatment at 100 °C, in other words, below the temperature of the exothermic pick.

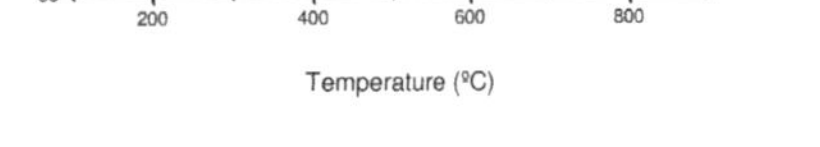

Figure 5. TG (bold line) and DTG of pure hydroxyapatite powder dried at 100 °C.

Figure 6. DTA of pure hydroxyapatite powder dried at 100 ° C.

Cr^{3+} -doped HAP calcined at 500 °C has good thermal stability, in agreement with figure 7 and 8. The crystallinity of the material is not altered along the process, being considered the absence of the exothermic pick (Figure 8). The mass loss starting from 750 °C is due to formation of TCP. This agreed with the research of Miyaji et al [17] when doping HAP with zinc, in which he indicated that Ca-deficient HAP tends to transform to TCP on heating at 800 °C depending on the deficiency of calcium.

Journal of Physics: Conference Series **249** (2010) 012049 doi:10.1088/1742-6596/249/1/012049

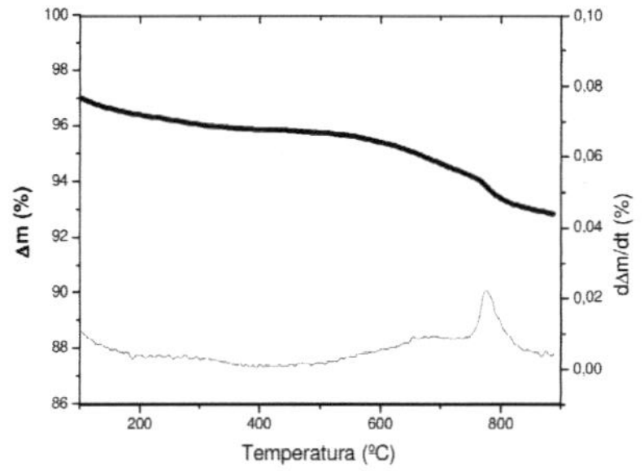

Figure 7. TG (bold line) and DTG of Cr^{3+} -doped hydroxyapatite powder calcined at 500 ºC.

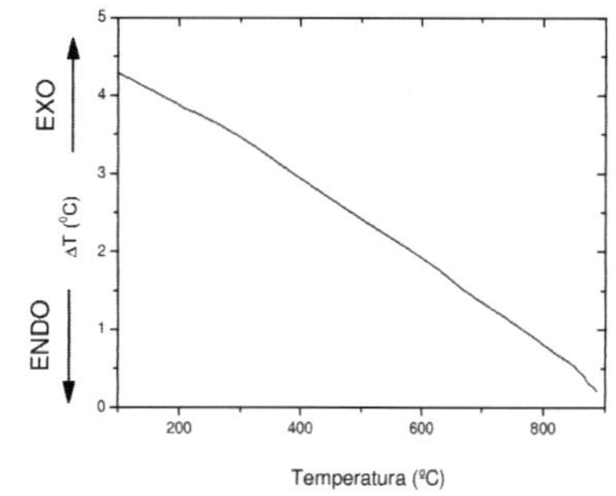

Figure 8. DTA of Cr^{3+} -doped hydroxyapatite powder calcined at 500 º C.

In Cr^{3+} -doped HAP dried to 100 ºC the elimination of the molecules of water is larger than in the pure sample in the same temperature, because this sample has amorphous structure (Figure 9). For it is reason the exothermic pick that is associated to the crystallization of HAP it is smaller compared with the pure sample (Figure 10).

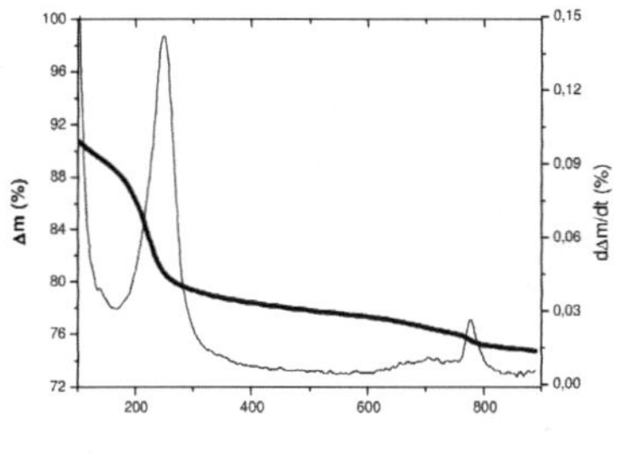

Figure 9. TG (bold line) and DTG of Cr^{3+} -doped hydroxyapatite powder dried at 100 ºC.

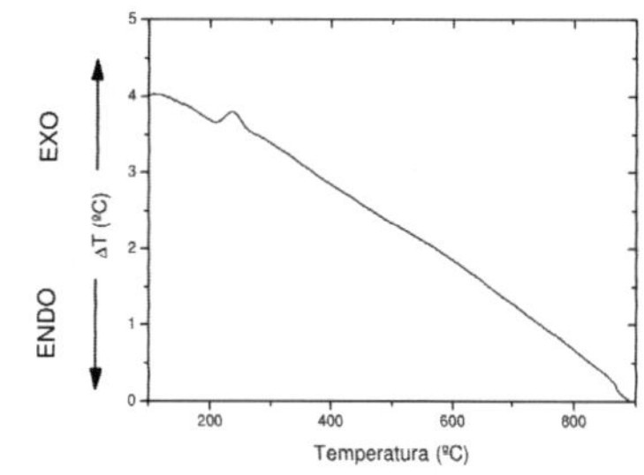

Figure 10. DTA of Cr^{3+} -doped hydroxyapatite powder dried at 100 º C.

4. Conclusions

Pure HAP and Cr-substituted Ca hydroxyapatites were synthesized by precipitation method under the specific conditions (pH 10.4, room temperature), and varying the temperature of thermal treatment of the samples, where the structural properties were investigated. The present study was successful in the preparation of chromium substituted hydroxyapatites. It was seen that in the production of pure and doped HAP using precipitation of $Ca(NO_3)_2.4H_2O$ with $(NH_4)_2HPO_4$ the best temperature of calcination of the samples in only phase nanocrystalline it is between 400 ºC and 530 ºC.

Acknowledgments
The authors wish to thank the Brazilian funding agencies CNPq and CAPES for financial support.

5. References

[1] Santos M H, Oliveira M, Souza L P F, Mansur H S, Vasconcelos W L 2004 *Materials Research* **7** 4 625– 630
[2] Araujo T S, Macedo Z S, Oliveira P A S C, Valerio M E G 2007 *Journal of Materials Science* **42** 7 2236-2243
[3] Araujo T S, Lima T A R M, Valerio M E G 2007 *Scientia Plena* **3** 7 285-291
[4] Araujo T S, Souza S O 2008 *Scientia Plena* **4** 11 1-7
[5] Suchanek W, Yoshimura M 1998 *J. Mater. Res.* **13** 94
[6] Liu Y K, Hou D D, Wang G H 2004 *Mater. Chem. Phys.* **86** 69
[7] Cao M H, Wang Y H, Guo C X, Wi Y J, Hu C W 2004 *Langmuir* **20** 4784
[8] Arends J, Christoffersen J, Christoffersen M R, Eckert H, Fowler B O, Heughebaert J C, Mancollas G H, Yesinowski J P, Zawacki S J 1987 *Journal of Cryst. Growth* **84** 515-532
[9] Garcia C, Paucar C, Gaviria J, Durán *Key Engineering Materials* **284-286** 47- 50
[10] Peña J, Vallet-Regi M 2003 *Journal of the European Ceramic Society* **23** 1687-1696
[11] Mostafa N Y 2005 *Materials chemistry and physics* 1-9
[12] Monma H, Goto M, Kajima H, Hashimoto H 1986 *Gypsum and Lime 202*
[13] Raynaud S, Champion E, Bernache-Assollant D, Thomas P 2002 *Biomaterials* **23** 1065–1072
[14] Tomida K, Kawano M, Shiraki K, Otsuka H *J.* 1996 *Min. Petr. Econ. Geol.* **91** 11-20
[15] Brown P W, Constantz B 1993 *Hydroxyapatite And Related Materials* 269-282
[16] Wakamura M, Kandori K, Ishikawa T 1997 *Polyhedron* **16** 2047-2053
[17] Miyaji F, Kono Y, Suyama Y 2005 *Materials Research Bulletin* **40** 209–220

Effect of PET on the crystallization of lithium/sodium acetate glasses studied by isothermal and non-isothermal kinetic methods

Silvanio S L da Costa[1], Marcelo L dos Santos[1], Marcelo A N Beltrao[1], Charlene R S Matos[1], Ana C S Alcantara[1], Iara F. Gimenez[1], Oswaldo L Alves[2], Ledjane S Barreto[1]

[1]Universidade Federal de Sergipe, DQI (Department of Chemistry),
[2]LQES (Solid State Chemistry Laboratory), IQ (Institute of Chemistry), UNICAMP

ledjane@ufs.br

Abstract. The kinetic parameters for the crystallization of CH_3COOLi/CH_3COONa glasses varying the Li:Na mole fraction LN21(2:1), LN31(3:1), LN41(4:1) and LN32(3:2) have been evaluated by isothermal and non-isothermal DSC measurements. Crystallization of LN21 samples containing variable amounts of poly(ethylene-terephthalate) (PET) has also been studied. Values of the Avrami exponent n varying between 2.05 and 2.33 were obtained from both isothermal and non-isothermal methods for all LN glasses, indicating a diffusion-controlled crystal growth with a decreasing nucleation rate. The more pronounced effect of PET on glass crystallization has been observed for low polymer amounts, being interpreted as due to polymer miscibility with the glass, which is absent for high PET amounts. The value of n = 4.75 for LN21+PET (3%) suggests a change in the crystallization mechanism from diffusion-controlled to interface-controlled crystal growth. The apparent activation energy (E) decreased with increasing PET amount, evidencing an improvement in glass stability against crystallization.

1. Introduction

Acetate glasses can be prepared by melting alkali metal acetates at relatively low temperatures [1-3]. Duffy and co-workers have shown that the structures of the resulting ionic glasses are comprised by M^+ and $C_2H_3O_2^-$ entities, the last one with two oxygen atoms able to coordinate metal ions [4,5]. To further understand the glass structure, we reported a computational study of structural and dynamic properties of crystalline and molten alkali acetates from molecular dynamics simulations [6-8]. We found that the crystalline phases are formed by layers of acetate anions and alkali cations and the structure of glassy phases can be described by the formation of deffects and expansion of the system, as a result of increasing thermal vibrations with heating. Glebovsky and co-workers described the structure of the first coordination sphere around lithium cations in glassy acetates by quantum chemical calculations [9].

Owing to their optical transparency (240 – 1400 nm) and the low melting temperatures, carboxylate glasses have been studied as hosts for luminescent species such as $Eu(fod)_3Phen\text{-}NO$ [10]. These glasses can also be useful as hosts for controlled release of pharmaceutical, biocidal, fungicidal, or hormonal species [11,12]. However, a major drawback to practical applications of these glasses is the high hygroscopicity, a common feature of many crystalline acetates of alkali and alkali earth elements. An alternative to improve the durability is the incorporation of polymeric additives to the glass composition. Important properties such as the optical transparency can be preserved by the proper choice of polymer, which must be thermally stable at the glass melting temperature.

Here we report a thermal analysis study of mixed Li/Na acetate glasses in the presence of poly(ethylene terephthalate) (PET) by isothermal and non-isothermal data. Empirical methods [13-15] have been used for determination of kinetic crystallization parameters.

2. Experimental

2.1. Materials

Anhydrous lithium acetate was obtained by the reaction of acetic acid and lithium carbonate. The final product was dissolved in water, filtered to eliminate residues, and recrystallized as the hydrated salt. The anhydrous phase was obtained by heating the corresponding hydrate under vacuum at 120 °C for 8 h. Reagent grade sodium acetate was purchased from Merck. Granular PET (poly(ethylene terephtalate)) was purchased from Aldrich.

2.2. Preparation and characterization of samples

The DSC measurements were carried out using aluminum cells under N_2 at 50 mL/min in a TA 2010 instrument, which was calibrated periodically using In and sapphire as standards. Glass samples were prepared by melting *in situ* mixtures of the raw materials of about 12 mg were in the DSC cell, followed by cooling with liquid nitrogen. Afterwards the isothermal and nonisothermal studies were carried out by reheating the freshly formed glass samples. Heating rates of 10, 12.5, 15, 17.5 and 20 °C/min were used for the non-isothermal study. The temperatures for isothermal studies were 109, 119 and 129 °C, and in this case the samples were heated to the chosen temperature and kept at this condition for 120 min. The samples were denoted LN41 (80:20 Li:Na mol%), LN31 (75:25 Li:Na mol%), LN21 (66:33 Li:Na mol%), LN32 (60:40 Li:Na mol%); LiAc (100 mol%). For the study of PET effect, the glass composition LN21 was chosen and, according to the polymer content (mass %), the sample nomenclature is as follows: LN21+PET (3%), and LN21+PET (30%).

For the measurements of optical microscopy images, a portion of the mixed raw materials was heated at 10 °C/min over glass plates in a Leica DMLM optical microscope coupled to a Metler Toledo FP90 heating accessory up to the melting temperature. Images were recorded at each 5 °C with a 20X magnifying lens.

3. Theory

Thermal analysis (TA) methods such as DTA or DSC are widely used for kinetic analysis of crystallization processes in amorphous solids. The crystallization kinetics based on these data is usually interpreted in terms of the standard nucleation-growth model formulated by Kolmogorov-Johnson-Mehl-Avrami [16-18]. All the theory can be found in references from [19-23].

This model describes the time dependence of the fractional crystallization x, usually written in the following form [19]

$$x = 1 - \exp[-(kt)^n] \tag{1}$$

where x is the volume fraction crystallized isothermally; t is the time and k is a constant while n is referred to as the Avrami exponent, a dimensionless parameter related to the morphology of the crystal growth. For the analysis of isothermal data, an equivalent expression of Eq. (1) can be used as follows [20]

$$\ln[-\ln(1-x)] = n \ln k + n \ln t \tag{2}$$

it follows from this expression that a plot of ln [-ln (1-x)] versus ln t gives n from the slope and k is calculated from the intercept. The crystallization activation energy E is obtained from the equation

$$\ln k = \ln k_0 - E/RT \tag{3}$$

where k_0 is the frequency factor (s^{-1}), E is the apparent activation energy (kJ/mol) representing the kinetic barrier to crystallization, and T is the temperature by plotting ln k as a function of $1/T$ for different temperatures.

The Avrami exponent n may also be obtained through non-isothermal measurements using the pseudo-isotherm method introduced by Ozawa [21]. In the non-isothermal method the sample is heated at a constant rate (α), and the heat flow is recorded as a function of temperature or time and the following expression can be used:

$$\ln[-\ln(1-x)] = n \ln[k(T-T_0)] - \ln \alpha \tag{4}$$

where x is the fraction crystallized at temperature T with heating rate α.

In a non-isothermal DSC experiment, the sample temperature is increased linearly versus time t. The heating rate is α (=dT/dt) and the temperature is expressed as $T = T_0 + \alpha t$, where T_0 is the initial temperature. In these conditions the rate constant, k, changes continually versus time and Eq. (1) must be written as follows [20]:

$$\left|-\ln(1-x)\right|^{1/n} = \int_0^t k(T(t'))dt'. \tag{5}$$

Solving this relation leads to the following equations [15]:

$$\frac{1}{n}\ln(-\ln(1-x)) = \ln\frac{k_0 R}{E} + \ln\frac{T^2}{\alpha} - \frac{E}{RT}, \tag{6}$$

$$\frac{1}{n}\ln(-\ln(1-x)) = \ln\frac{k_0 E}{\alpha R} - 2\ln\frac{E}{RT} - \frac{E}{RT}. \tag{7}$$

Consequently one may draw straight lines from which the value of E/R is deduced. Typical plots are those suggested by Ozawa [19]

$$\left|\frac{d(\ln\alpha)}{d(l/T_p)}\right|_x = -\frac{E}{R}, \tag{8}$$

and Chen [22]

$$\left|\frac{d(\ln T_p^2/\alpha)}{d(l/T_p)}\right|_x = -\frac{E}{R}. \tag{9}$$

The apparent activation energy has been calculated also from the constant fraction method (CFM) [23] using the equation:

$$\ln(t_x) = const - \ln A + E/R\,(1/T), \tag{10}$$

where t_x is the time for a transformed fraction for isothermal experiments, R is the gas constant, and A is a constant.

4. Results and discussion

Information related to nucleation and crystallization mechanisms for the LN glasses with different compositions as well as the LN21 composition with different PET contents were evaluated by the determination of the Avrami exponent (n), from both isothermal and non-isothermal methods applied to DSC data. The DSC curves in figure 1 ilustrate typical thermal profiles of the LN21 glass in the absence and in the presence of PET where the glass transition, crystallization and melting can be observed. In this work we report only the addition of PET to the LN21 composition because for the other compositions the presence of PET induced crystallization of the samples. It is worth mentioning that the DSC curves (not shown) were measured for all samples without PET, being very similar to the LN21curve.

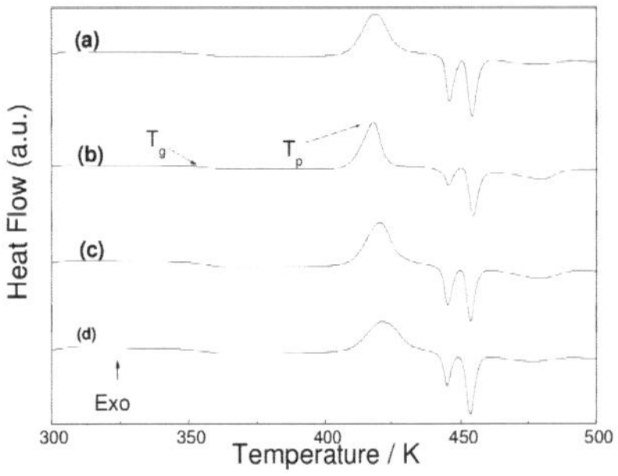

Figure 1- Differential scanning calorimetry curves (DSC): (a) LN21; (b) LN21+PET (3%); and (c) LN21+PET (30%).

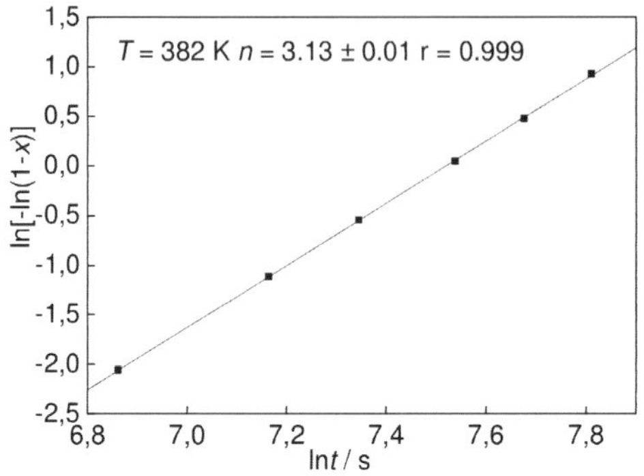

Figure 2- Plot of [-ln (1-x)] versus ln t for determination of of Avrami exponent from isothermal method for LN21+PET (3%) (a) and for LN21+PET (30%) (b).

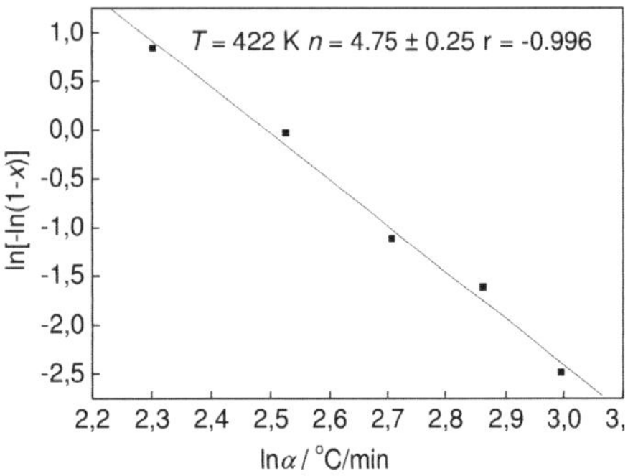

Figure 3- Plot of [-ln (1-x)] versus ln α for determination of of Avrami exponent from isothermal method for LN21+PET (3%) (a) and for LN21+PET (30%) (b).

For the determination of Avrami exponent, typical illustrative plots of ln[-ln (1-*x*)] versus ln *t* from isothermal data and ln[-ln (1-*x*)] versus ln *α* from non-isothermal methods are shown in figures 2 and 3. Similar plots (not shown) were obtained for all samples, with the resulting values of Avrami index *n* shown in table I. As expected, both plots resulted in straight lines. Table I gives the values of Avrami exponent *n* determined in this study from both methods, along with the ranges for the values of *n* presented in the literature [24] for interface-controlled processes and diffusion-controlled processes.

Table I- Values of the Avrami exponent for different samples and meaning of the values

Composition	n (Avrami)	n (Ozawa)	Transformation Condition (n)	
			DCG	ICG
LN41	2.33±0.06	1.99±0.31		
LN31	2.11±0.06	1.85±0.26	[a]>2.5	[a]> 4
LN21	2.05 ± 0.04	2.09±0.70	[b]2.5	[b]4
LN32	2.16±0.07	1.86±0.40	[c]1.5<n< 2.5	[c]3<n < 4
LiAc	2.24±0.07	2.17±0.52	[d]1.5	[d]3
LN21+PET(3%)	3.13 ± 0.01	4.75±0.25		
LN21+PET(30%)	2.16 ± 0.06	2.98 ± 0.55		

DCG – Diffusion-Controlled Growth, ICG – Interface-Controlled Growth, [a] Increasing nucleation rate, [b] Constant nucleation rate, [c] Decreasing nucleation rate, [d] Zero nucleation rate. *Source*: Reference [24]

In the absence of PET the results show a reasonable agreement between isothermal and non-isothermal studies. This finding suggests that the assumptions made for the extension to non-isothermal experiments of Avrami's equations can be considered acceptable. The values obtained indicate that the crystallization mechanism is barely sensitive to the sodium amount since the values obtained for the Avrami exponent were in the range 2.05-2.33 from isothermal method and 1.85-2.09 from nonisothermal method. Values in this range suggest a crystallization mechanism for the different LN glass compositions controlled by diffusion with a decreasing nucleation rate.

The values of the Avrami exponent for the LN21 glass composition increased in the presence of PET, which in general suggests a changing from diffusion-controlled to interface-controlled crystal growth. For the sample containing only 3 % of PET, the value of *n* from the isothermal method was 3.13 and from the non-isothermal measurement was 4.75, both relatively high in comparison to the values found for the PET-free LN21 composition. On the other hand, the values of *n* decreased with increasing PET proportion and the sample containing 30 % PET exhibited an Avrami exponent closer to that for PET-free sample, which could be related to an effect of the decreasing acetate percent. The general trend (increase in *n*) could be explained by phase segregation in the presence of PET, which is more pronounced for higher PET amounts. An examination of the melting process by optical microscopy images with *in situ* heating, figure 4, has been carried out to check for this hypothesis. The images evidenced that the LN21 glass containing 3 % PET appears homogeneous in this magnification and that for higher PET contents, the acetate glass-phase and PET apparently are segregated. Despite the low resolution compared to electron microscopy techniques, optical measurements were chosen due to the possibility of heating and forming in situ the low-melting glasses.

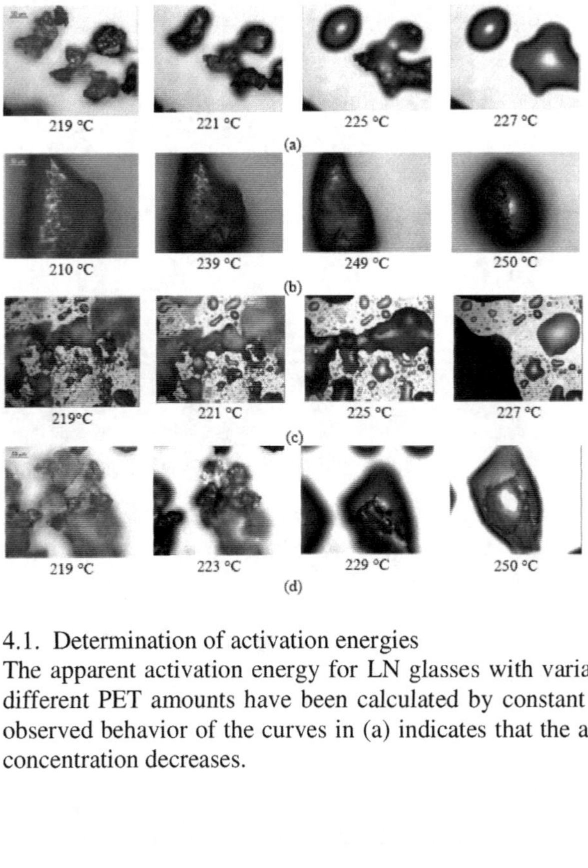

Figure 4- Optical microscopy images with 20X magnification obtained at increasing temperature up to melting for the compositions (a) LN21; (b) PET; (c) LN21 + PET3%; (d) LN21 + PET30%.

4.1. Determination of activation energies

The apparent activation energy for LN glasses with variable Li content and for the LN21 glass with different PET amounts have been calculated by constant fraction method (CFM) [23], figure 5. The observed behavior of the curves in (a) indicates that the activation energy decreases when the lithium concentration decreases.

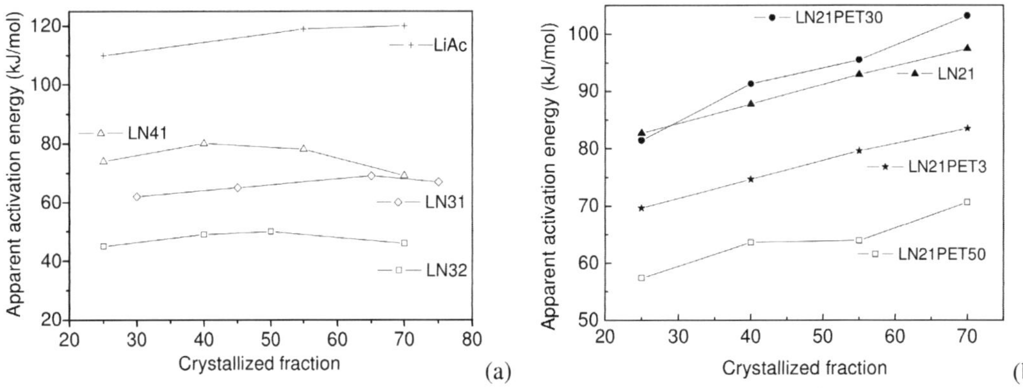

Figure 5- Variation of the activation energy with crystallized fraction for PET-free LN 31, 21 and 41 glasses (a) and for free and PET-containing LN21 glass (b).

In general the apparent activation energies (b) decrease in the presence of PET. For the evaluation of PET effect, activation energy has been determined by CFM method, also by the Arrhenius equation (3), and by the non-isothermal methods of Ozawa and Chen. Eq. (8) and Eq. (9). Typical plots are shown in figure 6 for LN21-PET 3%. All the values for apparent activation energy are summarized in table II.

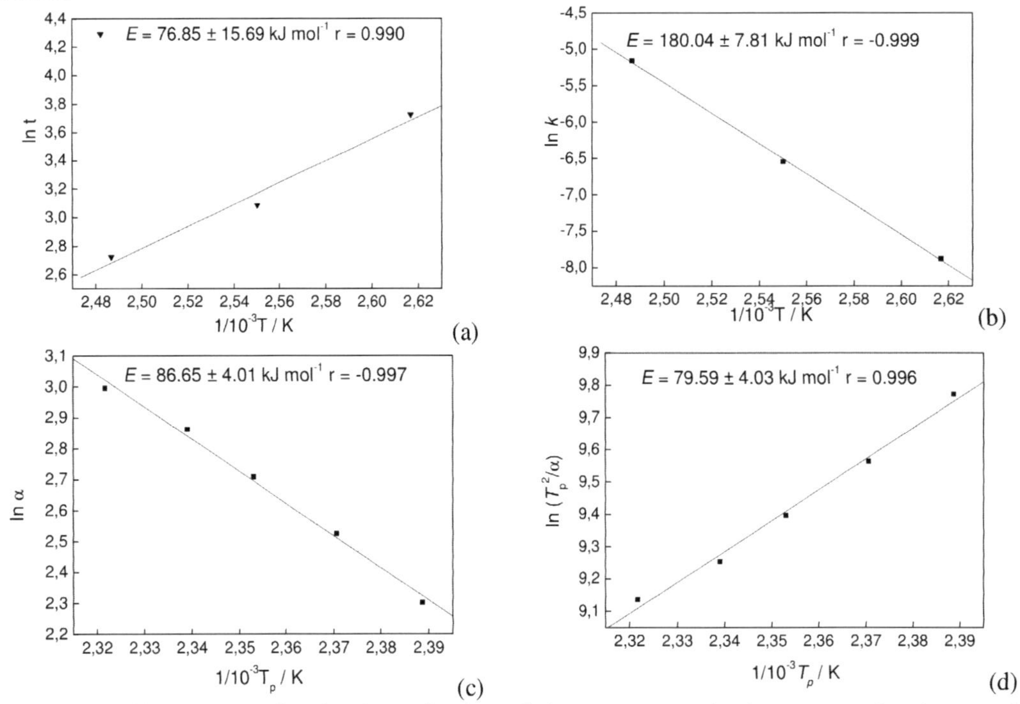

Figure 6- Different plots for the determination of the apparent activation energy for the sample LN21+PET3%: (a) by constant fraction method (CFM) (b) by the Arrhenius plot; (c) by the Ozawa method; (d) by the Chen method.

Table II- Apparent activation energy from isothermal and non-isothermal methods.

Samples	E / kJ mol^{-1} ln α Vs. $1/T_p$	E / kJ mol^{-1} ln (T_p^2/α) Vs. $1/T_p$	E / kJ mol^{-1} (CFM)	E / kJ mol^{-1} ln k Vs. $1/T$
LN21	111.94 ± 24.19	104.90 ± 24.19	90.22 ± 23.79	184.75 ± 7.98
LN21+PET (3%)	86.65 ± 4.01	79.59 ± 4.03	76.85 ± 15.69	180.04 ± 7.81
LN21+PET(30%)	89.12 ± 10.12	82.04 ± 10.12	92.87 ± 19.01	173.70 ± 4.77

In table II, the observed values of activation energy for glass crystallization decreased with the addition of PET, indicating that the glass stability toward crystallization increased. As stated previously by Poulain [28], the physical meaning of the activation energy of glass crystallization involves an apparent paradox. There is an empirical correlation between the activation energy of crystallization and glass stability, according to which a lower activation energy value may indicate a lower devitrification tendency [27-30], since this parameter also represents the activation energy of the viscous flow. It is not the purpose of this work to discuss this apparent paradox, but the observed results can be related to a decreased devitrification tendency with the addition of PET.

5. Conclusions

As conclusions of this work the values observed for the Avrami exponent indicated a changing in the crystallization mechanism upon addition of PET, from diffusion-controlled to a crystallization mechanism controlled by the interface. Finally the results indicated a decreasing in the apparent activation energy when increases PET composition, which could be related to an improvement of glass stability.

6. Acknowledgements

The authors are grateful to CNPq for financial support, to CNPq-PIBIC/UFS the fellowships and to Prof. G. Cardoso for the optical measurements.

7. References

[1] Uhlmann D R and Kreidl N J 1983 Glass-Forming Systems **V.1** Academic Press, New York
[2] Blair L A. and Duffy J A 1993 Phys Chem Glasses **34** 194
[3] Wilk Jr N R and Schreiber H D 1997 J Non-Cryst Solids **217** 189
[4] Baranova G I, Glebovsky D N, Zhuchkova T V and Panteleev I A 2001 Mol Cryst Liq Cryst **365** 1595
[5] Ingram M D, Lewis G G and Duffy J A 1972 J Phys Chem **76** 1035
[6] Wilk Jr N Rand Schreiber H D 1998 J Non-Cryst Solids **239** 192
[7] Barreto L S, Mort K A, Jackson R A and Alves O L 2000 J Phys Cond Matter **12** 9389
[8] Barreto L S, Mort K A, Jackson R A and Alves O L 2002 J Non-Cryst Solids **303** 281
[9] Barreto L S, Jackson R A and Alves O L 2002 Phys Chem Glasses **43C** 119
[10] Panteleev I A, Bedrina M E and Glebovsky D N 2002 Glass Sci Technol **75** 402
[11] Beltrao M A, Santos M L, Mesquita M E, Barreto L S, Costa Jr N B, Freire R O and Santos M A C 2006 J Lumin **116** 132
[12] Blair J A, Mordue A J, Duffy J A and Wardell J L 1994 J Control Release **31** 145
[13] Blair J A and Duffy J A 1992 Phys Chem Glasses **33** 191
[14] Taniguchi H, Yu J, Arai Y, Fu D, Yagi T and Itoh M 2008 Mater Sci Eng B **148** 48
[15] Camacho-Lopez M A, Haro-Paniatowski E, Lartundo-Rojas L, Livage J and Julien C M 2006 Mater Sci Eng B **135** 88
[16] Poulain M 1992 J Non-Cryst Solids **140** 1
[17] Avrami M 1939 J Chem Phys **7** 1103
[18] Avrami M 1940 J Chem Phys **8** 212
[19] Avrami M 1941 J Chem Phys **9** 177

[20] Cheng K 1999 Mater Sci Eng B **60** 194
[21] Tomasi C, Scavini M, Speghini A, Bettinelli M and Riccardi M P 2002 J Therm Anal Cal **70** 151
[22] Ozawa T 1965 Bull Chem Soc Jpn **38** 188
[23] Chen M S 1978 J Non-Cryst Solids **27** 257
[24] Putnis A 1995 Introduction to Mineral Sciences, Cambridge University Press, New York
[25] Christian JW 1975 The Theory of Transformations in Metals and Alloys, Pergamon, New York
[26] Boutarfaia A and Poulain M 1998 J Therm Anal Cal **51** 851
[27] Branda F, A. Marotta, A. Buri, J. Non-Cryst. Solids 134 (1991) 123-128.
[28] M. Matecki, Noiret-Chiaruttini I and Lucas J 1991 J Non-Cryst Solids **127** 136
[29] Poulain M 1996 Thermochim Acta **280-281** 343
[30] Costa B J, Poulain M, Messadeq Y, Ribeiro S J L 2000 J Non-Cryst Solids **273** 76
[31] Shiryaev V S, Adam J-L, Zhang X H and Churbanov M F 2005 Solid State Sci **7** 209

Temperature dependence of the electric field gradient at ^{181}Ta in nanostructured HfO_2 film

F.H.C. Cavalcante[1], A.W.Carbonari[1] and J.C. Soares[2]

[1]Instituto de Pesquisas Energéticas e Nucleares, IPEN-CNEN/SP, Brazil
[2]Centro de Física Nuclear da Universidade de Lisboa, Portugal

fabiohmc@ipen.br

Abstract. The hyperfine field at metal lattice sites in a nanostructured HfO_2 thin film was studied by the Perturbed Angular Correlation (PAC) technique using ^{181}Ta as probe nuclei. The thin oxide film was deposited by pulsed laser ablation on a silicon substrate kept at 673 K. The resulting film thickness was about 25 nm. The ^{181}Ta isotopes are formed in Hf sites of the sample as disintegration product of radioactive ^{181}Hf, which were produced by neutron activation of the very thin film by the reaction ^{180}Hf(n,γ)^{181}Hf.. PAC measurements of the electric field gradient at the ^{181}Ta probe sites .were performed with the thin film in the temperature range 295 – 1273 K in vacuum. Results showed that the film has a single crystalline structure and two different quadrupole frequencies could be observed corresponding to ^{181}Ta at monoclinic structure and with oxygen vacancy in its neighborhood.

1. Introduction

Hafnium Oxide (HfO_2) has been studied in the last years due to its high dielectric constant k ≈ 23 and interesting physics properties [1] that lead to use HfO_2 as a good candidate to replace SiO_2 in microelectronics. In order to be used in silicon electronic devices HfO_2 thin films must have superb quality with no defects. Therefore, it would be interesting to perform research using a very sensitive local technique in order to search for structural changes and the point defects behavior, getting a better understanding of the properties at the interface between the HfO_2 layer and the Si substrate. This work reports an investigation of HfO_2 thin films by measuring the temperature dependence of the electric field gradient (EFG) by means of the Perturbed Angular Correlation (PAC) technique. PAC is a powerful tool to study the hyperfine interactions as the EFG for instance using radioactive nuclei as probe [2]. The presence of hafnium in the films makes possible to use ^{181}Ta, an isotope formed in the beta decay of ^{181}Hf as probe nuclei.

2. Experimental procedure

The thin oxide film was deposited by pulsed laser ablation on a silicon substrate kept at 673 K. The film thickness was about 25 nm. The ^{181}Hf activity was produced by neutron activation of the thin film in the IEA-R1 Brazilian Reactor by the reaction ^{180}Hf(n, γ)^{181}Hf using a neutron flux of 2 x 10^{13} n . cm^{-2}. s^{-1} for 60 h. After irradiation the thin film was annealed in air at 1473 K and sealed in a small quartz tube in vacuum. PAC measurements were performed in the temperature range 295 – 1273 K. The PAC method is based on the hyperfine interaction of nuclear moments of the probe with extra nuclear magnetic fields or electric field gradients (EFGs). In the case of quadrupolar electric interaction, the experimental measurement gives the quadrupolar frequency ν_Q with respective distribution δ as well as the asymmetry parameter η of EFG. A detailed description of this method can be found elsewhere [2]. Results of previous PAC measurements in HfO_2 can be found in the literature [3,5]. The γ-γ cascade of (133 - 482) keV, populated in the β^- decay of ^{181}Hf, was used to measure the quadrupole interaction of the 482 keV ($5/2^+$) state of ^{181}Ta, with an anisotropy coefficient A_{22} = -0.288. The γ-γ PAC measurements have been carried out using a standard set up with four conical BaF_2 detector scintillators with a time resolution of 0.6 ns (FWHM). From the coincidence spectra N(θ, t), where θ is the angle between detectors and t is the time delay between events, the time differential anisotropy he first paragraph after a heading is not indented.

$$R(t)=2\frac{N(180^\circ,t)-N(90^\circ,t)}{N(180^\circ,t)+2N(90^\circ,t)}=A_{22}G_{22}(t)=\sum f_i G^i_{22}(t)$$

has been calculated. $G_{22}(t)$ is the perturbation function which describes the time modulation of the angular correlation perturbed by the hyperfine interaction and f_i are the fractional site populations of probe nuclei used to fit the data.

For a static quadrupole interaction, the perturbation function has the form

$$G_{22}(t)=\sum_{n=0}^{3} S_{2n}e^{-\delta\omega_n t}\cos(\omega_n t)$$

The frequencies ω_n are related to the quadrupole frequency $\nu_Q = eQV_{zz}/\hbar$ by $\omega_n = g_n(\eta)\nu_Q$. The coefficients $g_n(\eta)$ are known functions of the asymmetry parameter $\eta = (V_{xx} - V_{yy})/V_{zz}$, where V_{kk} (k = x,y,z) denote the principal components of the EFG tensor. The exponential function accounts for a lorentzian frequency distribution δ around ω_n. In the case of single crystals, the coefficients S_{2n} depend on both η and the angles between the emission direction of the γ-rays and the principal axes of the EFG tensor. Details of this dependence can be found elsewhere [2].

3. Results and discussion

Figure 1 shows the measured anisotropy function R(t) least squares fitted to the perturbation function $G_{22}(t)$ that describes the experimental data. The time spectra show the behavior of the quadrupole parameters corresponding to the monoclinic structure at different temperatures. The results correspond to measurements after annealing in air at 1473 K with the thin film sealed in vacuum. The orientation of the silicon crystal substrate during the measurements was such that the normal to the surface of the film was in the detectors plane, forming angles of 45° with the lines joining opposite detectors. Different orientations of the sample produced different PAC spectra which show that the thin film has a single crystalline structure [4].

The temperature dependence of the time zero anisotropy $A_{22}(t)$ shows different values for three temperature ranges, the first below 295 K in which A_{22} ~ 9 %, the second from 295 K to 500 K where A_{22} decrease with T and the third above 500 K when A_{22} increase with T. This behavior is likely caused by a reversible effect with the increase of temperature. A comparison of the A_{22} values at T = 295 K and for $T \geq 1200$ K shows the same A_{22}.

Table 1 shows some hyperfine parameters obtained in the best least square fit to the experimental data. It was observed that the best fit has been achieved using two different fractions characterized by different frequencies, η and δ.

The temperature dependence of quadrupole frequency (ν_Q) and asymmetry parameter (η) for the monoclinic lattice are shown in Fig. 2. The temperature dependence of the hyperfine parameters show that the major quadrupole frequency slowly decreases and the corresponding asymmetry parameter increases with the increase of temperature. This dependence agrees with PAC results reported for bulk HfO_2 [6] and show that the thin film is well structured and present the same behaviour as the bulk.

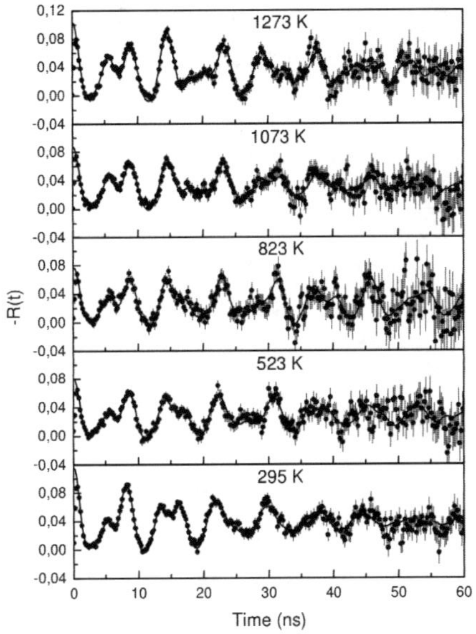

Figure 1. Perturbed R(t) functions for ^{181}Ta probe nuclei in HfO_2 thin film measured for five different temperatures. Solid lines are the least squares fit of the theoretical function to the experimental data.

Table 1. Best fit parameters for the spectra shown in figure 1.

Temp. Annealing (K)	ν_Q (MHz)	η	δ (%)	Fraction (%)
1273	720 (2)	0.49 (5)	3.51 (2)	73
	1181 (4)	0.46 (2)	0.02 (2)	27
823	753 (2)	0.40 (4)	1.73 (2)	82
	1175 (9)	0.43 (2)	0.03 (4)	18
523	770 (2)	0.37 (4)	2.99 (3)	84
	1134 (11)	0.35 (3)	0.01 (1)	16
295	800 (1)	0.33 (3)	4.44 (3)	95
	1052 (6)	0.39 (3)	0.02 (1)	5

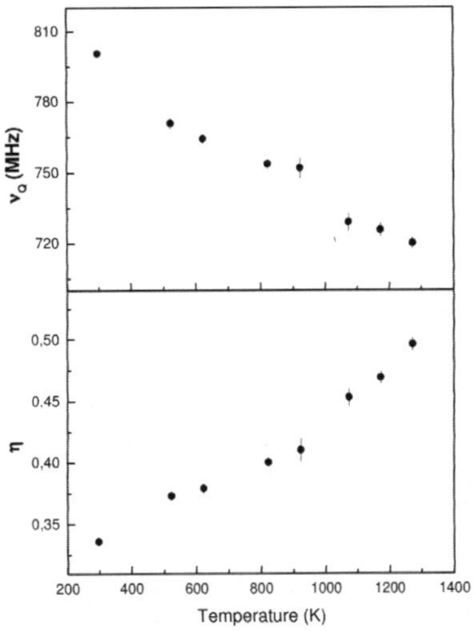

Figure 2. The temperature dependence of the quadrupole frequency (ν_Q) and asymmetry parameter (η) measured at the ^{181}Ta in the monoclinic site.

In Fig. 3 the fractions of two sites occupied by the probes are shown as a function of temperature. The major fraction (○) corresponds to ^{181}Ta at the monoclinic site and the minor fraction (●) has been ascribed to oxygen defect around ^{181}Ta. The monoclinic fraction decreases with the increase of temperature and consequently the second fraction increases. We believed that this behaviour can be explained by considering that as temperature increases oxygen defects are created. In order to test such consideration, first-principles calculations of EFG at Ta nuclei in Hf sites with oxygen defects (vacancy or extra oxygen) in their neighbourhood in HfO_2 are in progress.

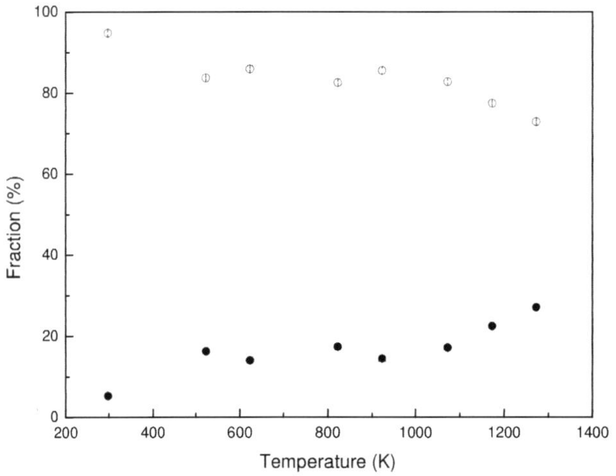

Figure 3. Temperature dependence of fractions of ^{181}Ta nuclei at Hf sites corresponding to: a) monoclinic sites (○) and b) oxygen-defect sites (●).

We have observed that only after annealing at very high temperatures it was obtained a very well-resolved electric quadrupole interaction[4]. The frequency distribution observed for the measurement taken just after the irradiation can be due to structural defects, very small nanocrystals or amorphous material, which have been removed with the high-temperature annealing. The time zero anisotropy A_{22} showed to be temperature dependent as observed in the measurements with HfO_2 powder samples [6]. The values of the quadrupole frequency and the asymmetry parameter obtained after annealing at 1473 K correspond to those expected for the quadrupole interactions of ^{181}Ta in the monoclinic phase of hafnium oxide: ν_Q = 800 MHz and η = 0.33 [6-9]. As also show by Ayala et al [6], the temperature dependence of ν_Q and η may be influenced by effects originated in lattice vibrations, these effects have distinct temperature behavior as the following: an increase in a, b and c crystal parameters what is likely to be correlated to the asymmetry parameter η. A different interpretation for the temperature dependence of EFG is also reported by Ayala et al [6] and it is related to the ^{181}Ta probe itself. HfO_2 is a insulator with a band gap of 5,5 eV. The decay of ^{181}Hf to ^{181}Ta creates an impurity atom in the lattice with a valence of 5^+ and n-type states could be created. Fig. 3 shows that the fractions of ^{181}Ta in HfO_2 thin film have an interesting behavior probably caused by oxygen vacancy. As the measurements have been carried out in vacuum and with the increase of temperature the lattice vibrations are stronger, oxygen ions could be removed provoking oxygen vacancies in the local neighborhood of the ^{181}Ta probes. As oxygen vacancies are more energetically favored in SiO_2 than in HfO_2 [10], local vacancy or an extra oxygen in the interface of the thin film and the substrate are more likely to be created. This effect may explain the observed increase of the second fraction in the PAC measurements.

4. Conclusions

The PAC measurements taken after the irradiation with neutrons and annealing showed that the film has a single crystalline structure with a preferred orientation regarding the main EFG axis which is essentially perpendicular to the surface of the substrate. A second conclusion is that we distinguish two fractions of probe atoms, one of ~ 80 % that can be associated to ^{181}Ta in monoclinic Hafnium oxide. The other fraction of ~ 20 % corresponds to the ^{181}Ta probes in a region of the film with oxygen defects probably near to the substrate interface.

Acknowledgments

FHMC thankfully acknowledges the support provided by Centro de Física Nuclear da Universidade de Lisboa (CFNUL) and CAPES in the form of research fellowships. Thanks are due also to José G. Marques and Olinda Conde for the irradiation of the sample and delivering the film respectively.

References

[1] Govindaraj R., Sundar C.S. and. Kesavamoorthy R 2006 *J. Appl. Phys.* **100** 084318
[2] Schatz G. and Weidenger A. 1996 *Nuclear Condensed Matter Physics* (Chichester: John Wiely \& Sons)
[3] Dogra R., Junqueira A. C., Saxena R. N., Carbonari A.W., Mestnik-Filho J., and Moralles M. 2002 Physical Review B 63 224104
[4] Pasquevich A.F., Cavalcante F.H.M. and Soares J.C. 2008 *Hyperfine Interactions,* 10.1007/s10751-008-9672-2
[5] Forker M. and Presa P. de la 2008 *Phys. Rev. B* 77 054108-1
[6] Ayala A., Alonso R. and López García A.1994 *Phys. Rev. B* 50 3547
[7] Lerf A and Butz T.1987 *Hyperfine interactions* 36 275
[8] Shitu J. and Pasquevich A. F. 1997 *J. Phys. Condens. Matter* 9 6313
[9] Broqvist P. and Pasquarello A. 2007 *J. Microelectronic Engineering* 84 2022
[10] Scopel W.L., Da Silva A.J.R., Orellana W., Fazzio A. 2004 *Applied Phys. Letters* 84 1492. n9

UV Thermoluminescence and Phosphorescence Properties of Mg^{2+} and Nd^{3+} Doped Nanostructured Al_2O_3

J F S Bitencourt[1], K A Gonçalves[1], S H Tatumi[2,4], P J B Marcos[3]

[1]Escola Politécnica da Universidade de São Paulo, Depto. de Engenharia de Sistemas Eletrônicos, av. Luciano Gualberto, 380, 05508-970, São Paulo, SP, Brazil.
[2]Faculdade de Tecnologia de São Paulo, Depto. de Ensino Geral, pça. Cel. Fernando Prestes, 30, 01124-060, São Paulo, Brazil.
[3]Faculdade de Tecnologia de São Paulo, Depto. de Mecânica de Precisão, pça. Cel. Fernando Prestes, 30, 01124-060, São Paulo, Brazil.

e-mail: tatumi@fatecsp.br

Abstract. Mg^{2+} and Nd^{3+} doped aluminium oxide samples were produced by polymer calcination method. Mg^{2+} doped samples did not exhibited significant fluorescence emission, using IR (LED, emission centered at 862nm) or green (Xe-lamp plus optical filter, emission centered at 520 nm) sources. Nonetheless, high thermostimulated luminescence was detected, with high emission peak at 190°C. A nanoscopic layer (about 50 nm width) of magnesium spinel was observed by Transmission Electronic Microscopy (TEM) for 2.61mol% doped sample; this layer can be the responsible for TL enhancement. Nd^{3+} doped sample exhibited low phosphorescence emission in the UV (Schott U-340) using IR source. TL peaks were detected at 185 and 265°C; the intermediary peak showed the highest emission. Occurrence of NdAl and $NdAl_2$ structures were detected in 5 mol% doped sample and $NdAl_2$ and $NdAl_4$ structures in 10 mol% doped sample.

1. Introduction

Since the 50's, aluminium oxide have been used as radiation detection and dosimetric material [1] using Thermoluminescence (TL) technique. Early samples exhibited high luminescent sensitivity but also high intensity fading, which prevent its use in a practical way. However, a lot of doped samples appeared in the last 25 years, presenting several new characteristics and improvements; the better example being the carbon doped alumina [2, 3, 4]. Although most of those materials did not exhibited such sensibility as carbon doped alumina, other characteristics can be obtained, such as high saturation point or low energy dependence.

Ceramic dosemeters, such as Al_2O_3, LiF and phosphate glasses, show advantages in comparison with polymeric films. Reproducibility, multi-readability [5], high-dose saturation point and low energy dependence are some of them.

Magnesium doped alumina is a well know material for dosimetry and radiation detection purposes; nonetheless, structural studies in order to explain luminescence enhancement had not been done. The

[4] To whom any correspondence should be addressed. tatumi@fatecsp.br

knowledge of the mechanisms of luminescence production will help the development of new materials and procedures.

Papin et al [6] studied Mg^{2+} doped alumina and found out TL peaks at, approximately, 190 and 350°C. He tried to explain the lower peak by the occurrence of oxygen vacancies. However, the emission intensity decreased for increasing dopant concentration and no explanation was given for high temperature peak.

Neodymium doped alumina is a novel material which application purposes are still in research. This dopant is easily found in photonic glasses, used in waveguides and optical fibres.

Thus, the objectives of the present work are to obtain characteristics TL curves from Mg^{2+} and Nd^{3+} gamma-irradiated samples, so as their OSL (Optically Stimulated Luminescence), in order to verify their use for radiation dosimetry and radiation detection. TEM (Transmission Electron Microscopy) images were obtained, trying to explain the change in luminescence properties from different compositions of samples.

2. Experimental Procedure

Magnesium and neodymium doped samples were produced using polymer calcination method (a variation from Pechini method). Four samples of neodymium doped material were produced (0.5, 2, 5 and 10 mol% of Nd_2O_3) and five ones doped with magnesium were obtained (0.47, 0.88, 1.33, 2.61 and 3.36mol%). Magnesium concentrations were determined by *Inductively Coupled Plasma Atomic Emission Spectroscopy* (ICP – AES).

Grain size study and crystals morphology were discussed previously at Bitencourt and Tatumi, 2009, achieving an average dimension of 8 μm for grain diameter.

The samples were irradiated using a ^{60}Co gamma radiation source (doses of 1, 2, 3, 4 and 5 Gy), which Kerma in air was 33.2 $Gy.h^{-1}$ at 40 cm from the source.

TL and OSL measurements were taken in a 1100-series model from Daybreak Medical and Nuclear Systems Inc. coupled with Schott U-340 optical filter (267-377.5 nm, 2 mm thickness) in the entrance of the photomultiplier tube (EMI 9235QA). Measured aliquot mass of magnesium doped samples was around 1.8 mg and 5.4 mg for neodymium doped samples. Heating rate was kept at $5°C.s^{-1}$ from RT to 300°C.

OSL measurements were carried out using two different sources of light: infrared (862 nm), provided by a LED, and green light from a xenon lamp coupled with one Schott VG-6 optical filter (transmission band centered at 520 nm). OSL decays were measured for the first 100 seconds under continuous excitation.

TEM images and Electron Diffraction (ED) were obtained using model CM200 from Philips Inc.

3. Results and Discussion

Figure 1 shows the average TL emission response variation for five different concentrations of magnesium (0.47, 0.88, 1.33, 2.61 and 3.36 mol%); in this case, the samples were irradiated with 3 Gy of gamma-radiation and the intensities values were calculated using the mean term of three TL maximum peaks intensities. It is possible to verify that the standard deviations from all the samples were kept low and the 2.61 mol% composition exhibited the greater sensibility to radiation exposure.

The luminescence decay observed for 3.36mol% composition is due to phase transition from alumina to magnesium spinel, which could be observed in TEM images and electron diffraction results, showed later in this section. Apparently, doping alumina with 3.36 mol% of magnesium converts most of the crystals into magnesium spinel, which TL response is not as high as alumina itself. Probably, the remaining emission is due to non-transformed crystals.

For Nd^{3+} doped samples (Figure 2), the data was obtained in the same way as the previous sample. The most sensitive sample was the 0.5 mol% Nd_2O_3 doped composition.

TL glow curves obtained from the best of magnesium and neodymium doped samples indicated above are shown, respectively, in **Figure 3** and Figure 4.

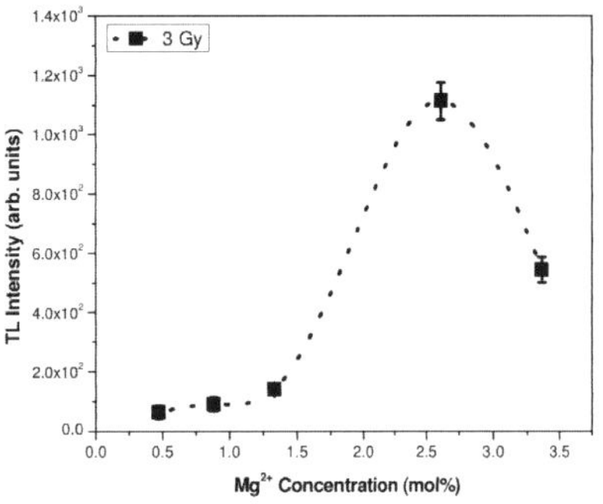

Figure 1. TL emission intensity in function of the magnesium concentration.

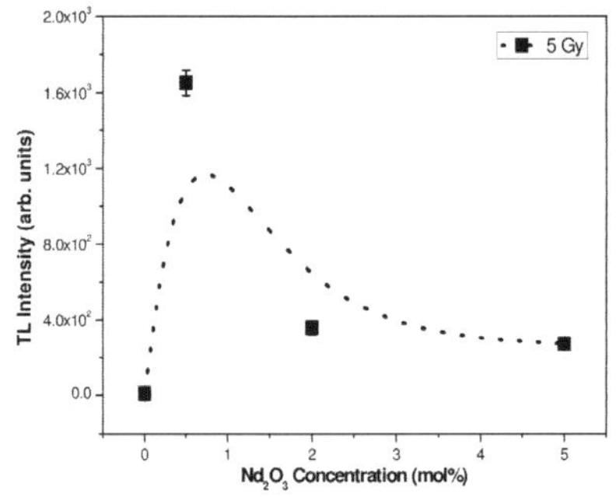

Figure 2. Curve of the TL emission intensity in function of the neodymium oxide concentration.

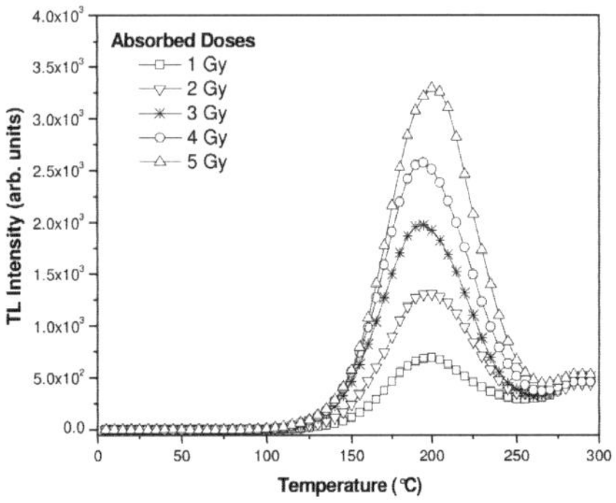

Figure 3. TL emission curves from Mg^{2+} doped α-alumina (2.61 mol% composition) in the UV region.

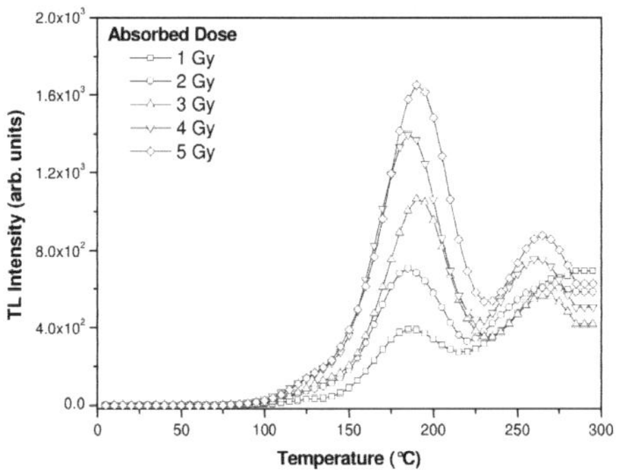

Figure 4. TL emission curves from Nd^{3+} doped samples in the UV band.

Magnesium doped sample produced high intensity TL peak at 190°C and its intensity is proportional to the absorbed radiation dose. The already mentioned work by Bitencourt and Tatumi showed the formation of higher temperature peaks, at 280, 360 and 432°C. The trapping centres responsible for these emissions are filled only when the material is exposed to even higher doses of radiation (above 100 Gy). For low doses, the emission of these centres is hidden behind the background.

Nd^{3+} doped samples, in the other hand, exhibited two TL peaks at 185 and 265°C for low radiation doses. Considering that the 190°C peak found for Mg^{2+} doped sample and 185°C peak detected in Nd^{3+} sample are produced by the same trapping and recombination centres, it can be said that the presence of neodymium atoms change significantly the concentration of such centres. Meanwhile, electronic transitions that produce luminescence at 265°C became proportionally significant. These two peaks were not visualised in Mg^{2+} doped sample, but it does not mean that they do not exist. TL intensity growth of 185°C TL peak from this sample was also proportional to the absorbed radiation dose.

Osvay and Deme [7], working with high-dose irradiated magnesium and yttrium co-doped alumina samples, found two TL peaks, one at 250 and another at 400°C. These peaks do not seem to be related to the presented results, despite of the similarity of both materials.

Unlike the first sample, Nd^{3+} doped material did exhibit OSL decay. The measurements were taken using ninety-nine channels of one second each. Due to low sensitivity or mismatch of the optical filter used, the decay was not abrupt and the signal was low. Nonetheless, it is possible to distinguish among the five different radiation doses applied to the samples. These decay curves are almost linear, allowing to be taken several times before a significant decrease of the luminescent intensity; that is the concept of multi-readability found in many commercial dosemeters, discussed in the first section.

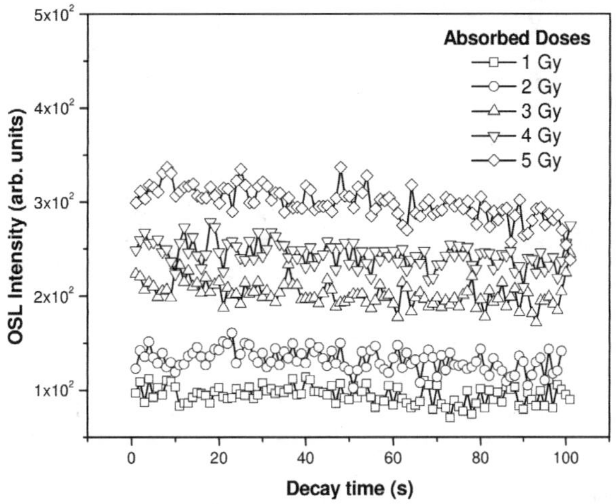

Figure 5. OSL from Nd^{3+} doped sample using IR stimulation source and detection in UV band.

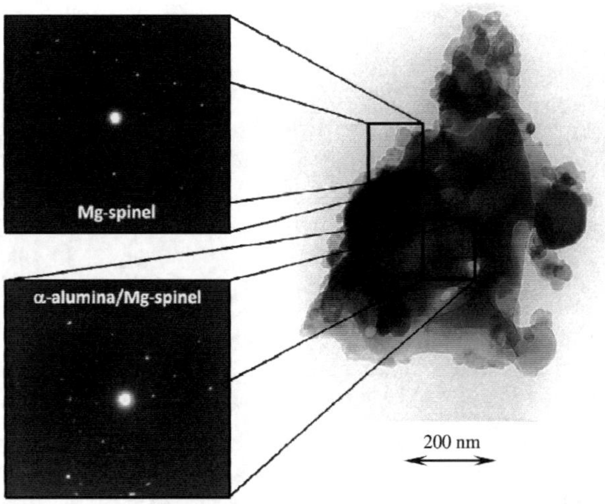

Figure 6. TEM image and Electron Diffraction from Mg^{2+} doped alumina. Superficial structure composed by magnesium spinel was generated on alumina clusters.

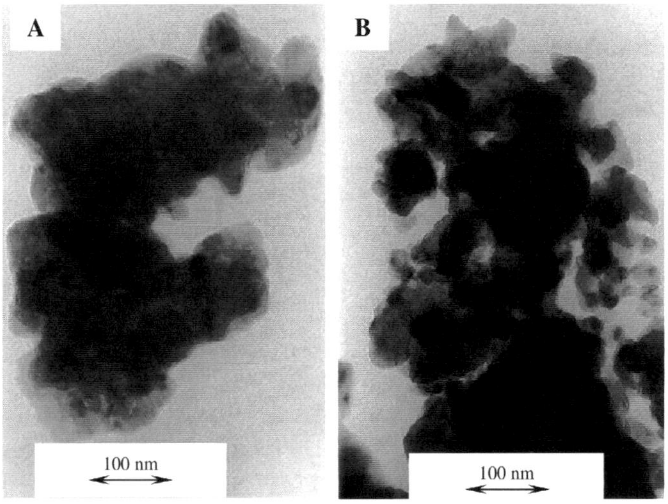

Figure 7. TEM images from Nd^{3+} doped alumina; 5 mol% (A) and 10 mol% (B). Electron Diffraction indicates the presence of NdAl and Nd Al_2 in (A) and $NdAl_2$ and $NdAl_4$ in (B).

TEM imaging and ED from Mg^{2+} doped sample detected a superficial structure composed by magnesium spinel (Figure 6). Such structure only appears in doped material and can be closely related to luminescence enhancement described previously. Also, the images showed a certain number of nanopores, i.e., nanometric pores, but they seem not to be linked to the luminescence properties of the material, for their dimensions do not change significantly when magnesium concentration is increased.

Probably, the spinel layer was created due to magnesium atoms diffusion to clusters surface, where its concentration was high enough to create another material composed by magnesium, aluminium and, of course, oxygen.

Alumina clusters are formed due to the coalescence among several nanometric single-crystals at high temperatures (above 1000°C). This means that the temperature is not enough to melt the substance, but only its surface, creating chemical bonds to nearby single-crystals.

The luminescence results from these samples are different from those published by Yoshimura & Yukihara, 2006 [8] and Lorincz et al., 1982 [9], that dealt with magnesium spinel. So, the luminescence properties must be originated by the junction, or interface, produced between spinel and alpha-alumina.

From Nd^{3+} doped alumina TEM images (Figure 7), it is not possible to discriminate the present phases, or how they interfere in luminescence behaviour. If neodymium atoms are changing places with aluminium in alumina molecules, just like what is occurring with magnesium doped samples, the local charge is not changed, once aluminium and neodymium have the same valence value (3+). Nonetheless, the atomic mass of neodymium (144.24 a.m.u.) is bigger than the mass of aluminium (26.98 amu); so, it is not expected that substitution of Al atoms by Nd atoms really occurs.

Furthermore, it was observed the formation of nanostructures composed by Nd and Al atoms; for 5 mol% doped samples, NdAl and $NdAl_2$ were observed and, for a higher dopant concentration sample, $NdAl_4$ nanoparticles, located at the surface of alumina clusters. Our results suggest that complex cluster of Nd and Al does not contribute at all to the luminescence efficiency of the sample.

4. Conclusions

TL measurements of irradiated Mg-doped samples indicate the presence of a high sensitive luminescent centre, producing a UV TL peak at 190°C. This emission is proportional to the magnesium concentration in the material. However, if the addition of magnesium surpasses the limit, alumina crystals turn into magnesium spinel crystals. So, the best composition found in this work is 2.61mol% magnesium doped alumina.

For Nd-doped alumina, the highest sensitivity was obtained with 0.5mol% Nd_2O_3 doped sample; two TL peaks were observed: 185 and 265°C. OSL measurement produced proportional emission intensity to the absorbed radiation dose, but a slow decay luminescence was observed.

Both samples exhibited proportional TL emission intensity growth within the radiation dose range, from 1 to 5 Gy, and no saturation point was reached in this dose interval.

Spinel nanolayer was observed on the surface of alumina aggregates in TEM images. This layer is highly probable to be the responsible for the luminescence enhancement observed in TL glow curves, because it is expected that the greatest concentration of luminescence defects or, in other words, anionic defects or F-centres, is located in the interface between the two compounds.

Acknowledgments
We would like to thank CNPq and FAPESP (Brazilian agencies) for financial support, Simone Perche from Laboratório de Cristalografia of Instituto de Física da Universidade de São Paulo (São Paulo University, Institute of Physics) for TEM images and Elizabeth Somessari from CTR, IPEN-CNEN for irradiating the samples analysed in this work.

References

[1] Cameron J R, Suntharalingam N and Kenney G N 1968 Thermoluminescent Dosimetry (London: The University of Wisconsin Press).
[2] Akselrod M S, Lucas A C, Polf J C and McKeever S W S 1998 Optically stimulated luminescence of Al_2O_3 Radiation Measurements vol. 29 No. 3-4 pp. 391-399.
[3] Molnár G, Benabdesselam M, Borossay J, Lapraz D, Iaconni P and Akselrod M 2001 Influence of the irradiation temperature on TL sensitivity of Al_2O_3: C Radiation Measurements vol. 33 pp. 619-623.
[4] Yukihara E G, Whitley V H, Polf J C, Klein D M, McKeever S W S, Akselrod A E and Akselrod M S 2003 The effects of deep trap population on the thermoluminescence of Al_2O_3: C Radiation Measurements vol. 37 pp. 627-638.
[5] Bulur E, Göksu H Y and Wahl W 1998 Infrared (IR) stimulated luminescence from alpha-Al_2O_3 Radiation Measurements vol. 29 No. 6 pp. 625-638.
[6] Papin E, Grosseau P, Guilhot B, Benabdesselam M and Iacconi P 1999 Influence of point defects on the thermoluminescence of alpha-Al2O3: application to dosimetry Radiation Protection Dosimetry vol. 84 No. 1-4 pp. 91-94.
[7] Osvay M and Deme S 2006 Application of TL dosemeters for dose distribution measurements at high temperatures in nuclear reactors Radiation Protection Dosimetry vol. 119 No. 1-4 pp. 271-275.
[8] Yoshimura E M and Yukihara E G 2006 Optically stimulated luminescence of magnesium aluminate ($MgAl_2O_4$) spinel Radiation Measurements vol. 41 pp. 163-169.
[9] Lorincz A, Puma M, James F J and Crawford, Jr. J H 1982 Thermally stimulated processes involving defects in gamma and x-irradiated spinel ($MgAl_2O_4$) J. Appl. Phys. Vol. 53 (2) pp. 927-932.
[10] Bitencourt J F S and Tatumi S H 2009 Synthesis and thermoluminescent properties of Mg^{2+} doped nanostructured aluminium oxide Journal of Luminescence ARTICLE IN PRESS.

Defect studies and optical activation of Yb doped GaN

K. Lorenz[1], E. Alves[1], S. Magalhães[2], M. Peres[2], T. Monteiro[2], A. Kozanecki[3], M. E. G. Valerio[4]

[1]Instituto Tecnológico e Nuclear and CFNUL, Estrada Nacional 10, 2686-953 Sacavém, Portugal
[2]Departamento de Física & I3N, Universidade de Aveiro, 3810-193 Aveiro, Portugal
[3]Institute of Physics, Polish Academy of Sciences, 02668 Warsaw, Poland
[4]Departamento de Física, Universidade Federal de Sergipe, 49100-000, São Cristóvão, SE, Brasil

Abstract. Wide band-gap semiconductors, particularly III-nitrides, became one of the most studied materials during the last decades. These compounds are the base of a new generation of optoelectronic devices operating in the UV-Blue region of the electromagnetic spectrum. Incorporation of rare-earth (RE) ions into nitrides creates new routes to build all-nitride electroluminescent devices, using the sharp intra-$4f^n$ transitions of these elements. The introduction of the RE ions in the nitride lattice during the growth or by ion implantation creates defects which influence the optical behaviour of the doped region. In this work we report the results on Yb implanted GaN. A combination of techniques (Rutherford backscattering/Channeling and Photoluminescence) was used to assess the mechanisms responsible for the optical and structural behaviour of the doped materials. Lattice site location experiments showed that Yb is incorporated into positions slightly displaced from the Ga-site. Clearly the optical activity of the RE could be enhanced by orders of magnitude reducing the number of non-radiative recombination paths related with defects.

1. Introduction

The incorporation of lanthanide ions in GaN-based materials has generated much interest lately due to the development of blue, green and red emitting devices based on the intra-$4f^n$ transitions of Tm^{3+}, Er^{3+} and Eu^{3+} ions [1-3]. Luminescence properties of GaN-based systems doped with rare earth (RE) ions have been widely studied on the basis of the crystal field splitting of the $^{2S+1}L_J$ manifolds of the $4f^n$ configurations. In nitride samples these ions mostly occupy the cation site symmetry (C_{3v}). Nevertheless they are often found in more than one distinct environment as observed using optical measurements. Extended X-ray absorption fine structure [4], emission channelling [5] and Rutherford backscattering spectrometry in the channelling geometry (RBS/C) [6,7] have shown that the majority of the RE ions occupy Ga-substitutional sites (C_{3v} symmetry) and some of the ions are slightly displaced from the Ga sites probably due to point defects in their vicinity.

In the visible spectral range other rare earth ions such as Tb^{3+}, Pr^{3+} and Sm^{3+} have also been studied by several research groups [8-12]. In the infrared Er^{3+} is known to be one of the most important ions as its $^4I_{13/2} \rightarrow {}^4I_{15/2}$ transition occurs at 1.54 μm which coincides with the minimum absorption wavelength region of silica based optical fibres. On the other hand Yb^{3+} ($4f^{13}$) despite the very simple scheme of levels with only the fundamental and excited multiplets ($^2F_{7/2}$ and $^2F_{5/2}$, respectively) which are separated by an energy of around 10000 cm^{-1} is one of the rare earth ions with scarce analysis in nitride hosts [13,14]. This RE ion besides the optical emissions can also be used as a sensitizer for other rare earth ions [15].

Here, we report on the optical activity of Yb^{3+} doped GaN layers produced using ion implantation and rapid thermal annealing treatments. After the implantation and each annealing step the samples were studied by RBS/C to asses the damage and lattice site location. Using above and below band gap photon energy excitation the intra-$4f^{13}$ recombination between the multiplets $^2F_{5/2} \rightarrow {}^2F_{7/2}$ of the Yb^{3+} was observed.

2. Experimental Details

GaN layers were implanted with an energy of 150 keV and fluences of 5×10^{14} cm^{-2} and 5×10^{15} cm^{-2}. The as-implanted samples were further submitted to rapid thermal annealing (RTA) in flowing nitrogen during 120 s at 1000 °C using a proximity cap.

Ion beam analysis (RBS/C) was performed with 2.0 MeV He^+ ions and the backscattered particles were detected with solid state detectors placed at 140° and close to 180° with resolutions of 13 and 16 keV, respectively. A high precision (~0.01°) computer controlled goniometer was used to place the samples in position and perform the channelling measurements. The PL studies of the annealed samples were measured with above and below band gap excitation using as excitation sources the 325 nm laser line of a cw He-Cd laser, a 450 W Xe arc lamp coupled to a monochromator and the 488 nm laser line of a cw Ar^+ laser. The luminescence was dispersed using a Spex 1704 monochromator or a Fluorolog-3 Jobin Yvon-Spex. A Brucker 66V Fourier-transform spectrometer was also used for the infrared spectral region PL measurements. The signal was detected either by cooled Hamamatsu photomultipliers (R928, H9170-75) or by a North-Coast EO-817 liquid nitrogen cooled germanium detector.

3. Results and discussion

The aligned RBS/C spectra of the ytterbium as implanted GaN clearly show the damage produced by the implanted ions, Fig.1.

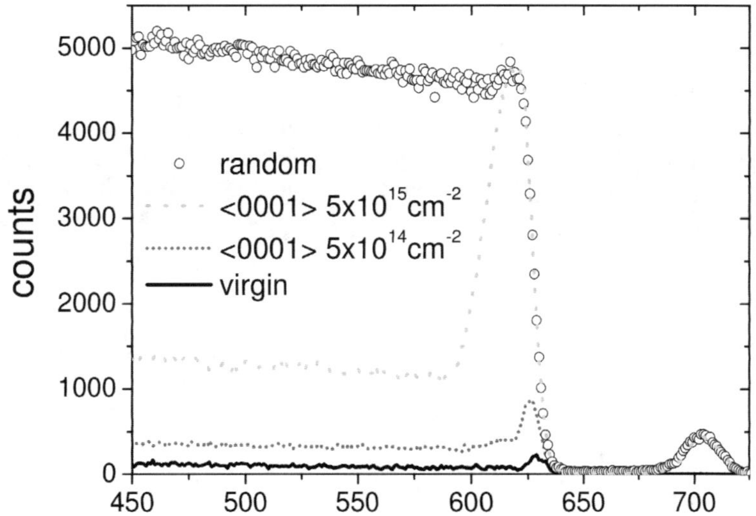

Figure 1. Random and <0001>-axis aligned RBS spectra obtained with 2.0 MeV He^+ before (virgin) and after implantation with 150 keV Yb ions to a fluence of 5×10^{14} and 5×0^{15} at/cm^2, at room temperature.

The backscattering yield of the aligned spectrum for the sample implanted with the highest fluence overlaps in the implanted region with the random spectrum revealing the complete loss of long range order typical for high fluence implantation [16,17]. For the lower fluence, besides the surface peak a second peak is visible close the end of range of the implanted ions in agreement with the behaviour observed for other RE in GaN [6, 18]. The low damage level in the implanted region in this sample allows the lattice site location of the Yb ions in the GaN structure. Detailed angular scans along the <0001> and <10-11> axial directions are shown in Figure 2. The close agreement between the Ga and Yb curves demonstrates the incorporation of the Yb ions in Ga sites. The Yb scan is found slightly narrower than the one for Ga indicating a small displacement from the perfect substitutional site. Similar results were reported for measurements using the emission channelling technique [14].

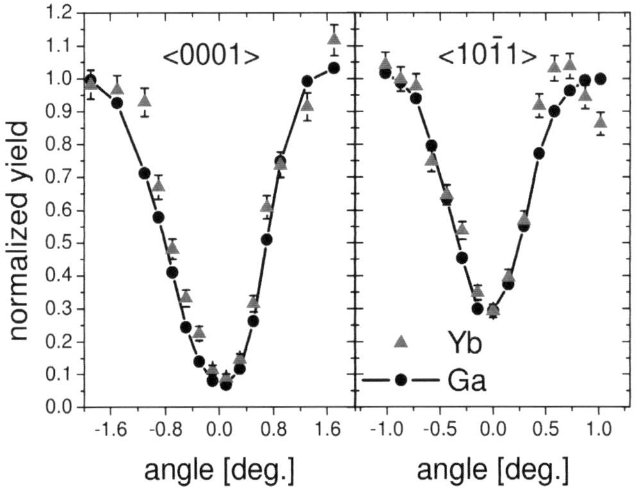

Figure 2. Yb- and Ga-Angular scans for the <0001> and <10-11> axes obtained directly after the implantation with 5×10^{14} at/cm^{-2}..

After annealing we observe a small decrease of the implantation damage and the angular scan along the <0001> axis seems to indicate an increase of the displacement of the Yb ions perpendicular to the c-axis, Fig.3. This displacement is responsible for the narrowing of the Yb curve with respect to the GaN one. Emission channelling [14] did not show an increased displacement after annealing. This difference could be attributed to the much higher ion fluence used in the present study and suggests that implantation defects are responsible for the displacement of Yb. The sample implanted on the higher fluence showed a complete decomposition of the samples' surface after annealing. The highly damaged surface layer sublimates and Yb segregates at the surface upon high temperature annealing.

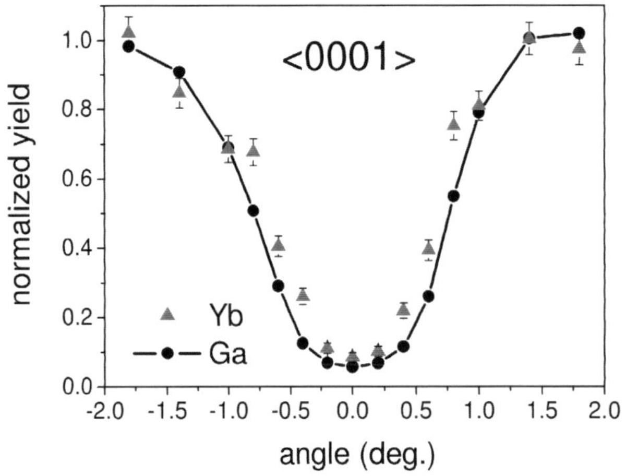

Figure 3. Angular scans obtained for the sample implanted with 5x10^{14} cm^{-2} after annealing for the <0001> axes.

Despite the relatively low damage level and the high fraction of Yb in regular sites in the low-fluence sample no PL signal was observed directly after the implantation. Figure 4 shows the low temperature PL spectra of annealed GaN:Yb samples obtained with above band gap excitation. The spectrum shows the typical D^0X feature of GaN layers at ~357 nm followed by broad emission bands. As observed the dominant optical active defect is different in both samples being the sample implanted with $5x10^{15}$ cm^{-2} dominated by a violet recombination with maxima at 382 nm and 414 nm.

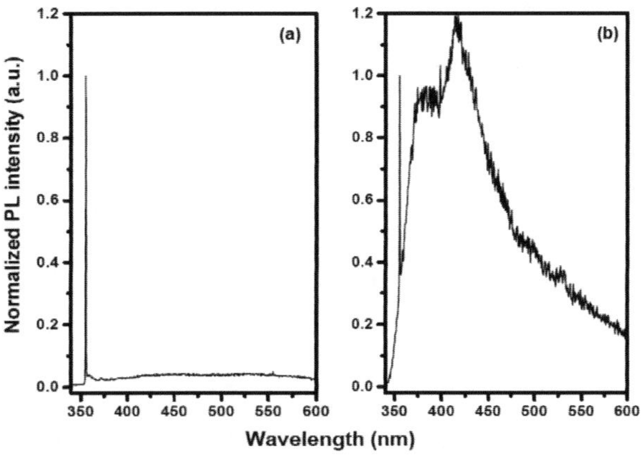

Figure 4. 12 K PL spectra of samples implanted with $5x10^{14}$ cm^{-2} (a) and $5x10^{15}$ cm^{-2} (b).

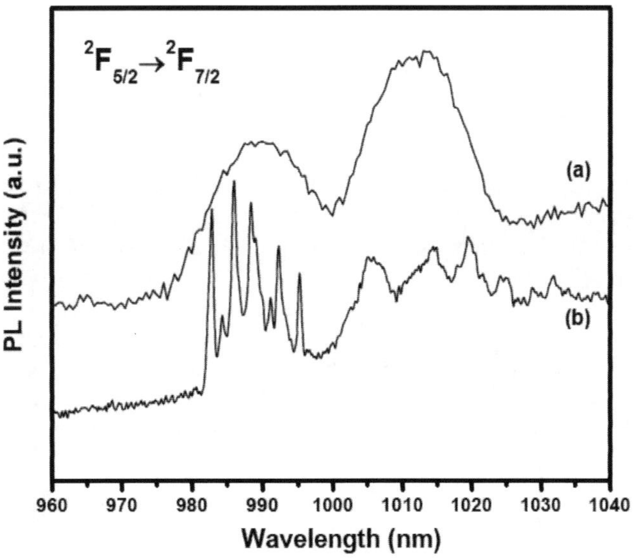

Figure 5. (a) 12 K low resolution PL spectra obtained with 357 nm photons using a Xe lamp coupled to a monochromator and (b) 70 K high resolved spectra obtained with a 488 nm excitation line of an Ar^+ laser using a FTIR system.

Figure 5 shows the $^2F_{5/2} \rightarrow {}^2F_{7/2}$ transition of Yb^{3+} ion observed in our doped and annealed low-fluence sample. The spectra were taken under different excitation conditions and at different temperatures. They were also measured in different set-ups with different resolution conditions.

One of the main characteristics corresponds to the excitation paths for the observed emission. PLE monitored at the intraionic emission maxima shown in Figure 6 indicates that besides the above band gap excitation the intraionic Yb^{3+} luminescence is excited via band gap electronic levels that match both the donor bound exciton recombination and the violet band. Furthermore, and in accordance with the data shown in Figure 5, the intraionic emission can also be excited by photons with lower energy such that ones corresponding to the 488 nm wavelength.

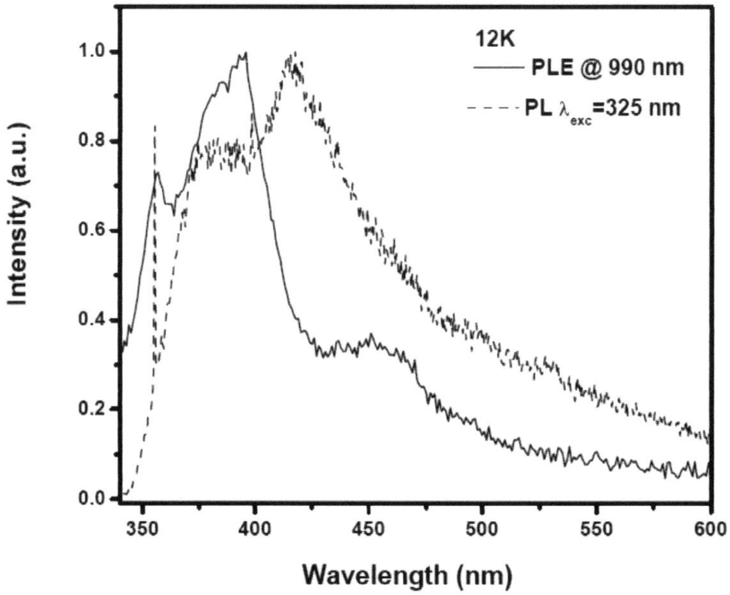

Figure 6. 12 K PL spectra (dash line) and PLE spectra (full line) monitored at 990 nm.

RBS/C measurements clearly evidence that the Yb^{3+} occupies near-substitutional Ga sites. In C_{3v} site symmetry the manifolds $^2F_{7/2}$ and $^2F_{5/2}$ split into four and three Kramers doublets, respectively. At lower temperatures and assuming that only the lowest energy level of the excited $^2F_{5/2}$ multiplet is populated four transitions are expected for the intraionic recombination. However, in accordance with previous observations [13] the observed spectra (Fig. 5) are rather complex which cannot be accounted for this simple explanation. Thermally populated stark levels, vibronic coupling effects and/or multiple optically active centres cannot be excluded. The slight displacement from the perfect substitutional site evidenced by RBS/C may point to the interaction of Yb with defects which could be the origin of multiple optically active centres. Furthermore, a detailed analysis of the influence of the different excitation mechanisms on the resolved PL must be realized.

4. Conclusions

Implantation of GaN with Yb leads to the incorporation of Yb into Ga-sites. After annealing a clear displacement of the Yb from the regular lattice site ions is observed occurring at the same time as the partial damage recovery of the GaN crystal indicating an interaction of Yb ions with defects. The optical activity of the samples after implantation was totally quenched. The annealed GaN:Yb samples studied using low temperature PL and PLE spectra reveal that optical activation of Yb is achieved. The infrared emission characterized by multiple narrow features which cannot be explained by a simple model assuming that the emission is provided by the lowest energy level from the excited multiplet. The analysed data also indicate that the intraionic recombination can be excited using above or below band gap excitation being some of the energy levels coincident with the donor bound exciton recombination and violet emission.

Acknowledgments
The authors want to thank to R. Ferreira for helping PLE experiments. We acknowledge the support by FCT, Portugal (PPCDT/FIS/57550/2004) and bilateral collaboration Portugal-Poland (GRICES).

References

[1] A. J. Steckl, J. C. Heikenfeld, Dong-Seon Lee, M. J. Garter, C.C. Baker, Y. Wang, R. Jones, IEEE Journal of Select Topics in Quantum Electronics, **8**, 749 (2002).
[2] T. Monteiro, J. Soares, M. R. Correia and E. Alves, J. Appl. Phys. **89**, 6183-6188 (2001).
[3] T. Andreev, N. Q. Liem, Y. Hori, M. Tanaka, O. Oda, D. L. S. Dang, B. Daudin, Phys. Rev. B, **73**, 195203 (2006).
[4] K. P. O'Donnell, B. Hourahine, Eur Phys. J. Appl. Phys. **36**, 91 (2006).
[5] U. Wahl, E. Alves, K. Lorenz, J. G. Correia, T. Monteiro, B. De Vries, A. Vantomme, R. Vianden, Mat. Sci. Eng. B **105**, 132 (2003).
[6] K. Lorenz, U. Wahl, E. Alves, E. Nogales, S. Dalmasso, R. W. Martin, K. P. O'Donnell, M. Wojdak, A. Braud, T. Monteiro, T. Wojtowicz, P. Ruterana, S. Ruffenach, O. Briot, Optical Materials **28**, 750 (2006).
[7] E. Alves, K. Lorenz, R. Vianden, C. Boemare, M. J. Soares, T. Monteiro, Mod. Phys. Lett. B **15**, 1281 (2001).
[8] J.B. Gruber, B. Zandi, H.J. Lozykowski, W.M. Jadwisienczak, J. Appl. Phys. **91**, 2929 (2002).
[9] Y. Hori , T. Andreev , D. Jalabert , E. Monroy , L.S. Dang , B. Daudin , M. Tanaka , O. Oda , Appl. Phys. Lett. **88**, 053102 (2006).
[10] A. Wakahara, Optical Materials **28**, 731 (2006).
[11] H.J. Lozykowski , W.M. Jadwisienczak , I Brown_, J. Appl. Phys. **88**, 210 (2000) .
[12] T. Monteiro, C. Boemare, M.J. Soares, R.A. Sá Ferreira, L.D. Carlos, K. Lorenz, R. Vianden, E. Alves, Physica B **308–310**, 22, (2001).
[13] W.M. Jadwisienczak, H.J. Lozykowski, Optical Materials **23**, 175 (2003).
[14] M. Dalmer, M. Restle, A. Stotzler, U. Vetter, H. Hofsass, M.D. Bremser, C. Ronning, R.F. Davis, Mat. Res. Soc. Symp. Proc. **482**, 1021 (1998).
[15] A. Kozanecki, K. Homewood, B.J. Sealy, Appl. Phys. Lett. **75**, 793 (1999).
[16] S. O. Kucheyev, J. S. Williams, and S. J. Pearton, Mater. Sci. Eng. **R33**, 107 (2001).
[17] F. Gloux, T. Wojtowicz, P. Ruterana, K. Lorenz, E. Alves, J. Appl. Phys. **100**, 073520 (2006).
[18] B. Pipeleers, S. M. Hogg, A. Vantomme, J. Appl. Phys. **98**, 123504 (2005).

Journal of Physics: Conference Series **249** (2010) 012054 doi:10.1088/1742-6596/249/1/012054

EXAFS and Raman scattering studies of Y and Zr doped nano-crystalline tin oxide

KE Rammutla[1], AV Chadwick[2], J Harding[3], DC Sayle[4] and RM Erasmus[5]

[1]University of Limpopo, Department of Physics, P/Bag X1106, Sovenga, 0727, RSA
[2]University of Kent, Centre for Materials Research, Canterbury, UK
[3]University of Sheffield, Department of Engineering Materials, Sheffield, UK
[4]Cranfield University, Centre for Materials Science, Swindon, UK
[5]University of the Witwatersrand, School of Physics, Johannesburg, 2050, RSA

Email: rammutlae@ul.ac.za

Abstract. Nanocrystals of Y and Zr doped SnO_2 have been prepared by sol-gel route and annealed at 200, 400, 600, 800 and 1000 ^{0}C. The X-ray diffraction (XRD) results showed the average size of the particles in the freshly prepared samples to be ~ 3 nm. The Extended Absorption Fine Structure (EXAFS) technique was used to study the dopant environments in nanocrystalline tin oxide. In all Y-doped samples, except the one annealed at 1000 ^{0}C, there is clear evidence that Y has not entered the SnO_2 lattice. This is clearly supported by the Raman scattering results. In all Zr-doped samples, there is a simple substitution for Sn by Zr.

1. Introduction

The unique physical and chemical properties of nano-crystalline materials in comparison to their bulk counterparts [1-4] have recently aroused a great deal of interest. Reasons for such interest are, for example, that they can adopt different morphologies to bulk crystals leading to extraordinary surface chemistry and can dissolve higher concentrations of impurities and this offers further scope for tailoring new catalytic materials. Furthermore, semiconductors made from nanoparticles have unusual optical, electrical and magnetic properties and nanocrystalline ceramics have greater hardness and plasticity. Tin oxide occurs naturally as cassiterite and has the rutile structure. It is an n-type semiconductor which is widely used in flammable gas sensors. Cation doping is widely used to improve the sensing. In the present study EXAFS, XRD and Raman scattering techniques were used to study the Y^{3+} and Zr^{4+} doped nanocrystalline SnO_2. The objective of the studies was to locate the dopant environments and in particular to discover whether they were in the crystal matrix or at the surfaces.

2. Experiments

To prepare Y and Zr doped nanocrystalline tin oxide, appropriate amounts of Tin chloride pentahydride and water soluble metal salts (Yttrium chloride hexahydrate and Zirconyl chloride octahydrate, respectively) were dissolved in 40 ml of distilled water to prepare a 0.1M solution. To this solution, 5 ml of aqueous ammonia was added and the resulting white precipitate recovered by evaporation. The byproduct of this reaction is ammonium chloride and was removed by extensive washing with distilled water. The addition of a small quantity of nitric acid also resulted in the

Journal of Physics: Conference Series **249** (2010) 012054 doi:10.1088/1742-6596/249/1/012054

removal of this byproduct. The material was then oven dried at 100 ^{0}C overnight and thoroughly ground in an agate pestle and mortar. The absence of ammonium chloride peaks in the XRD patterns confirmed the successful removal of all the ammonium chloride. These samples were prepared with nominal dopant concentrations of 1 and 10 mol% for yttrium and zirconium.

The XRD of all the samples were performed on a Phillips Diffractometer on a Phillips Generator PW1130. The average grain size of the samples, L, was determined by the broadening of peaks on the XRPD pattern, using the Scherrer equation [5]; i.e.

$$L = \frac{K.\lambda}{\beta . cos\theta}$$

where K is the constant dependent on the particle shape (0.89 for spherical particles), λ is the wavelength of the X-ray radiation (Cu K_α λ=1.5406 Å), θ is the diffraction angle and β is the peak width (corrected for instrumental broadening).

All EXAFS measurements were carried out at the CCLRC Daresbury Synchrotron Radiation Source on station 9.3. This station is equipped with a specially designed furnace so that samples can be heated *in situ* to 1100 °C. Data were collected at room temperature. The synchrotron has an electron energy of 2 GeV and an average current during the measurements was 150 mA. For dilute samples spectra were collected in fluorescence mode. In some cases we were able to use some samples with a higher concentration of dopant (10 atom per cent) and use the more rapid transmission mode. The EXAFS samples were prepared by thoroughly mixing the powders with fumed silica and pressing into pellets in a 13 mm IR press. Data analysis was performed with interactive computer programmes, developed at the Daresbury SRS [6]. For each spectrum a theoretical fit was obtained by adding shells of atoms around the central excited atom and least-squares iterating the Fermi energy, E_o, the radial distances, *RD*, and the Debye-Waller type factors, $A\ (= 2\sigma^2)$. This latter factor will contain contributions from both thermal disorder and static variations in *RD*. The programme also allowed iteration of the co-ordination number, *CN*, but this option was not used in the present work. The data were Fourier filtered to include the region of 1 – 5 Å of the Fourier transform containing the dominant peaks.

The Raman spectra of the samples were collected using a JY – T64000 (Jobin Yvon) Raman spectrometer equipped with a liquid nitrogen cooled CCD detector. An Argon ion laser with the wavelength of 514.5 nm was used as the excitation source. All the spectra were recorded at room temperature.

3. Results and discussion

The samples investigated were Y and Zr doped tin oxide at levels of 1 and 10 atom per cent. The as-prepared samples had crystallite sizes of ~3 nm. Samples were heated at 200, 400, 600, 800 and 1000 °C for 60 minutes. This caused particle growth and after the 1000 °C anneal the particles had grown to ~40 nm.

3.1. Y-doped SnO_2

K edge EXAFS spectra were collected for 10 atom percent samples. The Fourier transforms (FT's) of the spectra are shown in figure 1. In all the samples, except the one annealed at 1000°C there is clear evidence for only one shell around 2.3 Å. This is the Y-O correlation and is close to the Y-O distance in Y_2O_3. From previous studies of doped tin oxide [7,8] the EXAFS are interpreted, i.e. the lack of a second peak, as evidence, that the Y has not entered the SnO_2 lattice. The samples consist of nanocrystals of SnO_2 with Y in the surface layers. The EXAFS of the sample annealed at 1000 °C shows well-defined peaks out to 5 Å. The spectrum shows an excellent fit (table 1) to the crystallographic data for $Y_2Sn_2O_7$ [9], resulting from a high temperature solid-state reaction.

Journal of Physics: Conference Series **249** (2010) 012054 doi:10.1088/1742-6596/249/1/012054

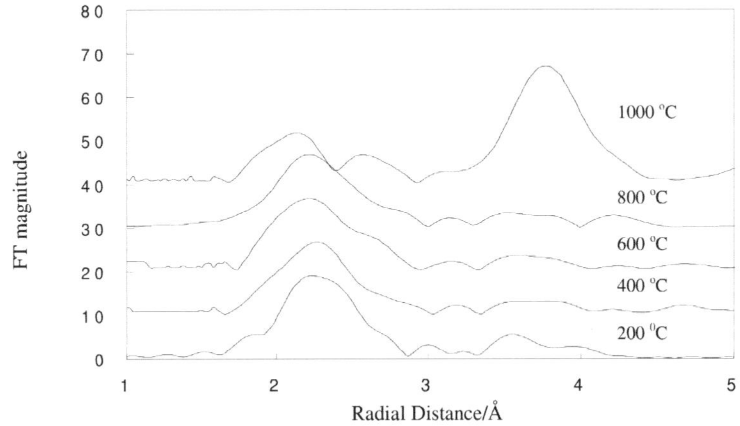

Figure 1: FT of the Y K edge EXAFS for 10% Y doped SnO_2.

Table 1. EXAFS for Y doped SnO_2 after 1000 °C

XRD data for $Y_2Sn_2O_7$			Y K-edge EXAFS	
Atom	**CN**	**R/Å**	**R/Å**	**A/Å^2**
O	2	2.246	2.233	0.012
O	6	2.495	2.427	0.037
Y	6	3.667	3.636	0.017
Sn	6	3.667	3.655	0.026
O	12	4.198	4.087	0.034
O	6	4.660	4.970	0.037
Sn	6	5.186	5.185	0.020

The Raman spectra were also collected for Y doped SnO_2. For SnO_2, three main Raman bands, A_{1g} (630 cm^{-1}), B_{2g} (774 cm^{-1}) and E_g (472 cm^{-1}) are expected as predicted by group theory [10]. A_{1g} and B_{2g} are related to the expansion and contraction vibration mode of the Sn-O bonds whereas the E_g is associated with vibration of oxygen in the oxygen plan. The Raman spectra of yttrium doped nanocrystalline SnO_2 are shown in figure 2. The Raman spectrum (not shown) of the as-prepared sample shows a broad Raman band at 572 cm^{-1} and this is ascribed to the amorphous tin(IV)-hydrous oxide [11]. For the sample heated at 400 ^{0}C the band associated with the amorphous phase is still observed but new peaks start to develop at 630 cm^{-1} and 774 cm^{-1}. At 600 ^{0}C the 572 cm^{-1} peak is no longer visible and at 800 ^{0}C the three Raman active modes i.e. A_{1g}, B_{2g} and E_g are observed although only the A_{1g} band is clearly visible due the particle size being small. For the sample heated at 1000 ^{0}C, all the three bands are clearly visible with the A_{1g} being the most intense. In addition to those associated with SnO_2, more bands at 508 cm^{-1}, 412 cm^{-1} and 311cm^{-1} are observed. These bands resemble the Raman vibrational modes of yttrium stannate ($Y_2Sn_2O_7$) [12]. These results show that at 1000 ^{0}C, SnO_2 is starting to react with Yttrium (Y) to form $Y_2Sn_2O_7$ and are consistent with the EXAFS results.

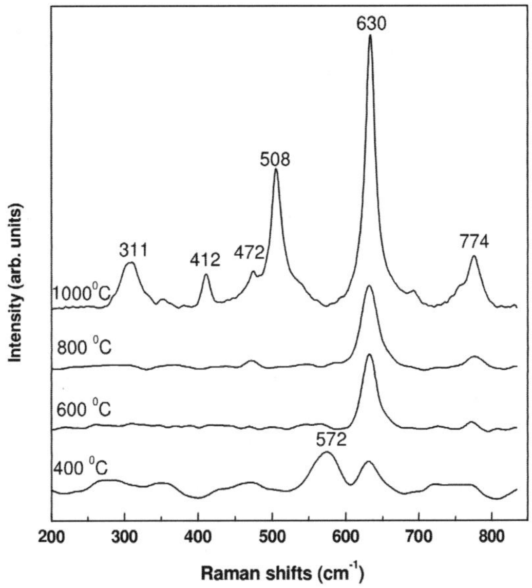

Figure 2. Raman spectra of Y doped nano-SnO_2

3.2. Zr-doped SnO_2 .

K edge EXAFS spectra were collected for 10 atom percent samples. The FT's of the spectra are shown in figure 3. In this case there are clearly resolved peaks out to 5 Å. The spectra show an excellent fit to Zr substituting directly for Sn in the lattice. This is expected given that both cations have 4^+ charge and are similar in size. The Zr-Sn correlations at ~3.2 and 4 Å clearly grow in size with heating. This is directly related to the growth in the particle size on heating [7,8]. The structural parameters for pure SnO_2 are shown in table 2. The results of fitting the Zr K-edge EXAFS for the samples heated at 200 and 1000 °C are shown in table 3 and the similarities with table 2 confirm that Zr is on the Sn site. The differences in the A parameters reflect the particle sizes [7,8].

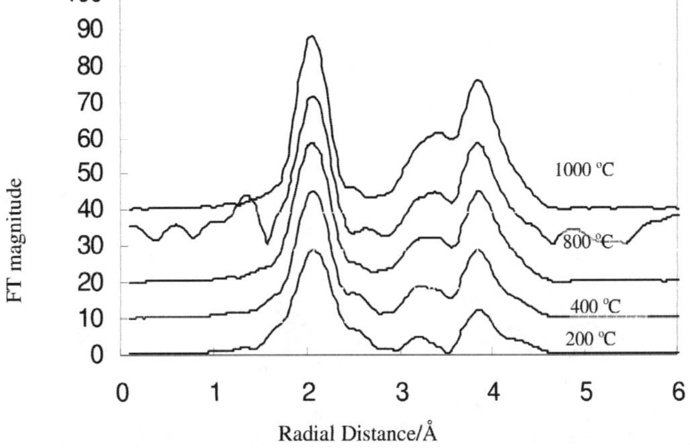

Figure 3: FT of the Zr K edge EXAFS for 10% Zr doped SnO_2.

Table 2. Structural parameters for bulk SnO_2

XRD data for SnO_2			Sn K-edge EXAFS		
Atom	CN	R/Å	CN	R/Å	A/Å^2
O	2	2.045	6	2.050	0.005
O	4	2.088			
Sn	2	3.185	2	3.201	0.007
O	4	3.594	4	3.571	0.001
Sn	8	3.708	8	3.730	0.007

Table 3. Zr K-edge EXAFS for 10 % Zr doped SnO_2

Model		200 °C		1000 °C	
Atom	CN	R/Å	A/Å^2	R/Å	A/Å^2
O	6	2.108	0.017	2.091	0.009
Sn/Zr	2	3.252	0.019	3.213	0.007
O	4	3.597	0.009	3.600	0.011
Sn/Zr	8	3.751	0.036	3.731	0.014

4. Conclusions

There is a clear evidence that Y has not entered the SnO_2 lattice in all Y-doped samples up to the one annealed at 900 ^{0}C. At 1000 ^{0}C, SnO_2 starts to react with Yttrium (Y) to form $Y_2Sn_2O_7$. This is clearly shown by both the EXAFS and Raman scattering results. In all Zr-doped samples, there is a simple substitution for Sn by Zr, even at 10 atom percent doping.

Acknowledgements

We wish to thank the CLRC for providing support for the use of the Daresbury facilities and the Daresbury staff, particularly Dr. Ian Harvey and Dr. Fred Mosselmans for assistance with the synchrotron experiments. KER is grateful for the National Research Foundation for funding.

References

[1] Gleiter H 2000 *Advanced Materials* **48** 1.
[2] Seigel RW and Fougere GE 1995 *Nanostructured Materials* **6** 205.
[3] Henglein A 1989 *Chem. Rev.* **89** 1061.
[4] Efros AL and Rosen M 2000 *Ann. Rev. Mat. Sci.* **30** 475.
[5] Klug HP and Alexander LE 1974 *X-ray Diffraction Procedures* (Wiley: New York).
[6] Binsted N, Campbell JW, Gurman SJ and Stephenson PC 1992 *SERC Daresbury Program Library*, (Daresbury Laboratory, Warrington, Cheshire, UK).
[7] Chadwick AV and Rush GE 2001 '*Nanocrystalline Metals and Oxides: Selected Properties and Applications*', eds P. Knauth and J. Schoonman, (Kluwer, Boston) chapter 5.
[8] Davis SR, Chadwick AV and Wright JD 1997 *J. Phys. Chem. B* **101** 9901.
[9] Facer GR, Howard CJ and Kennedy BJ 1993, *Powder Diffraction* **8** 245.
[10] Yu KN, Xiong Y, Liu Y and Xiong C 1997 *Phys. Rev.* B **55** 2666.
[11] Ristic M, Ivanda M, Popovic S and Music S 2002 *J. Non-Crystalline Solids* **303** 270.
[12] Li K, Li H, Zhang H, Yu R, Wang H and Yan H 2006 *Materials Bulletin* **41** 191.

Holographic recording of surface relief gratings in stilbene azobenzene derivatives at 633 nm

A.Ozols[1], D.Saharov, V.Kokars, V.Kampars, A.Maleckis, G.Mezinskis, A.Pludons

Faculty of Material Science and Applied Chemistry, Riga Technical University, Azenes iela 14/24, LV-1048, Riga, Latvia

E-mail: aozols@latnet.lv

Abstract. Holographic recording in stilbene azobenzene derivatives by He-Ne 633 nm laser light has been experimentally studied. It was found that surface relief gratings (SRG) can be recorded by red light. Usually shorter wavelengths are used to induce the *trans-cis* photo-isomerization in organic materials. SRG with 2 μm period and an amplitude of 130 nm have been recorded with 0.88 W/cm^2 light in about 20 minutes in amorphous films of 3-(4-(bis(2-(trityloxy)ethyl)amino)phenyl)-2-(4-(2-bromo-4-nitrophenyl)diazenyl)phenyl)acrylonitrile spin-coated on glass substrates. Self-diffraction efficiency up to 17.4% and specific recording energy down to 114 J/(cm^2%) were measured. The recorded SRG were stable as proved by subsequent AFM measurements. The photo-induced changes in absorption spectra did not reveal noticeable signs of *trans-cis* transformations. Rather, spectrally uniform bleaching of the films took place. We conclude that a photothermally stimulated photo-destruction of chromophores is responsible for the SRG recording. The recording of stable SRG in the stilbene azobenzene derivatives we studied is accompanied by the recording of relaxing volume-phase gratings due to the photo-orientation of chromophores by the linearly polarized recording light. It should also be noted that holographic recording efficiency in stilbene azobenzene derivatives exhibit an unusual non-monotonic sample storage-time dependence presumably caused by the peculiarities of structural relaxation of the films.

1. Introduction

Holographic recording in stilbene azobenzene derivatives by He-Ne 633 nm laser light has been experimentally studied. Azocompounds are among the most efficient materials for holographic recording [1,2,3,4] that we have studied [5,6]. Commonly, blue and green light is used for recording in these materials because the absorption spectra of azocompounds have maxima at these wavelengths [1]. In this case a *trans-cis-trans* transformation [1,2] of the material under laser radiation takes place. In the red spectral range absorption is considerably lower yet recording is possible [5, 6]. The mechanism of such recording is unclear and this fact creates the motivation for further researche. Discovery of efficient materials for optical recording of information in the red part of spectrum will allow using cheaper red lasers for recording instead of more expensive UV, blue and green lasers. As is known from laser theory, the lasing processes dictates that energy requirements scales as ~λ^{-4} where λ is the wavelength of the laser radiation. Besides, with red light higher-quality of volume-

[1] To whom any correspondence should be addressed.

holographic optical elements can be achieved because of the more homogeneous exposure through the volume.

In this paper, holographic grating (HG) recording at 633 nm has been experimentally studied in organic molecular glasses of stilbene azobenzene derivatives. The HG recording efficiency was measured as a function of the sample storage time before recording. The photo-induced changes in absorption spectra were measured. Atomic force microscope (AFM) measurements were made as well. It was found that efficient surface-relief grating (SRG) recording is possible and that material relaxation has a strong and non-monotonic influence on the holographic recording efficiency.

2. Samples

We have studied different chemical compositions of stilbene azobenzene derivatives. However, in this paper we describe only the results for the most efficient composition whose full chemical designation and chemical structure is given in figure 1. Stilbene azobenzene dervative films were applied to glass substrates by spin-coating. The chromophore substance was dissolved in dichloromethane. A drop of the solution was placed on the rotating glass plate where it was distributed rather evenly. After evaporation of dichloromethane the film formed. Absorption spectrum of the sample is shown in figure 2.

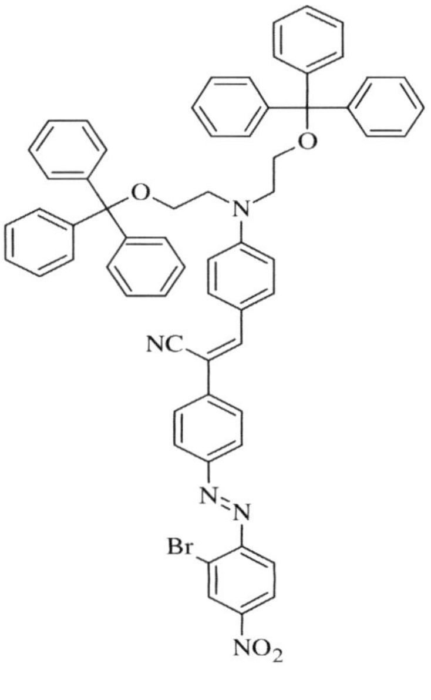

Figure 1. The structural formula of the studied most efficient composition 3-{4-[bis(2-trityloxyethyl)amino]fenyl}-2-{4-[(2-bromo-4-nitrophenyl)diazenyl]phenyl}acrylonitrile.

3. Experimental measurements

The holographic measurements were made with the setup shown in figure 3. Two equally strong He-Ne laser beams P_R and P_S symmetrically incident on the sample recorded the transmission HG.A LGN-222 type He-Ne laser(produced in Ukraine) operating in TEM_{00} mode with horizontal linear polarization was used for recording and readout at λ=632.8 nm.The diffraction efficiency (DE) could not be measured directly due to the fast oscillations of diffracted-beam power that were observed during exposure. Therefore, the measurements had to be made without interruption of one beam. Therefore, we have measured self-diffraction efficiency (SDE) at PR$\neq$ 0, PS $\neq$ 0 instead of DE. Light power measurements were made using a Gentec Solo power meter (Canada). The measurement results were stored in its memory and processed after the recording. This registration method allowed us to measure fast SDE changes during the exposure.

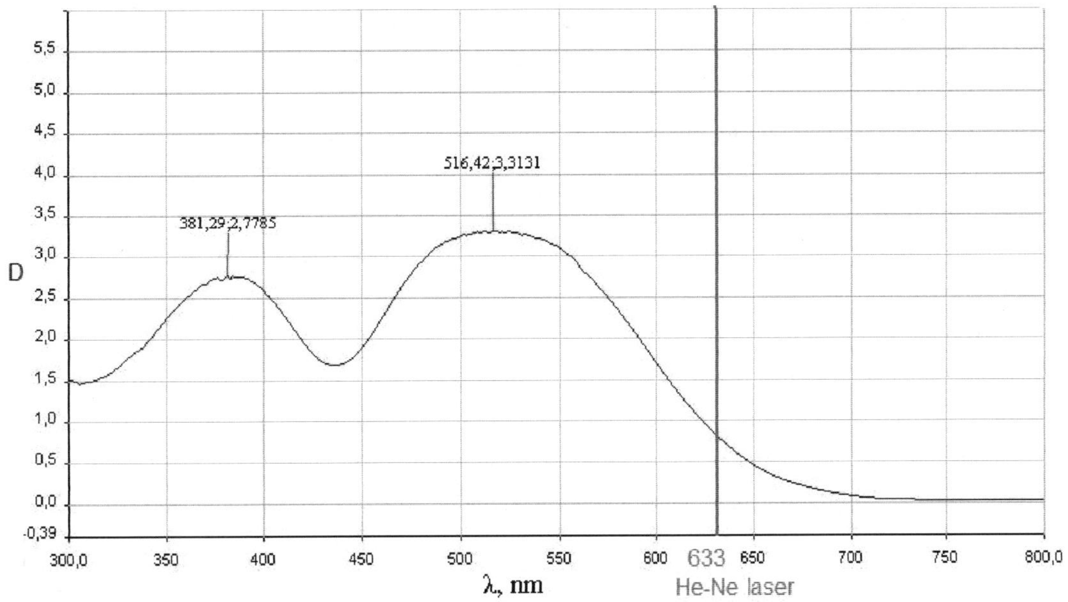

Figure 2. Absorption spectrum of the sample with chemical composition shown in figure 1.*D* is optical density, λ is wavelength.Vertical thick line shows the 632.8 nm line of the He-Ne laser by which the recording and readout of HG were made. Absorption maxima are red-shifted in comparison with azobenzene [1] to be more red- light sensitive.

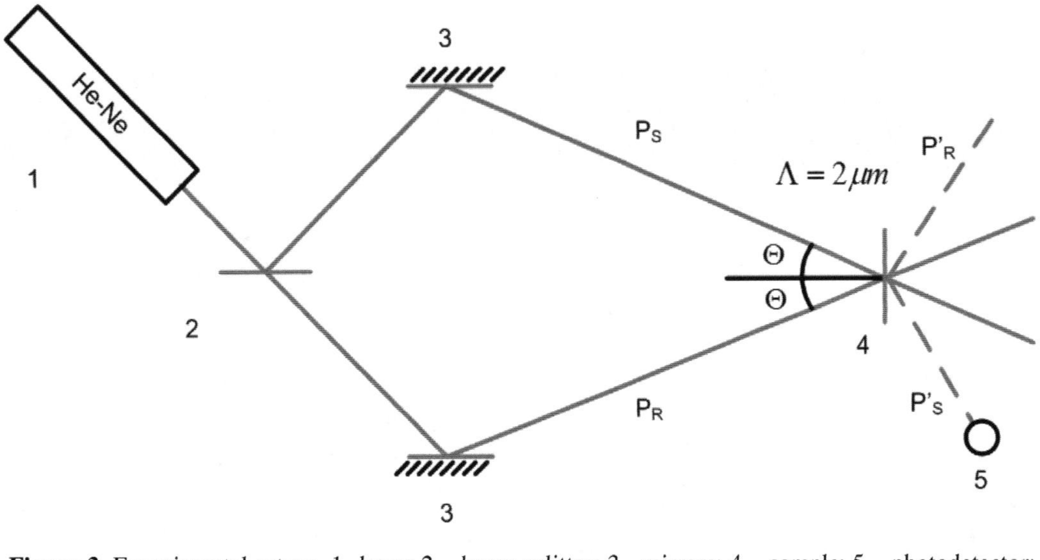

Figure 3. Experimental set-up. 1- laser; 2 – beam splitter; 3 - mirrors; 4 – sample; 5 – photodetector; P_S, P_R – incident beams; Θ – angle of incidence; P'_S, P'_R – diffracted beams; Λ – period of the recording interference pattern and the HG.

We defined SDE according to the formula:

$$SDE = \frac{P_d(k=-2, l=-1)}{P_s}, \tag{1}$$

where $P_d(k=-2,l=-1)$ is the sum of the minus-first-order diffracted power of the P_S beam and of the minus-second-order diffracted power of the P_R beam , P_S is the incident readout light power. If P_R=0, SDE=DE. A detailed optical scheme explaining the recording and readout processes is given in figure 4. The connection between SDE and DE is discussed below.

Non-focused laser beams with a $1/e^2$ (defined by the intensity decrease from the centre) diameter of 1.5 mm were directed to the same point on the sample and formed interference pattern there. The holographic grating period was 2 μm. It was calculated according to the formula

$$\Lambda = \frac{\lambda}{2\sin\Theta}, \tag{2}$$

where Λ is the holographic grating period, λ the recording light wavelength, Θ the angle of incidence of the recording beams. The graphs of the SDE exposure time dependences were created using measurements results.

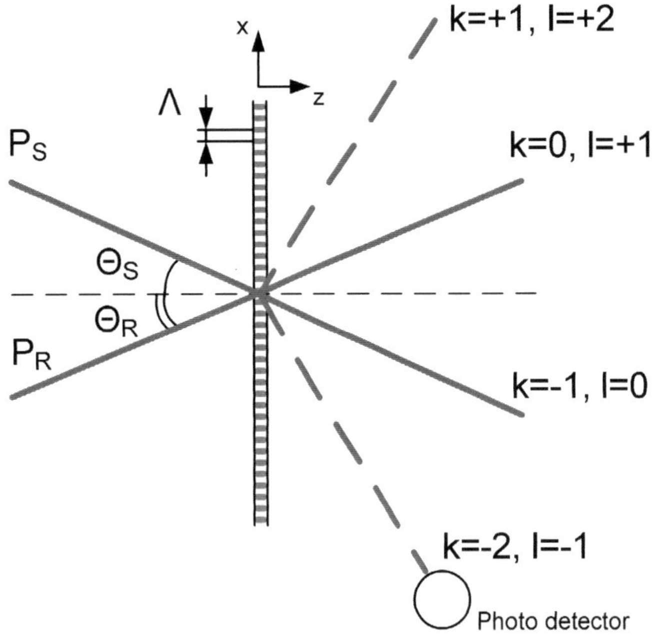

Figure 4. Detailed optical scheme of the recording and readout processes. P_S and P_R are the incident laser beam powers, $\Theta_S = \Theta_R = \Theta$ the angles of incidence of the beams, k the different diffraction orders of the P_R beam, l the different diffraction orders of the P_S beam, Λ=2μm the period of the holographic grating, Δ the grating displacement due to setup instability and/or movement of molecules. $\Delta\frac{2\pi}{\Lambda} = \Phi_S - \Phi_R$, where Φ_S and Φ_R are the phases of the zeroth-order diffracted waves of the P_S and the P_R beam, respectively.

For each SDE graph the specific recording energy W_{max} [J/cm^2%] was calculated according to the formula

$$W_{max} = \frac{It}{\eta_{max}}, \tag{3}$$

where I is the average recording light intensity, t is the exposure time corresponding to the maximal DE η_{max}. (actually, maximal SDE, see below).The average recording light intensity was defined by the formula

$$I = \frac{4(P_R + P_S)}{\pi d^2}, \quad (4)$$

where P_S and P_R are the powers of the recording beams, d is the $1/e^2$ laser beam diameter. The recording light intensity varied from 0.74 to 0.98 W/cm^2.

As mentioned, the self-diffraction efficiency (SDE) was used to estimate the diffraction efficiency (DE). According to calculations made on the basis of thin-hologram theory [7] SDE and DE are related by the expression

$$SDE = DE_{(-1)} + \frac{P_R}{P_S} DE_{(-2)} + 2\left(\frac{P_R}{P_S} DE_{(-1)} DE_{(-2)}\right)^{1/2} \cos(\Phi_S - \Phi_R + \Phi_{S(-1)} - \Phi_{R(-2)}), \quad (5)$$

where P_R and P_S are the incident beams with equal power, $DE_{(-1)}$ is the minus first-order diffraction efficiency, $DE_{(-2)}$ is the minus second-order diffraction efficiency of P_R beam, Φ_S and Φ_R are the phases of the zeroth-order diffracted P_S and P_R beams respectively, $\Phi_{S(-1)}$ and $\Phi_{R(-2)}$ are the phase of the first-order diffracted wave P_S and the phase of the minus-second-order diffracted wave P_R. The first term of the sum in the expression is the diffraction efficiency that needs to be estimated. The third term describes the instability of the induced grating, that can be caused by equipment instability or (and) molecules moving under the influence of light . We have estimated that in our experiments the SDE values were in the range from 0.5 DE(-1) to 1.6 DE(-1).

The photo-induced changes of absorption spectra were measured to elucidate the recording mechanism. Measurements were performed with PERKIN ELMER Instruments Lambda35 UV-VIS spectrometer. There were 25 records made with a diameter of 1.93 mm in the sample area of 5x5 mm2 to fit the entrance slit width of the spectrometer. Apart from holographic and spectral measurements the surface relief of the samples was studied with a scanning atomic force microscope, VEECO model SPMII (USA).

4. Results and discussion

The measurement results are presented in figures 5 to 7. The maximum SDE value was 17.4%. Small dips in SDE time dependences (figure 5) are due to the non-monotonic nature of the chromophore photo-orientation counteracted by thermal back reactions. Large dips are due to the small instabilities of the recording setup causing small shifts of the recording interference pattern with respect to the recorded grating. This behavior was repeatable and follows from the expression (5).

It is also seen from the figure 5 that the sample storage time before recording significantly influences the recording efficiency. There exists an optimal storage time that enables the maximum efficiency, i.e., the DE and SDE maximum, and the specific recording energy minimum. This optimal sample storage time is about 50 days for sample No2 for which the data are presented. At 66 days the efficiency has decreased noticeably. The optimal storage time varies even for different samples of the same chemical composition. For the composition studied in this paper (of the figure 1) it varied from 6 days to 51 day. The physical mechanism of the storage- time effect is not yet clear. We believe that it is due to the peculiarities of structural relaxation of stilbene azobenzene derivatives which form organic molecular glasses consisting of pure chromophores having a large free volume.

AFM measurements (figure 6) show that a surface- relief grating (SRG) is recorded. SRG amplitudes as high as 113 nm has been achieved. The maximum SRG amplitude achieved (in epoxy-based polymers) to-date is 160 nm [9].SRG was stable during a year as shown by repeated AFM measurements. To our knowledge this is the first SRG recording using red light. It should also be noted that HG recording results in a large increse in surface roughness. The average roughness increased from 11 nm to 85 nm for the samples we studied.

In contrast to SRG, DE and SDE decreased in the course of storage. Thus we have concluded that complementary volume phase gratings are recorded simultaneously with SRG and relax in the course of the HG storage over several months. The phase character of these additional volume HGs is shown below by the absorption spectra. The conclusion about the volume- phase gratings is supported by the DE calculations of SRG according to the formula

$$\eta = \frac{\tau}{3}\left[\frac{\pi(n-1)d_{1\max}}{\lambda\cos\theta_i}\right]^2, \tag{6}$$

where τ is the transmittance of the grating, n its average refractive index, d_{1max} is the amplitude of thickness modulation, λ the readout wavelength, θ_i is the readout beam incidence angle inside the medium. This formula (6) was derived by us on the basis of thin- hologram theory [7]. The calculation at $\tau = 0.398$, $n = 1.5$, $d_{1max} = 113$ nm, $\lambda = 633$ nm, $\cos\theta_i = 1$ gives $\eta = 1.04$ % ,which is much less than the maximum SDE value of 17.4 %.

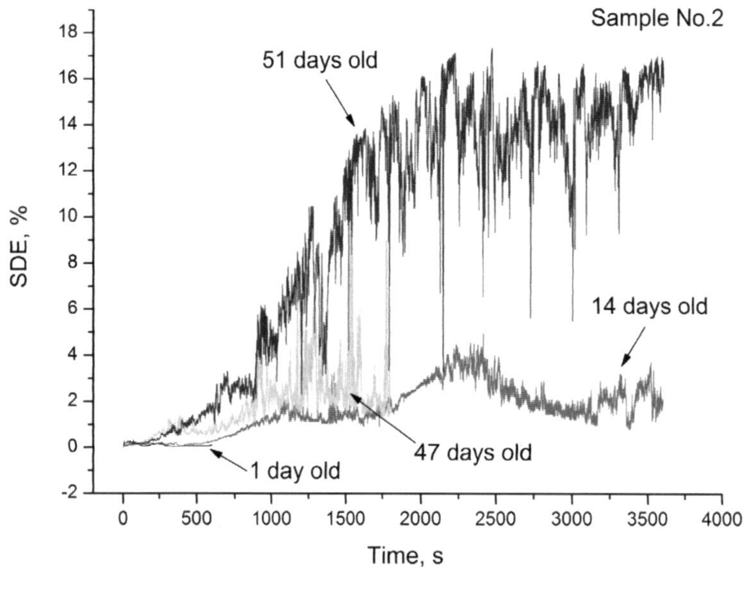

Figure 5. Self-diffraction efficiency (SDE) exposure- time dependences for different sample No2 storage times (marked by arrows) before the recording. The recording light intensity was 0.88 W/cm^2.

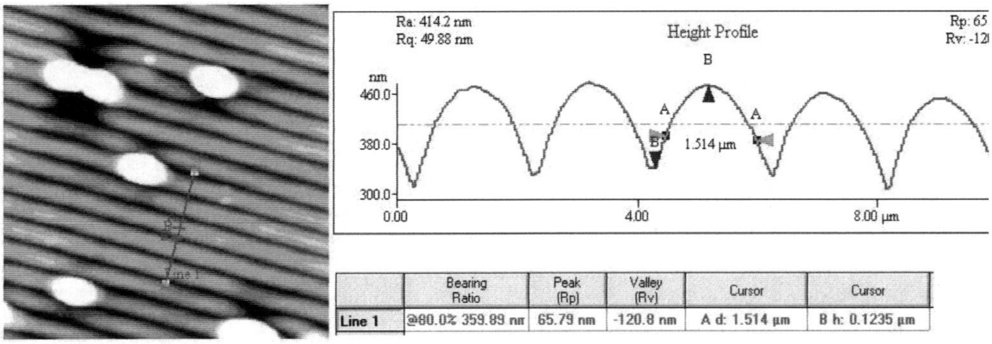

Figure 6. Surface relief and its profile in the central area of the recording. The SRG amplitude is 113 nm, the maximum surface- relief change is 226 nm.

The photo-induced changes in absorption of the sample during holographic recording are shown in figure 7. Photo-induced bleaching is mainly observed. As can be seen, the absorption changes at 633 nm are negligible and a pure phase recording took place. At shorter wavelengths amplitude recording can take place as well.

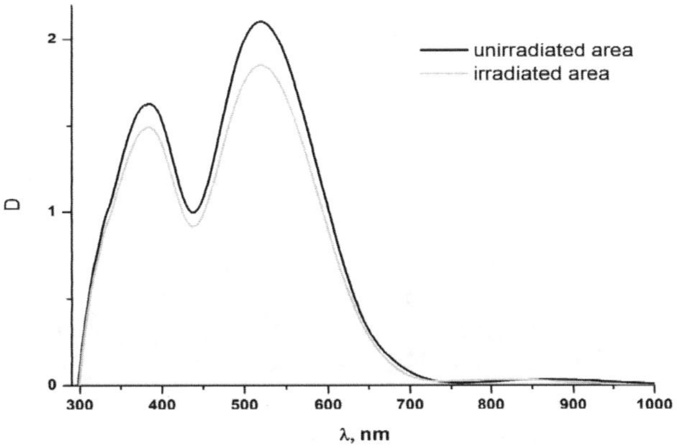

Figure 7. Photo-induced changes of absorption spectrum produced by 633 nm light.

Figure 7 shows that optical density has decreased almost uniformly over the spectrum, except for a a little increase at long wavelengths beyond 700 nm. This is in contrast with known *trans-cis* spectral transformations in UV and visible bands [8]. Besides,the *cis-trans* back reaction usually proceeds on time scales from seconds to hours [1,8], and, therefore, it was finished when the absorption spectra were measured about two months after recording. Thus, we cannot exclude the *trans-cis* photo-isomerization and the mass transfer by electric gradient force [1] or other mechanisms but these seem unlikely at 633 nm. Yet, further experiments are needed to draw a conclusion about the contribution of photo-isomerization. Mostly uniform reduction of the absorbance after recording can be explained by the photo-degradation, i.e., by chromophore destruction under irradiation during the recording process. For example, N-N bonds can be broken by 633 nm photons [9]. Photoablation is also possible, but it needs recording intensities higher than 1 W/cm^2 [9].

Calculations show that photo-degradation in our experiments can be photo-thermally stimulated. The photoinduced temperature increase of a thin film on a bulk substrate by a CW laser in the center of the beam is [10]

$$\Delta T = \frac{KIr}{k}, \quad (7)$$

where K is the absolute absorption coefficient, I is light intensity, $r=d/2$ is the laser beam radius, k is the heat conductivity of the film. In our case $K = 0.64$, $I = 0.88$ W/cm^2, $r = 0.965$ mm, $k = 0.08$ W/mK (the value for polystyrene which we assumed to be approximately equal to the unknown heat conductivity of our film) and $\Delta T = 69$ K. Thus the temperature of sample can be close to the glass transition temperature $T_g = 94^0$C of the film if the room temperature is 20^0C.

There are different possible recording mechanisms of the complementary volume- phase HG. The most probable is the photo-orientation mechanism [5,6,9]. Chromophore groups and fragments of molecules that have dipole moments can change their orientations and spatial positions under the influence of linearly polarized light. As a result, the refraction and absorption indices change. A spatially periodic density modulation accompanying photo-degradation can also contribute. Practical aspects of efficient red- light SRG recording include simpler and cheaper production of holographic optical elements – gratings, couplers, waveplates, antireflection coatings, etc.

5. Conclusions

Efficient stable SRG recording with the surface relief modulation amplitudes up to 113 nm has been achieved in stilbene azobenzene derivatives at 633 nm. To our knowledge, this is the first direct SRG recording with red light. Usually, UV, blue or green light is used.

Photo-induced absorption spectra changes are consistent with assuming that photo-thermally stimulated degradation of the films is the main recording mechanism. However, the contribution of *trans-cis* photo-isomerization cannot be excluded.

SRG recording is accompanied by volume-phase holographic grating (HG) recording that slowly relaxes in the course of the sample storage over several months. These volume HGs are, most probably, recorded due to the photo-orientation of chromophores by linearly polarized *p*-light.

Holographic recording efficiency in stilbene azobenzene derivatives is a non-monotonic function of the sample storage time. An optimal storage time exists after sample preparation that enables the maximum efficiency. This optimal storage-time is not uniquely determined by the chemical composition because even for the same composition it can be different. Most probably, this storage time effect is caused by peculiarities in the structural relaxation of stilbene azobenzene derivatives which form organic molecular glasses consisting of pure chromophores with a large free volume.

6. References

[1] Sekkat Z and Knoll W 2002 *Photoreactive Organic Thin Films* (Orlando, FL: Elsevier Science)
[2] Schwartz K 1993 *The Physics of Optical Recording* (Berlin: Springer Verlag)
[3] Delaire J A and Nakatani K 2000 Linear and nonlinear optical properties of photochromic molecules and materials *Chem Rev.* **100** 1817-1845
[4] Rasmussen P H, Ramanujam P S, Hvilsted S and Berg R H 1999 A remarkably efficient azobenzene peptide for holographic information storage *J.Am. Chem.* **121** 4738-4743
[5] Ozols A, Kampars V, Reinfelde M, Kokars V 2003 Hologram recording in azobenzene oligomers *Proc. SPIE* **5123** 102-109
[6] Ozols A, Kampars V, Reinfelde M, Kokars V 2005 Structure optimization of azobenzene oligomers for holography *Phys. Stat. Sol.(c)* **1** 673-676
[7]Ozols A 977 Diffraction efficiency of thin amplitude-phase holograms. *Opt. Spektrosk.* **42**, 168-174
[8] Rau H. Photoisomerization and photoorientation of azobenzenes, in *Photoreactive Organic Thin Films* , edited by Sekkat Z and Knoll W 2002(California, Academic Press) pp.3-40
[9] Oliveira O N, Kumar L L, Tripathy S K Surface-relief gratings on azobenzene-containing films, in *Photoreactive Organic Thin Films* , edited by Sekkat Z and Knoll W 2002 (California, Academic Press) pp.429-483
[10] Stradins P, Shvarts K, Teteris J 1989 The relation of photo and thermal components of photoinduced changes in amorphous semiconductors *J. of Non-Cryst. Solids* **114** , 79-81

Journal of Physics: Conference Series **249** (2010) 012056 doi:10.1088/1742-6596/249/1/012056

A Combined Conductivity and XAS Study of Plastically Crystalline Electrolytes

A.V. Chadwick, A. Berko. A.N. Blacklocks and W. Edwards

Functional Materials Group, School of Physical Sciences, University of Kent, Canterbury, Kent CT2 7NH, UK.

a.v.chadwick@kent.ac.uk

Abstract. We report a study of the conductivity of LiBr and copper salt doped succinonitrile along with X-ray absorption measurements for the bromide and copper ions. The work shows that the ions are in well-ordered sites for most of the plastic phase and relatively high conductivity is due to a small fraction of the ions which are mobile.

1. Introduction

Plastic crystals are a class of molecular materials [1, 2] that consist of roughly spherical or globular molecules; thus the simplest examples are methane (CH_4) and carbon tetrachloride (CCl_4) and more complex systems are the bridged cyclic terpenoids, such as camphor and camphene. The materials are characterized by a solid state phase transition or transitions. In the lowest temperature non-plastic phase the materials are typical organic solids with low crystallographic symmetries and are mechanical brittle and friable. In the solid phase prior to melting the solids are soft, hence the name plastic crystals, and show relatively rapid translational diffusion [2]. In the plastic phase the molecules are undergoing rapid endospherical reorientation and are often referred to as rotator phase solids. In addition, the plastic phases tend to be high symmetry, either fcc, hcp or bcc. On melting there are relatively small increases in the entropy as the only gain is translational freedom.

Two types of plastically crystalline electrolyte are possible. One type is when the ion or ions of the salt are exhibiting plastic behaviour, i.e. one or both ions are undergoing rapid rotation, the conductivity being improved by doping with a salt [3, 4]. The second is when a plastic solid with a high dielectric constant dissolves a salt [5, 6]. Both types can have solid state conductivities at room temperature in excess of 10^{-3} S cm-1, and coupled with appropriate mechanical properties, have potential applications as the electrolytes in solid-state batteries. They offer an alternative to solid polymer electrolytes as the conductivities are higher and the cation transport numbers appear higher.

We have just initiated a programme of XAS studies on plastically crystalline electrolytes and the results presented here are for succinonitrile electrolytes. In the pure material the plastic phase extends from 238 to 335K and the dielectric constant is 55, i.e. higher than that for many organic solvents (e.g. acetonitrile has a dielectric constant of 36.6). Succinonitrile will dissolve a wide range of salts and in the non-plastic phase the ionic conductivity is low, typically ~ 10^{-6} S cm^{-1}. On heating the conductivity of the solutions rapidly rises to 10^{-4} to 10^{-3} S cm^{-1} at 300K and then there is a further less pronounced increase up to the melting point. This indicates a link between the molecular motion of the organic molecule, rotation and possibly translational diffusion, with the migration of the ions. We will report ionic conductivity and X-ray absorption spectroscopy (XAS) results for lithium and copper salt doped succinonitrile.

2. Experimental

2.1. Materials

All preparation work was carried out in a glove box under inert nitrogen atmosphere (water ≤ 1ppm). The basic components for the plastic crystalline electrolytes were mixed in the appropriate compositions. The composite electrolytes were loaded into 2 × 4 in poly bags (Aldrich Z183377-1PAK) cut into 20 × 20mm squares and heat-sealed before removal from the glove box for XAS measurements.

Impure succinonitrile (Aldrich, 99%) was purified by zone-refing and dried with molecular sieve. The salts used were all purchased from Aldrich, except for the copper(I) trifluoromethanesulfonimide and the copper(II) trifluoromethanesulfonimide (Cu(I) TFSI and Cu(II) TFSI, respectively), and are listed below:-

Lithium bromide, LiBr (99.995+%, Aldrich).

Copper(I) trifluoromethanesulfonate, available as a 1:1 complex with toluene, copper(I) triflate (99.99%, Aldrich).

Copper(II) trifluoromethanesulfonate, copper(II) triflate (≥97%, Aldrich).

The preparation of Cu(I) TFSI and Cu(II) TFSI followed the procedures used to synthesise these materials for conductivity studies [3]. Detailed reaction conditions were supplied by Professor Armand [7]. The Cu(II) TFSI was synthesised by boiling copper(II) oxide, CuO (99.99%, Aldrich) under reflux with bis(trifluoromethane)sulfonimide (≥95.0%, Aldrich) in distilled water. The water was then carefully distilled off and the Cu(II) TFSI dried under vacuum. The Cu(I) TFSI was synthesised by boiling Cu(II) TFSI under reflux with copper metal-60-100 mesh.

The samples discussed here contained 1 mole per cent dopant in succinonitrile.

2.2. Conductivity measurements

The conductivity cell was constructed from brass and PTFE. The sample was sandwiched between two carbon discs separated by a small PTFE ring producing a sample disc of a consistent geometric dimensions (thickness = 0.88 mm, diameter = 12.75 mm). The cell was contained within a Pyrex tube with electrical feeds. Temperature was measured with a chrome-alumel thermocouple. Conductivity measurements were carried out at 50 kHz using a 4192ALF Impedance Analyzer 5Hz-13MHz, Hewlett-Packard. Tests showed that there was no appreciable frequency dependence of the conductivity up to 1MHz.

The samples were melted and loaded between the two carbon discs in a glove box to avoid moisture. This was then loaded into the cell, which was in turn loaded into the Pyrex tube and all openings sealed. The entire setup was then removed from the glove box and connected to a nitrogen supply. The sample was then heated to above the melting point of the pure plastic crystal under investigation and then left to cool for a day. Conductivity measurements were then made at approximately every 10K from room temperature (in the case of NPG) to above the melting point of the pure crystal and then back again. For succinonitrile, which has its plastic phase transition at 238K, measurements were made from approximately 210K. The sample was left to equilibrate for 1 hour before a measurement was made at each temperature. The sample temperature was then recycled to check reproducibility.

2.3. XAS measurements

Measurements were made on station 9.3 at the CLRC Daresbury Synchrotron Radiation Source. The synchrotron has electron energy of 2 GeV and the average current during the measurements was 150 mA. Station 9.3 is equipped with a double Si (220) or Si (111) crystal monochromator that can be offset from the Bragg angle to reject harmonic contaminants from the monochromatic beam. The Si (111) was used for the softer edges (3d metals) and the Si (220) for the harder edges (Rb and Br). In the present work the harmonic rejection was set at 50%. The detectors were standard ion chambers for transmission experiments and a 13 element solid state detector for fluorescence experiments. The poly bag sample containers were mounted in an Oxford cryostat specially designed for XAS experiments. There are Mylar windows on four sides of the sample; an entry and exit for the X-ray beam and windows to detect the fluorescence X-rays. Spectra were collected at various temperatures, depending on the system. The cryostat was capable of maintaining samples to ±0.1°C over the temperature range -193 to 100°C.

The data were processed in the conventional manner using the Daresbury suite of EXAFS programmes: EXCALIB, EXBACK (or EXBROOK) and EXCURV98 [8,9]. Phase shifts were derived from *ab initio* calculations within EXCURV98. This code also includes routines to treat multiple scattering effects in highly symmetric structures. For each spectrum a theoretical fit was obtained by adding shells of atoms around the central excited atom and least-squares iterating the Fermi energy (E_o), the radial distances (*RD*), and the Debye-Waller type factors, A ($=2\sigma^2$). This latter factor will contain contributions from both thermal disorder and static variations in *RD*. The programme also allowed iteration of the co-ordination number (*CN*). The quality of the fit is measured by an *R*-factor and the errors in *RD* are ~± 0.02 Å and ~± 20% in *A* and *CN*.

3. Results and Discussion

As will be seen later the results for the lithium bromide and copper salt doped succinonitrile presented significant differences. We will begin by presenting the results for the lithium bromide doped system.

The conductivity results for 1 mole per cent LiBr doped succinonitrile are shown in Figure 1. The plots show similar features to the data reported by Long et al [5] for a series of lithium salts dissolved

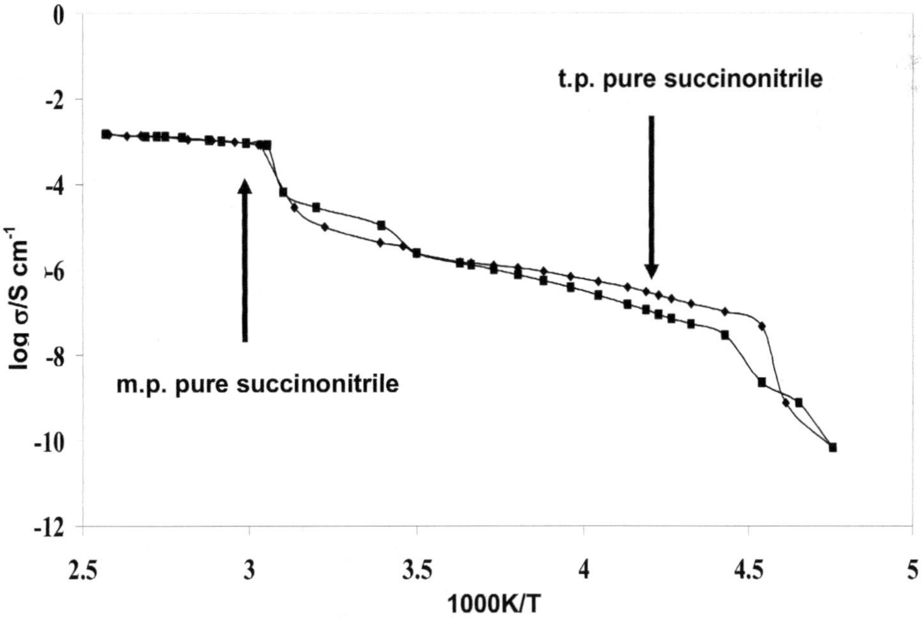

Figure1 Conductivity plots of 1 mole per cent LiBr in succinonitrile.

in succinonitrile. There is clearly a change in conductivity on changing from the plastic to non-plastic phase, although the transition temperature appears to be considerably reduced in the conductivity measurements. In most of the plastic phase the conductivity plot is roughly linear; i.e. 1000K/T from 4.5 to 3.3. What is striking is the rapid rise in the conductivity in the temperature regime prior to melting. This suggests that there is some onset of disorder prior to melting.

Br K-edge XAS data were collected in transmission mode at four temperatures, namely 200K (non-plastic phase), 273K (plastic phase), 323K (plastic phase) and 343K (melt) for the 1 mole per cent doped sample. The EXAFS and Fourier transform for the sample measured at 200K are shown in Figure 2. The spectrum was fitted to a simple model where the Br ion was coordinated to four succinonitrile molecules through the CN grouping. A comparison of the spectra at different temperatures is shown in Figure 3 which plots the Fourier transforms of the spectra. It can be seen that

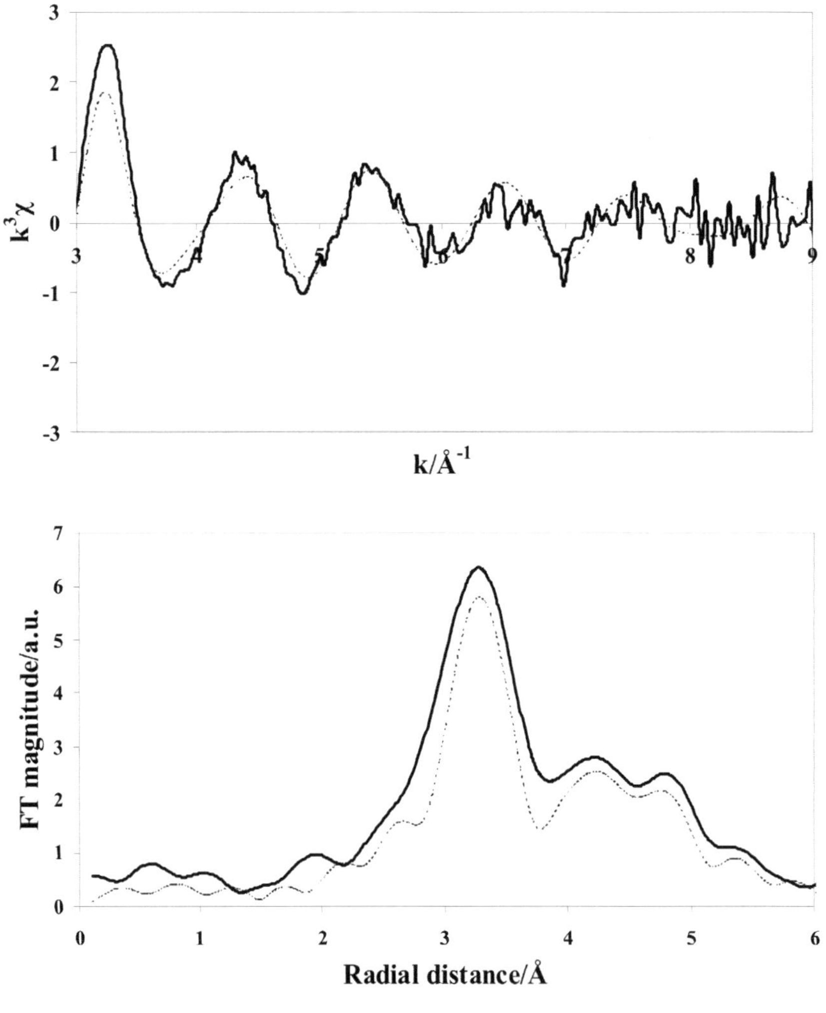

Figure 2 The Br K-edge EXAFS of 1 mole percent LiBr in succinonitrile at 200K; (a) normalised EXAFS, (b) the Fourier transform. The solid line is experiment and the dotted line the best-fit.

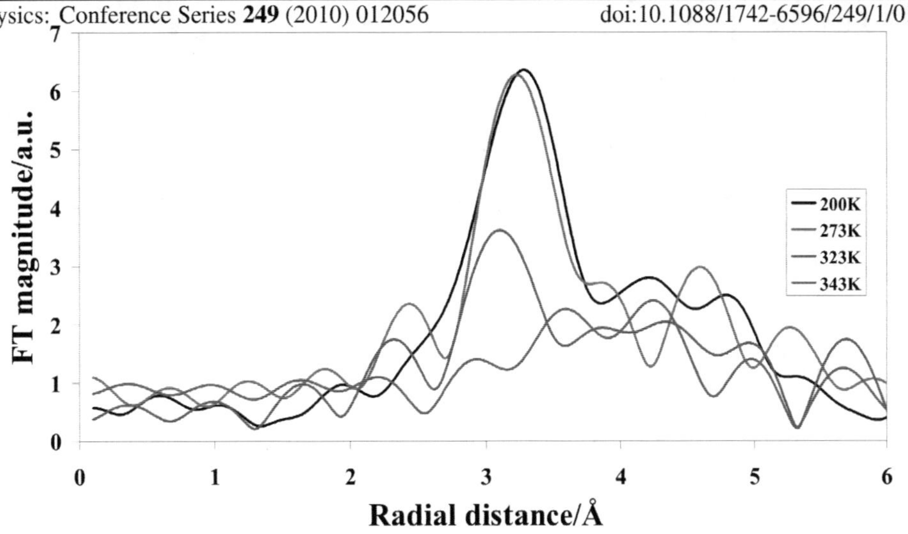

Figure 3 The Br K-edge EXAFS of 1 mole percent LiBr in succinonitrile at a range of temperatures

there is no major change in going from non-plastic to plastic phase. However, the peaks are attenuated at 323K (1000K/T = 3.1) and are lost in the molten sample. We have performed a detailed fitting of the spectra but only a qualitative interpretation is needed here. The attenuation of the peaks is due to a loss of order in the system and is therefore consistent with the conductivity results.

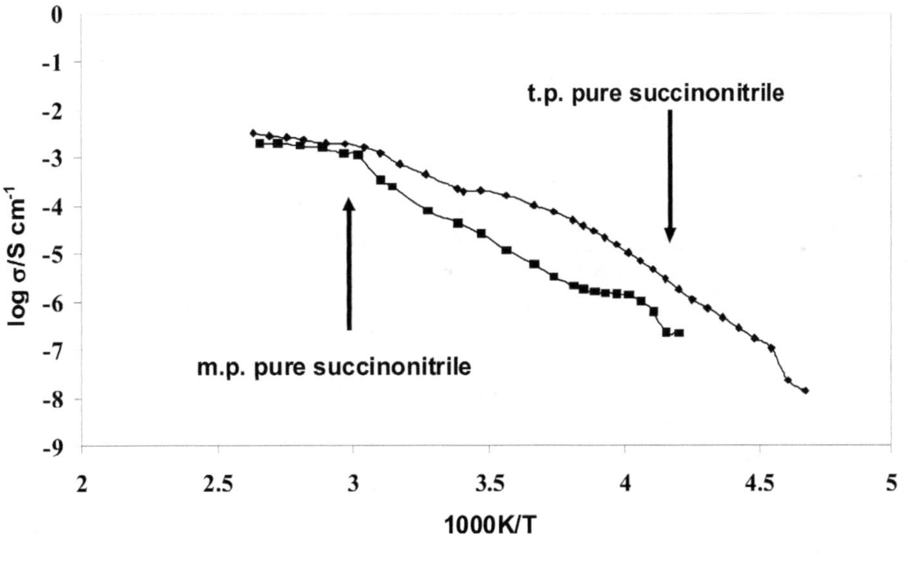

Figure 4. Conductivity plots of copper salts in succinonitrile; 1 mole percent Cu(I) TFSI (upper) and Cu(II) TFSI (lower) in succinonitrile.

The conductivity results for some of the copper salt doped succinonitrile samples are shown in Figure 4. The results are in reasonable agreement with the published data for these systems [3]. In these systems the plots were roughly linear in the plastic phase, unlike the plots for lithium bromide.

The Cu K-edge EXAFS of all the copper salt doped samples showed remarkably similar spectra. The results for 1 mole per cent Cu(II)TFSI sample are shown in Figure 5. All the systems showed three dominant peaks in the Fourier transforms which were modeled with four succinonitrile molecules coordinated to the copper ion. In addition there was remarkably little attenuation of the spectra on heating and even on melting. This indicates that there is a strong complex formed by the copper ion and the host. This is consistent with the linearity of the conductivity plots

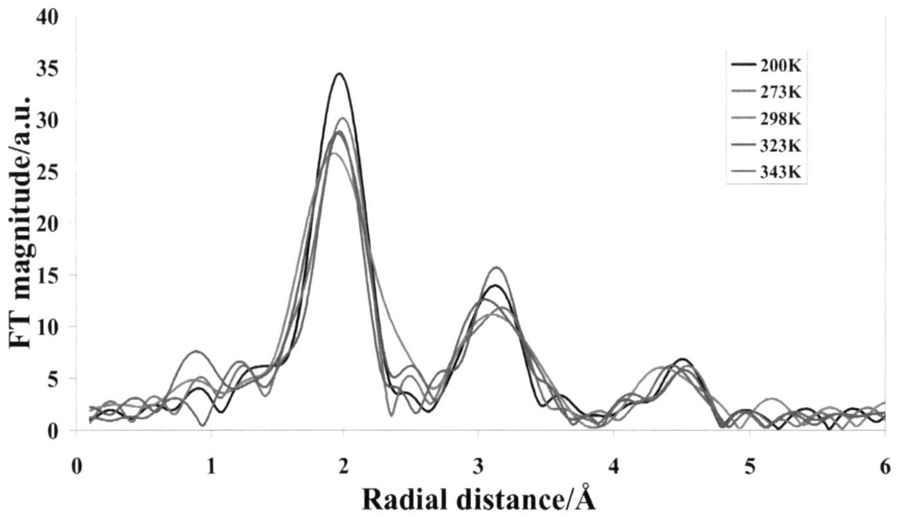

Figure 5 The Fourier transform of the Cu K-edge EXAFS of 1 mole percent Cu(II)TFSI in succinonitrile at a range of temperatures.

The present results for both the LiBr and the copper salt doped succinonitrile suggest that the systems are highly ordered for most of the plastic phase; in the former case to within ~ 50K of the melting point and in the latter case the whole of the plastic phase. Unfortunately, we could not explore the Li^+ or anion environments with XAS and the differences may reflect the stronger binding between copper ions and the host compared to bromide ions and the host. The overall order in the systems raises interesting questions concerning the mechanism of the ionic transport. The intuitive assumption would be assign the relatively high conductivities to the high levels of rotational disorder and the rapid diffusion of the host molecules; i.e. similar to a molten salt. Indeed the diffusion studies show for lithium salts in succinonitrile that the diffusion of anions, cations and host molecules, albeit at temperature within 40K of the melting point, are similar in magnitude. However, succinonitrile is a complex material showing 'extrinsic' rather than intrinsic plastic crystal behaviour [2]. Succinonitrile molecules have *trans* and *gauche* forms and can be considered as 'self-doped'. The current XAS results suggest that in the region of the dopant ions there is a high degree of order, except at temperatures close to the melting point for the case of lithium bromide doping. In this case the thermal disorder is presumably sufficient to 'break' the interactions between ion and host. Hence the majority of the ions, as XAS monitors the average structure of the local environment, is in highly ordered sites and is relatively immobile. This suggests that conductivity may be due to a small fraction of ions (<10%) that dissociate from the host molecules and are mobile.

4. Conclusions

We have reported conductivity and XAS results for LiBr and copper salt doped succinonitrile. The conductivity results are in reasonable agreement with previous studies [3,5] of the systems and are sufficiently high at ambient temperatures (10^{-4} to 10^{-3} S cm^{-1}) to be considered for application as electrolytes in solid state batteries [3,10]. The XAS results show that the ions studied here, Br^- in LiBr doped succinonitrile and Cu^{+} and Cu^{2+} in copper salt doped succinonitrile, are in highly ordered environments for the bulk of the plastic phase. This suggests that only a small fraction of the ions in the systems are responsible for the relatively high ionic conductivity.

Acknowledgements
We wish to thank the EPRSRC for studentship support (ANB and WE) and to staff at the Daresbury SRS, notably Dr Ian Harvey, for help with the XAS experiments.

References

[1] Timmermans, J. J. Phys. Chem. Solids,**18** (1961) 1.
[2] Sherwood, J.N., (Ed.), The Plastically Crystalline State, Wiley-Interscience, 1979.
[3] Alarco, P.J, Abu-Lebdeh, Y., Abouimrane, A., Armand, M. Nature Materials **3** (2004) 476.
[4] MacFarlane, D.R., Forsyth, M., Adv. Mater. **13** (2001) 957.
[5] Long, S., MacFarlane, D.R. Forsyth, M., Solid State Ionics **175** (2004) 733.
[6] MacFarlane, D.R, Huang, J.H., Forsyth, M., Nature **402** (1999) 792.
[7] Armand, M.A., private communication.
[8] Gurman, S. J., Binsted, N. and Ross, I., J. Phys. C: Solid State Phys., **17**, (1984) 143
[9] Binsted, N., EXCURV98: CCLRC Daresbury Laboratory computer program (1998).
[10] Abouimrane, A., Abu-Lebdeb, Y., Alarco, P.J. and Armand, M., J. Electrochem. Soc., **151**, (2004) A1028.

Applicability of Topaz Composites to Electron Dosimetry

K S Bomfim and D N Souza
Departamento de Física, Universidade Federal de Sergipe
São Cristóvão, SE, Brazil, 49100 000

E-mail: dnsouza@fisica.ufs.br

Abstract. Thermoluminescent dosimetric topaz properties have been investigated and the results have shown that this mineral presents characteristics of a good dosimeter mainly in doses evaluation in radiotherapy with photons beams in radiotherapy. Typical applications of thermoluminescent dosimeters in radiotherapy are: *in vivo* dosimetry on patients (either as a routine quality assurance procedure or for dose monitoring in special cases); verification of treatment techniques; dosimetry audits; and comparisons among hospitals. The mean aim of this work was to evaluate the efficiency of topaz-Teflon pellets as thermoluminescent dosimeters in high-energy electron beams used to radiotherapy. Topaz-Teflon pellets were used as TLD.

1. Introduction

Thermoluminescent dosimetric topaz properties have been investigated for some years and the results have shown that this mineral presents characteristics of a good dosimeter mainly in doses evaluation in radiotherapy with photons beams in radiotherapy [1-4]. In a comparative analysis between the TL response presented by topaz pellets and the conventional dosimetric procedures in radiotherapy of a 6 MV x-ray beams performed with a ionizing chamber Magalhães et al. observed that the topaz-Teflon pellets show a similar behavior to the results obtained by the conventional dosimetry, because the dose profiles obtained in equivalent tissue material had a good agreement [4].

Nowadays, typical applications of thermoluminescent dosimeters (TLD) in radiotherapy are: *in vivo* dosimetry on patients (either as a routine quality assurance procedure or for dose monitoring in special cases); verification of treatment techniques; dosimetry audits; and comparisons among hospitals [5, 6].

Megavoltage electron beams represent an important treatment modality in modern radiotherapy, often providing a unique option in the treatment of superficial tumors [5]. As the TLD are already widely employed in dosimetry to photons beams is expect the interest in these dosimeters electron beams monitoring. However, unlike in MV photon beams over regions of electron equilibrium, the response of a TL detector placed in a phantom exposed to a beam of electrons in the megavoltage energy range may depend on several factors, such as the mean electron energy at the phantom and detector surfaces, detector size, density of the detector and of the phantom medium or depth in the phantom at which the detector is irradiated [6].

The mean aim of this work was to evaluate the efficiency of topaz-Teflon pellets as thermoluminescent dosimeters in high-energy electron beams.

2. Materials and Methods

Topaz-Teflon pellets of size 6 mm diameter × 1 mm thickness were used as thermoluminescent dosimeters. The preparation of topaz-Teflon was as described by Souza et al. in a previous paper [3].

The simulations in electron beams were performed on a linear accelerator from Radiotherapy Service of the Hospital de Urgência de Sergipe, using megavoltage electron beams (8, 10, 12 and 14 MeV) at the source surface distance of 100 cm for 10×10 cm^2 for the 5x5 cm^2 radiation fields.

The electron beam central axis depth dose curves were obtained by irradiation of topaz-Teflon on diverse depths under acrylic plates. To each irradiation a group of 5 detectors were irradiated with absorbed doses in the range from 10 to 200cGy.

The TL measurements were performed in Harshaw 3500 equipment using a linear heating program with a rate of 10°C/s between 25°C up to 300°C.

3. Results

A good thermoluminescent dosimeter (TLD) must show a thermoluminescent peak at high enough temperature so as not to be affected by room temperature and still low enough so as not to interfere with black body emission from the heating planchet of the thermoluminescent equipment that occurs near at 400°C.

Figure 1 shows the typical thermoluminescent glow curve of the topaz-Teflon after irradiation with dose of 2 Gy from a 10 MeV electron beam at the depth of maximum dose (22 mm) under acrylic plates. We can observe that the maximum TL peak intensity occurs on 200°C. The glow curves exhibited by other pellets irradiated using 10, 12 and 14 MeV they were similar to this curve.

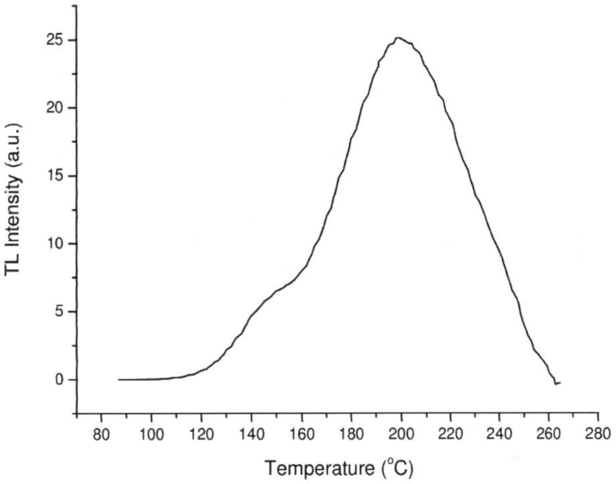

Figure 1. Thermoluminescent glow curve of Topaz-Teflon pellets irradiated with electron beams (8 MeV).

Figure 2 shows the TL response of the topaz-Teflon dosimeters as a function of absorbed dose from electron beams of 8 and 14MeV, respectively. The figure 3 shows 2 shows the TL response for the energies of 12 and 14 MeV. In both figures, the calibration curves presented responses proportional to absorbed dose between 10^{-1} Gy and 2 Gy. The curves are useful in the whole tested dose range and no dose saturation was observed. Meanwhile, a strong energy dependence can be also observed, the samples irradiated with 10 MeV showed a more intense TL response. Ideally, the energy response should be flat; the system calibration should be independent of energy over a certain range of radiation qualities, consequently for use of topaz-Teflon in monitoring doses from electron beams the energy correction must be included. The measurement uncertainties in those studies were always below 5%.

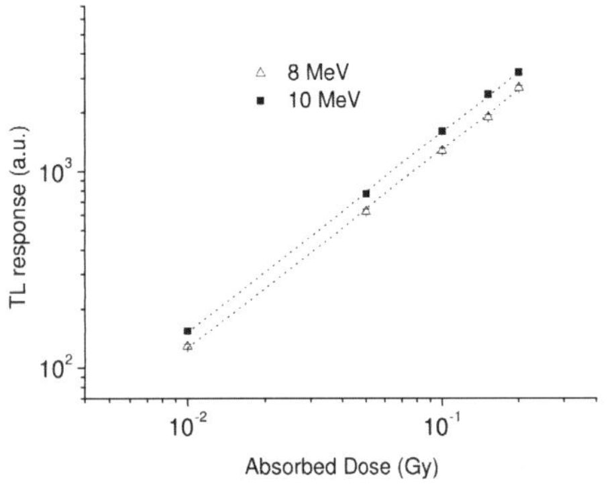

Figure 3. Dose response of Topaz-Teflon pellets irradiated by electron beams (8 and 10 MeV).

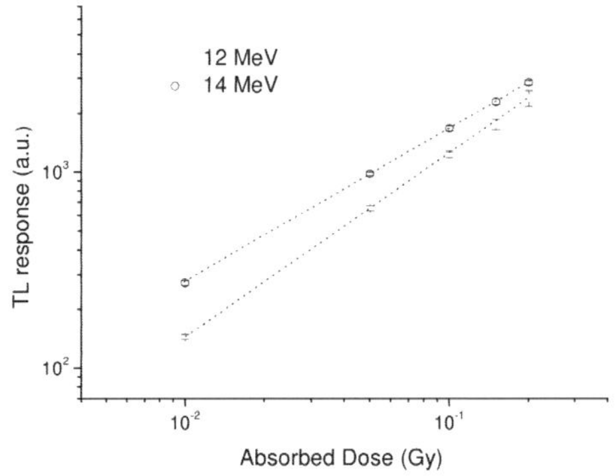

Figure 3. Dose response of Topaz-Teflon pellets irradiated by electron beams (12 and 14 MeV).

Figure 3 shows the result of energy dependence. The values were normalized to the TL response of the samples irradiated with 2 Gy in 8 MeV. The dependence of the TL and on the energy of X-rays between 8 and 14 MeV were less than 20% Samples exposed to the beam of 12 MeV showed the TL response of lesser intensity and when the pellets were exposed at 10 MeV beam they exhibit the highest intensity, 96% and 120% of the TL intensity when compared to dosimeters irradiated at the 8 MeV, respectively. The uncertainties associated with the measures were no more than 6.2%.

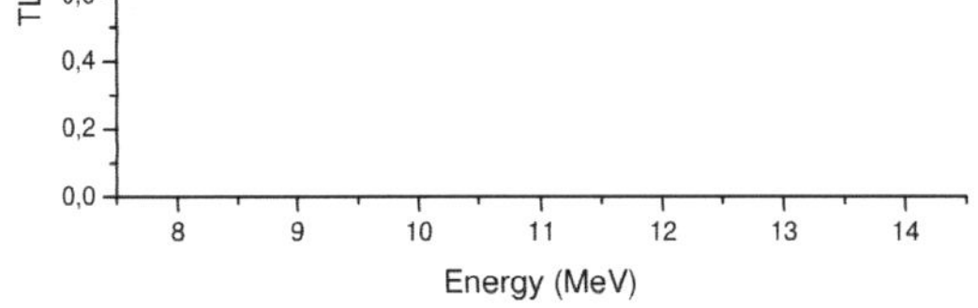

Figure 4. Dependence of the TL response topaz–Teflon pellets on the energy of electron beams. All the samples were irradiated to 2 Gy.

Typically, the electron beams central axis depth dose curve exhibits a high surface dose (compared with megavoltage photon beams), and the dose then build up to a maximum at a certain depth referred

Journal of Physics: Conference Series **249** (2010) 012057 doi:10.1088/1742-6596/249/1/012057

to as the electron beams depth of dose maximum [5]. In order to analyse the TL response the detectors were irradiated with doses between 0.1cGy and 200cGy under different thickness of acrylic plates.

The TL response of detectors irradiated with electron beams were similar to central axis percentage depth dose (PDD) curve obtained in conventional dosimetric procedures performed with ionizing chamber (calibration in routine procedures). TLD results showed that the absorbed doses presents around 80% in surface and increased up to 100% in depth of dose maximum (build-up region) for each evaluated energy and it was reduced in the high depths, as can be seen in figures 4 to 7.

To 8 MeV beams the maximum dose was obtained in the depth of 17 mm, in one of 26 mm the PPD result was of 77.1± 3.0% showing a 3% low then the value obtained by conventional procedures. When the depth increases more than 30 mm was observed divergences if compared with results obtained with ionizing chamber (figure 4).

Figure 4 shows the PDD to 10 MeV irradiation. In this figure it is possible observe that the maximum absorbed dose occurs at 22 mm. Moreover, we can say that the average values of PDP found with the TLD and the one obtained by calibration are in accordance up to the depth of 30 mm. In depth of 35 mm the PDP obtained with the pellets was equal to 57.6 ± 2.8%, 12.8% lower than the value found by the calibration, 64.7%. For depths higher than 35 mm the PDP values seemed quite distinct from obtained by the calibration, although by calibration PDD have not been obtained in depth of 40 mm.

The topaz-Teflon detectors when irradiated with 12 MeV, showed a accordance to the results of calibration, with an PDD of approximately 83% in surface witch increased by 100% to a depth of 28 mm and it being similar up to the depth of 35 mm in depth, the PDP obtained with TLD was 89%, only 3.4% lower than the obtained with calibration in water, which was equal to 92%. From the depth of 40 mm the differences between the responses to PDP were higher reaching up to 38% in depth of 45 mm (figure 5).

In the energy of 14 MeV the maximum dose depth was 29 mm according to calibration results obtained in routine procedures. The PPD results obtained with topaz-Teflon pellets has been showed lower than calibration in the interval of depth to range from 15 to 35mm, to depth of 15 mm this difference was 4.2%; to 45mm the result was similar to the one of calibration and to a depth of 60 mm the difference was 45% (figure 6).

The depth of the 90% dose level beyond the depth max is defined as the therapeutic range for electron beam therapy. To the energies studied the PDD obtained with topaz-Teflon pellets were almost similar to PDD obtained on calibration procedures.

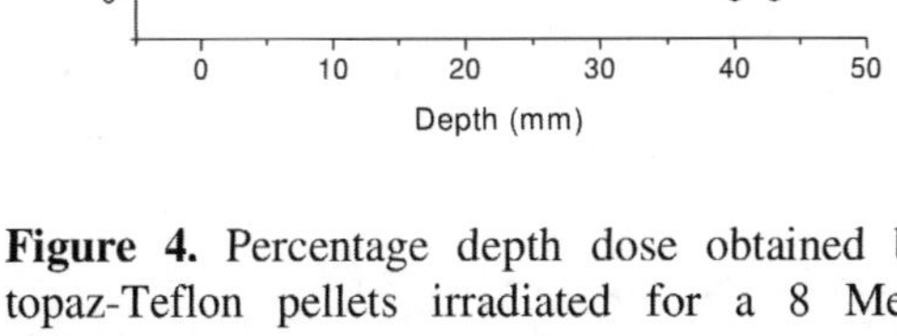

Figure 4. Percentage depth dose obtained by topaz-Teflon pellets irradiated for a 8 MeV electron beams.

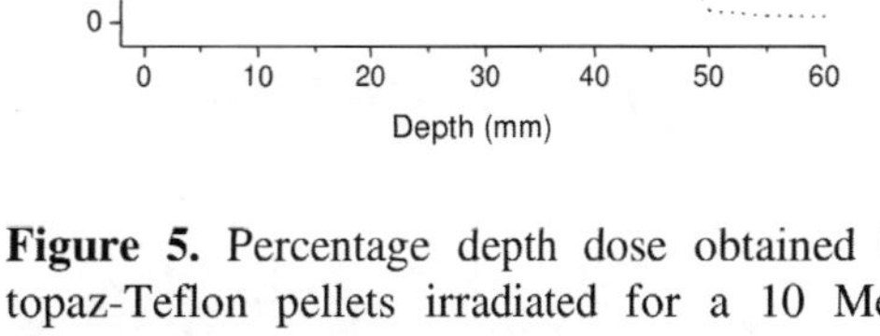

Figure 5. Percentage depth dose obtained by topaz-Teflon pellets irradiated for a 10 MeV electron beams.

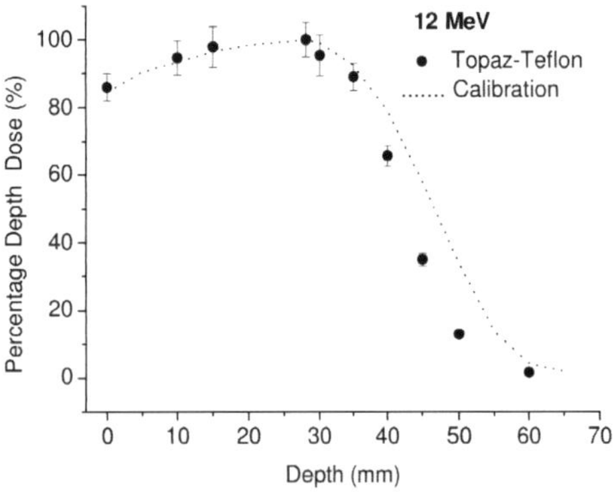

Figure 6. Percentage depth dose obtained by topaz-Teflon pellets irradiated for a 12 MeV electron beams.

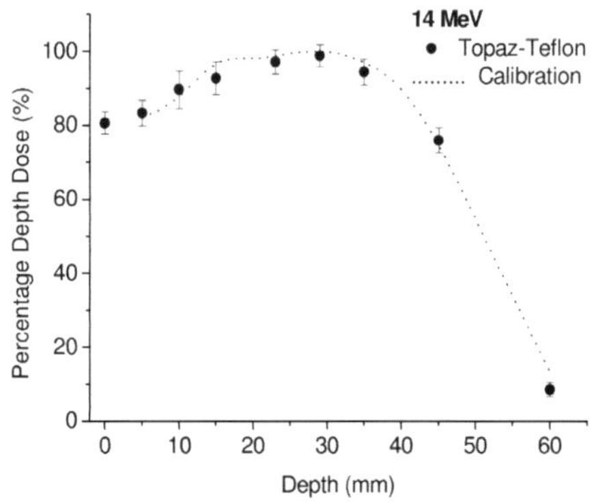

Figure 7. Percentage depth dose obtained by topaz-Teflon pellets irradiated for a 14 MeV electron beams.

4. Conclusions

The topaz-Teflon detectors showed TL response dependent to absorbed dose when irradiated in megavoltage electron beams, indicating that those TLD can be used for dosimetric monitoring in electron radiotherapy procedures mainly in the therapeutic range.

Ideally, the energy response should be flat; the system calibration should be independent of energy over a certain range of radiation qualities, consequently for use of topaz-Teflon in calibration procedures of electron beams the energy correction has to be included in the determination of absorbed doses.

5. Acknowledgments

The authors wish to thank Coordenação de Aperfeiçoamento de Pessoal Superior (CAPES) and Conselho Nacional de Desenvolvimento Científico e Tecnológico (CNPq), Brazilian agencies, for parcial financial support.

References

[1] Souza D N, Valerio, M E G , Lima, J F, Caldas, L V E 2001 *Rad. Eff. Def .Sol* .**156** 325
[2] Souza D N, Valerio, MEG, Lima, JF and Caldas, LVE 2002 *Radiat. Prot. Dosim.* **100** 413
[3] Souza D N, Meira, R A, Valerio, MEG, Lima, JF and Caldas, LVE 2003 *App Radiat Isot* **58** 489
[4] Magalhães C M S, Macedo, Z S, Valério, M E G, Hernandes, A C and Caldas, L V E 2004 *Nucl Instr and Meth B* **218** 277
[5] Podgorsac E B Radiation Oncology Physics: A Handbook for teachers and students Viena: IAEA, 2005
[6] Nahajowski D, Góra E, Rozwadowska-Bogusz B, Lesiak J, Polak B, Kabat D, Zawadzki P and Waligórski M P R 2008 *Radiat.. and Measur.* 879

Development of Sintered Composites of $CaSO_4$ Doped With Eu^{3+}

M A P Chagas, L J Parente and D N Souza
Departamento de Física, Universidade Federal de Sergipe
São Cristóvão, SE, Brazil, 49100 000

E-mail: mchagas@fisica.ufs.br

Abstract. The main purpose of this work is the developing of sintered composites in order to allow their use as thermoluminescent dosimeters using a simple route to preparation, employing high temperature. The composites were prepared with $CaSO_4.2H_2O$ doped with EuO_3. Initially $CaSO_4.2H_2O$ was dehydrated in oven at 100ºC for 1h. After addition of dopants, glass and homogenization, PVA was incorporated into the compounds to make their compaction easier. The pellets with 6 mm of diameter and 1 mm thick were sintered at 750ºC for 6 h. The TL emission curves showed two peaks when the samples were irradiated with beta source ($^{90}Sr+^{90}Y$) and X-ray. During calibration, TL responses were found to be proportional to the absorbed dose. The results showed that this material can potentially to be used in dosimetry.

1. Introduction

Many authors have shown results from studies on the subject of doped detectors, revealing variousof their properties. Those studies includes the thermoluminescent (TL) response to dose; TL emission kinetic; specific ionization density; enabled dopants and their concentrations, co-dopants; processes of energy transfer; thermally stimulated exoeletronic emission, results from of optical absorption and paramagnetic resonance measures[1-5]. The elements from the group of rare earth (RE) are frequently used as dopants of the CaSO4 for the production of thermoluminescent detectors; amongst them, the most often used are: Dy, Tm and the Eu [5-10]. The model for the thermoluminescent emission of the doped $CaSO_4$ with RE is the one of energy transference. According to this model, there is no direct recombination of charge carriers due to the small quantity of the dopant (RE), but the energy of the electron-hole recombination is transferred to the RE^{+3}. However, no direct evidence has been published so far [3]. The europium is a soft, malleable and ductile silvery-looking metal that shows the electronic configuration [Xe] $4f^4$ $6s^2$ [10]. The europium presents electronic configuration [Xe] $4f^7$ $6s^2$, it is classified as a metal of internal transition and it is solid in its natural state [10].

In this work we are considering a new composite to be employed in the dosimetry of the ionizing radiations to the based on CaSO4, using the europium and glass for use in thermoluminescent of ionizing radiation dosimetry, mainly. We decided to use the glass in order to sintering the composites at a temperature lower than the melting point (1450°C). Normally, the thermoluminescent dosimeters based on $CaSO_4$ are produced with Teflon. However, there is a limitation for the use of this material as a binding because Teflon volatilizes at temperatures higher than 300°C.

2. Methodology and Materials

The samples were prepared in the Laboratory of Materials Preparation and Characterization Department of Physics of Federal University of Sergipe. It was used $CaSO_4$ with 98% of pureness (Labsynth Ltda) doped with 0.1, 1 and 2% of EuO_3 with 99.9% of pureness (Aldrich Chemical Company Inc.) and recycled window glass samples classified as transparent for industrial usage were utilized. The glass sample was crushed into pieces with diameter of 63 μm. After the milling step, the glass powder was dried and mixed with the topaz crystals (50% in mass). The samples were mixed with 10 ml of distilled water and taken to a magnetic agitator where they remained for 30 minutes. After that, drying took place in oven at a temperature of 100°C for 24 h. The dried powder was homogenized again in mortar with the polyvinilic alcohol (PVA) addition. The pellets were submitted to a uniaxial presage of 100 kgf/cm^2 and sintered at temperatures of 750°C for 6h, with a rate of heating of approximately 10°C per minute, and a free rate cooling down to room temperature in order to obtain pellets with 1.0mm thickness and diameter of 6.0mm.

A beta source ($^{90}Sr+^{90}Y$) of the dermatological applicator type and an x-ray device of clinical application (VMI Compact Plus - 500) was used to irradiate the pellets. The beta source had a 6.2 10^{-3} Gy/s dose rate. The x-ray device has a 40 at 125 kV tube voltage and a 0.2 mAs to 150 mAs at 70 kV tube current time products. The x-ray beam was previously calibrated using an ionizing chamber. Before each exposition, the pellets were submitted at thermal annealing at the temperature of 400ºC to 30min.

The measures of thermoluminescence were carried out at interval up to 6 h after irradiation. Measurements were made with a Harshaw 3500 thermoluminescent reader using a heating rate of 10°C/s . The maximum deviation found in the TL intensity was of ± 6.5%.

3. Results

Initially, was made a study of the possibilities of preparation of $CaSO_4:Eu^{3+}$ with different concentrations of dopants. The best results were showed to the composites doped with 2%. Figure 1 shows the TL glow curves of $CaSO_4:Eu^{3+}$ + glass performed using 2% of Eu^{3+} whiter that has been irradiated with 22.6 Gy from a beta source. TL emission exhibits two peaks with different intensities, one high TL peak with maximum of intensity at 195°C and a short one at 270ºC. The other ones prepared with less dopant present similar TL peaks, but with reduced intensity. The glow curves of $CaSO_4:Eu^{3+}$ + glass are similar to the observed by Bernal at al [11]. The TL peaks there are at a high enough temperate so that it is not necessary a pre-annealing in the using of dosimetric procedures.

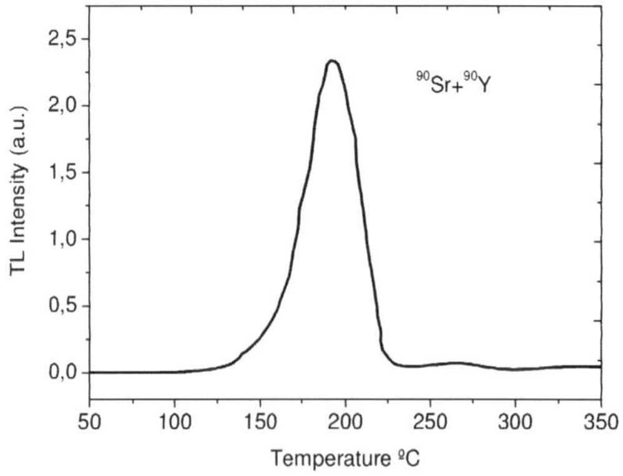

Figure 1. Typical TL emission curve of a $CaSO_4:Eu^{3+}$ + glass composites irradiated with $^{90}Sr+^{90}Y$.

The pellets exhibits a good rigidity and they showed no break or weight loss during the routine experimentation.

Pellets of $CaSO_4:Eu^{3+}$ + glass can be used in successive irradiation–reading–annealing cycles without change in their sensitivity.

Figure 2 shows the TL response of the $CaSO_4:Eu^{3+}$ + glass pellets irradiated with doses between 2.6 and 26.2 Gy of a $^{90}Sr+^{90}Y$ source. The TL intensity was proportional to the dose in all this range. The measurement uncertainties in those studies were always below 5%.

Figure 3 shows the glow curves of the pellets which were ***x-irradiated*** with 1.65 10^{-4} Gy. The glow curve was similar at that one irradiated with beta source.

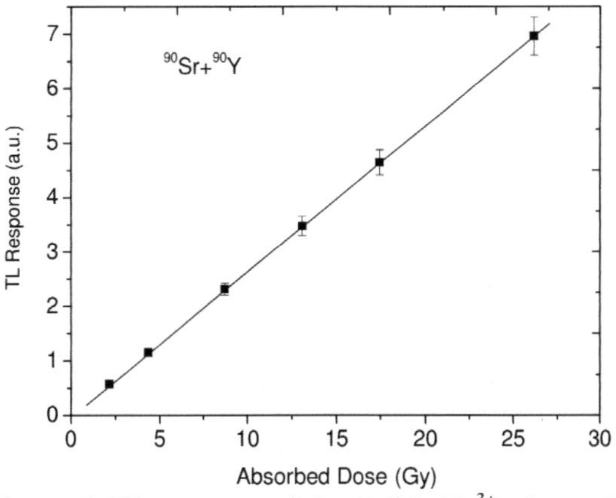

Figure 2. TL response of the $CaSO_4:Eu^{3+}$+glass pellets as a function of absorbed dose in the range of 3.27 to 22.6 Gy ($^{90}Sr+^{90}Y$).

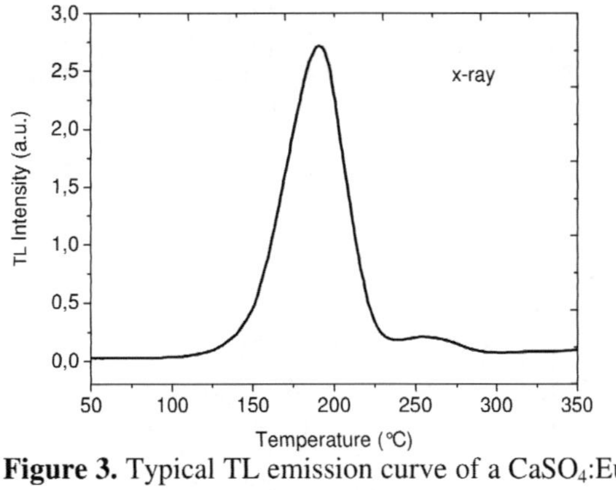

Figure 3. Typical TL emission curve of a $CaSO_4:Eu^{3+}$ + glass composites irradiated with x-ray.

In the figure 4 the TL response of the pellets was plotted as a function of the absorbed dose of x-rays. Calibration curves presented responses proportional to absorbed dose between 1.5 Gy and 6.0 Gy for the x-radiation.

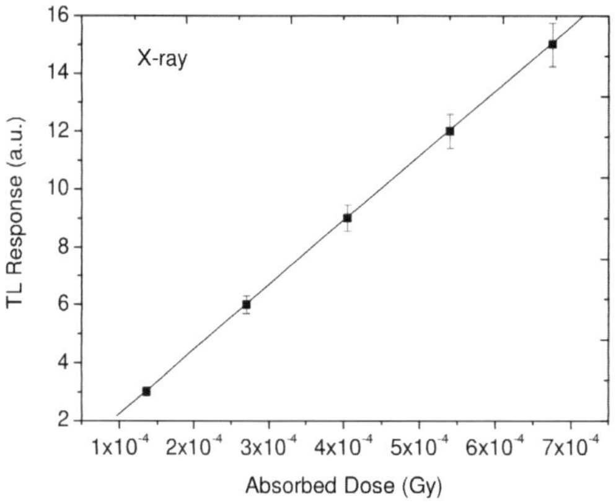

Figure 4. TL response of the $CaSO_4$:Eu^{3+}+glass pellets as a function of absorbed dose from x-ray.

4. Conclusions

The currently accepted procedures of radiological protection aim at reducing the exposition associated with practices that make use ionizing radiation, both for the occupationally exposed workers and for the environment. Therefore it is important to use detectors of high sensitivity in the personal and the environmental dosimetry. According to this study, it could be observed that the $CaSO_4$ detector doped with europium is appropriate for dosimetric purposes. Due the previously mentioned properties associated with their low cost, the described detectors could be used in teaching laboratories as a complement of didactic practices involving the use of other kinds of radiation detectors. Further techniques for preparation of the described composites, as well as other radiation sources for testing the pellets, could be within the scope of future works.

Acknowledgments

The authors are thankful to Conselho Nacional de Desenvolvimento Científico e Tecnológico (CNPq) and Coordenação de Aperfeiçoamento de Pessoal de Nível Superior (CAPES) for partial financial support.

References

[1]. Cristovan F H, Eiras S P, W. Cruz O and Dias I. F. L 2005 *Quim. Nova,* **28**, (6) 964
[2]. Hinatsu Y, Doi Y, Ito K 2003. *J. Solid. State Chem.* **172** (2) 438
[3]. Lakshmanan A R 1999 *Progr. In Mater. Sc.* **44**: 1
[4]. Q. Tang, C.X., Zhang, P.L., Leung, M.,Li; D.L., Luo, 2005. *Acta Phys. Sinica* **54** (1): 64-69.
[5]. K.N.R., Taylor; M.I.,Darby, 1972. *Physics of Rare Earth Solids Chapman and Hall LTD.*
[6]. Patil R R, Muthal P L, Dhopt S M, Kondawar V K and Moharil S V 2006 J. Lumines. **126** 517
[7]. Cameron J R, N.;Suntharalingam, G. N.,Kenney, 1968. *Thermoluminescent Dosimetry.* The University of Wisconsin Press.
[8]. L. L.,Campos; M. F, Lima, 1983. *J. Lumin.* **28** 3.
[9]. M. A. P. Chagas, 2005. *Estudos espectroscópicos de cristais dopados com o íon* Nd^{3+}. *Dissertação de Mestrado.* Universidade Federal de Sergipe.
[10]. D.R, Lide, 2006-2007. *Handbook of Chemistry and Physics* **87** Edition.
[11]. Bernal R, García-Haro A R, Machi L, Brown F, Pérez-Salas R, Castaño V M and Cruz-Vázquez C, 2008. *Rad. Meas.* **43** 371.

Virtual restoration of cracks in digitized image of paintings

G Schirripa Spagnolo[1], **F Somma**[2]
[1] Dipartimento di Ingegneria Elettronica, Università degli Studi "Roma Tre"
[2] Dipartimento di Fisica, Università degli Studi "Roma Tre"
Via della Vasca Navale n.84, 00146 Roma, Italy

Abstract. An integrated methodology for the detection and removal of cracks on digitized image is presented in this paper. Crack-like pattern detection have been a matter of high concern among researchers mostly for its useful contribution to a variety of applications. The results presented here regard the craquelure of old paintings, however, the same methodology can be used for a much wider set of application. Many images contain similar patterns: crack in protective coating for polymers and other surfaces; fatigue crack in MEMS/NEMS; crack in epoxies used for underfill and encapsulation microelectronics components; etc. In this paper the cracks are detected by thresholding the output of the morphological top-hat transform. Afterwards, the thin dark brush strokes which have been misidentified as cracks are removed using automatic procedure. Finally, crack filling using texture synthesis algorithms. The methodology has been shown to perform very well on digitized images suffering from cracks.

1. Introduction

Every day, digital image processing techniques are used in all science fields. In digital image analysis, crack-like pattern detection, better known as ridge-valley structure extraction in some literature [1-4], has been a matter of high concern among researchers, mostly for its potentially useful contribution to a variety of applications. Many old painting suffer from breaks in substrate, the paint, or varnish. These patterns are usually called craquelure and can be caused by aging, drying, and mechanical factors. Many other images contain similar patterns, such as fatigue crack in MEMS/NEMS, crack in epoxies used for underfill and encapsulation microelectronics components,etc. All these examples are current areas of research in the modeling and classification fields for which the results in this dissertation can find an application.

Image processing techniques used to detect and remove crack can be employed to inspection and/or diagnosis. Besides it can be useful for, improve her perceptiveness (legibility) of the digital images. In particularly this is true for the digital images used for didactic or multimedia applications.

An integrated methodology for the detection and removal of cracks on digitized paintings is presented in this paper. Here we focus the attention on the possibility of classification and virtual cancellation of cracks on digitized old paintings. In this way, a "virtual" restoration is performed. It can provide clues to art historians, museum curators and the general public on how the painting would look like in its initial state, i.e., without the cracks. Furthermore, it can be used as a nondestructive tool for the planning of the actual restoration [5].

A methodology to identify and to restore the present cracks on digital images is introduced in this paper. The technique consists of the following stages:

- Detection of the cracks.
- Separation of the brush strokes, which have been misidentified as cracks.
- Implementation of the crack filling procedure.

The outline of the paper is as follows. Section 2 describes the craquelure formation. The crack-detection procedure is presented in Section 3. The method for filling the craks is propossed in Section 4. Section 5 concludes the dissertation by outlining the contributions of this work and discussing potentials for the future.

2. Craquelure Formation

Craquelure is the pattern of cracks that develop across a painting with age [6]. It is apparent in all older pictures and influences their appearance to greater or lesser extent. Figure 1 shows an example of craquelure in a painting.

Figure 1. Example of a painting with cracks.

This network of fine cracks depends on the materials used for the painting, the painting technique of the artist, the atmospheric variations the painting has been exposed to, and, of course, manipulation and/ or storage conditions [7].Generally, the paint layer is protected by the varnish and problems occur when the originally transparent varnish becomes yellowish or greenish or simply loses its transparency during ageing. As the paint layer ages, the solvent is no longer capable of keeping the paint layer intact, thus cracks begin to form. This is the most common reason behind the forming of cracks. Other known reasons include physical tensions within the structure of the painting and external impacts such as in human handling.

The pattern of cracks is a visible record of the physical tensions within the structure of the painting. The ways in which tensions are generated and dissipated are dependent upon the choice of materials and methods of construction employed by the artist [8].

Craquelure can be a very important element in judging authenticity, use of material or environmental and physical impact because these can lead to different craquelure patterns. Although most conservation of fine artwork relies on manual inspection of deterioration signs, the ability to screen the whole collection semi-automatically is believed to be a useful contribution to preservation. It is hoped that the mass screening of craquelure patterns will help to establish a better platform for conservators to identify the cause of damage [9].

3. Detection of the Crack

Crack-like pattern detection, also known in some literature as ridge-valley structure extraction, has been a matter of high concern among researchers mostly for its useful contribution to a variety of

applications [10-12]. In this paper, we present a method to identify and extract the craquelure pattern based on the algorithms developed by Giakoumis et al. [5], Abas [8, 13], Iyer et al. [14].
Crack usually have low luminance (bright cracks) o high luminance (light cracks) and, thus can be considered as local intensity minima (or maxima) with rather elongated structural characteristics. Therefore, a crack detection can be applied on the luminance component of an image should be able to identify such minima (or maxima).
For our reference, we will define a two-dimensional (2D) grayscale image (luminance component) having a range of $[I_{\min}, I_{\max}]$ as a functional $F: \mathbb{R}^2 \rightarrow [I_{\min}, I_{\max}]$ and a 2D structural element as a functional $B: \mathbb{R}^2 \rightarrow \mathbf{B}$ where **B** is the set of the neighborhoods of the origin.
Gray-scale morphological dilation and erosion can be visualized as working with two images namely the image being processed F and the structuring element **B**. Each structuring element has a specific shape that act as a probe. Basic morphological operators with respect with respect to the structuring element **B**, a scaling factor q, image F and a processing point $P_0 \in \mathbb{R}^2$ can be defined as:
Gray-scale *erosion* and *dilation* are defined as [10]:

$$\text{erosion:} \quad \varepsilon_B^q(F)(P_0) = MIN_{P \in P_0 + q \cdot B(P_0)}\left(F(P)\right) \tag{3.1}$$

$$\text{dilation:} \quad \delta_B^q(F)(P_0) = MAX_{P \in P_0 + q \cdot B(P_0)}\left(F(P)\right) \tag{3.2}$$

$$\text{opening:} \quad \gamma_B^q(F) = \delta_B^q\left(\varepsilon_B^q(F)\right) \tag{3.3}$$

$$\text{closing:} \quad \phi_B^q(F) = \varepsilon_B^q\left(\delta_B^q(F)\right) \tag{3.4}$$

Dilatation, in general, causes objects to dilate or grow in size; erosion cause objects to shrink. The amount and the way that they grow or shrink depend upon the choice of the structuring element **B**. The opening operation is somewhat like erosion in that it tends to remove some of the foreground (bright) pixels from the edges of regions of foreground pixels. However, it is less destructive than erosion in general. As with other morphological operators, the exact operation is determined by a structuring element. The effect of this operator is to preserve foreground regions that have a similar shape to the structuring element, or that can completely contain the structuring element, while eliminating all other regions of foreground pixels. However, "closing" is similar in some ways to dilation in that it tends to enlarge boundaries of foreground (bright) regions in an image (and shrink background color holes in such regions), but it is less destructive of the original boundary shape. The effect of this operator is to preserve background regions that have a similar shape to this structuring element, or that can completely contain the structuring element, while eliminating all other regions of background pixels.
Cracks can be detected with the implementation of a very useful morphological filter known as the top-hat transformation developed by Meyer [15]. These details can be lines or areas with particular sizes. Top-hat operators can function as a closing or opening operator based on the features to extract from an image. *Opening* top-hat operators (OTH) will detect bright details in an image while *closing* top-hat operators (CTH) (or also sometimes known as the bottom-hat operator) are designed to detect dark details. Formulation for OTH and CTH are shown denoted by Equations (3.5) and (3.6) respectively.

$$\text{Opening Top-Hat:} \quad OTH_B^q(F) = F - \gamma_B^q(F) \tag{3.5}$$

$$\text{Closing Top-Hat:} \quad CTH_B^q(F) = \phi_B^q(F) - F \tag{3.6}$$

The top-hat operator can be tuned for detection of specific features by modifying two important parameters [16]:

- The shape and the size of the structuring element S. A square-shaped or disk-shaped structuring element may be used. The size must be carefully selected according to the thickness of the crack to be detected.
- The number of times in which erosion or dilation is performed.

The transformation produces a grey-scale image with the desired features enhanced to a certain level.

Opening top-hat operators will detect bright details in an image while closing top-hat operators (or also sometimes known as the bottom-hat operator) are designed to detect dark details [8, 14].
Figure 2 shows an example of craquelure identification obtained with *opening* top-hat operator.

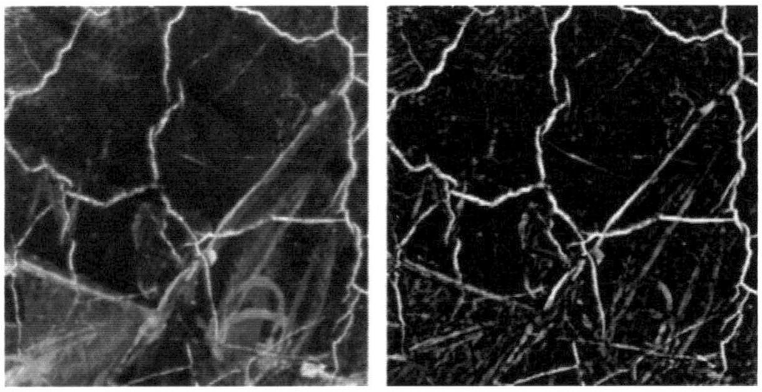

Figure 2. Example of craquelure identification by *OTH*.

Figure 3 shows an example of craquelure identification obtained with *closing* top-hat operator.

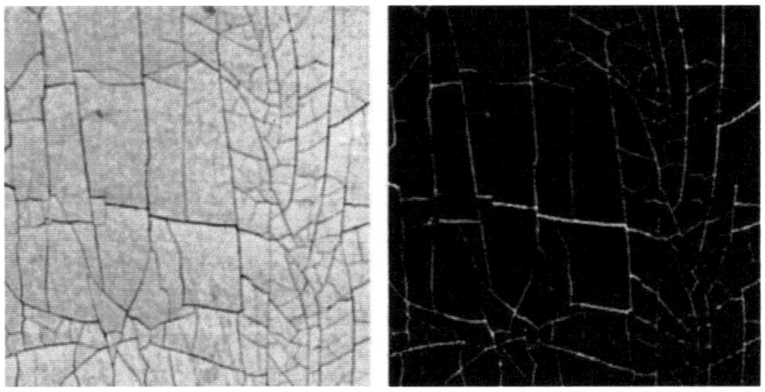

Figure 3. Example of craquelure identification by *CTH*.

The top-hat transformation produces a grayscale output image t(k; l). If its value is large, the corresponding pixel belongs to a crack (or crack-like element). Otherwise, this pixel location corresponds to background. Therefore, a thresholding operation is required to separate cracks from the background.

3.1. Separation of the brush strokes from the cracks
In some paintings, thin dark brush strokes exist (e.g. in hair), which have almost the same features (thickness, luminance) as cracks. Therefore, it is possible that the top-hat transform misclassifies these brush strokes as cracks. Thus, in order to avoid any undesirable alterations to original image, it is very important to separate these brush strokes from actual cracks, before the implementation of the crack filling procedure is necessary to prevent undesirable alterations to the image. Such a separation can be done on the basis of the hue and saturation values of the corresponding pixels. In the HSV color model (also called HSB - Hue, Saturation, Brightness), hue (H) is associated with the dominant wavelength in a mixture of light wavelengths and represents the color type (such as red, green). It ranges from 0 to 360 degree, with red at 0 degree, green at 120 degree, blue at 240 degree and so on. Saturation (S) refers to amount of white light mixed with a certain hue. The lower the saturation of a color, the more "grayness" is present and the more faded the color will appear. Saturation ranges are

from 0 to 100%. In this work saturation is presenting in range 0-255. Value (V) refers to the brightness of the color. Value ranges are from 0-100% [5].

By statistical analysis of typical digitized old painting images, it has been concluded that the hue of the cracks usually ranges from to 0° to 60°. On the contrary, we observed that the hue of dark brush strokes varies, as expected, in the entire gamut [0°, 360°]. Furthermore, crack saturation ranges usually from 0.3 to 0.7, while brush stroke saturation ranges from 0 to 0.4. Thus, on the basis of these differences, a great portion of the dark brush strokes, falsely detected by the top-hat transform, can be separated from the cracks. This separation can be achieved by classification using a median radial basis functions (MRBF) neural network, which is a robust version of radial basis functions (RBF) network [17]. The input vectors of the network consist of the hue and saturation values of pixels identified as cracks by the top hat transform. During the recall phase each input is assigned to one of the two available output classes (cracks and thin dark brush strokes).

4. Reconstruction of defective pixels (crack-filling methods)

After identifying cracks, the next task is to restore the image using local image information to fill (interpolate) the cracks.

The correction of defective pixels can be viewed as an image reconstruction task. Within it, the pixel being repaired, as well as much of its neighboring pixels, may be defective, i.e., their values are only loosely related to the respective original values. The values at these pixels should thus be ignored by the reconstruction algorithm.

At this stage of the solution, we assume that the defects are already detected. Two approaches to image restoration of missing regions, in literature, are image inpainting and texture synthesis.

Image inpainting attempts to fill in, a defective image region, in a natural way. The pixels in the defective image are treated as missing. The algorithm iteratively solves some partial differential equations that smoothly propagate the information to the surrounding missing region preserving the gradients apparent in the boundary.

Texture synthesis algorithms are concerned with producing a large texture image from a small sample of a texture. The resulting texture image should appear to arise from the same texture to a human observer. The texture should not appear to be duplicated or artificial. In our work we have used this approach to reconstruct defective pixels. Particularly, we have used the algorithms developed by Bergman et al. [18]

The used reconstruction algorithm uses a defect map, and is intended to overcome the shortcomings of local reconstruction, in particular in textured areas. Rather than averaging, it copies like textures from elsewhere in the image.

The reconstruction algorithms, developed in this paper, works according to the following procedure.

Let D and R be positive integers such that $D < R$. Typical values for D and R, for example, are $D = 5$ and $R = 100$. At each defective pixel i,

1. Let C_i be the context of the defective pixel i. The context includes all color channels of the pixels in a $D \times D$ neighborhood around pixel i.
2. Look at all the $D \times D$ neighborhoods, N_j , in an $R \times R$ region around pixel i. Neighborhoods with defective pixels are excluded from the search.
3. Find the neighborhood N_j which is most similar to C_i. Specifically, for every neighborhood N_j, compute the sum of squared differences between the non-defective pixels of C_i and the corresponding pixels of N_j. Let N_{j*} be the neighborhood with the lowest sum of squared differences.
4. If the sum of squared differences is below a pre-defined threshold, replace the defective pixels in C_i with the corresponding pixel values from N_{j*}.
5. Otherwise replace pixel i with the median value of its $D \times D$ neighborhood.

Note that in step 4 several pixels are typically replaced. Replacing several pixels together has two advantages. The first advantage is speed, because the number of searches for matching neighborhoods is reduced. The second is, perhaps surprisingly, in image quality. Work in texture synthesis [11] has shown that texture is duplicated better when groups of pixels are copied rather than single pixels. In step 5, when no "good enough" context is found, the algorithm defaults to the median solution.
Figure 4 shows the results obtained with the used algorithm of reconstruction.

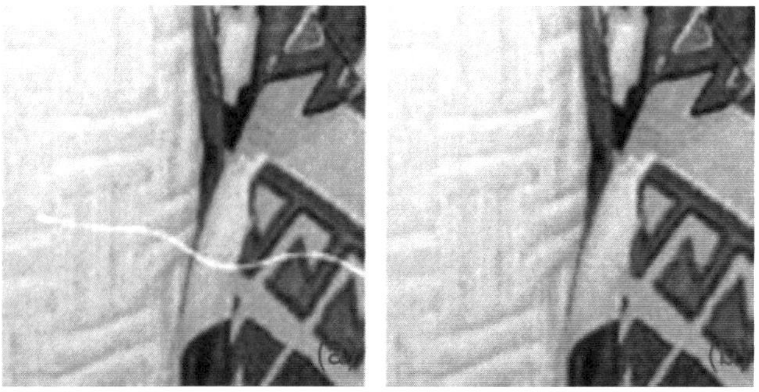

Figure 4. Example of reconstruction results. The resulting repair is shown in (b).

To evaluate the performance of the proposed methodology, we have tested the algorithms with varying crack pattern, color, and background. In our tests the size of the images is 1024 × 1024 [pixels].
The first test is performed (see Figure 5(a)) on a particular of the fresco "*The expulsion from the Eden*" (Sistine Chapel – The Vatican City). Figure 5(b) shows cracks enhanced by a closing top-hat operator (CTH) operator, while figure 5(c) shows the thresholding operation. Finally, figure 5(d) shows the reconstruction of defective pixels.

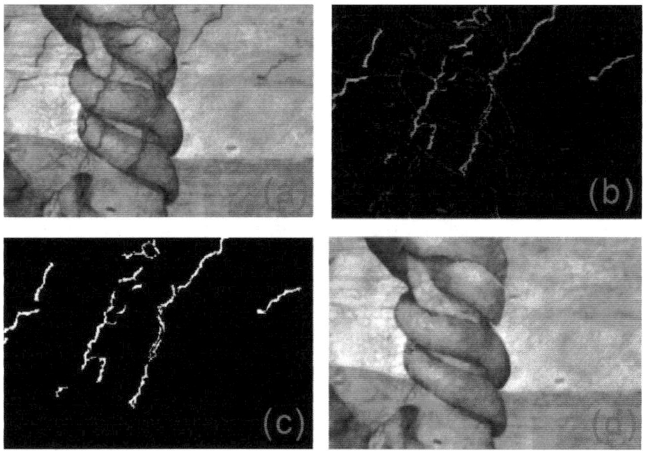

Figure 5. Virtual restoration of the fresco: particular of "*The expulsion from the Eden*".

A second test has been realized on the image of an old painting with bright cracks (Figure 6(a)). Figure 6(b) shows the crack enhancement obtained with opening top-hat operator (OTH), while the figure 6(c) displays the results thresholding. The result of the crack filling is shows in figure 6(d).

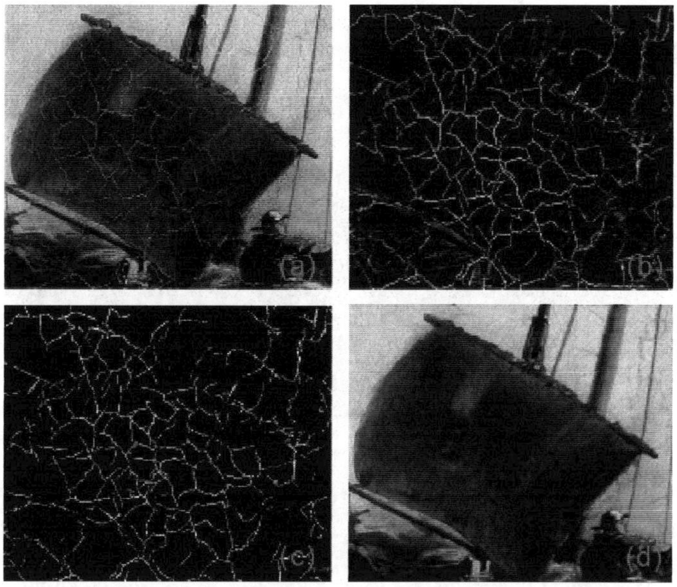

Figure 6. Virtual restoration of old painting with bright cracks.

5. Conclusion

This paper presented a new full automatic methodology for detection and removal cracks on digital image. The results presented in this paper regard the craquelure of old paintings, however, the same algorithms can be used for a much wider set of application. Many images contain similar patterns: crack in protective coating for polymers and other surfaces; fatigue crack in MEMS/NEMS; crack in epoxies used for underfill and encapsulation microelectronics components; etc.

Image processing techniques used to detect and remove crack can be employed to inspection and/or diagnosis in many scientific field. Besides it can be useful for, improve her perceptiveness (legibility) of the digital images. In particularly this is true for the digital images used for didactic or multimedia applications.

The software works well in the individualization and reconstruction of cracks. However, future works have to study other types of defects that are present in the digital images, such as dust, scratch, spot light, etc.

6. References

[1] A.M. Lòpez A M, Lumbreras F 1999 Evaluation of Methods for ridge and valley detection *IEEE Trans. Pattern Anal. Mach. Intell.* **21** 327–335

[2] Elberly D, Gardner R, Morse B, Pizer S, Scharlach C 1994 Ridges for image analysis, *J. Math. Imaging Vis.* **4** 353–373

[3] Gaugh J, Pizer S 1993 Multiresolution Analysis of Ridges and Valleys in Grey-Scale Images *IEEE Trans. Pattern Anal. Mach. Intell.* 15(6) 635–646

[4] Haralick R 1983 Ridges and Valleys on Digital Images *Computer Vision, Graphics and Image Processing* 22(10) 28–38

[5] Giakoumis I, Nikolaidis N, Pitas I 2006 Digital image processing techniques for the detection and removal of cracks in digitized paintings, *IEEE Transactions on Image Processing* **15** (1) 178 – 188

[6] Brucklow S L 1998 A Stylometric Analysis of Craquelure *Computers and the Humanities* **31** 503–521
[7] Knut Nicolaus 1999 *The restoration of Paintings* Könemann Veragsgesellschaft mbH (Cologne), ISBN 3-89508-922-2
[8] F.S. Abas 2004 Analysis of Craquelure Patterns for Content-Based Retrieval *Thesis of PhD University of Southampton, Faculty of Engineering, Science and Mathematics School of Electronics and Computer Science*
[9] Brucklow S L 1996 The description of craquelure patterns *Studies in Conservation* **42** 129-140
[10] Zana F, Klein J C 2001 Segmentation of Vessel-Like Patterns Using Mathematical Morphology and Curvature Evaluation *IEEE Transactions on Image Processing* **10** 1010–1019
[11] A.M. Lopez, F. Lumbreras 1999 Evaluation of Methods for ridge and valley detection, *IEEE Trans. Pattern Mach. Intell.* **21** 327–335.
[12] Elberly D, Gardner R, Morse B, Pizer S, Scharlach C 1994 Ridges for image analysis, *J. Math. Imaging Vis.* **4** 353–373.
[13] Abas F S and K. Martinez K 2003 Classification of painting cracks for content-based analysis In: *IS&T/SPIE's 15th Annual Symposium Electronic Imaging 2003: Machine Vision Applications in Industrial Inspection XI* 149-160, ISBN: 0-8194-4811-7
[14] Iyer S, Sinha S K 2005 A robust approach for automatic detection and segmentation of cracks in underground pipeline images *Image and Vision Computing* **23** 921–933
[15] Meyer F 1979 Iterative image transforms for an automatic screening of cervical smears *J. Histoch. Cytochem.* **27** 128–135.
[16] Giakoumis I, Pitas I 1998 Digital Restoration of Painting Cracks," in *ISCAS '98, Proceedings of the IEEE International Symposium on Circuits and Signals* 269–272
[17] A. G. Bors and I. Pitas, "Median radial basis function neural network 1996 *IEEE Trans. Neural Netw.* **7**(6) 1351–1364
[18] Bergman R, Maurer R, Nachlieli H, Ruckenstein, Chase P, Greig D, 2008 Comprehensive Solutions for Removal of Dust and Scraches from Image *J. Electron. Imaging* **17** 013010 DOI:10.1117/1.2899845

Journal of Physics: Conference Series **249** (2010) 012060 doi:10.1088/1742-6596/249/1/012060

Defects in UV-vis-NIR reflectance spectra as method for forgery detections in writing documents

F Somma[1], P Aloe[1] and G Schirripa Spagnolo[2]
[1] Dipartimento di Fisica, Università degli Studi "Roma Tre"
[2] Dipartimento di Ingegneria Elettronica, Università degli Studi "Roma Tre"
Via della Vasca Navale n.84, 00146 Roma, Italy

Abstract. Documents have taken up a very important place in our society. Frauds committed in connection with documents are not at all uncommon, and, in fact, represent a very large domain of the forensic science called "questioned documents". In the field of forensic examination of questioned documents, the legitimacy of an ink entry is often an essential question. A common type of forgery consists in materially altering an existing writing or adding a new writing. These changes can be characterized by means of optical spectroscopy. The aim of this work is to perform the UV-vis-NIR reflectance spectrophotometry to analyze a range of blue and black commercial ballpoint pens, in order to investigate the discriminating abilities of the different inks found on the same document.

1.Introduction

Ever since paper was invented, documents have taken up a very important place in our society. They play a decisive role in fields such as communication, education, culture, art, sanitation, hygiene, or packaging, and it is almost impossible to imagine life without them. Today, the advances of computer science have put the digital exchange of information into a privileged position in all our societies, yet paper is still a preferred medium in many applications: books, notes, contracts, testaments, receipts, letters, tickets, banknotes, and so on. A great number of transactions can be digitally executed, but in many cases a signature is required for proof of consent. Therefore, frauds committed in connection with documents are not at all uncommon, and in fact represent a very large domain of forensic science called "questioned documents". A common type of forgery consists in materially altering an existing writing or adding a new writing. Therefore, ink analysis is an important forensic procedure that can reveal useful information about questioned document [1]. In the last years, numerous techniques have been proposed for forensic ink examination [2]. The wide array of materials used in inks, coupled with possible contamination from the writing surface confronts forensic ink chemists with a complex analytical challenge to carry out this type of analysis. But the aim of most analyses is to determine whether two pieces of written text originated from the same ink, therefore, comparison of different writing inks on a document is the main goal of most investigations. In these cases, inks can be differentiated by determining their macro and/or micro optical properties. In particular, for ballpoint pen ink the UV–vis–NIR absorption/reflection spectrum is immensely informative and, though features may be highly convoluted, the spectrum from around 300 nm to 1200 nm is a reliable quantitative measure of the mix of dyes in that sample [3-6]. This technique is straightforward to execute and provides fairly rapid and repeatable results, performed in scientific laboratory or also in field with portable spectrophotometer.

Journal of Physics: Conference Series **249** (2010) 012060 doi:10.1088/1742-6596/249/1/012060

2.Material and methods

In order to assess the ability of UV-vis-NIR spectroscopy to distinguish between different ballpoint pen inks a Perkin Elmer Lambda 900 Spectrophotometer was used. In Figure 1 we present the reflectance spectra obtained directly from the ink lines of three different ballpoint.

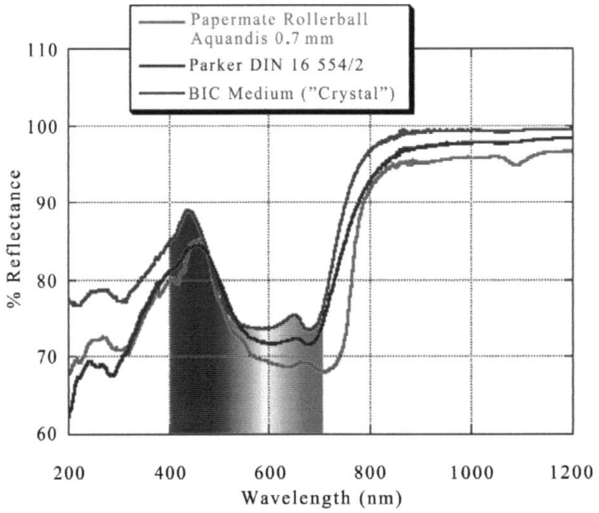

Figure 1. UV-Vis-NIR reflectance spectra relative to three different ballpoint pen inks. Small shift from the base line (100 % reflectance) have been introduced for improving the visibility of spectra.

Observing Figure 1 it is easy to see that the spectrophotometry has the potentiality to distinguish different inks. The different molecules used to obtain inks cause different absorption bands.
The infrared reflectance with illumination in the near infra-red region (700–1000 nm) can be used to detect whether two pieces of the same text are originated by different pen.
Figure 2(a) shows an example of the check with fraudulent changes of the amount: the addition of "1" in front of 100€ and "*Mille*" in front of "*cento*".

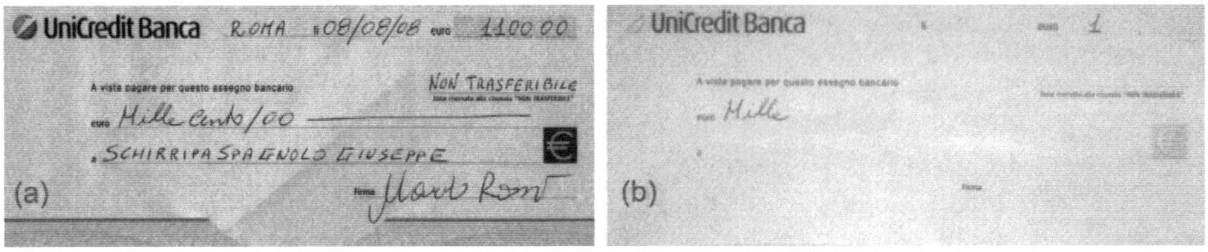

Figure 2. Example of check with fraudulent changes. Two black inks of different manufacture have been used. (a) Photo obtained using normal light. (b) Photo obtained in IR light. The difference absorption in the infrared region denote the use of two different pens.

The forgery can be detected also using a portable fiber optics spectrophotometer (see Figure 3). Figure 4 shows the UV-vis-NIR reflectance spectra of the two inks. The visible range exhibit no differences, on the contrary in the IR zone they have meaningful differences.

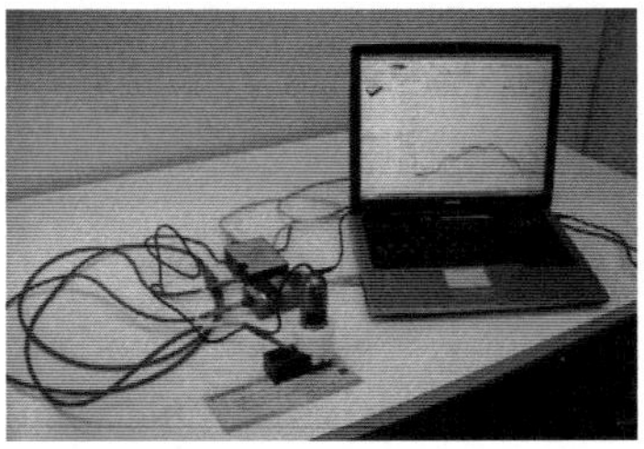

Figure 3. Portable fiber optics spectrophotometer.

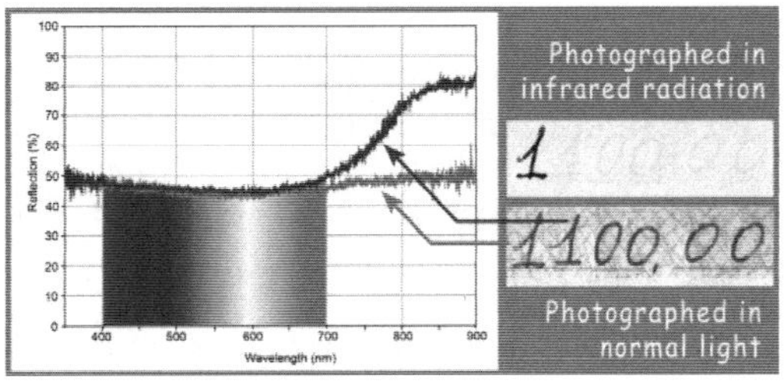

Figure 4. UV-vis-NIR reflectance spectra of the two different pen inks.

Not always the IR photo is able to identify if two pieces of written text was originated by different pens. The Figure 5(a) shows another check, compiled by ballpoint pen with blue ink, with the same kind of change found in the first example. The alteration was made with a second pen having, in visible range, a color "very similar" to that of the original text. In the IR photo (Figure 5(b)) it is not possible to identify any forgery.

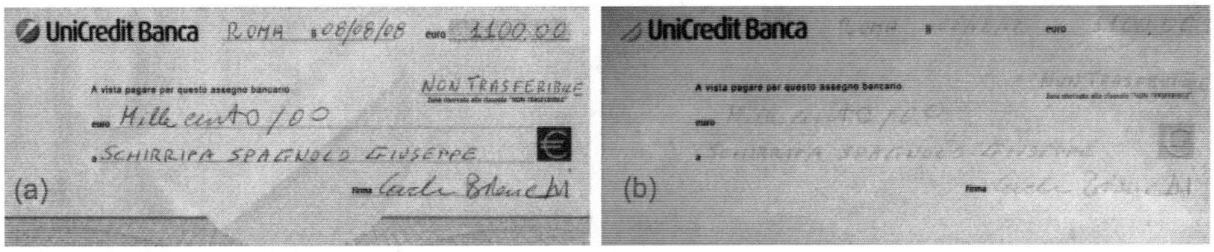

Figure 5. Example of check with fraudulent changes. Two blue inks of different manufacture have been used. (a) Photo obtained with white light. (b) Photo in IR light. In this case the IR photo isn't able to enhance the use of two different pens.

Also on the second check we are performed the UV-vis-NIR reflectance spectra with the portable fiber optics spectrophotometer (see Figure 6). As in the previous case, the spectra related to the two different ink lines, in the visible region, are similar. Instead, the measurements in the IR range are more determinant to distinguish between pens from different manufacturers.

Modern demands for specialized writing and printing instruments have resulted in an explosion of ink formulations, each of which may contain dozens of chemical components (e.g., acid or basic dyes, organic or inorganic color pigments, surfactants, antioxidants, viscosity adjusters, resins, glycol and glycerol, waxes, oils, and pigments). The organic components in the ink, generally, can be monitored studying NIR and IR spectra.

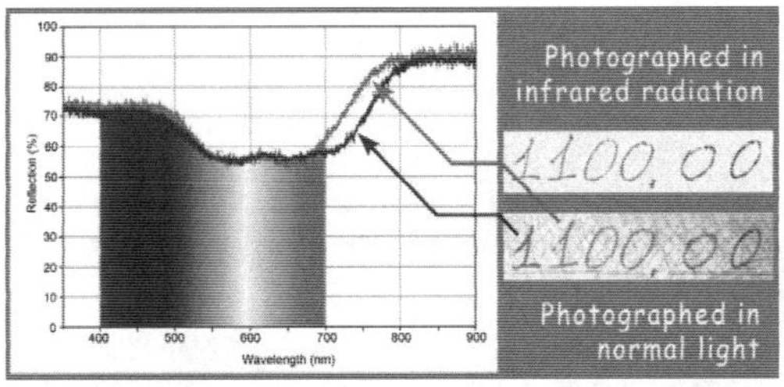

Figure 6. UV-vis-NIR reflectance spectra of the two pen inks used in above check.

3.Conclusion

In this paper the UV-vis-NIR reflectance spectrophotometry has been applied to determine whether two pieces of written text are originated from the same ink. Furthermore, we have set a fiber optics spectophotometer to obtain UV-vis-NIR spectra directly on the ink line of a document. Our experimental approach has the advantage of being non-destructive. The experimental set up has shown the possibility to characterize optically and differentiate the different inks in a fast and straightforward way.

In forensic examination of inks, one of the most challenging investigation is the dating of handwritten documents. With elapsing of time, the inks undergo to some change in their chemist/physical characteristics, related with the variations of document storage. Spectrophotometric measurements, on ink line to determine the age of handwritten documents, are in progress.

4.References

[1] Kelly J S, Lindblom Schientific 2006 *Examination of Questioned Documents* (CRC Press: Boca Raton) ISBN 978-0-8493-2044-6

[2] ASTM E1442-01 *Standard Guide for Test Methods for Forensic Writing Ink Comparison.*

[3] Roux C, Novotny M, Evans I, Lennard C 1999 A study to investigate the evidential value of blue and black ball-point pen inks in Australia, *Forensic Sci. Int.* **101** 167–176.

[4] Vila A, Ferrer N, Garcia J F 2007 Chemical composition of contemporary black printing inks based on infrared spectroscopy: basic information for the characterization and discrimination of artistic prints, *Anal. Chim. Acta* **591** 97–105.

[5] Adam C D, Sherratt S L, Zholobenko V L 2008 Classification and individualisation of black ballpoint pen inks using principal component analysis of UV–vis absorption spectra *Forensic Sci. Int.* **174** 16-25.

[6] Weyermann C 2005 Mass spectrometric investigation of the aging processes of ballpoint ink for the examination of questioned documents *Ph. D. Thesis , Faculty of Biology and Chemistry, Justus-Liebig-University Giessen.*

Comparative study of the dose rate influence in the radiochromic polymeric film dosimeter response

A M S Galante and L L Campos
Radiation Metrology Centre, Institute of Energetic and Nuclear Research,
Av. Prof. Lineu Prestes, 2242, 05508-900, Cidade Universitária, Sao Paulo, Brazil

E-mail: sgalante@ipen.br

Abstract. Most of film dosimetry is based on optical signal measurements since the spectrophotometric technique is cheap and widespread. Radiation induced chemical reactions take place in the material creating and/or enhancing absorption bands in the visible and/or ultraviolet regions of the spectrum. The response of most films used in routine dosimetry is usually influenced by the environmental conditions (temperature, humidity, ligth), dose rate and post irradiation storage. Several film dosimeters are commercially available for the absorbed dose measurements, in this work the effects of dose rate changes in optical density (OD) of film dosimeters (polymethylmetacrylate – PMMA) and polycarbonate – PC) were investigated and compared. Polycarbonates has been investigated as a new dosimetric material for industrial radiation processing aiming low cost, easy handle, fast and accurate response. PMMA dosimeters has been used routinely in dosimetry. The dose rate (kGy h^{-1}) may influence the dosimeter response and appropriate corrections are necessary. The dosimeters were irradiated with ^{60}Co gamma-radiation at dose rate in the range 0.26 – 2.6 kGy h^{-1} and with electrons from accelerator with maximum energy of 1.5 MeV and dose rate between 4.19 and 33.59 kGy s^{-1} to absorbed doses of 5 and 15 kGy, respectively. The irradiated dosimeters were measured using Shimadzu UV-2101PC spectrophotometer. It can observed that the films shows moderate dose-rate dependence (about 10%) in the analised range.

1. Introduction

Effects induced by ionizing radiation (accelerated electrons, ion and protons, gamma and X rays) have been largely used to modify the chemical and physical properties of the polymers [1,2]. Some of the changes have been atributed to the scissioning of the polymer chains, breaking of covalente bonds, promotion of cross-linkages, formation of carbon clusters, liberation of volatile species and, in some cases, even formation of new chemical bonds [3-5].

Polycarbonates are usually applied in neutron and alpha particles detection using nuclear tracks detection technique now a days has been studied as a dosimeter to measure gamma-rays doses [6]. The ionizing radiation changes the color of polycarbonates of colourless to yellow-green. The radiolysis mechanisms of polycarbonates explain the formation of isolated phenoxy radicals (responsible for the yellowing) and two different radical pairs as well as the evolution of large amounts of CO_2 and CO [7-9].

Several film dosimeters are commercially available and the PMMA is widely used as ionizing radiation dosimeter in industrial aplications. The interesting effect induced radiation in this material is the color or optical density change that can be related to the absorbed dose [10-12].

Optical absorption is a useful method for investigations of the induced changes in the polymeric material. The dosimetric system is based on the measurement of specific optical absorbance k at 412 nm and 640 nm wavelength for PC and PMMA, respectively.

In this work the dose rate dependence response for PC and PMMA red 4034 perspex detectors was studied for photons and particles radiations using in gamma source and accelerated electrons of 1.5 MeV maximum.

2. Experimental

Polycarbonate (PC) sheets (2 x 1 m) manufactured by Policarbonatos do Brasil and commercially available 3 mm thick (crystal) and density 1.2 $g.cm^{-3}$ was used for this work. The PC sheets were cut into strips 10 mm width and 30 mm length. Polymethylmethacrylate (PMMA), Harwell type Red 4034, is manufactured with a red dye, 3 mm thick, cut into ships 30 mm length, 10 mm width and sealed in laminated foil sachets.

Both samples were irradiated with ^{60}Co gamma-rays using the Gammacell 220 source installed at the Radiation Technology Center – CTR / IPEN-CNEN/SP. The absorbed dose rate correspondig to 2.55 kGy h^{-1} in march/2008, was detemined by Fricke dosimeter and certified by IAEA. The PC and PMMA films were tested for dose rate dependence for total doses up to 5 kGy.

To evaluated the electron dose-rate dependence of the OA response, both samples were irradiated with 1.25, 1.311, 1.40 and 1.499 MeV electrons at doses rate between 4.19 and 33.59 kGy s^{-1} varying the current between 0.97 and 7.77 mA. The irradiation process was performed in a JOB-188 Dynamitron Inc. Electron Accelerator, IPEN-CNEN/SP.

The UV-Vis measurements were performed in the maximun wavelength using the Shimadzu UV-2101PC. Each data point represents an average of three measurements.

3. Results

Radiation induced damage in polymers depend on accumulated absorbed dose and can result in dramatic changes in the polymer structure on relatively short time scales.

The response of PC and PMMA samples irradiated with ^{60}Co gamma rays with different doses rates, 0.255, 0.765, 1.275 and 2.55 kGy h^{-1} and 5 kGy is reported, figure 1 and 2, respectively. The dose rate at the studied range had no effect on the change of specific absorvance of the PMMA red 4034 perspex detectors, however, significant difference in the optical response was observed for differences of 10 times, in the rate dose at low doses for PC detectors; the influence at higher doses is being studied.

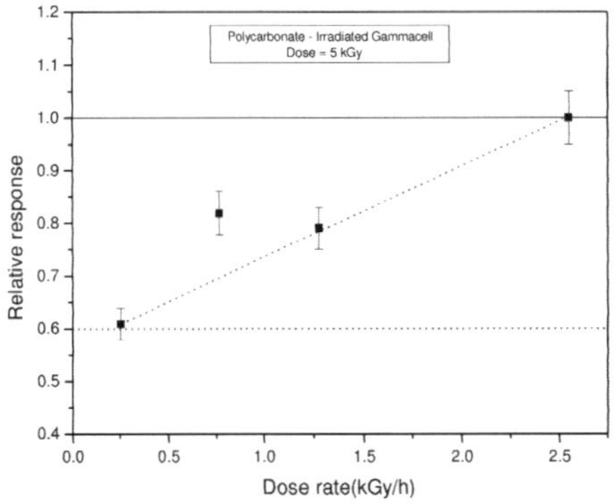

Figure 1. Dose rate dependence response for PC samples irradiated with ^{60}Co gamma radiation, measured at 412 nm wavelength.

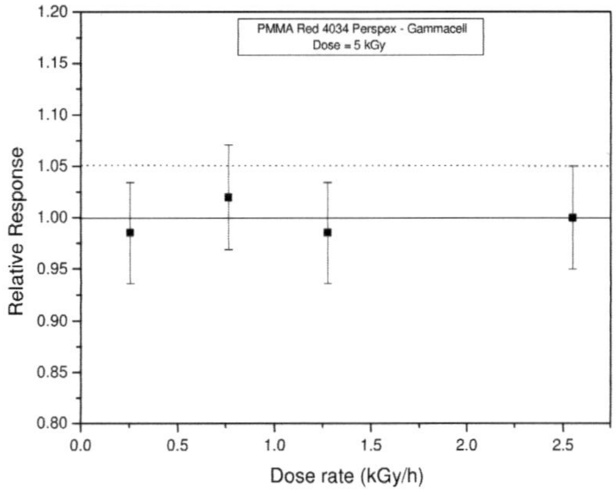

Figure 2. Dose rate dependence response for PMMA red 4034 Perspex irradiated with ^{60}Co gamma radiation, measured at 640 nm wavelength.

The optical response of PC and PMMA samples irradiated with electrons accelerated (absorbed dose of 15 kGy) in the conditions listed in the table 1 is showed in figure 3 and 4 for PC and PMMA, respectively. The change in the detectors response is less 10% between the maximum and minimun obtained values and 6% in average, demonstrating that there is no significant dependence of the response at different energies, currents and dose rates.

Table 1: Irradiation parameters – PC and PMMA samples irradiated in the JOB 188 eletrons accelerator, IPEN-CNEN/SP.

Irradiations Conditions	Energy (MeV)	Current (mA)	Dose rate (kGy/s)
1	1.250	0.97	4.21
2	1.250	1.89	8.39
3	1.250	3.79	16.81
4	1.250	7.57	33.59
5	1.311	0.97	4.19
6	1.311	1.94	8.39
7	1.311	3.89	16.82
8	1.311	7.77	33.59
9	1.400	1.01	4.20
10	1.400	2.02	8.41
11	1.400	4.04	16.82
12	1.400	8.07	33.59
13	1.499	1.05	4.19
14	1.499	2.11	8.42
15	1.499	4.21	16.79
16	1.499	8.42	33.59

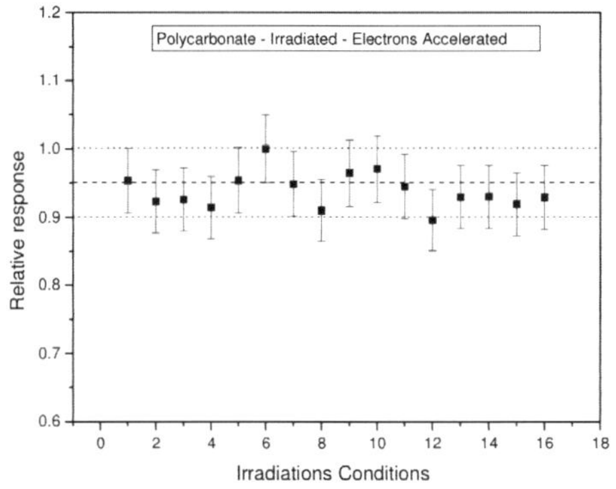

Figure 3. PC detectors irradiated with electron accelerated, dose 15 kGy.

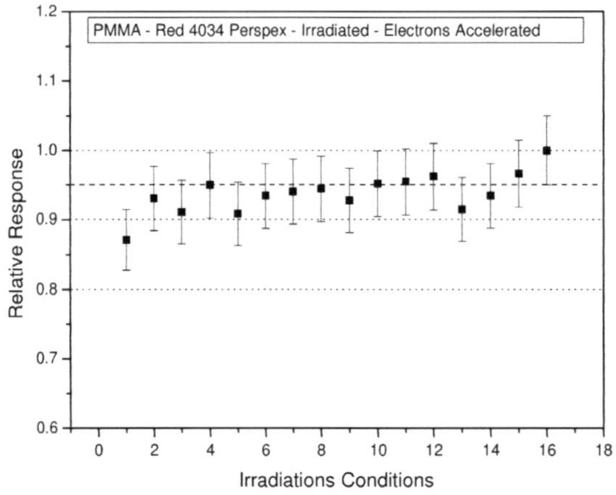

Figure 4. PMMA red 4034 Perspex detectors irradiated with electron accelerated, dose 15 kGy.

4. Conclusions

Comparisons between Polycarbonate and PMMA detectors has showed that both presents convinient dosimetric response when irradiated with ionizing radiation, the changes in optical density are visible, measurable and can be related with absorbed dose. These dosimeters are easy to handle, low cost and can be used in a wide dose range.

Dose rate dependence may introduce large errors to the dose measured. To minimize these errors the detectors can be calibrate in the uses conditions or determinated these errors and made appropriate corrections of the reading.

The dose rate effect (gamma and accelerated electrons) in the studied range for PMMA detectors presents no significative dependence, so it can be used in varied conditions of irradiation.

PC detectors presents significative dependence in response in differents gamma dose rates in low dose. This suggests that it should be used in a range of higher doses, research to determine this effect are being made. No significative effect was observed in irradiations with accelerated electrons.

Journal of Physics: Conference Series **249** (2010) 012061 doi:10.1088/1742-6596/249/1/012061

Acknowledgments
The authors are thankful to FAPESP (Fundação de Amparo à Pesquisa do Estado de São Paulo) and to CNPq (Conselho Nacional de Desenvolvimento Científico e Tecnológico) for the financial support and to Radiation Technology Center - CTR of IPEN for irradiations.

References
[1] *American Society for Testing and Materials* 1996 Standard practice for use of a polymethylmethacrylate dosimetry system (*ASTM E 1276 – 96*) Annual Book 12.02
[2] *American Society for Testing and Materials* 2005 Standard practice for use of cellulose acetate dosimetry systems (*ASTM 51650-05*)
[3] Chung J Y J 1997 Stabilization of gamma-irradiated polycarbonate *Medical Plastics and Biomaterial Magazine*. Technical Paper Series
[4] Kumar R, Virk H S, Verma KC, De U, Saha A and Prasad R 2006 Physico-chemical modifications induced in Makrofol N-polycarbonate by swift heavy ions. *Nuclear Instrumensts and methods in Physics Research B* **251** 163 – 166
[5] Sinha D, Sahoo K L, Sinha U B, Swu T, Chemseddine A and Fink D 2004 Gamma-induced modifications of polycarbonate polymer *Radiation Effects & Defects in Solids* **159** 587-595
[6] Souto E B 2007 Projeto, montagem e caracterização de um dosímetro para radiação de nêutrons. Dissertação (Mestrado) - Instituto de Pesquisas Energéticas e Nucleares - IPEN/CNEN-SP São Paulo
[7] Ilić R, Skvarc J and Golovchenko A N 2003 Nuclear tracks: present and future perspectives *Radiat. Meas.* **36** 83 - 88
[8] Nouh S A, Morsy A A and El-Husseiny H M 2004 Radiation-induced changes in some SSNTDs. *Radiation Effects & Defects in Solids* **159** 115 - 124
[9] Saad A F, Atwa S T, Yokota R and Fujii M 2005Radiation-induced modifications on spectroscopic and thermal properties of CR-39 and SR-90 nuclear tracks detectors *Radiat. Meas.* **40** 780 - 84.
[10] Galante A M S, Villavicencio A L C H and Campos L L 2004 Preliminary investigations of several new dyed PMMA dosimeters *Radiat. Phys. Chem.* **71** 391- 394.
[11] Galante A M S and Campos L L 2006 Electron dose radiation response of dyed PMMA detectors developed at IPEN. *Radiat. Prot. Dosim.***120** 1-4 113-116
[12] Whittaker B and Watts M F 2001 The influence of dose rate, ambient temperature and time on the radiation response of Harwell PMMA dosimeter. *Radiat. Phys. Chem.* **60** 101-110

Journal of Physics: Conference Series **249** (2010) 012062 doi:10.1088/1742-6596/249/1/012062

ESR dating of teeth from Brazilian megafauna

L C Oliveira[1], A Kinoshita[1,2], A M F Barreto[3], A M Figueiredo[4], J L L Silva[5] and O Baffa[1]

[1]Departamento de Física e Matemática, FFCLRP, USP Ribeirão Preto, SP, Brazil.
[2]Universidade do Sagrado Coração, Bauru, SP, Brazil.
[3]Universidade Federal de Pernambuco, Recife, PE, Brazil
[4]Instituto de Pesquisas Energéticas e Nucleares (IPEN), São Paulo, SP, Brazil
[5]Universidade Federal de Alagoas, Al, Brazil

E-mail: lcoliveira@biomag.usp.br

Abstract. The study of radiation defects created in biomaterials, such as bone and teeth, can be used in dating with importance to palaeontology and archaeology. Two Stegomastodon teeth (AL1 and AL2) from north-eastern Brazilian megafauna were studied by electron spin resonance (ESR) spectroscopy. The samples were collected in Fazenda Ovo da Ema, (913349 / 3714965) UTM, Alagoas state, Brazil. The dating of these samples can contribute to the better knowledge of megafauna presence in this region as well as to the events associated to the extinction of these species.

1. Introduction

Electron spin resonance (ESR) is a non-destructive method of measuring the concentration of paramagnetic centres and free radicals in solids and liquids. Its application to dating technique is based on the fact that fossil material received radiation from uranium, thorium, potassium and their decay products plus cosmic radiation during the period that it was buried. The absorbed dose can be converted in age through the annual dose rate of the place where the sample was collected and the internal dose rate given by the radioisotopes concentration. For teeth the ROSY ESR software has been successfully used to this calculation [1].

The ESR dating signal arises from carbonate impurities in the dahllite lattice [2]. It exhibits an axial powder pattern ESR spectrum with g-value near 2.0018 and 1.997 (at 337.6 mT and 338 mT respectively in figure 1). Dose response of the peak-to-peak intensity of the signal at g=2.0018 (signal A) is routinely used in ESR dating of tooth enamel to determine the equivalent or archaeological dose (AD).

The ESR measurement is relatively simple, requires a small quantity of material, and the ESR technique has a wide range of age applicability (from a few hundreds to million years). The amount of material to be dated can be an impediment for the use of others conventional techniques. Higher microwave frequency has been used as an alternative in these cases, allowing the use of ESR dating for small sample masses and also getting spectrum with higher resolution and sensitivity [3]. Kinoshita et al, 2005 [4] demonstrated that the K-Band (24GHz) ESR spectra of fossil teeth presents similar resolution obtained using Q-Band (35GHz) with advantage in the signal quantification that is more precise in comparison with 35GHz. The ESR dating has generated relevant data for archaeological studies, especially in cases where fossilization process does not allow the use of the conventional C-14. These results are in agreement with data from others techniques [5,6,7,8].

It is important to stress that this dosimetric/dating technique not destroy the paramagnetic centers during the measurement process and in situ measurements can also be performed with appropriate instrumentation [9].

In the north-eastern Brazil countryside a case of extinct wildlife Pleistocene of large mammals, known as megafauna, has been documented. This fauna, tiger-teeth-of sable, mastodontes, giant armadillos, giant Sloths, toxodon, paleolhama, among others, are found mainly in deposits of tanks, which are depressions in crystalline embasement of the Pre-Cambrian, which formed old lagoons. It is common assign to these animals an age which is linked to the end of the Pleistocene, or approximately 10 thousand years before the present. After the interstadial stage (+ / - 60,000 years), a warmer phase of the last glacial period in the Late Pleistocene. Possibly climate changes with a tendency for more dry and warm periods, leading to reduction in the supply of water in the region caused changes in vegetation composition that nowadays have the predominance of plants with leaf that has little mass and little nutritional value. The extinction of these mammals and fauna are associated mainly to these changes.

2. Materials and methods

2.1. Sample preparation and measurement set up

Two Stegomastodon teeth samples were obtained from Fazenda Ovo da Ema, (913349 / 3714965) UTM, Alagoas state, Brazil. The enamel was mechanically separated from dentin and chemically treated with a saturated solution of NaOH (30%) in an ultrasound bath, in order to remove the remaining dentin. The samples were etched in an HCl solution (1:10) and an external layer of ~ 500μm was eliminated. Later, the enamel was powdered in fine particles ($\phi < 0.5$mm), divided in 14 aliquots of about 100 mg and a set of additive dose, up to 2500 Gy was given. The enamel powders were irradiated by ^{60}Co γ-rays at room temperature with a dose rate of 4.47 kGy/h with an appropriate build up layer. The doses were 0, 10, 20, 50, 100, 200, 300, 400, 600, 800, 1000, 1500, 2000 and 2500Gy respectively, for each aliquot.

ESR measurements were performed at room temperature, after two months, with a Varian E-4 X-band spectrometer. The magnetic field was modulated at 100 kHz with peak-to-peak amplitude of 0.2 mT. A 20 mW microwave power was used in order to avoid saturation of the ESR signals. A scan range of 10 mT with a scan time of one minute and time constant of 200 ms was taken, given 300 data points resolution.

2.2. g factor measurement

The absolute measurement of g factor requires knowledge of the microwave frequency and the resonant magnetic field. A number of systematic errors can enter [10] in this determination; hence absolute measurements are rarely attempted. Instead, it is usually employed standards for which the g factors are well known. In this work, the DPPH (α, α-diphenyl-β-picrylhydrazyl) powder with g factor of 2.00037±0.0002 was used as standard. The unknown g factor g_X is then calculated from

$$g_X = \frac{g_S H_S}{H_X} = \frac{g_s H_S}{H_s - \Delta H} \cong g_s\left(1 + \frac{\Delta H}{H_s}\right) \quad (1)$$

here g_S is the g factor of the standard and H_S and H_X are the resonant magnetic fields for standard and unknown, respectively, and $\Delta H = H_S - H_X$.

2.3 Additive dose method and age

In this work the peak-to-peak method was used. This method consists on taking the amplitude between the top to the bottom of a chosen ESR line, as showed in Figure 1 represented by the letter A. A sample irradiated with known doses (Q) yields a growth curve. For each aliquot measurement the distance A is taken and plotted against the received dose. Later, the experimental data points are fitted to a mathematical expression. The model that best express the growth curve as function of received dose was discussed by Ikeya to be an exponential that is expressed in equation (2).The following equation was used to relate the natural dose or archaeological dose (AD) received by the sample up to the present:

$$I(Q) = I_s(1 - e^{-(Q+AD)/SD}) \tag{2}$$

where I_S is the saturation intensity, SD is the saturation dose in a laboratory irradiation. Thus, by fitting the experimental data (dose response) to equation 2, the equivalent/archaeological dose (AD) can be found out.

The age t of a tooth is determined by the ratio of the equivalent dose to the effective annual dose rate d, which is composed of external (d_{ext}) and internal (d_{int}) components. This can be done under the hypothesis that the annual dose rate presently determined is the same as in previous time and that the fading of signal is negligible. Therefore, an integral expression is [11] is used to solve for t:

$$AD(t) = \int_0^t d\,dt \tag{3}$$

where, AD is obtained through the additive dose method. The value for d as well as for t was found out by using the ROSY software through the radioisotope's concentration in the sediment and in the sample [3]. The concentration of ^{238}U and ^{232}Th present in the samples and from the soil where the samples were buried were obtained by neutron activation analysis (NAA) and the concentration of potassium in the samples was determined by atomic absorption spectroscopy.

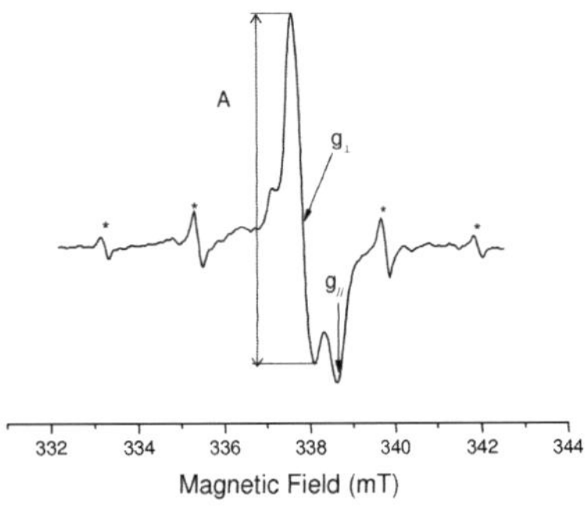

Figure 1. ESR spectrum of tooth enamel (AL2). Signal amplitude at g⊥ was used to determine the AD. * indicates the isopropyl ($(CH_3)_2C$.-R) signal.

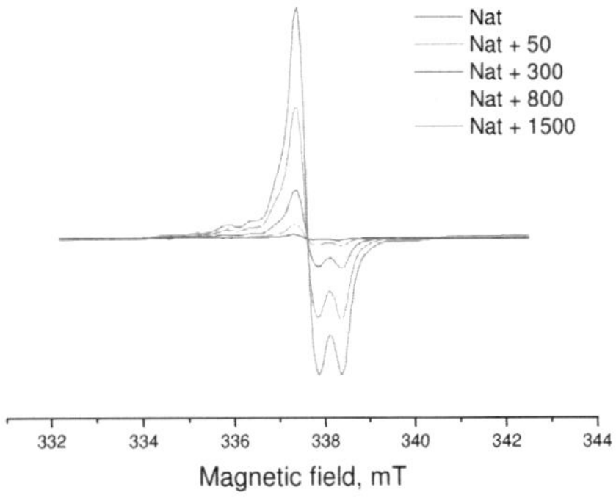

Figure 2. Enhancement of the ESR signal of the AL2 tooth sample obtained after 50, 300, 800 and 1500Gy, respectively of ^{60}Co irradiation.

3. Results and discussions

The native ESR spectrum of these samples shows a strong signal with spectroscopic factors at $g_\perp$ = 2.0025 and $g_{//}$ = 1.9973 related to CO_2^- radical in hydroxyapatite. Figure 1 shows the native spectrum of sample AL2. We can also verify the presence of the signal of isopropyl radical with hyperfine splitting of 2.17 mT. This septet signal was observed by others authors in middle Pleistocene tooth samples [11].

The additive dose method was used to plot the dose-response curve (figure 3) and an archaeological dose (AD) of 24.5 ± 0.5 Gy and 87.8 ± 0.5 Gy, for AL1 and AL2, respectively, were found out by using exponential fitting, through the equation (2). Carefully measurements made it possible to find a correlation level higher than 0.999. Through the figure it is possible see that the fitting is visually very good.

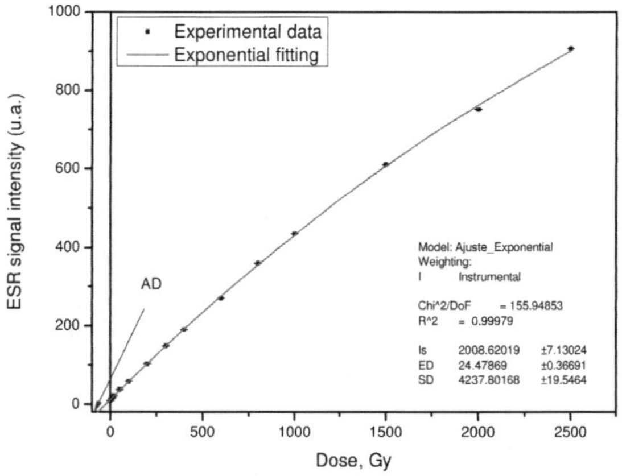

Figure 3. Dose–response curves of sample AL1 fitted by Equation (2).

The total annual dose rate and the sample age were found by using the ROSY software [1] where we introduced the concentration of uranium-238, thorium-232 and potassium-40 as well as the cosmic rays rate. The concentration of these radioactive elements was found by NAA technique. The results of this analysis are reported in Table 1. With these data on hand and the archaeological dose the age of the two samples were calculated. In Table 2 the age of the two teeth sample are presented as well as the environmental annual dose rate, where the samples were buried. The values, for the annual dose rate, are relatively high when they are compared with the earth's dose rate that is generally between 0.5 and 1.0 mGy/a. However, other author has found value of 1.7 mGy/a (for more details see reference [5]). Nevertheless, values lower than 0.5 mGy/a or higher than 5mGy/a are rarely to be found at an ordinary place. This fact, allow us to conclude that these values are acceptable.

Our results indicate dates that are coherent with other archaeological and paleontological findings by others authors in Brazil [12].

Table 1. Radioisotope concentrations in the tooth (enamel and dentine) samples and soil obtained by NAA (μg g-1).

Sample	**[^{238}U] (ppm)**	**[^{232}Th] (ppm)**	**[^{40}K] (ppm)**
AL1- Enamel	< 0.05	<0.01	<750
AL2- Enamel	< 0.05	<0.01	<750
AL1- Dentine	10.9 ± 0.5	<0.01	<750
AL2- Dentine	2.9 ± 0.5	<0.01	<750
Soil	4.1±0.8	22.7±0.2	17837±444

Table 2. Ages and total dose rate (D_t) of the tooth samples. The cosmic rays dose of 250 μGy/a was considered.

Sample	**AD**	**EU age**	**EU D_t**	**LU age**	**LU D_t**

	(Gy)	(ka)	(mGy/a)	(ka)	(mGy/a)
AL1	24.5±0.5	10±0.5	2.355	10±0.5	2.340
AL2	87.8±0.5	39.5±1	2.220	39.8±1	2.202

Conclusions

Based on knowledge of the physics of stable radiation defects created in hydroxyapatite, present in enamel, it has been possible to determine the accumulated dose in the material and by modeling the dose deposition the age can be calculated. Our results indicate dates that are coherent with other archaeological and paleontological findings in Brazil.

The ESR dating of Stegomastodon waringi tooth are the first ones that were done on mammals fossils on Alagoas state. The ages obtained by ESR suggest that this animal had lived on this region between the interstadial and on the end of the Pleistocene period. This warmer period was indicated by the appearing and predominance of plants with little leaf mass on the semi-arid region of the Brazilian northeast.

Acknowledgments

To Brazilian agencies Fundação de Amparo a Pesquisa do Estado de São Paulo (FAPESP), Conselho Nacional de Desenvolvimento Científico e Tecnológico (CNPq) and Coordenação Aperfeiçoamento do Pessoal de Ensino Superior (CAPES) for partial financial support. IPEN for the irradiation of the samples and C.A. Brunello and Lourenço Rocha for technical support.

References

[1] Brennan B, Rink W, Rule E, Schwarcz H, Prestwich W. The Rosy ESR Dating Program. Ancient TL. 1999;17:45–53.

[2] Rink W. Electron spin resonance (ESR) dating and ESR applications in quaternary science and archaeometry. Radiation Measurements. 1997;27(5-6):975–1025.

[3] Kinoshita A, Karmann I, Cruz Jr, FW, Graeff CFO, Baffa O, K-band ESR spectra of calcite talagmites from southeast and south Brazil. Applied Radiation and Isotopes. 2005;62:247–250.

[4] Kinoshita A, Franca AM, Almeida FAC, Figueiredo AM, Nicolucci P, Graeff CFO, Baffa O. ESR dating at K and X band of northeastern Brazilian megafauna. Applied Radiation and Isotopes . 2005;62:225–229.

[5] Kinoshita A, Figueiredo AMG, Felice GD, Lage MCSM, Guidon N, Baffa O. Electron spin resonance dating of human teeth from Toca da Santa shelter of Sa˜o Raimundo Nonato, Piauí, Brazil. Nuclear Instruments and Methods in Physics Research B. 2008;266:635–639.

[6] Watanabe S, Ayta WEF, Hamaguchi H, Guidon N, La Salvia ES, Maranca S, Baffa O. Some evidence of a Date of First Humans to Arrive in Brazil. Journal of archaeological Science. 2003; 22:1361-1366.

[7] Skinner AR, Hay RL, Masao F, Blackwell BAB. Dating the Naisiusiu Beds, Olduvai Gorge, by electron spin resonance. Quaternary Science Reviews. 2003;22:1361–1366.

[8] Mascarenhas S, Baffa O, Ikeya M. ESR Dating of Human Bones from Brazilian Shell-Mounds (Sambaquis). American Journal of Physical Anthropology. 1982;59:413-417.

[9] Furusawa M, Ikeya M. Electron spin resonance imaging utilizing localized microwave magnetic field. Jpn J Appl Phys. 1990;29:270–276.

[10] Swartz HM, Bolton JR, Borg DC. Biological applications of electron spin resonance. John Wiley & Sons; 1972.

[11] Ikeya M. New Applications of Electron Spin Resonance: Dating, Dosimetry and Microscopy. World Scientific; 1993.

[12] Kinoshita A, Barreto A, Alves R, Maria Figueiredo A, Eduardo de Souza Sarkis J, Dias ML, et al. ESR dating of teeth from northeastern Brazilian megafauna. Radiation Measurements. 2008;43(2-6):809–812.

Journal of Physics: Conference Series **249** (2010) 012063 doi:10.1088/1742-6596/249/1/012063

Thin-film structures with nanocrystals: an origin of enhanced photo-response

O Goncharova[1, 2, 3], **R M Montereali**[2] and **G Baldacchini**[2]
[1] Institute of Physics, National Academy of Sciences, 68 Independence Avenue, Minsk 220072, Belarus
[2] Frascati Research Center, Ente per le Nuove Technologie, l'Energia e l'Ambiente, 45 E. Fermi Street, Frascati 00044, Rome, Italy

E-mail: OlgaGoncharova08@gmail.com

Abstract. To discover well the properties of nanothin crystalline layers and nanometer-sized crystals, we investigated the relatively thick multilayer structures composed of high quantity of nanothin layers with nanocrystals. Alternate nanolayers of 150–10 nm thicknesses with LiF, CaF_2 and CdS nanocrystals have been deposited onto irradiation-resistive substrates by thermal evaporation of pure crystals. Some multilayers were γ-irradiated in air at room temperature with dose of 83 kGy. X-ray diffraction and microscopy studies reveal that the multilayers consist of nanocrystals with cubic phase and defined size. Thin-film structures were oriented along the (111) plane. Absorption spectra of non-irradiated LiF nanocrystals of 100 nm size suggest evidence of metal colloids presence. We find that photoluminescence spectra of γ-irradiated nanostructures with metal colloids and various LiF contents show the enhancement of F_3^+-colour centres excitation in the region of metal colloids absorption and the increase is observed of the emission intensities ratio of F_3^+ and F_2 centers with respect to initial crystals γ-coloured in identical conditions. Emission intensities of both centers under excitation at 458 nm correlate with LiF content. These effects, which are related to high-quality nanostructures, but at the same time depend strongly on the defect content, especially as far as their 1–2 ps nonlinearities are concerned, could depend on nanocrystal purity and metal excess collection in their boundaries regions. Our results provide an original contribution to the understanding of the influence of the nanolayer-by-nanolayer deposition, γ-irradiation on these specific structures, and of the metal aggregates on the properties of nanocrystals and nanolayers.

1. Introduction

Nano-sized layers, films, crystals and multilayers play an important role in future technology as they exhibit different and often unique properties with respect to the initial macroscopic materials [1–6]. Film structures with nanocrystals have attracted much interest for the possibility to grow them in their final position within switching devices, detectors, efficient dosimeters, emitters and solar cells. The possibility of using these nanostructures not only for fundamental studies, but also for novel devices fabrication, is stimulating the efforts to control the nanocrystal size, shape, structure, space arranging as well as their composition and structural defects. Unfortunately, not only bulk, but boundary (surface) states of nanolayers and nanocrystals affect strongly their properties as their size reduces [1, 7]. The properties of the nanoelements as prepared and post-growth processed by different methods

[3] To whom any correspondence should be addressed.

can be affected by the presence of various defects and also by aggregation of impurities and intrinsic defects to the colloids in the bulk and surface regions related to the used preparation technique.

Several physical deposition methods are usable for preparation of thin films and nanostructures such as laser-assisted, electron-beam or thermal evaporation from composite targets and undoped bulks, sequential sputtering or evaporation [1–6, 8]. Ion implantation [9], thermal treatments [10] and ionising radiation bombardment (combined with special annealing [11–14]) are established tools for metal nanoparticles formation in solids. Understanding the structural evolution during the preparation of nanocomposites can be of great help in understanding how to control the characteristics of the final products readily.

Lithium fluoride is a well-known material which, after irradiation by ionizing radiation (electrons, ions, γ-rays and X-rays) or laser light, is characterized by *F*-colour centres (CCs) aggregates [3–5] and metal colloids formation sensitive to the content of oxygen, hydroxyl and divalent metal ions in initial crystals [11–14]. Clusters of *F*-CCs (lithium colloids) and impurity colloids can be created in the irradiated regions of LiF and at temperatures where diffusion of the defects is possible [10–12]. Even though the actual mechanism of the formation of such colloids is still an open question, a lot of attention was devoted to study optical properties of the crystals and their near surface nanothin layers containing these metallic aggregates. Since thin-film nanolayers and nanocrystals have been studied only in a limited way, in this paper, we carried out the following measurements to investigate them.

2. Experimental details and measurements

Microstructural and optical characteristics of the investigated samples are summarized in figures 1–8. The experimental samples are multilayers and thin-film interferometers (TFIs) with the intermediate multilayers [15–17]. Each structure unit is a thin-film nanolayer with a nanocrystal array inside it (figures 1, 2). The geometry of the nanostructures can be tuned by controlling the thicknesses and the compositions of the "sensitizing" and complementary, buffer, nanolayers. Every nanolayer can be designed of one material [15] and as composition of nanocrystals, clusters of two materials [16].

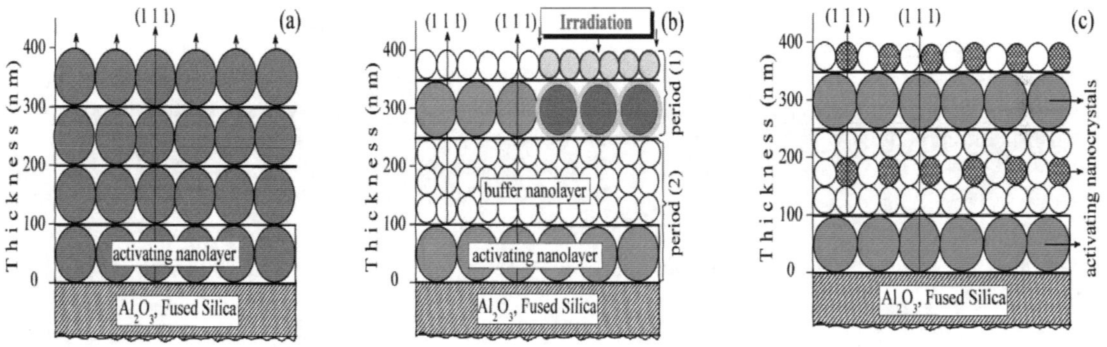

Figure 1. Schematic representation of the periodic multilayer nanostructures with *c*-oriented nanocrystals prepared from one (a), two (b) and more than two initial materials (c).

2.1. Samples preparation

The elaborated samples consist of dielectric (LiF, CaF_2) or/and semiconductor (CdS) nanocrystals. Samples with various "sensitizing" nanocrystals are grown on radiation-resistant, flat substrates in the layer-by-layer manner of the reproducing of thin-film nanolayers with defining compositions [15, 16]. "Sensitizing" nanolayers are prepared with thicknesses equal to nanocrystal sizes [15], and by using special techniques for nanocrystal ordering [17]. The approach presented in this paper can be extended to fabricate thin-film multilayer structures of a wide range of materials. To avoid radiative coloration, thermal evaporation was the preferred fabrication technique. Undoped crystals, 99.999% pure, are used as starting materials. The clean substrates were kept at room temperature (RT) on a copper holder, and the vacuum pressure in the evaporation chamber was about 10^{-6} Torr. The films thicknesses were measured by interference method. As-prepared (fresh) samples are kept for seven days in a dry box before starting the optical measurements in the open atmosphere at RT at about 50% humidity. Crystals and films produced in the manner of continuous growth of the result thicknesses were used as reference samples. After the growth some samples were γ-irradiated at RT up to 83 kGy [19]. Self-assembled nanolayers of LiF and CaF_2 nanocrystals were used as templates. In the case of CdSe/CaF_2, CdS/CaF_2 [20] and LiF/CaF_2 nanostructures (figure 2), more radiation-resistant crystalline CaF_2 nanolayers were used to promote the *c*-axis ordering and the size restriction of "sensitizing" CdSe, CdS and LiF nanocrystals. After thermal evaporation and post-growth γ-colouration, spatially ordered arrays of CCs could be fabricated in the periodic multilayer nanostructures (figure 1b) [6].

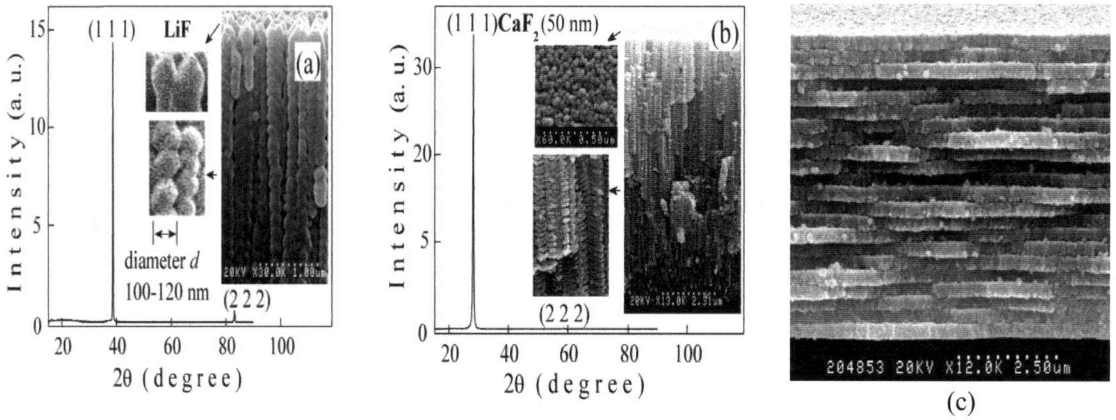

Figure 2. XRD and SEM images of non-irradiated thin-film structures with *c*-oriented nanocrystals of cubic phase and defined sizes, prepared from: LiF (a), CaF_2 (b), and LiF and CaF_2 initial crystals (c).

2.2. Samples characterization

The microstructure of thin-film samples is determined by X-ray phase analysis, scanning electron microscopy (SEM) and atomic force microscopy (AFM). The *c*-axis ordering (texture) analysis of thin-film nanostructures was performed by means of pole figures and θ-2θ diffraction patterns (figures 2–6). A Seifert XRD *3003 PTS* four circles diffractometer employing CuK_α (λ= 1.54059 Å) radiation was used for pole figure measurements. The primary beam was collimated by a 0.5 mm pinhole and a 1 mm slit was placed in front of the detector. The accelerating voltage and current was fixed at 40 kV and 30 mA, respectively. Pole figures were recorded varying ϕ between 0° and 360° with a sampling interval of 4°, and varying χ between 0° and 75° with $\Delta\chi$=2.5°. The acquisition time was 5 seconds. X-ray θ-2θ diffraction pattern were collected by using a Rigaku Geigerflex diffractometer with CuK_α radiation. The diffractometer was equipped with a graphite monochromator in order to suppress K_β radiation and to increase signal to noise ratio. The divergence slit, scattering slit and receiving slit were 0.5°, 0.45° and 0.3°, respectively. The measurements were performed with a sampling interval of 0.05° and a fixed time of 10 seconds.

Optical transmittance (*T*) and reflectance (*R*) spectra of samples are measured with a dual beam *Cary 500 Scan* and a *Lambda 19DM* Perkin-Elmer spectrophotometers in the spectral range 190–3300 nm (figures 6, 8). Optical absorption (OA) (figure 7) is derived from transmittance and is used to estimate the defects in nanocrystals with respect to initial crystals. The photoluminescence (PL) measurements were carried out by using a Jobin Yvon *Fluorolog-3* spectrofluorometer with the front-face detecting geometry, by exciting the samples at the wavelength of 458 nm (*M*-band, LiF), see figures 7b,c. Nonlinear properties were detected using a pump-and-probe technique and 3-ps [11–15] and 150-fs laser pulses [3, 5], and reported in figure 8. All measurements were performed at RT.

3. Thin-film structures with dielectric nanocrystals

3.1. Thin-film structures with nanocrystals of one fluoride material

X-ray diffraction (XRD)-, SEM- and AFM-results show that non-irradiated multilayers are structures with definite size of nanocrystals. (Final thickness of LiF, CaF_2 and LiF/CaF_2 structures is approximately 6.5 μm.) Presence of (111) and (222) peaks in XRD data of the multilayers indicates that they have *c*-axis orientation perpendicular to the substrate surfaces. The XRD patterns of LiF and CaF_2 multilayers, see figure 2, show intense (111) peaks at $2\theta = 38.7°$, $28.3°$ and a considerably less intense (222) peaks at $2\theta = 83.1°$, $58.5°$, respectively. Not-presented here XRD-, SEM- and AFM-results validate the fact that irradiation neither causes a change in the packing nor increases the size of nanocrystals. The (111) peaks prevail in the XRD-patterns of samples, and γ-irradiation with the dose of 83 kGy only improves the crystallinity and the orientation of nanostructures (increase in the intensities of (111) peaks). The XRD results correlate well with the texture analysis (figures 3–5).

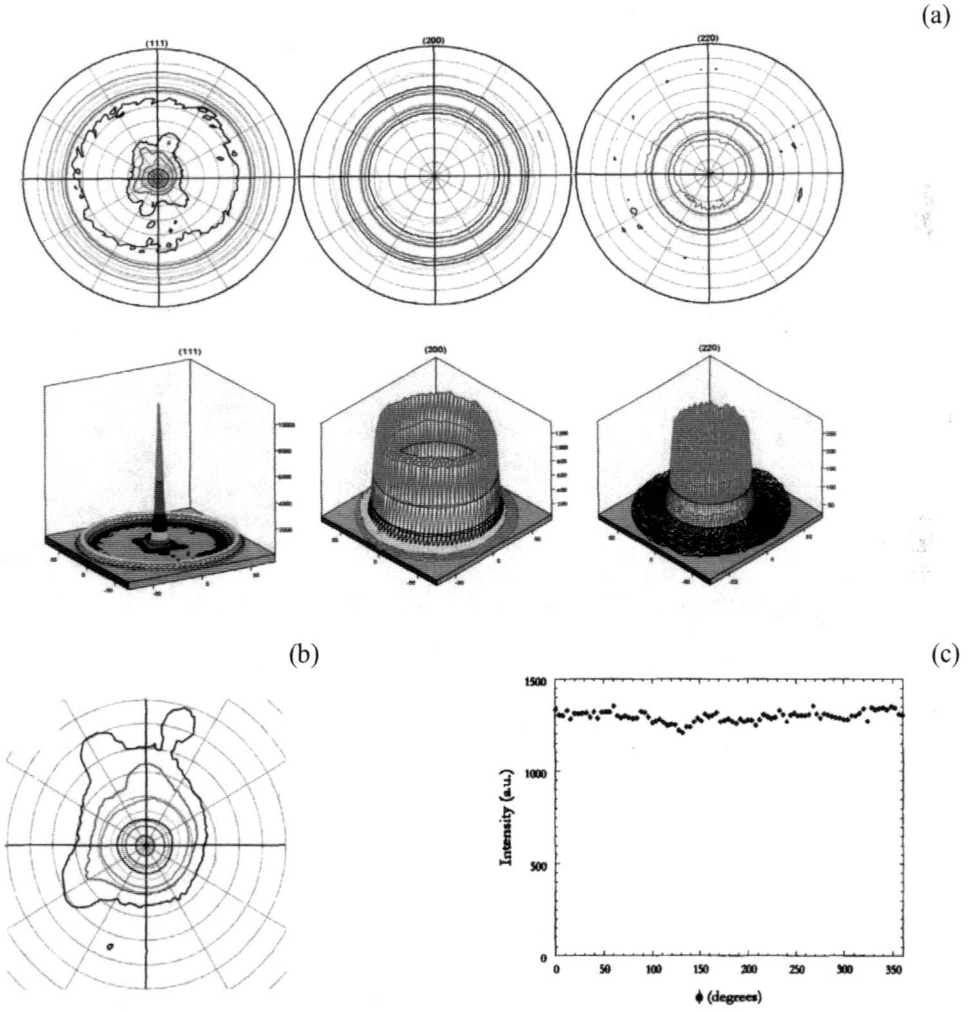

Figure 3. 2d and 3d (111), (200) and (220) LiF pole figures of the non-irradiated LiF nanostructures (a). Magnification of (111) LiF pole figure in the region $\chi < 30°$ (b) and ϕ-scan of the (200) LiF pole figure demonstrating the in-plane isotropy (c).

In figure 3a two (2d) and three dimensional (3d) (111), (200) and (220) LiF pole figures of the non-irradiated LiF sample are reported. A sharp (111) texture is present. The Full Width at Half Maximum (FWHM) of the pole located at χ=0 in the (111) pole figure is about 6°, and the FWHMs of both (200) and (220) poles are 8°. This is due to a very low mosaic spread in the growth direction. Figure 3b shows a magnification of the (111) pole figure in the range χ <30°. The isointensity lines were chosen in the following way: the intensity of the nth line is $I_n = I_{max}/2^n$. The circular shape of the pole indicates that the in-plane nanocrystal distribution is uniform. This hypothesis is supported by the analysis of the cut of the (200) pole image at χ=55, shown in figure 3c: no difference in intensity is observable varying ϕ angle, confirming that any preferential in-plane direction does not exist.
The (111), (200) and (220) LiF pole figures for the γ-irradiated LiF sample are shown in figure 4. For this sample perfect (111) texture is presented (a). The FWHM of the (111), (200) and (220) LiF poles are 5°, 8° and 7°, respectively. The detailed (111) pole in the range χ<30° (b) and the cut of the (200) pole figure at χ~55° (c) are also reported. The (111) LiF pole profile is perfectly circular and the (200) pole ϕ-scan is constant indicating a high in-plane isotropy, differently from what we observed for non-irradiated LiF sample in which a low fraction of slightly misoriented nanocrystals can be inferred.

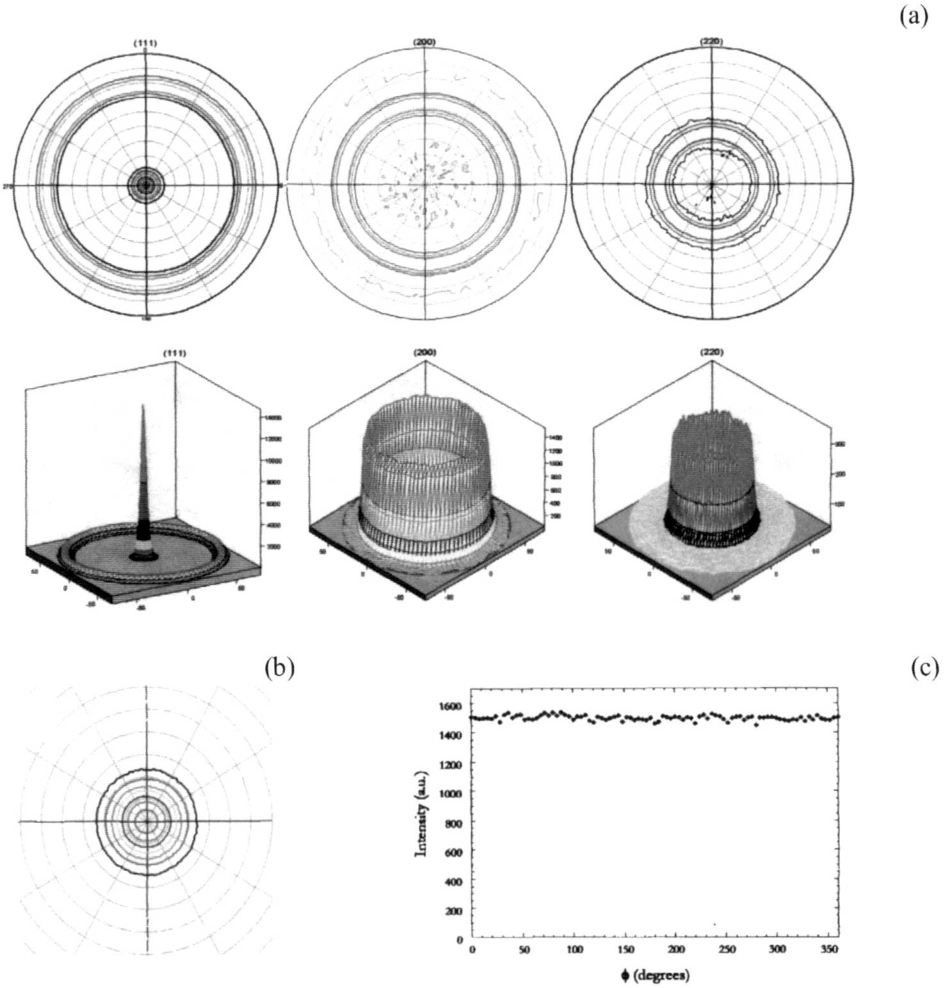

Figure 4. 2d and 3d (111), (200) and (220) LiF pole figures of the γ-irradiated LiF nanostructures (a). The perfect texture is shown. Magnification of (111) LiF pole figure, χ <30° (b) and ϕ-scan of (200) figure demonstrating in-plane isotropy (c).

It must be emphasized that the difference is very small: an analysis of the (111) LiF pole figures revealed that the integrated intensity of the range χ<10° ($I_{\chi<10}$) to the integrated intensity of the total (111) pole figure (I_{total}) ratio varies from 14.95% (for the non-irradiated LiF sample) to 15.05% (for the sample γ-irradiated at RT with dose 83 kGy). Conversely, the pole maximum is incresed of about 20% in the irradiated sample confirming the effect of the structure improving due to the γ-irradiation.
In figure 5 the (111), (400) and (220) CaF_2 pole images of non-irradiated CaF_2 sample are reported (a). Also in this case, the presence of a good (111) texture is evident. The FWHM of (111), (400) and (220) poles are 5°, 7° and 7°. The (111) peak profile is circular down to the lowest intensity and no evidence of peak broadening was observed. The pole is misaligned respect to the centre of about 0.5°, as can be seen in figure 5b. Since the measurement was repeated several times, an error in the mounting of the sample can be excluded. A different profile of the ϕ-scan performed on the (400) and (220) poles was recorded (figure 5c), characterised by a weak intensity dependence vs ϕ. This behaviour is due to the (111) CaF_2 direction not exactly perpendicular to the substrate surface.

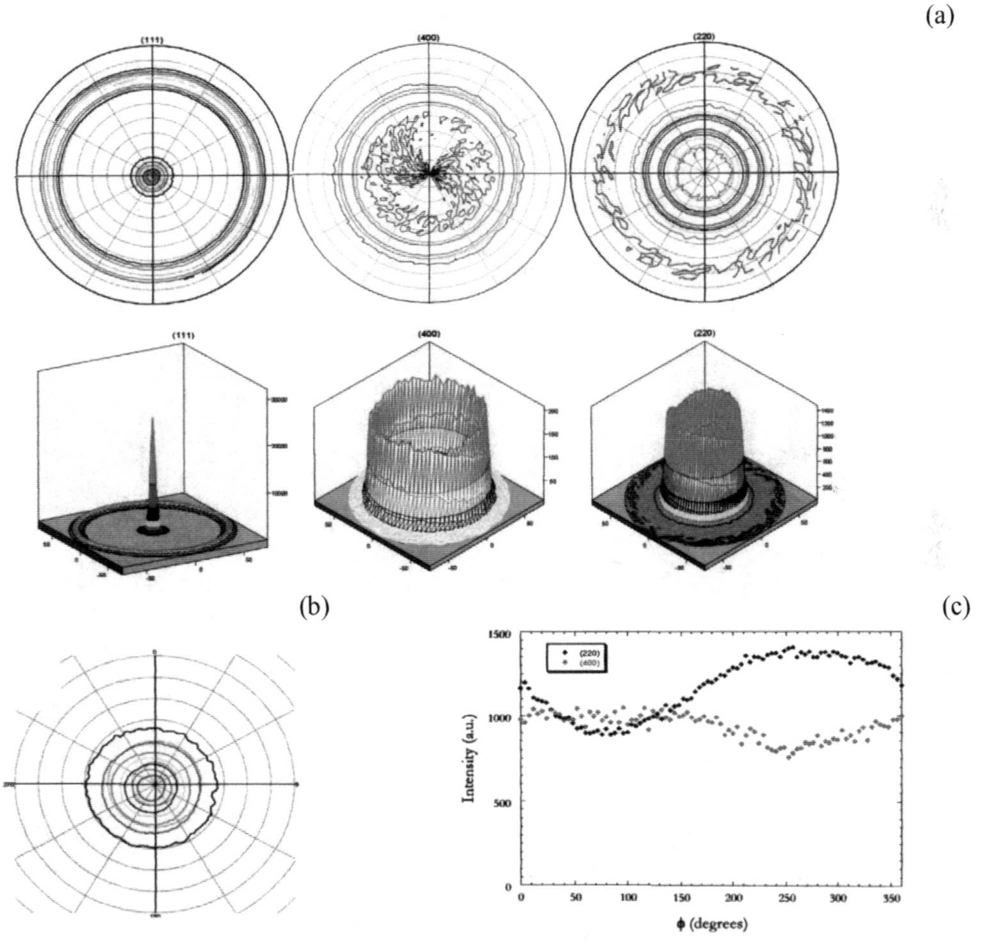

Figure 5. 2d and 3d (111), (200) and (220) CaF_2 pole figures of non-irradiated CaF_2 nanostructures (a). Magnification of (111) CaF_2 pole figure, χ <30° (b). A misalignment of ~0.5° is present and is consistent with the (400), (220) ϕ-scan (c).

Not reported here the poles for the γ-irradiated CaF_2 sample revealed the development of more strong (111) texture. The FWHM of the (111) pole is about 4° indicating a very small mosaic spread.

3.2. Thin-film structures with nanocrystals of various fluoride materials

To establish the influence of content of LiF nanocrystals with given size on PL spectra of multilayers, two LiF/CaF_2 structures have been prepared. The first, LiF/CaF_2(1) (figure 1b, period 1), is the 48-period structure of 100 nm LiF-nanolayers and 50 nm CaF_2-nanolayers. The next, LiF/CaF_2(2), is the 24-period structure consisting of 100 nm LiF- and 220 nm CaF_2-nanolayers (figure 1b, period 2).

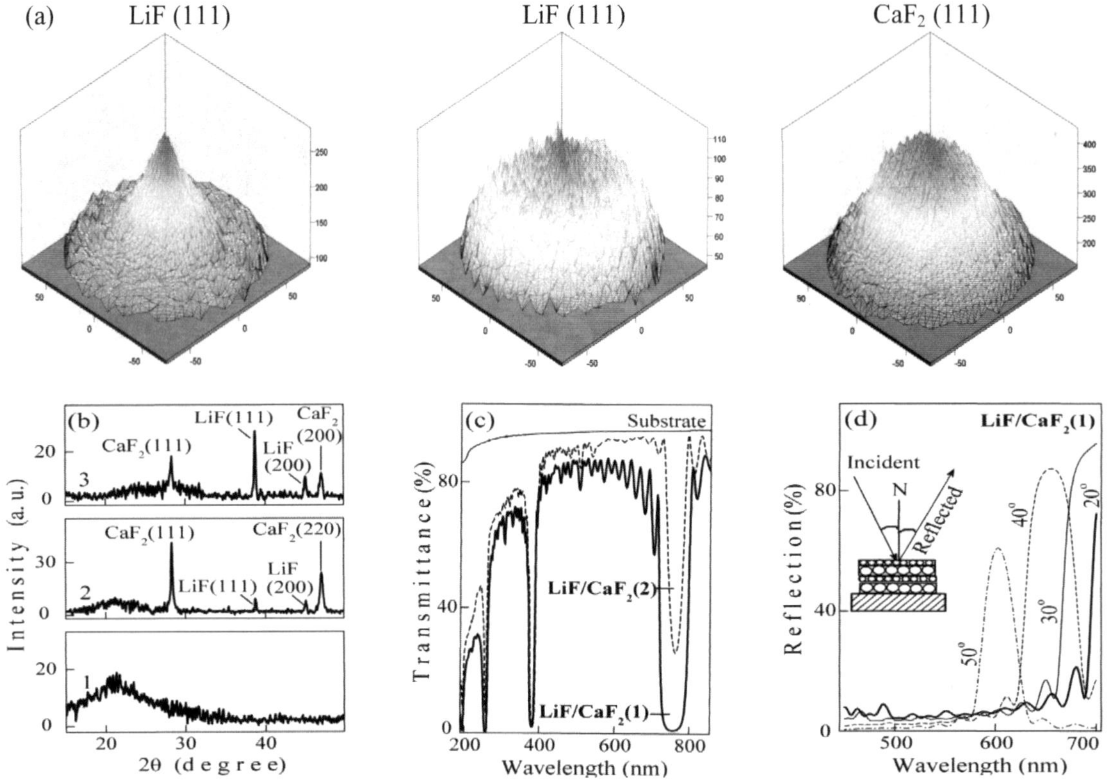

Figure 6. 3d (111) LiF, CaF_2 pole images of fresh LiF/CaF_2(1) (left) and LiF/CaF_2(2) multilayers (middle, right) (a). XRD patterns of LiF/CaF_2(2) before (1), after irradiation (2) and fresh LiF/CaF_2(1) (3) (b). Transmission spectra of the multilayers and substrate (c). Reflection spectra of LiF/CaF_2(1) vs. angles of incidence in the range 20°–50° (d). The graphs are *p*-polarized light.

Both multilayers were composed of ~100 nm LiF and ~50 nm CaF_2 nanocrystals and were *c*-oriented structures (figures 6a,b). In figure 6a the (111) LiF and CaF_2 pole images of as-prepared LiF/CaF_2(1) (left) and LiF/CaF_2(2) samples (middle, right) are reported. A well-resolved (111) LiF texture is present on both cases. The quantity of used isolines are 128, and the minimal and maximal intensities in the (111) LiF pole figures are about 91.0 and 283 cps for the LiF/CaF_2(1) samples, and 45.0 and 115.4 cps for the LiF/CaF_2(2) samples. This is due to the result LiF thickness in the LiF/CaF_2(1) sample (4.5 μm) 2.7 time larger with respect to the LiF/CaF_2(2) sample (1.7 μm). The CaF_2 pole image (figure 6a, right) indicates that the preferential nanocrystal orientation is (111) direction. The crystallinity and (111) texture improving for the γ-irradiated LiF/CaF_2 samples is supported by the XRD patterns shown in figure 6b: strong difference is observable between the data measured for LiF/CaF_2(2) before (curve 1) and after γ-irradiation at RT with dose 83 kGy (curve 2).

High-reflectivity and transmissivity bands are provided by the LiF/CaF_2 structures (figure 2c). Spectral positions of the bands were not changed after γ-irradiation but were sensitive to the polarization state and angle of incidence of light, see figure 2d. The LiF/CaF_2 structures, covered at the surface by CaF_2 nanolayers, are more mechanically stable and resistant to external influences than LiF nanostructures.

Journal of Physics: Conference Series **249** (2010) 012063 doi:10.1088/1742-6596/249/1/012063

3.3. Optical absorption features of non-irradiated thin-film structures

OA spectra of non-irradiated LiF nanostructures and initial crystals with small content of Mg^{2+}, O^{2-} and OH^- ions [13] show well-resolved absorption of Mg^{2+} colloids [11, 12] (figure 7a). (The non-irradiated and γ-irradiated substrates remain non-absorbing in the above spectral range.) According to the literature, besides Mg colloids with the absorption band peaked at 4.4–4.6 eV, the low-wavelength absorption band at 3.6–4.1 eV, related to the formation of more large Mg colloids or intrinsic Li colloids [11, 12], are reliably distinguished in non-irradiated LiF nanostructures (figure 7a) [21] and initial crystals γ-irradiated with various doses and bleached by annealing after irradiation (figure 7a, insert) [13, 14].

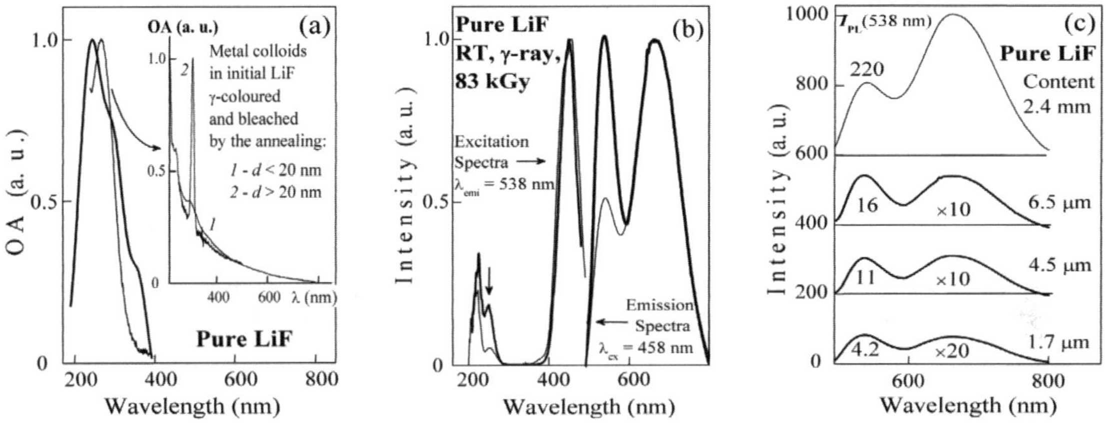

Figure 7. OA spectra of non-coloured (a) and PL spectra of γ-coloured nano (bold curve) and bulk (solid curve) crystalline structures (b) and the structures with various LiF contents (c).

3.4. PL spectra of the irradiated thin-film structures

Non-presented here PL spectra of γ-irradiated CaF_2 nanocrystals are composed of the reference 425 nm band excited at the wavelength of 370 nm. The PL spectra of initial LiF crystal, LiF and LiF/CaF_2 nanostructures having the same size of LiF nanocrystals and irradiated in analogous conditions are given in figures 7b,c. In comparison with the initial crystals, an increase of F_3^+-CCs excitation in the region of Mg colloids absorption in nanostructures is observed, together with an enhanced emission in the green region with respect to the red one for γ-irradiated film structures (figures 7b). The normalized emission spectra of nanostructures with various LiF content (bold) have the same shape: the F_3^+ photoluminescence band intensity at λ_{max} = 538 nm increases of the same factor in comparison with the F_2 emission at 670 nm with respect to the spectrum of the crystal (solid curve). The PL spectra show an increase (correlated with LiF nanocrystal content) of the intensities of F_3^+ and F_2 bands (figure 7c). These results could be explained by the high structural and optical qualities of these nanostructures (the overall emission is proportional to the quantity of luminescent LiF nanocrystals) and by the nanocrystal purity with metal excess collection in their boundaries regions [22].

4. Thin-film structures with semiconductor and dielectric nanocrystals

We measured nonlinear optical (NLO) effects for the reference films (figure 8a, curve 1), elaborated multilayers (figure 8a, curves 2, 3) and specially-elaborated multipeak TFIs, containing *c*-oriented cubic CdS nanocrystals of defined sizes, as reported in figure 8. Sample 3 and TFIs intermediate layers are composite film structures with semiconductor (CdS) and dielectric (CaF_2) nanocrystals (figure 1c). Separation of the transient shift and absorption saturation in TFIs gives us the features of both effects. Characteristic 1–2 ps decay of the shift confirms the metal clusters nature of the effect [23, 24]. Maximal NLO effects in nanocrystals are increased with size reduction, and are enhanced by extra excitation power, pulse durability [20, 22] and γ-irradiation [19]. These effects can be explained by accumulation of colloids and electrons at nanocrystal boundaries.

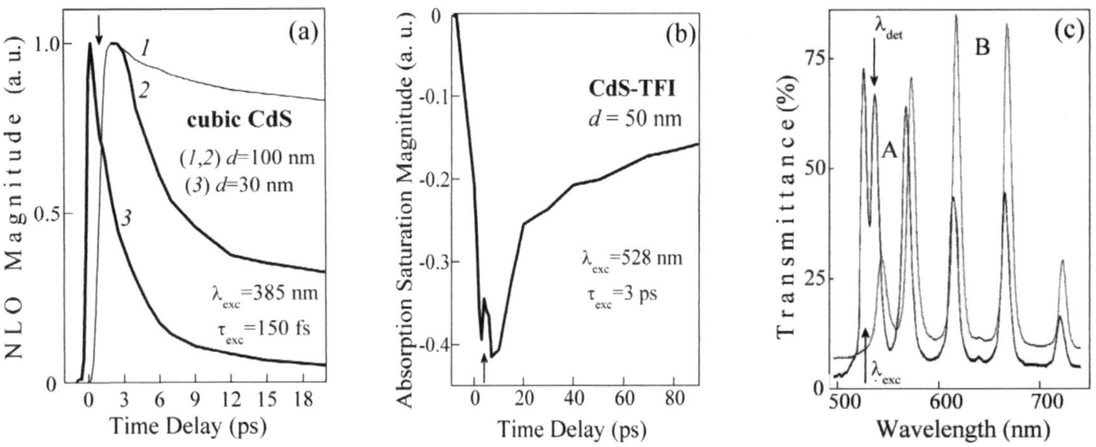

Figure 8. NLO characteristics detected for: (a) reference films (1), multilayers (2, 3), (b) multipeak TFIs with *c*-oriented cubic CdS nanocrystals of defined sizes. (c) Transmission spectra of the TFI before (B) and at time of excitation (time delay between the excitation and registration is zero) (A).

Impurities and nonstoichiometric excess tend to be expelled from the small crystalline core, as the distance a defect or impurity must move to reach the surface of a nanocrystal is very small.
Self-purification of the nanocrystals can be explained through technological arguments and is an intrinsic property of defects in nanocrystals, prepared by the methods of alternative deposition of nanolayers with thicknesses 150–10 nm. Post-growth thermal treatments, ionising radiation bombardment and light irradiation of the nanocrystals are the next tools for the nanocrystal purification, metal colloids formation and the nanocrystal properties modification.

5. Conclusion

The *c*-axis highly oriented thin-film multilayer structures composed of dielectric (LiF, CaF_2) and semiconductor (CdS) nanocrystals prepared by thermal evaporation show enhanced photo-response, which is related to high-quality nanocrystals, but at the same time depends strongly on the defect content, especially as far as their 1–2 ps nonlinearities are concerned. Critical factors are nanocrystal purity as well as metal impurities and nonstoichiometric excess collection in their boundaries regions.
The nanostructures (more high-quality than original bulk structures) can be part of efficient emitting elements, detectors and modulators of light, as well as a novel type of dosimeters. Such optically stimulated luminescence dosimeters (integrable with SiO_2:OH fibres) provide capability for remote monitoring radiation locations which are difficult to access and hazardous. These devices are relatively simple, small in size, cheap in production and have low power consumption. Moreover, they are also suitable for space radiation dose exploration. In addition, they can be used in other radiation measurements, and have interesting perspectives in the case of periodical nanostructures composed of the nanolayers of CdS and LiF nanocrystals with high-contrast indices of refraction (figure 1c).

Acknowledgments

The authors are indebted to O Bilan, A Pace and V Gremenok for their help in irradiation and microstructure testing of the samples. Many thanks are due to M A Vincenti, S Tikhomirov and O Buganov for their assistance during PL and NLO experiments. This work is supported by the Belarusian National Program for Crystalline and Molecular Structures Research, Project CMS-35, and by the grants of the ENEA Fellowships Programme (Frascati Research Center, Rome, Italy).

6. References

[1] Poole Ch and Owens F 2003 *Introduction to Nanotechnology* (New York: John Wiley)
[2] Goncharova O 1994 Semiconductor microcrystallites in thin-film matrix: fabrication methods, optical properties and application prospects *New Materials for Thin Film Functional Electronic-Technological Components* ed V Labunov (Minsk: Nauka i Tekhnika) chapter 6, pp 99–176
[3] Baldacchini G, Cremona M, d'Auria G, Martelli S, Montereali R M, Montecchi M, Burattini E, Grilli A and Raco A 1996 *Nucl. Instr. Meth.* **B 116** 447–51
[4] Montereali R M 2002 Point Defects in Thin Insulating films of Lithium Fluoride for Optical Microsystems *Ferroelectric and Dielectric Thin Films* (*Handbook of Thin Films Materials* vol 3) ed H S Nalwa (New York: Academic Press) chapter 7, pp 399–431
[5] Kumar M *et al* 2005 *J. Phys. D: Appl. Phys.* **38** 637–41
[6] Goncharova O 2005 Microstructure and properties of periodic multilayer thin-film structures with nanocrystals *Advances in Spectroscopy for Lasers and Sensing: Proc. NATO Advanced Study Institute (Erice, Sicily, Italy, 6–21 June 2005) (NATO Science, Series II: Mathematics, Physics and Chemistry* vol 231*)* ed B Di Bartolo and O Forte (Berlin: Springer) p 556
[7] Goncharova O, Karpushko F V and Sinitsyn G V 1983 *Technical Physics* **28** 1142–44
[8] Pulker H K 1979 *Appl. Optics* **18** 1969–77
[9] Wang Y H, Lu J D, Wang R W, Mao Y L and Cheng Y G 2008 *Vacuum* **82** 1220–23
[10] Bryukvina L I, Ermolaeva E A, Pidgurskii S N, Suvorova L F and Khulugurov V M 2006 *Phys. Sol. State.* **48** 68–72
[11] Schwartz K K, Lyushina A F and Vitol A Ya 1969 *Sov. Phys. Sol. State* **11** 1885–90
[12] Lushchik A, Lushchik Ch, Schwartz K, Vasil'chenko E, Papaleo R, Sorokin M, Volkov A E, Neumann R and Trautmann C 2007 *Phys. Rev.* B **76** 054114
[13] Baldacchini G, Goncharova O, Kalinov V S, Montereali R M, Nichelatti E, Vincenti A and Voitovich A P 2007 *Phys. Status Solidi* **C 4** 744–48
[14] Baldacchini G, Goncharova O, Kalinov V S, Montereali R M, Vincenti A and Voitovich A P 2007 *Phys. Status Solidi* **C 4** 1134–38
[15] Goncharova O and Demin V 1993 Photosensitive, resistive, nonlinear optical thin-film heterostructures *RF Patent* 2089656
[16] Goncharova O and Demin V 1994 Photosensitive, resistive, nonlinear optical composite films *RF Patent* 2103846
[17] Goncharova O and Demin V 1994 Narrow-band thin-film Fabry–Perot interferometer *RF Patent* 2078358
[18] Goncharova O 1998 Light-induced ultrafast bleaching in films of different arranging of semiconductor nanocrystals: materials *Proc. Int. Conf. on Coherent and Nonlinear Optics (Moscow, Russia, 29 June – 3 July 1998)* (Moscow: Moscow State University Press) p 246
[19] Goncharova O, Kalinov V and Voitovich A 2004 *Radiat. Meas.* **38** 775–79
[20] Goncharova O 1996 Picosecond all-optical switching in thin-film Fabry–Perot interferometers and cavityless devices *Notes and Perspectives of Nonlinear Optics* (*Nonlinear Optics* vol 3) ed O Keller (London: World Scientific) pp 636–42
[21] Goncharova O and Gremenok V 2007 Comparative study of electronic structure of thin film nanocrystals prepared by low-temperature vacuum deposition *Proc. Int. Conf. CLEO/Europe-2007 (Munich, Germany, 17–22 June 2007*) p 1-1
[22] Goncharova O , Montereali R M and Baldacchini G 2008 *Abstr. Int. Conf. ICL-2008 Lyon, France, 7–11 July 2008*) p 160
[23] Goncharova O 1996 Studies on a Nature of the Picosecond Transient Absorption in Thin Films Activated by Semiconductor or Metal Nanocrystals *Proc. Int. Conf. EQEC-1996 (Hamburg, Germany, 8–13 Sept. 1996*) p 87
[24] Z. Lu and K. Zhu 2008 *J. Phys. B: At. Mol. Opt. Phys.* **41** 185503–09

Journal of Physics: Conference Series **249** (2010) 012064 doi:10.1088/1742-6596/249/1/012064

Study of Fricke gel dosimeter response for different gel quality

C C Cavinato* and L L Campos
Instituto de Pesquisas Energéticas e Nucleares, IPEN-CNEN/SP, Av. Prof. Lineu Prestes, 2242, Cidade Universitária, Sao Paulo, SP, Brazil

E-mail: ccavinato@ipen.br

Abstract. The Fricke xylenol gel (FXG) dosimeter has been studied for application in radiotherapy because it is capable of to measure the spatial distribution of radiation doses. The dosimetry is based on the oxidation of ferrous (Fe^{2+}) to ferric (Fe^{3+}) ions radiation induced, related to the radiation dose. The gel material usually employed is the 300 Bloom gelatin, which is imported and very expensive in Brazil. Aiming to analyze the viability of to use a locally produced and low cost gel material, in this work the spectrophotometric responses of FXG solutions prepared using 270 Bloom gelatin commercially available and 300 Bloom gelatin imported were compared. The absorption spectra of solutions prepared with 5% by weight 270 and 300 Bloom gelatins non-irradiated and irradiated with ^{60}Co gamma radiation in the dose range between 0.5 and 100 Gy were analysed, the dose-response curves were evaluated and the useful dose range was established. The obtained results indicate that the FXG solution prepared with 270 Bloom gelatin presents good performance, similar to that presented by the FXG solution prepared with 300 Bloom gelatin and its use can be recommended owing to the low cost and the availability in local market.

1. Introduction

In radiotherapy, radiations sensitive gels have potential as a three-dimensional (3D) dosimeter [1] because are capable of to measure the spatial distribution of radiation doses using the magnetic resonance imaging (MRI) technique [2, 3]. Among others [4, 5, 6, 7] the Fricke xylenol gel (FXG) dosimeter has been very studied by several researches [8, 9, 10, 11] since 1984 [12], when was developed. The FXG dosimetry is based in oxidative conversion of ferrous (Fe^{2+}) to ferric (Fe^{3+}) ions by ionizing radiation [12].

The spatial integrity of the dose distribution in the FXG dosimeter is maintained by gel matrix present in the dosimeter because it restricts the diffusion of the ferrous and ferric ions [3, 13]. Usually, the gel matrix employed in gel dosimeters is the 300 Bloom gelatin [2, 14, 15], imported and very expensive in Brazil.

The purpose of this work is to analyze the spectrophotometric response of FXG solution prepared with 270 Bloom gelatin commercially available, locally produced and of low cost compared to FXG solution prepared with 300 Bloom gelatin imported.

---

* To whom any correspondence should be addressed.

2. Materials and methods

2.1. FXG preparation

The Fricke gel solutions were prepared using 5% by weight 270 or 300 Bloom gelatins, tri-distilled water, 50 mM H_2SO_4, 1 mM NaCl, 1 mM ferrous ammonium sulphate hexahydrate or Mohr salt $[Fe(NH_4)_2(SO_4)_2 \cdot 6H_2O]$ and 0.1 mM ferric ions indicator xylenol orange ($C_{31}H_{28}N_2Na_4O_{13}S$) [8, 10].

Immediately after preparation the non-irradiated Fricke gel solutions were conditioned in acrylic cuvettes (10 mm of the optical path length) for the spectrophotometric measurements. The samples were maintained under refrigeration (4° C ± 1° C) and light protected during 12 h before irradiation.

2.2. FXG samples irradiation

Thirty minutes before irradiation the dosimeters were maintained at room temperature (~ 25° C) and light protected. The FXG samples were irradiated in two different ^{60}Co gamma radiation sources (*Panoramic* 20 TBq and *Gammacell 220* 110 TBq) pertaining to Radiation Technology Centre of IPEN-CNEN/SP with absorbed doses between 0.5 and 100 Gy, dose rates of 0.031, 0.0307, 2.47 and 2.44 kGy/h, free air and electronic equilibrium conditions.

The optical measurements were performed thirty minutes after irradiation.

2.3. Evaluation

The spectrophotometry technique [9, 16, 17] was used to evaluate the optical responses. The absorption measurements were performed in the optical range between 190 and 900 nm using SHIMADZU UV-2101PC spectrophotometer.

2.4. Parameters studied

The absorption spectra of solutions prepared with 5% by weight 270 and 300 Bloom gelatins non-irradiated and irradiated with ^{60}Co gamma radiation in the dose range between 0.5 and 100 Gy were analysed, the dose-response curves were evaluated and the useful dose range was established.

3. Results and discussions

The presented data in all dose-response curves correspond to the arithmetic mean of three FXG samples for each absorbed dose studied and they present uncertainties that were calculated based in the standard deviation times coverage factor k = 3 providing a reliable level of 99%.

3.1. FXG – 270 Bloom

The FXG solution prepared with 270 Bloom gelatin presented one absorption band at 441 nm related to Fe^{2+} ions and other absorption band at 585 nm [2] related to Fe^{3+} ions (figure 1).

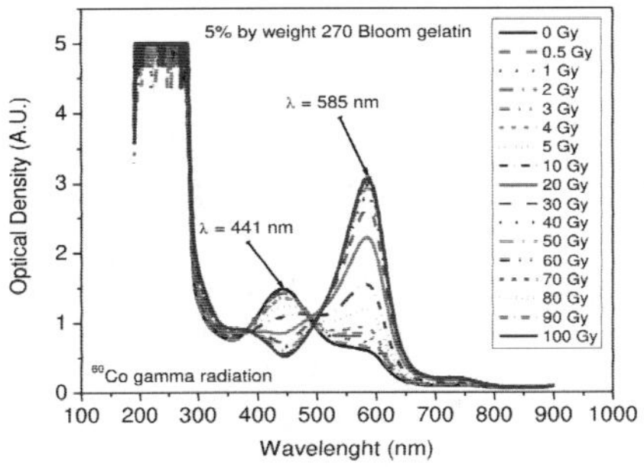

Figure 1. Optical absorption spectra of Fricke gel solution prepared with 270 Bloom gelatin.

The lower dose detection limit was calculated using the equation (1):

$$D_L = \left[\left(\overline{OA_0} - b\right) + (3\sigma)\right] \times f_{cal} \quad (1)$$

were D_L = Lower dose detection limit, $\overline{OA_0}$ = Mean optical absorption corresponding to thirty non-irradiated FXG samples, b = Linear coefficient of the line set, σ = Standard deviation of the mean and f_{cal} = Dose calibration factor. The calculated D_L value was found to be equal to 0.5 Gy (figure 2a).

The upper dose limit adopted can be extended up to 30 Gy (figure 2b), the maximum value before saturation of the spectrophotometric response. The results presented correspond to the wavelength of 585 nm.

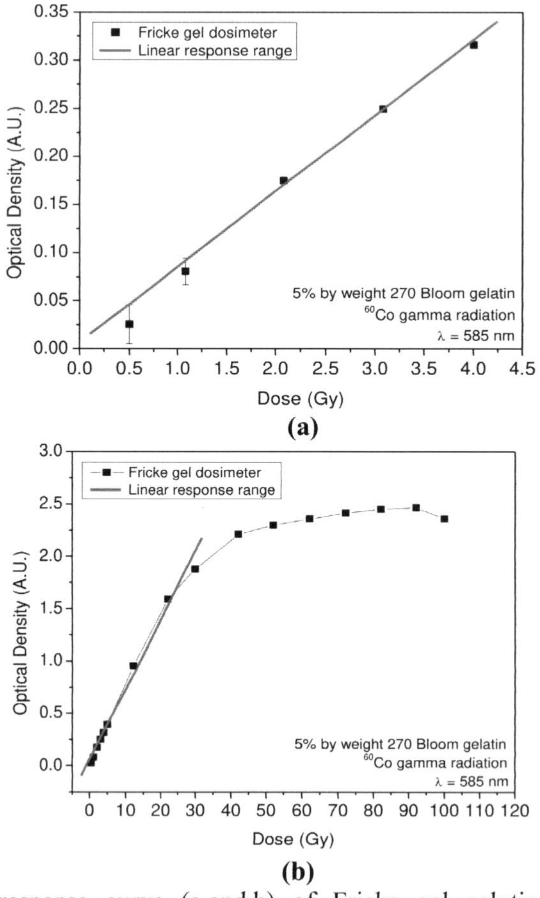

Figure 2. Dose-response curve (a and b) of Fricke gel solution prepared with 270 Bloom gelatin.

3.2. FXG – 300 Bloom

The FXG solution prepared with 300 Bloom gelatin presented one absorption band at 438 nm related to Fe^{2+} ions and other absorption band at 585 nm [2] related to Fe^{3+} ions (figure 3).

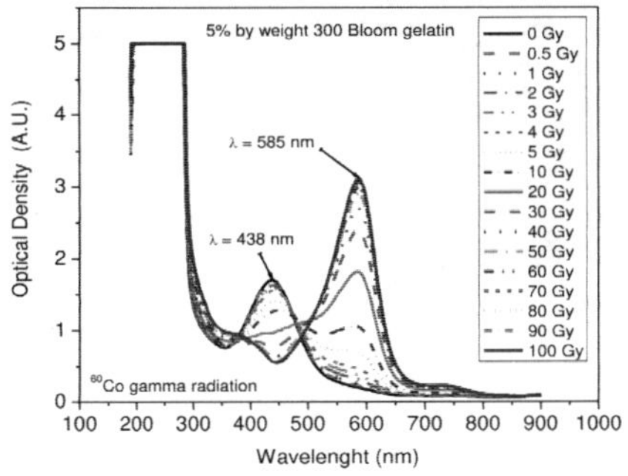

Figure 3. Optical absorption spectra of Fricke gel solution prepared with 300 Bloom gelatin.

The lower dose detection limit was calculated using the equation (1) and was found to be equal to 0.5 Gy (figure 4a).

The upper dose limit adopted can be extended up to 30 Gy (figure 4b) the maximum value before saturation of the spectrophotometric response. The results presented correspond to the wavelength of 585 nm.

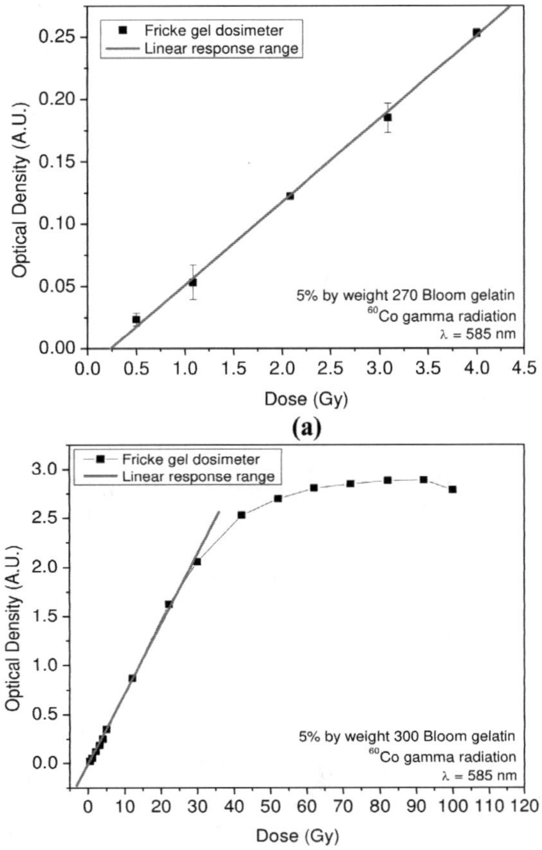

Figure 4. Dose-response curve (a and b) of Fricke gel solution prepared with 300 Bloom gelatin.

3.3. Comparison – 270 and 300 Bloom

In the figure 5 the spectra of non-irradiated samples prepared using 270 and 300 Bloom gelatins are compared. It can be observed the presence of small pre-existing Fe^{3+} ions concentration in both Fricke gel solutions which must be subtracted of irradiated samples. In this work the pre-existing Fe^{3+} ions concentration in different Fricke gel solutions (270 and 300 Bloom) was always subtracted.

The similar dose-response behavior in the dose ranges between 0.5 and 30 Gy of both solutions is presented in figure 6.

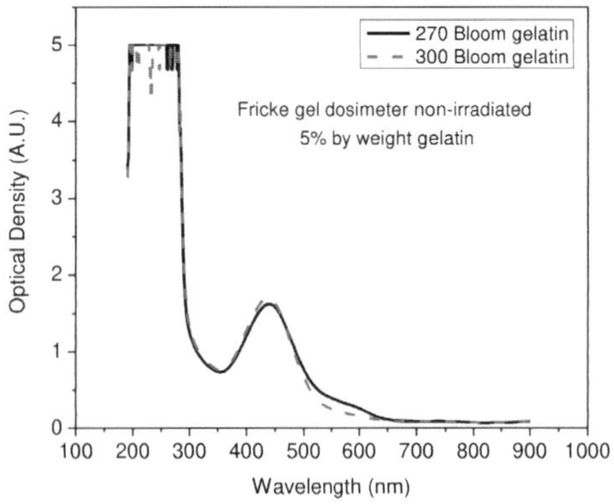

Figure 5. Comparison between the spectra of Fricke gel solutions prepared with 270 and 300 Bloom gelatins.

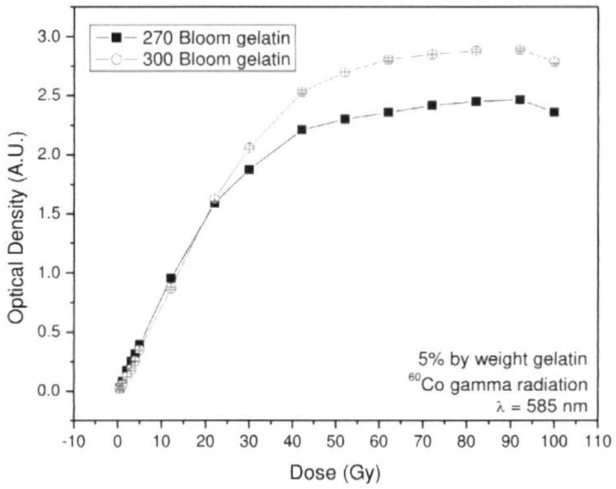

Figure 6. Comparison between the dose-response curves of Fricke gel solutions prepared with 270 and 300 Bloom gelatins.

4. Conclusions

The obtained results indicate that the 270 Bloom gelatin presents good performance, similar to that presented by solution prepared with 300 Bloom gelatin and its use can be recommended owing to the low cost and the availability in local market.

Acknowledgments

The authors are grateful to the FAPESP, CNPq, CAPES, CNEN and IPEN by the financial support.

References

[1] Healy B J, Zahmatkesh M H, Nitschke K N and Baldock C 2003 *Med. Phys.* **30** 2282-91
[2] Bero M A, Gilboy W B and Glover P M 2001 *Rad. Phys. Chem.* **61** 433-5
[3] Audet C and Schreiner L J 1997 *Med. Phys.* **24** 201-9
[4] Baldock C, Lepage M, Bäck S A J, Murry P J, Jayasekera P M, Porter D and Kron T 2001 *Phys. Med. Biol.* **46** 449-460
[5] Maryanski M J, Schulz R J, Ibbott G S, Gatenby J C, Xie J, Horton D and Gore J C 1994 *Phys. Med. Biol.* **39** 1437-55
[6] Karaiskos P, Petrokokkinos L, Tatsis E, Angelopoulos A, Baras P, Kozicki M, Papagiannis P, Rosiak J M, Sakelliou L, Sandilos P and Vlachos L 2005 *Phys. Med. Biol.* **50** 1235-50
[7] Papagiannis P, Karaiskos P, Kozicki M, Rosiak J M, Sakelliou L, Sandilos P, Seimenis I and Torrens M 2005 *Phys. Med. Biol.* **50** 1979-90
[8] Olsson L E, Petersson S, Ahlgren L and Mattsson S 1989 *Phys. Med. Biol.* **34** 43-52
[9] Bero M A and Kharita M H 2004 *J. Phys. CS* **3** 236-9
[10] Bero M A 2007 *Nucl. Instr. and Meth. in Phys. Res.* A **580** 186-9
[11] Galante A M S, Cervantes H J, Cavinato C C, Campos L L and Rabbani S R 2008 *Rad. Meas.* **43** 550-3
[12] Gore J C, Kang Y S and Schulz R J 1984 *Phys. Med. Biol.* **29** 1189-97
[13] Chu W C 2001 *Proc. Natl. Sci. Counc. ROC(B)* **25** 1-11

[14] Bero M A, Gilboy W B, Glover P M and El-masri H M 2000 *Nucl. Instr. and Meth. in Phys. Res.* B **166-167** 820-5
[15] Schreiner L J 2004 *J. Phys. CS* **3** 9-21
[16] Podgorsak M B and Schreiner L J 1992 *Med. Phys.* **19** 87-95
[17] Mattsson L O, Johnson K A and Svensson H 1982 *Acta Radiol. Oncol.* **21** 139-144

AUTHOR INDEX

AUTHOR INDEX

AUTHOR INDEX